NATURE GUIDING

WILLIAM GOULD VINAL

*Formerly Professor of Nature-Study at the Rhode Island College of Education;
now in the New York State College of Forestry
at Syracuse University.*

ITHACA, N. Y.
THE COMSTOCK PUBLISHING CO.
1926

NATURE GUIDING

WILLIAM GOULD VINAL

Formerly Professor of Nature-Study at the Rhode Island College of Education;
now in the New York State College of Forestry
at Syracuse University.

ITHACA, N. Y.
THE COMSTOCK PUBLISHING CO.
1926

Open access edition funded by the National Endowment for the
Humanities/Andrew W. Mellon Foundation Humanities Open Book
Program.

First paperback printing 2019

ISBN 978-1-5017-4085-5 (pbk.)
ISBN 978-1-5017-4086-2 (pdf)
ISBN 978-1-5017-4087-9 (epub/mobi)

Librarians: A CIP catalog record for this book is available from the
Library of Congress

PREFACE

THE PERSON who trains for Nature Guiding is dealing with one of the most difficult and at the same time one of the most enjoyable professions of the world. Nature Guiding is old but not traditional. It does not have a rut. The Nature Guide, therefore, must be resourceful. This book aims to provide a point of view and considerable material on methods of Nature Guiding. The next decade should witness a marked elaboration in the ways of Nature Guiding.

During the past fifteen years the writer has been trying out plans of presenting Nature-study in schools and camps. This book is an outgrowth of these efforts. The author is deeply indebted to the hundreds of students, campers, and teachers, who have cooperated throughout the growth of these ideas which have culminated in the publishing of NATURE GUIDING.

The term Nature Guide includes not only the government Nature Guide but the Nature Counsellor of the summer camp, the scout naturalist, the many parents who are making an earnest effort to direct their children in nature interests, and the school teacher who is guiding, but not drilling, in Nature-study.

Most of the chapters in this book are reproductions of articles published in various educational magazines. The author wishes to make the following acknowledgments: To the *Nature-Study Review* for Chapters XII, XXIII, XXIV, XXVI, XXXV, XLIV and parts of Chapters I, II, and XI. To the *Rhode Island Arbor Day Publication* for Chapters XXV, XL, XLII, and XLIII, and part of Chapter XX. To *School News and Practical Educator* for Chapters XVI, XVII, XX, XXII, and part of Chapters II, XIX, and XXI. To the *Popular Educator* for Chapters XXX, XXXI, XXXII, and part of Chapter XXVIII and XXIX. To the *Primary Educator* for parts of Chapters XXVIII and XXIX. To the *Journal of Geography* for Chapter XXXVII; *Bird Lore* for Chapters XLVI, XLVII, and XLVIII; to the *General Science Quarterly* for Chapter XXXIII; the *Boys Life Magazine* for Chapter V; The *Nature Magazine* for part of Chapter I; The *Rhode Island Educational Circular* for Chapter L; *Camps and Camping* for Chapter X; The *Camp Fire Guardian* for Chapter IV; The *Playground* for Chapter XLIV and LIII. *Yosemite Nature Notes* for most of Chapter LI; and the *Educational Bimonthly* for part of Chapter XII.

The author also wishes to thank Dr. Walter E. Ranger, Commissioner of Education for the State of Rhode Island, and Mr. W. A. Slingerland, Manager of the Comstock Publishing Company, for the loan of cuts used in their publications, and Mr. William Wessel, Assistant Camp Director of the Boy Scouts of America, for the privilege of taking the pictures of fire building, camp craft, and the outdoor bathtub.

Chapter XXVIII, Nature-Study by Grades, was prepared in collaboration with a special committee on science in elementary schools. The writer is indebted to the following who aided with their counsel: Emerson L. Adams, Assistant Commissioner of Public Schools for State of Rhode Island; Dr. John L. Alger, President of Rhode Island College of Education; Harold L. Madison, Curator Children's Museum, Cleveland; and Isaac O. Winslow, Superintendent of Schools, Providence.

I cannot conclude this preface without thanking the teachers of the Henry Barnard School, of the Rhode Island College of Education, whose courtesy was unfailing during the many years I worked with them. I wish to express my gratitude to their principal, Professor Clara E. Craig, who was ever ready to aid me with her counsel and experience. Her development of the Americanized-Montesorri system has been a great inspiration. I also wish to thank the members of the staff of the Nature Lore School who have so generously given of their time and knowledge to make this important movement a success. Appreciation is especially extended to Professor Anna Botsford Comstock, who is always an inspiring leader in Nature-study work.

WILLIAM GOULD VINAL

Syracuse University, February 1, 1926.

TABLE OF CONTENTS

PUBLISHERS' ACKNOWLEDGMENTS

I appreciate the courtesy of the publishers who have allowed me to use the following quotations and poems: Harper and Bros., poem, FRIENDS, by John Kendrick Bangs; The Bobbs-Merrill Company, for quotations from James Whitcomb Riley as follows—from RHYMES OF CHILDHOOD, copyright 1890, from GREEN FIELDS AND RUNNING BROOKS, copyright 1892, and other quotations; the *Boston Transcript* and Miss Ruth Hall for the poem TO AN ENGLISH SPARROW by Hazel Hall; The Century Company for Blanche E. Wade's NATURE STUDY AND TEACHER, and Carolyn Wells' AN APPLE LESSON which appeared in the *St. Nicholas Magazine*; *Chicago Tribune* for THE LINE O' TYPE OR TWO; Henry Turner Bailey for quotation from THE CHILDREN'S BIRTHRIGHT; Henry Van Dyke for quotations on pages 53 and 322; The Science Press for quotation from the *Scientific Monthly* for April, 1923, page 492; The Comstock Publishing Company for various quotations from Professor Anna Botsford Comstock and Dr. L. H. Bailey; Houghton Mifflin Co., Riverside Press, and Mrs. Enos Mills for IMAGINATION GUIDES THE RACE which is in part used in YOUR NATIONAL PARKS by Enos Mills; the Oswego Normal School for the picture of Dr. Sheldon.

FOREWORD

A MAN may be a scientist, of high attainment, but it does not follow that he shall successfully lead the uninitiated into an understanding of the natural world, nor yet give a child an interest in and a love for his natural environment. To do this requires not alone scientific attainment but also a comprehension of the absence of knowledge and lack of training in observing natural phenomena of the ordinary adult, and an understanding of the mind and interest of the child.

The author of this volume is a man of science and has shown to a marked degree his ability to interest his pupils in science and to give to them careful methods of investigation and show to them many ways of interesting the children in the out-of-doors when they should become teachers. His long experience as a successful head of a Girls' Camp and his active interest in the Boy Scouts have given him the opportunity to know our boys and girls of to-day as they are, and this knowledge has been of great use to him in preparing teachers of nature-study. His field classes have been conducted with remarkable success, with the invariable result that his pupils become interested as well as learned in the ways of nature.

The author has been exceptionally broad in his training. He began with work for the Fish Commission and later became deeply interested in Botany and finally centered his interest in Forestry. His Arbor Day manuals for Rhode Island were remarkable for their excellence, and brought him wide recognition. It is, perhaps, because of the diversity of his scientific training that the author became interested in the nature lore movement. For years he has been the moving spirit in preparing Nature Counsellors for the Summer camps and he has undoubtedly done more to bring children of the camps into true companionship with their natural surroundings, than has anyone else.

It is fortunate that the author in the midst of his multiplicity of duties has found time to write the lessons contained in this volume and thus increase the radius of his inspiration and beneficient influence.

ANNA BOTSFORD COMSTOCK,
Professor of Nature-Study,
Emeritus, in Cornell University

There was a child went forth every day;

And the first object he look'd upon, that object he became;

And that object became part of him for the day, or a certain part of the day, or for many years, or stretching cycles of years.

The early lilacs became part of this child,

And grass, and white and red morning-glories, and white and red clover, and the song of the phœbe-bird,

And the third-month lambs, and the sow's pink-faint litter, and the mare's foal, and the cow's calf,

And the noisy brood of the barn-yard * * * * * *

And the apple-trees cover'd with blossoms, and the fruit afterward, and woodberries, and the commonest weeds by the road; * * * * * *

And the school-mistress that pass'd on her way to the school, * * * * * *

The village on the highland, seen from afar at sunset—the river between,

Shadows, aureola, and mist, the light falling on roofs and gables of white or brown, three miles off, * * * * * *

These became part of that child who went forth every day, and who now goes, and will always go forth every day.

—*Walt Whitman.*

INTRODUCTION

I. There is a Need for Nature Leaders

One of the difficulties in the establishment of nature-study has been that there is no field for the work. This is no longer true. There is an awakening throughout the country.

1. *Summer Camps need Nature Counsellors.* It is admitted that nature lore is the most important and the most difficult position to fill in the summer camp. There are over 5000 camps.

2. *Scouting and Camp Fire Organizations*, have the same difficulty. A few scout councils have their naturalist and others will take them on when they can find competent leaders.

3. *The Nature Guide Movement* in Yosemite National Park started in 1920. In 1925 the number in Yosemite increased from six to ten guides and had spread to many other parks. The time is near at hand when every community will have its Nature Guide.

4. *The Public School* officials are realizing that the grade teacher cannot teach nature-study without help. They are meeting the problem with nature-study supervisors. The study is a vital source for project work. Although camping is in the "private school" stage it will undoubtedly become a part of the public school system. Nature leaders are being demanded more and more.

5. *The Playgrounds* need leaders who know the nature-study method.

6. *The Country Day School* movement is one toward the nature-study idea.

7. *Universities* are introducing nature-study departments.

8. *Normal Schools* are changing to Teachers Colleges and in the expansion of their work are planning intensive nature training.

9. *From the National Conference on Out-door Recreation* called by President Coolidge: "That the conference endorse Nature-Study in schools and the extension of the Nature-Study Idea to every American school and family; . . . . . that the establishment of museums of natural history in National Parks will increase the educational recreational value of the parks."—*Resolution of the Conference.*

II. "Nature Guiding" is a Book of Methods for Nature Leaders

Nature Guiding is the science of inoculating nature enthusiasm, nature principles, and incidentally nature facts into the spirit of individuals. When these facts have accumulated to such an extent that the owner needs to organize them it becomes the science of Geology, Botany, Zoology, or Agriculture rather than nature-study. Everyone has use for nature-study and only a minority attain the need of organizing their knowledge into science.

The greatest handicap to effective and successful nature-study is the lack of teachers trained in the methods of nature-study. A university professor may know Phanerogamic Botany yet fail to give his own child a sustained enthusiasm for flowers. Scout leaders who have never studied science are more frequently successful in nature guiding then the college graduate in science. This is why the nature-study teacher must first of all catch the method. It is a matter of contagion. The method of nature-study is the spirit of nature-study.

The nature-study method is simple if followed in its desirable details. If teachers are willing to study the analysis of the situation and be painstaking enough to apply these rules until they become habitual they have laid the foundation for successful nature-study. This is the function of a nature-study department in our Teachers' Colleges. When the pupil teacher can apply these same steps and principles to nature conditions characteristic of a new community— that point is the beginning of professional growth.

Nature knowledge is unlimited. It will develop through experience. For the young teacher it is not so important how much he knows as what he will do with what he knows. Nature laws, like moral laws, develop best when lived rather then when preached about.

Nature-study is in a healthy, experimental stage with several distinct lines of development. The purpose of this book is to make clear the methods in these various phases. Good results may be obtained by any one of these methods but the optimum is a combination and adaptation of all. We want men and women who have an abiding and pleasurable interest and enjoyment in their forests, gardens, orchards, home surroundings, and in the immediate life which these areas support.

PART I
NATURE-LORE FOR CAMPS

WANDERLUST

Beyond the East the sunrise, beyond the West the sea,
And East and West the wanderlust that will not let me be;
It works in me like madness, dear, to bid me say good-bye!
For the seas call and the stars call, and oh, the call of the sky!

I know not where the white road runs, nor what the blue hills are,
But man can have the sun for friend, and for his guide a star;
And there's no end of voyaging when once the voice is heard,
For the river calls and the road calls, and oh, the call of a bird!

Yonder the long horizon lies, and there by night and day
The old ships draw to home again, the young ships sail away;
And come I may, but go I must, and if men ask you why,
You may put the blame on the stars and the sun and the white road and the sky!
—*Gerald Gould, in Stevenson's "Home Book of Verse."*

CHAPTER I

Nature Leaders for Camps

I. *The Call of the Camps*

When your grandmother was a girl the neighbors used to shake their heads doubtfully and say "Nothing but a regular tomboy, anyway." She could play "Four-old-cat" or climb a tree with skill that was envied by many of the boys. If the up-streeters wanted to go 'cross lots they always waited for "Jimmie" Morrill. Grandmother was a live girl. She use to pick huckleberries, spin cloth, romp the fields, feed the cattle, and tramp the roads. From these activities she gained health of mind and body.

But times have changed. A new kind of girlhood has appeared. The ordinary girl is contented to sit in a stuffy school room, flat-chested and sallow skinned. She exercises her tongue and finger tips but the body muscles remain flabby. Her circulation has become sluggish, her nerves shriek, she gorges herself with chocolates, takes piano and violin lessons, goes to the movies twice a week, dancing school every Friday, and to various parties on the holidays. Do you wonder at the pallid cheeks, pale eyes, and need of after dinner pills to assist digestion? This type of girl has not learned to live.

It is now spring and that peristent desire to get into the fields, to eat green herbs, to scale the wall, to wear old clothes, to sleep in a cabin, is a call of nature. It is a desire to leap back into the good old days. You want to get away from the crowded streets, the school room, and the moving picture shows. You wish to meet untamed nature, as did your grandmother, to stride through the rough forest and to quench your thirst from clear pools. This is a natural feeling. It is a call that may be satisfied by the summer camp. It is a positive necessity to go? Then look into a directory for girls' camps and take your pick. Mountain or plain, sea-shore or lake, boating or horse-back riding, it is all there to satisfy your craving for the great out-of-doors.

The opportunities at a girls' camp are manifold. In the first place there is contact with congenial councilors who are true and tried in character. They are usually college graduates who have had experience dealing with girls. The girls and their sympathetic leaders store up energy together for the coming year.

In a camp one sleeps in a tent or cabin with three or four other girls and a councilor. In the days of large families the girls had to learn to consider the happiness of others. One gets the same sort of training at camp for there the elbows touch and they cannot be

Nothing is lost on him who sees
    With an eye that genius gave;
For him there's a story in every breeze
    And a picture in every wave.'',
                                —*Moore.*

too sharp. The out-door air makes red blood and healthy appetites, and in camp one not only eats but sleeps in the open.

The first order of the day is reveille which means to get ready for setting-up drill. Indisposition is not a feminine grace in a girls' camp, and every one hustles out for morning exercises. The working of the big muscles—the trunk and leg muscles—builds up health. It gives arterial tone and prevents kidney or heart disease which are increasing so rapidly under the nerve racking pace of today.

The morning air whets the appetite. There is a rush for the table where Miss Camper finds eggs, milk, and johnny cake. As time goes on her mania for candy diminishes. She begins to eat to live, instead of live to eat. The response in her digestion and general health are good omens of right eating.

The morning activities may consist of athletics. In a baseball game a girl is able to acquire that general sturdiness so characteristic of boys. She learns to sacrifice her own wishes for the best interest of the team and besides, it takes courage to slide bases. This spirit of self-sacrifice and courage will help greatly in after life.

The greatest fun of the day is swimming time. As everyone should learn swimming before stepping into a canoe there is a big incentive to acquire this important art. In Japan every boy and girl is taught how to swim but the American girl is just beginning to inherit her right to aquatic sports.

After the mid-day meal comes a period of rest and quietness. This may be the time of arts and crafts. It was the spinning and weaving industries of the home that developed the real artistic sense of our grandmothers. These industries have been taken from the homes and put in the factories. In camp, however, this work is lifted from the plane of necessity to that of opportunity.

After the rest hour the campers may take a hike. They learn hiking rules by experience, such as, never walk over anything that you can walk around, or never step on anything that you can step over. The hike may be in the form of a scouting party. The scouts must learn how to find their way and be able to interpret all signs and tracks. There is nothing more fascinating than to sit around the camp-fire after a long ramble, you are tired but it is a "good feeling" fatigue.

The camp-fire is started, not "um big fire" but "um little fire" as the Injun did,—a fire over which one could cook a supper. The girls are bubbling over with good health and appetites. The fish, just landed from the lake, are baked in a hole in the ground and the baked potatoes are poked out of the hot embers. The ends of the corn husks are tied over the cob, soaked in water, and placed in the hot coals for twenty minutes. Melted butter is then put on with a brush. And then the feast comes and it will be remembered longer than any banquet in a marbled hall hotel. Such occasions form a sunny spot in ones' memory and are pleasant to look back upon as the years roll along.

Cooking breakfast on the Atlantic shore.

"Here is the spring of springs, and the waterfall of waterfalls. A man may stand here and put all America behind him."
—*Henry D. Thoreau's "Cape Cod"*

In the evening the camp community gathers around the fire place (not gas logs) and makes merry in song and story. It was the same in colonial days when the neighbors met and made their own fun. More people should learn to play rather than hire others. There are those who have become so fixed in mind and character that they are not able to learn to play. It is claimed that all the virtues of the human race are brought out in play and if this is true, the play element is an important element in the camp activities.

And so the days go—all too quickly—for the girl who is enjoying the fascination of living with Nature all summer long, hiking through quiet woods and paddling along clear streams. She should have learned the natural pleasures of the great out-of-doors. Let us hope that she is more tom-boyish and in the best sense of the word. She should have red blood, sound nerves, a quick ear, keen sight, a quick step, and many other of the good characteristics of our grandmother—that good old lady who lived so long ago.

II Counsellor Guidance

Getting a counsellor's job has been a mad rush and gamble. Someone hears of a friend who has such a position and it seems quite appropriate that he should also have a share in the spoils. The idea makes a healthy growth during the night and the candidate makes an onset on Kamp Klondike. He must have a job "immediatment." One letter this past week announces a friend who is to graduate from St. Paul's Hospital in June and would like to rest up during the summer. Would I please tell her how to go to work to get a counsellor's position? Another is teaching biology in the Smith Hall School and is now prepared to take a position as nature counsellor. There is a dire need of a counselor's clearing house.

This chapter is primarily an analytical study of the vague idea counselorship and, secondly, a guide to would be counselors in summer camps. A camp counsellor is not an errand boy nor simply an employee with wages and time off, nor is he taken for a rest cure. A counsellor is a red-blooded leader who is interested in children and the out-of-doors, and seeing the challenge of this new profession, has enrolled in the camp program to do his bit in child development.

The usual leaders in camps are art-craft, camp-craft, dancing, dramatic, song, swimming, outdoor cooking, nurse, and nature-lore. Because one is accomplished in the technique of one of these specialties is no guarantee that he is a good counsellor for a camp.

I know of a song counsellor—a graduate of a well known Conservatory of Music—who on the first day of camp examined every camper as to voice and ability to sing the chromatic scale. His summer's program failed the first day. The next summer a college trio came from Dixie. They just sang and sang. The campers caught the spirit and the camp became known as a "singing camp." Everyone sang just for the love of singing. Singing became a tremendous power in developing the spirit of that camp. Every camp has its songs,—old favorites that are sung best about the camp fire. Then there are the marching songs that sort of swing along on a hike. And in the nautical camp the chanteys are given their sea way. Give us singing counsellors—those who are versatile in camping and can sing in rhythm with their camp work.

In cleaning out my letter files the other day I came across a written application for a position as swimming counsellor. I re-read the letter with a mingled feeling—I did not know whether to laugh or cry. This applicant not being accepted as a counsellor, had come as a camper. She was one of our poorest swimmers. To swim was painful, if one were to judge by her facial contortions as she churned the bay with her laborious breast stroke. Poor Girl! She did not realize that 90% of the girls going to a private camp are already good swimmers. She had not heard of the Pre-camp School of the Camp Directors' Association where a rigid examination for a diploma is given to swimming counsellors nor did she know of the American Red Cross Training Camp for Swimming Counsellors and her responsibility for forty young lives, to say nothing of her own life,—that had never dawned on her inexperienced mind. To her swimming meant to be able to keep above water and to get somehow, somewhere, sometime.

A young lady appeared at my office door recently (May she read this little sketch for I did not tell her). Possibly she had just had her face in a flour barrel but I doubt if she ever deals with such practical things; possibly she painted her cheeks from within but I doubt if she knows the meaning of that expression; possibly her cousin is one of her folks but I doubt if she be more than a cousin. "My cousin is in a camp and I have applied for a job too. I used your name for reference and hope that you don't mind." I assured her that I should be glad to write "something" should the director inquire. I suppose that her specialty is dramatics although I did not inquire. Would she understand—could she understand if I had asked her to express her views concerning campers emulating their counsellors? In camp one has to be "sized up." No "tinkling cymbals" for real girls of the woods. That is the last place in which one can get by with "veneer."

But I am just giving real instances of honest-to-goodness applications. All directors know that they are typical. I must now get down to brass tacks and tell some of the things that directors look for in choosing a counsellor. Grandfatherly advice—words from Aunt Minnie some will say—but nevertheless they are the things that count when one is seeking counselorship.

Counsellorship Inventory. The following list of the ten qualities of a counsellor is presented so that a wouldbe counsellor can take account of his stock on hand. The best measurement is obtained by having several honest friends write down a per cent for each quality on the basis of ten. The sum of the average of such an analysis is a certain plan of getting acquainted with one's own characteristics and often clears the way for a marked improvement. If the total adds up to 90–100% it shows a very promising outlook but of course does not guarantee success for it is by their results that we shall know them. If the average is 80–90% the chances of success may even then be fairly good. In this case there should be a second reckoning to discover the possibility of cultivating new habits. For example: the candidate who scores 4% in social background can by conscious effort learn correct English, acquire clean habits, and practice common courtesies. The remedying of a bad habit requires intelligent treatment and the time required may be a matter of years—depending upon the malignancy of the case and the persistency of the sufferer.

1. *Good Social Background:* By social background is meant the "everyday" qualifications. Is *personal cleanliness* habitual? Does the candidate use *correct English?* I know of one counsellor who is high grade in everything but his speech. That is decidedly colloquial. He says "yeah" for yes and "Don't yer know." This discounts him considerably in the eyes of his director and by the campers who know the difference. Does he know and practice the *common courtesies* of life? Is he *companionable* and unselfish or is he "the only child" kind? Is he *humble* or conceited?

2. *Love of Children:* Counsellors are sometimes a misfit because they do not distinguish between a mere interest and a love of children. In camp the counsellor lives with the boys or girls. They are playmates. How many children look up to you as a pal? If they do not do that now, what reason do you have to suspect that they will when you get to camp?

3. *Power of Moral Guidance:* A counsellor must have the character that discriminating parents wish their children to emulate. It is not enough to be physically wholesome and mentally clean. A counsellor must be a genuine character builder. He must have rock-bottom courage to say "no" or "yes" at the right time. His trustworthiness stands the test even when acting quickly as in a game. His decisions are fundamentally just and firm.

4. *Mature Judgment:* Counsellors should be mature enough to guide yet young enough to enjoy participating in the various camp activities. Campers go to him as an advisor and he respects their confidences.

5. *Attractive Personality:* As important as is knowledge, I would rather have a leader with an attractive personality than a person with a whole encyclopedia of knowledge and no magnetism. A person is worthless about camp if he cannot attract young people. The most important is the *sunny smile*—sort of a "laugh and the world laughs with you" personality, the ability to enjoy a hike when caught in the rain—to come through joyful with a blister on your foot. And then a *keen sense of humor*—the sort of a person who can

take as well as give. And he must have a *sympathetic heart*—generous at the right time and firm when necessary. And a generous supply of *loyalty*—loyal to tentmates, to the team, to the director, to the camp, to other counsellors. Does not cause cliques, crushes, or factitious taste. Can stand the test of being *adaptable*, i.e. can change plan from sunshine to rain or daylight to darkness.

6. *Physically Fit:* The physical obligations of the counsellor are clear cut. He must have the ability to lead on hikes, to run in play, to climb hills, to paddle a canoe, to assist in swimming. It is potent for him to set the example of erect carriage, and good health habits. He must know how to adapt hikes and games to the individual needs, the dangers of violent exercise, the relation of exercise to meals, the value of corrective work. The physical qualities of a counsellor are clean, sound health habits, tested endurance, reserve energy and a will to safeguard the same.

7. *Outdoor Minded:* A person who does not have an enthusiastic love of the out-of-doors has no place in camp. He becomes a liability rather than an asset. He should have a real interest in plants and animals. He should love to assist the swimming counsellor—the nature counsellor—and so on—each in turn. He should know the common camp laws of the trail. He must have an honest belief in nature ideals. He must be a protector of all wild life.

8. *Camping Knowledge:* First it should be announced that directors have unanimously decided that the best counsellors come from the ranks of the campers. It is estimated that there were 150,000 boys and girls in private camps last summer and there were many more in organization camps such as scouts, Camp Fire, Y. M. C. A., and Y. W. C. A., and there is a long procession of experienced campers back of last summer's list. It is equally true that the majority of applications come from those who have never camped. The first preparation for counselorship is to attend a well organized camp.

The knowledge of camping is an indispensable asset. The counsellor, above all, must be perfectly at home in all kinds of weather. He must possess a certainty in picking a camp site, making a fire, and organizing a meal, for his assurance radiates on the group. If he is not a master of fire building and first aid he should never be given the charge of a trip. He must recognize these responsibilities.

A good counsellor should be at least as accomplished as a good camper in any department of camping. He should have a fund of nature enjoyment and at least an amateur understanding. He should be able to point out some nature things well. He should realize that the artcraft department presents an opportunity and be eager, if his time from duties permit, to utilize the equipment. He should recognize that games have a place in the program and whenever convenient should enter in for the sportsmanship of playing. He should have a few camp-fire stories which he can tell in his own words. He should join in the camp songs. He should be in for the fun of "getting up a show." His camp training of earlier days has made him an all round camper.

9. *Specially Gifted:* Many would-be counsellors, in their own estimation, can fit in anywhere. All they need is the opportunity. It is true that a good counsellor is an all-round camper. It is equally true that he should have the gift of doing something unusual. Every group of young people finds someone toward whom they gravitate with the usual request "to do something." This counsellor is a good entertainer. He directs conversation. He possesses originality. And along with this should go the unusual knowledge and ability in one of the nine departments of camping already named. He is counsellor in nature-lore, camp-craft, or camp music—that in which he is well trained.

10. *Genuine Leader:* It is well known that a person may have knowledge but does not know how to get rid of it. This is a frequent criticism of college professors. His methods are just as important as his subject. If this is important in a teacher it is even more important in the case of a camp counsellor. In camp there are no walls to obscure the view—the chairs are not nailed to the floor, there is no filing to class. One does not have to take this and that. He is expected to do but he does that which he prefers.

The counsellor must have the power to organize. He plans in cooperation with the other leaders. He is punctual—not only at meals, but in arising, and bed making. He is versatile in his own work and a good suggester. He has the ability to put it across to young people. He knows child nature. He is so enthusiastic about his own work that the campers get it by contagion. Everyone about him feels a thrill. He is enterprising. He is firm. The group obeys. He develops leaders within the group. He has initiative and a goodly measure of tact. He knows what to do in an emergency. He changes his plans cheerfully.

* * * *

By now you may have decided your course. If you are not fitted for the work you cannot afford to waste the camper's time and opportunity. If it is still your chosen service and you are "looking ahead" to actually entering this new profession— I am going to entrust you with the name and address of people who are giving their time to the cause: Camp Supplies, Inc., 52 Chauncey St. Boston. We need to mobilize the best leaders for the great educational program of camping. They will cooperate in helping you to attain this end.

* * * *

A nature counsellor has to have all the requisites of other counsellors plus those desirable to make him a good nature leader.

III Counsel for Nature Counsellors

It has almost become a maxim that a college graduate will fail as a nature counsellor. In the same way that he tried to teach college science in the high school he is now trying to transplant classroom science into the woods, and it does not work.

One counsellor gave up her work at the end of two weeks last summer. There were many things that she did not know "because she had only required thirty flowers in her high school botany class." From her point of view it was due to her number being thirty instead of a hundred or more. From the director's point of view it was due to her "high school botany class" methods, and the director was right. "Camp Botany," where one learns thirty trees and knows not one, cannot compete with "woodcraft," where the hickory will bend just right for a rustic toaster and the sassafras gives an aroma to the woodsman's tea. In botany one looks at nature, but in camp one becomes a part of nature—as much difference as between a lecture on swimming and swimming.

The question often arises,—Am I suited to leading nature-study trips? or, How can I train myself to become a nature leader? The following notes will help individuals solve the problem.

Do you take walks in the fields and woods? Or do you prefer the movies?

Are you willing to say "I do not know"? If not you will not enjoy leading children. Those who know most say "I do not know" with most ease.

Do you use "big words"? Names are only a means to an end. Never require the learning of scientific names. Use a popular language.

Do you tell true to nature stories? Has your experience been rich enough to enable you to round out a study now and then with a true story? Or is your story telling limited to nursery rhyme and faking.

What kind of heels do you wear? The kind of heels one wears proclaims deep facts about one's innermost self. People with common sense wear low heels. Out-door leaders should have common sense.

Do you carry a jack-knife? Guides, tramps, fishermen, boys, scouts, do.

Will a dog follow you? There are people whom a dog will not follow. A dog makes friends. He knows them. Do you make friends with dogs?

Will a chicken eat out of your hand? This is a true test of worth.

Can you whistle? This joyful pastime is not a monopoly for boys. A whistling girl is an asset to any community. Can you imitate bird-calls?

Do you teach nature-study? Are you so impressed with its value that you cannot help imparting it to others? "All foreigners" you say? "All Americans" I say. Have you tried it? By what right have you not?

Did you ever see the sun rise? Probably not if you went somewhere every night last week, and the same the week before that, ad infinitum. A regular "mania" for going. Sharp contrast to "early to bed and rise." To see the sun rise once and then again because it is so beautiful is true culture.

Did you ever sleep out in the open? Fearlessness of nightly spooks is absolutely essential to good sense in the day. If you can have a good time sleeping on the rough ground you have one quality of a naturalist.

Do you attract groups of people around you? A councilor must attract. It is not sufficient "to get along" with people. How do you attract?

How far can you go into the woods without discovering something? Thoreau used to worry if he had gone a few feet without seeing news.

Are you a campfire leader? If so, you are interested in girls and nature.

Do you shrink from cows, toads, snakes, and earthworms? Do you add school girl superlatives when you retreat? If you feign fear it is because adults taught you the example. To pass this along is a crime.

Can you throw a ball like a boy? This is a test to your early love for the out-doors. I wish there were more tomboys. It is a compliment.

Do you ever feed the birds? This shows an active or superficial interest.

In what way do you protect wild life? Do you leave the wood-lily where it grew? Do you go around or through a beautiful plant collection? Do you pick bunches or sprays? Have you stepped over or on worms? Why?

Are you always dwelling on the beauties of some other place? You should not to do this unless you can see something beautiful where you are?

Did you ever climb a tree? You should once even after twenty.

What pets have you had? What became of them? Why?

Did you ever dig in the soil with your hands? A child has this interest.

What collections have you made? And why did you make them?

Are you a college graduate? It is not necessary, to be a successful nature leader. Many of the most successful ones are successful in spite of their education in school. They were taught in a woods-school.

Do you object to a tan? How much talcum powder do you use to avoid this?

Did you ever go out into the rain for a walk? Did you really enjoy it?

Will weeds or seeds or slips grow for you? Explain your answer?

How would you dress for a hike? No hat hugh muff, low shoes, muffler, hobble skirts, bright colors, etc? etc? etc?

TRAINING LITTLE CITIZENS

THE CHILDREN'S BIRTHRIGHT

By Henry Turner Bailey, Director, Cleveland School of Art.

All children ought to be familiar with the open country. They should know the joy of playing in healthful mud, of paddling in clean water, of hearing roosters call up the sun, and birds sing praises to God for the new day.

They should have the vision of pure skies enriched at dawn and sunset with unspeakable glory; of dew drenched mornings flashing with priceless gems; of grain fields and woodlands yielding to the feet of the wind; of the vast night sky "all throbbing and panting with stars."

They should feel the joy of seed time and harvest, of dazzling summer noons, and of creaking, glittering winter nights. They should live with flowers and butterflies, with the wild things that have made possible the world of fable.

They should experience the thrill of going barefoot, of being out in the rain, without umbrellas and rubber coats and buckled overshoes; of riding a white birch, of sliding down pine boughs, of climbing ledges and tall trees, of diving head first into a transparent pool.

They ought to know the smell of wet earth, of new mown hay; of the blossoming wild grape and eglantine; of an apple orchard in May and of a pine forest in July; of the crushed leaves of wax myrtle, sweet fern, mint and fir; of the breath of cattle and of fog blown inland from the sea.

They should hear the answer the trees make to the rain, and to the wind; the sound of rippling and falling water; the muffled roar of the sea in a storm, and its lisping and laughing and clapping of hands in a stiff breeze. They should know the sound of the bees in a plum tree in May, of frogs in a bog in April, of grasshoppers along the roadsides in June, of crickets out in the dark in September. They should hear a leafless ash hum, a pine tree sigh, old trees groan in the forest, and the floating ice in a brook making its incomparable music beneath the frozen crystal roof of some flooded glade.

They should have a chance to chase butterflies, to catch fish, to ride on a load of hay, to camp out, to cook over an open fire, to tramp through new country, and to sleep under the open sky. They should have the fun of driving a horse, paddling a canoe, and sailing a boat, and of discovering that Nature will honor the humblest seed they plant.

Things that children can do in cities are not to be compared with such country activities. Out of the country and its experiences has come and always will come the most stimulating and healthful art of the world. One cannot appreciate and enjoy to the full extent nature-books, novels, histories, poems, pictures, or even musical compositions, who has not had in his youth the blessed contact with the world of nature.

CHAPTER II

Nature-Lore for Camps

I. Organization

Nature-Lore originated with the pioneer who loved his woodsy home. It was the way of Thoreau and the training school for Lincoln. Perhaps Enos Mills was the first to start an organized school to teach it to others. He called it *Nature Guide* and *The Trail School*. And now thousands of boys and girls are experiencing the lore of

Long's Peak Inn, Colorado, where Enos Mills created Nature Guiding. The Trail from the Inn to the summit of Long's Peak is a distance of seven miles.

nature in summer camps. The *Nature-Lore School* was organized in 1920 to train nature leaders for the summer camps. This chapter is an attempt to present this point of view to those who are facing the unexplored possibilities of an outdoor education.

The study of nature has been constant but the ideals of nature-study have been continuously growing and changing. Our first text book to deal with nature-study objects was the New England Primer with its religious precepts. Then there was Poor Richard's Almanac which was first published by Franklin in 1732. Every household had a copy of the Old Farmer's Almanac which was used faithfully in weather prediction. Even today, in rural districts, it is strung on a string and hung on a nail for ready reference. But we do not need to learn the signs of the zodiac and neither is it necessary to hew

13

logs and thatch the roof. From rail splitter to horse trader, we were emphasizing the variations in social needs. In colonial days—when certain medicines depended on the knowing of particular weeds it was important to distinguish these weeds by name. Today it is important to know how these same weeds reproduce in order to prevent them in our gardens. New opportunities are now offered for the enjoyment of nature through such factors as automobile-camping and leisure time caused by modern industrial society. Social needs change and with this change there should be a change in subject matter. Often times, however, subject matter which was a need a long time ago, is inherited and the social needs of the day remain unrecognized by the school.

The social aim of Nature-Lore is Nature Service. Nature Service is efficiency in supplying the nature needs and wants of a community. It should contribute enormously to the leisure time. Community Nature Service includes such activities as the development and enjoyment of parks, the beautifying of the streets and commons by the Village Improvement Society, the planting and conserving of shade trees, the encouragement of beautiful school grounds, the distribution of shrubs and plants by the Chamber of Commerce, horticultural exhibitions and flower shows, nature guides for field trips and outings, scouting, planning and equipping a camping park, etc. Home Nature Service includes opportunities and guidance in gardening, ornamentation of grounds, care of house plants, the fun of having pets, the family outing, auto-camping. Any of these activities suggest worthy projects for a school.

Park citizenship requires more than a passing word. Without education our parks become a farce or tragedy. Our parks should have a far-reaching influence upon education. The fact that the maintenance and character of these parks are now in the hands of the visitors make it very important that all citizens be trained to understand the duties and opportunities in the use of the parks. Bolsheviki Russia shows what a tragedy may come from the hands of a large uneducated people. The same thing can happen in a park republic.

Nature recreation is one of the greatest contributions that Nature-Lore can make to society. In rural homes the boys and girls were busy in the woods and fields. They were living with living things. Playtime meant picking flowers, chestnuting, berrying, fishing, boating, exploring. In cities, and our population is largely urban, boys and girls do not have these natural enjoyments. The factory system and the industrial revolution of organized labor have greatly reduced human labor and increased their leisure. Our educational training must provide for complete living and for the enjoyment of the increased social leisure. Nature recreation teaches happiness in the woods, on the hike, and when camping out.

The social needs of different communities are quite different. Obviously the nature problem in the school filled with children from an Italian settlement with vineyards is very different from the problem of teaching in a country day school. The teacher must first know what types of problems occur in the lives of ordinary citizens of the community. If it is a case of pruning the present nature-study course, begin by eliminating facts which are socially of little value.

The psychological aims of Nature-Lore relate to the individual and should contribute to the broader social aim. Of the psychological aims of nature-study information has commonly been emphasized at the expense of ideals and interests. It is a great deal easier for the teacher to ask the children to learn a list and give an examination than it is to develop ideals and interests and test the results. The pendulum swings from the extreme method of learning lists to the opposite extreme of not learning any facts. The knowledge of certain facts is necessary for the welfare of human society. One ignorant person in a community may be responsible for the house flies that exist in that locality. One house fly can carry enough typhoid germs to innoculate the entire population. The sanitation of a group of people depends upon the membership of that group and can only be safe-guarded by universal education. In the case of forests, the appalling loss from preventable fires, and the economic distress from the lack of conservation are striking examples of the need of a nature education which emphasizes scientific information.

The other psychological aims of Nature-Lore which are fundamental are: (1) *Nature habits* such as kindness to animals, protection of native plants, feeding hungry birds, and leaving a clean picnic ground, must come through practice and not by sermons. (2) *Nature-Lore ideals,*—everyday beauty, "all's right with the world," sportsmanship, keeping fit, cooperation, "The world is too much with us." (3) *Abiding Interests*—current events, civic problems, and scientific progress in nature-study. The school that is disposed to fence itself off from these aims—social and psychological—is giving its citizens a one-sided training. A child in this system is in his environment but not of it. He is unable to share in the outdoor responsibilities and enjoyments about him because of his one-sided development. A host of boys and girls are realizing these aims not because of the schools but in spite of them. Their nature-lore experiences come through such agencies as the summer camp and scouting organizations.

Nature-Lore is more than knowledge, more than utility, more than discipline, more than good citizenship, more than appreciation. It is all of these interests interwoven. Our horizon is narrow if it provides growth in one direction and sets up barriers in other directions. If there is a one-sided emphasis on learning the names of birds or on gardening, or on experiments, the capacity to appreciate other things may be lost. I know of a school where the pupils obtained credit for collecting fifty insects and for pressing and mounting fifty flowers. The pupils had learned that if one had forty-nine insects that the required number could be secured by using colored ink and when the array was complete it often served for several generations in a family, being handed down from brother to sister in turn. I know of other pupils who had a genuine appreciation of nature but were killed off by the teacher's system of taxation. The tax might be to learn to recognize a tree list. But the aims and ideals of democracy have grown from taxation without representation to the application of the Golden Rule in all places. Pupils should have a much larger share in the planning of projects and in devising means of solving the problems which the general project presents.

As a further help in organizing Nature-Lore one should use the intensive rather than the extensive method. The encyclopedic method of teaching the whole field of nature—the teaching of *every* large unit as though on a Cook's Tour, such as,—ferns, shells, trees, birds, flowers, will not do. Also the holding of the telescope on one of these groups to the exclusion of all others will not do. The waste is too great.

There is also a tendency to follow the "order of evolution" beginning with the protozoa and proceeding in an ascending order to man. A zoologist almost always organizes his material in an evolutionary order. The ways in which certain structures appear to have grown out of other structures seem to demand this order of procedure. For example, the breathing of frogs by gills was probably inherited from a fish-like ancestor. Consequently in order to present the frog properly it would seem necessary to know the fish. Little children have no understanding of evolution. This arrangement in Nature-Lore is not necessary or desirable. Nature-Lore should be adapted to the capacity and interest of the pupils. The large unit comes according to the season and the details are determined by varying circumstances or happenings—never the same twice.

Let us start nature lore impromptu this fall. Impromptu nature-lore has this in its favor—the teacher has nothing at stake as a result of preparation and the subject does not put up an umbrella as though being administered a dose. An "impromptu" does not seem as important as a prepared for lesson. It is like the college student who had just been studying the use of the camera metre in photography. He tried it out with great precision on his grandmother. She looked at the print and said,—"Isn't it dreadful!" He had his picture but he did not have his grandmother. College graduates usually fail as a nature guide as they cannot make nature study natural. I asked one the other day if it were a hundred miles to Boston and she said, "No, it is ninety-nine and three-tenths." The best impromptu nature lesson is a field trip and it need not be called a lesson at all. You may be interested in the happenings on such an occasion.

The particular excursion which I am about to describe was introduced by the following challenge: "On the Great Island, just across the bay from camp, is a black-crowned night-heron rookery. The rookery is of such a character that it offends the five senses. Mosquitoes infest the pines, the day is hot and these pine wood thickets are sultry, the dead fish are unsightly and have a strong odor, and the herons often throw their last meal at you. Their cry has been likened to the Indian war-whoop. The interesting feature of the trip is that this particular colony is the farthest out on Cape Cod and the object of the trip is to band the grown-up birds. These bands are furnished by the Biological Survey at Washington and whoever finds one of these birds, no matter where, he is supposed to report to the government headquarters. There are some indications that they go to the coast of Maine before going south for the winter. It requires a great deal of skill and bravery to run down one of these birds. Whoever catches one can have the honor of naming the bird after themselves How many wish to go on this excursion?" It is needless to say that a full quota accepted the challenge.

A Fair Wind

The next step was to equip the expedition. It was decided to have Goulash for food. This meant a little arithmetic which worked out as follows for the party of twelve: Two quart cans of tomato, two one-quart cans of corn, six boiled potatoes, three onions, and a half-pound of bacon. The onions and bacon were fried brown and then added to the mixture. The goulash was served in tin dippers. The individual equipment simply consisted of a tin dipper and spoon. The Cap'n carried the canned things. Dinner was cooked and served on the shore of the island before diving into the forest.

At the rookery we did just what one would expect from the challenge. There was much scurrying through the underbrush and loud peals of laughter as the herons bluffed their pursuers by a loud squawk or by a wide open defiant mouth. The hiding of the heron's head under a sweater sort of put a quietus on the bird and he peacefully succumbed to being banded. Then it was a regular "gym" exhibition when he was placed on a lower limb of the tree. He would balance with his wings, grab hold with his beak, and clutch with his dangling legs. By trapeze performances he finally gained the uppermost branches where he again felt safe and secure. Thousands (So it seemed) of questions were shot at the leader of the party. Many were answered and many could not be. What good are herons? Why does the government protect them? etc. etc. ad liberatum. One learns more by questioning than by being questioned. Note that

this plan is the reverse to the usual "lesson" in the school room.
The enthusiasm of the trip culminated in composing an original
song which was given before the "stay at camp" people.

> The herons are lengthy birds,
> And they're wise in their way,
> With their long bills and longer legs,
> They fish in the bay.
> Trousers turned up above the knees,
> They fish without a line,
> For meal time is anytime,
> To their young in the pines.
> —*To the tune of "Father Time."* (*A Vassar Song*)

What book work was connected with this trip? I can hear this
unspoken query. My answer is *none* concerning the black-crowned
night heron. Not but what there might be many occasions for it,
and rightly. The thing that excited our curiosity enough for a little
research in the camp library was the appearance of the lighthouse
and buildings on Billingsgate Island. On the way across the bay
someone noticed that Billingsgate was unusually clear and the houses
seemed to stand up out of the water. It was pronounced a mirage.
A girl from Alabama remembered that the mirage on the Sahara
Desert made things seem to be bottom side up. Much discussion
followed but it was not settled until arriving back at camp. A book
was finally found that described what we had seen as a looming.
Everyone now knows the difference and the cause of looming and of
a mirage. This part was not planned and is one of those things that
makes every nature lore experience different.

As we look back over this experience what is there that we have
asked for as the ideals of nature lore? First there is the nature service
for the Biological Survey. The benefits of such a survey need not be
gone into at the present writing. There is the realization of the necessity of bird protection and bird reservations—a very essential thing
when it comes to appropriations to save some of our disappearing
species. It was a lesson in having a good time under difficulties—
perhaps not an aim but rather of a good training in sportsmanship.
And who is there that shall say that the information was any the
less or that the experience was less rich than an assignment within
the four walls of the class room? As to out door cooking, singing for
enjoyment, knowledge of the tides, and the hundred and one things
that are a part of the trip,—how shall we value them? We who have
tried it have unlimited faith in the Nature-Lore method.

II. *Nature-Lore vs. Nature-Study*

In 1920 the National Association of the Directors of Girls Camps
voted to hold a Nature Training School for nature councillors. This
school is now a permanent institution.

It is a significant thing that the course has been termed *Nature-
Lore* rather than *Nature-Study*. The words study, teacher, class,
lessons, etc. are tabooed in camp. Not that study is less—it is

deeper. The aims of camp cannot be discounted because of this change in vocabulary. It is a challenge for the teacher to investigate the *why* of the change.

Nature-lore is possibly the goal but hardly ever the realization of Nature-study. Both are organized nature-learning. Just as Nature-study is not Elementary Science,—so too, Nature-lore is not Nature-study. The point of attack and the results can never be the same, until the schools have their camps and their opportunities of forest recreation. Detailed comparisons will be given to show the present differences in method and what we may expect as results.

NATURE-STUDY

1. Nature-study is mainly aquaria, cages, flower-pots, and pictures in the schoolroom.

For example: In perch study the fish is placed in an aquarium in the schoolroom. Questions to direct the observation of the pupil are written on the board, such as: What is the shape? Describe the tail fin. What is the color along the side? How does this help him? Compare with the picture of the fish from Japan. Perch study in New York City or Toronto is the same. Since it is a schoolroom lesson all the pupils are doing the same thing at the same time. One must either fish or be a "clam"—there is no other opportunity. The project method may alleviate the situation but how many teachers are there who can use this method successfully in nature-study in a graded room with the hurly-burly of a modern curriculum. Most so-called project lessons are old lessons made over with more attractive bait.

2. Nature-study is mainly studying living things or about living things.

For example: A Nature-study lesson on the red squirrel means a live animal in the cage, a stuffed specimen or a picture. Let us choose the best conditions—a live animal. The pupil stands off to one side and observes. The questions may bring out a few isolated facts. Other questions require reading: Has it cheek pouches? When are the young born? Describe the front teeth. These answers are artificially produced. The pupil now has generalized statements about red squirrels which are true anywhere. The knowledge is pigeonholed with

NATURE-LORE

1. Nature-lore is mainly swimming, fishing, foraging, and photographing out-of-doors.

For example: In the camp dining hall announcement is made that we are going to Gull Pond. Those who want to go fishing may meet "A.B." on the beach to get bait. As this part of Cape Cod is sandy there are no earthworms except in a few gardens. The bait is, therefore, sandfleas. Right from the beginning perch-lore differs from perch-study. By 9 a. m. the whole camp is "en route," a four mile hike o'er hill and dune. Unlike the school one may fish or not. Those who fish acquire real knowledge of perch. Others may pick blue-berries for the flapjacks or have a shampoo in the lake. In any case one can not be busy around Gull Pond without learning much of nature even though that be not the aim of the expedition. In the group are enthusiastic leaders, chosen because they believe in their work, big sisters but not one "teacher," so-called.

2. Nature-lore is mainly living with living things.

For example: Last summer a pair of red squirrels took possession of the mail box in the pines. Some of the Campers became too enthusiastic in peeking at the "cute" little fellows and the squirrels had a moving day. The mother squirrel carried each wee baby by the nape of the neck to a safer height—a bird house in the top of a pitch pine. The campers looked on in breathless suspense— a nature-lore lesson without words. The observers had been partners in the group. They had had an experience with a living animal under other daily collections from History, Grammar, etc. More interesting than these? Yes, but pigeonholed nature-knowledge is not what we want.

3. Nature-study is mainly an outer urge. The *teacher* says: Next week we will *study* the frog. Who *will* catch one for me? *I* will put a list of questions on the board which *I* want you to answer. You may *have to* read some of the library books on frogs in order to find some of the answers. *Do not* remove the wire screen as the frog may get away. *I* will put this sign "Please give me a fly!" on the side of the jar so that you will not forget to feed it. Such a lesson is full of devices and persuasions. It is a teacher-made plan.

4. Nature-study in the traditional school is individualized.
By a traditional school is meant the formal, arbitrary school that has been in vogue since the colonial Dame School; 90% of our nature-study lessons are taught on this plan. The criticism is of the *method* of nature-study rather than of nature-study. The procedure is as follows: To-morrow, John will bring his cat to school. For thirty minutes each one of us will think about the cat. For one or two minutes each one will be thinking,—What is the advantage of the cat having retractile claws? While Mary is trying to find how many toes the cat has on its front foot the others may be wondering what Mary is going to say. Helen may give the cat some milk. (Probably all need to stretch their muscles by now.) And so it goes for the assigned time. There is no doubt in the mind of the visitor as to which person is the teacher. As to the pupils—they are all worker bees cramped in the same kind of a cell.

a natural setting. Not learned all about red squirrels? No, for a nature-lore lesson is never complete, but possibly a keener enjoyment and one to be remembered longer.

3. Nature-lore is mainly an inner urge. The camper says: Why can we not catch a frog and watch it swim? The swimming councillor has previously told them that the frog does the breast stroke perfectly. However this was not bait for a lesson on frogs. They discovered that the frog executed his strokes so rapidly that they couldn't see how he did it. A toad was then suggested. Soon some one wished to see the circulation in the foot and finally the inner organs. Only those staid who wanted to see the dissection. In fact those who thought that they might squeal or have hysteria were asked to go away. Knowledge sought by the pupil is more lasting. It is a self-assigned learning and is carried to a purposeful conclusion by the initiative of the learner.

4. Nature-lore in the summer Camp and a few schools is socialized.
The summer camp is a small democracy. Its life is one of spontaneity,—free but orderly. If John is fishing there is no class to fold their hands and watch him. One may fish or not as he choses. If someone else decides to prepare the fish for dinner all well and good. There is team work but not class work. Someone remembers having rolled a fish in clay and baking it. Volunteers want to hunt for clay. Others like the fish broiled. A competition is started in the preparation of fish,—others join the game. Professor Palmer has described in detail many socialized nature activities in his Cornell leaflets. Many such games arise impromptu. Every member of the hive is like the queen bee with plenty of room and food for growth. The period of growth is not arbitrarily put down as thirty minutes, or at 10 a. m. or for the last week in January in the third grade. There is doubt in the mind of the visitor as to which one is the leader. And to the life of the camper has been added another of those never to be forgotten experiences.

5. Nature-study is something that is taught.

In this respect it is not unlike the three R's. The same dose is prescribed in all cases with absolutely no regard for the requirements. Yet nothing has been suggested in the nature-lore method that could not be applied in the school. Peter Bell never went to school in the woods as did Thoreau. Peter Bell went to the three R's every day. And he went into a nature-study class. And when he came out he thought that a primrose was a primrose. How could he think otherwise? If there was nothing contagious how could he catch anything? And who is taking Peter Bell, and your boy and my boy to the woods as we were taken?

5. Nature-lore is something that is caught.

This audience is supposedly interested in nature. You were born naturalists. Everyone is a born naturalist but is usually killed as a naturalist before he is ten. You escaped the killing-off process. Right here I wish to ask a personal question: Was your early interest for nature kindled in the four walls of a school room or did you catch the enthusiasm outside? I have my mother to thank for the song of the robin in the old "high-top" tree. And one of the greatest days of my life was when Dad took me "bobbing eels." A nature-study lesson could never take the place of these experiences. These experiences are nature-lore.

6. Nature-study starts and stops with a bell.

Arithmetic is arithmetic no matter whether the bluebird flies South or the bluebird flies North, and arithmetic it must be.

6. Nature-lore comes at any time.

It is not scheduled and may not be planned. It may happen on the way to breakfast or during a mid-day rest. It may interrupt a baseball game or it may tip over a canoe.

7. Who are the graduates of the Nature-study School? One time I took a large party of nature-study students on a carefully planned outing. When we arrived several came up and wanted to know how long they had to stay. They did not see any fun in a wood-frolic. They wanted to get back to the "Movies." You too may have had disappointments in the lack of interest in the nature which you appreciate so well. My experience is that the nature student hardly ever gathers what may be called momentum in nature appreciation or accumulative enthusiasm.

7. Who are the graduates of the Nature-lore School?

All country boys and girls. All pioneers. It was the school of Lincoln, John Muir, Thoreau, Burroughs, Enos Mills, and Dallas Lore Sharpe. It was the school of every one who has a woodsy spot where he loves to return. The place where he feels peculiarly at home. The locality that he can write best about in his letters. It is the sea captain who longs for another trip; it is the old man who dreams of a "barefoot boy with cheeks of tan;" it is the mountaineer who returns to his West Virginia Home.

III. What can We Expect from Nature-Lore Experiences?

There are many evidences that the summer camp is not the only institution which expects results from nature experiences. In Detroit the public schools have been placed on an eleven month system, so that all the children of all the people may go to an organized camp. For the business firm we may mention the General Electric Company which maintains a camp for employees during vacations. Leaders in Social Settlement Work and Community Centers, missionaries and preachers, are seeking the why and how of nature recreation. Along with the attempt to develop the music resources of our communities is coming the idea of recreation in nature. Vacations can no longer be vacancies. What are we doing to meet the situation?

It is stated that 20,000,000 people in this country attend the motion picture show daily. They pay $4,000,000 at 18,000 theatres. This means that one-fifth of our population have the "movie mania." The moving picture man has a means of amusement (not synonymous with recreation) which he has sold for a good price. We have a better proposition which we have not been able to give away. The Sunday supplement is equally popular. We must reorganize. Now is an opportune time to place our wares on the market. Nature-lore gives experiences for the individual or for the group. It gives another wholesome opportunity for spending leisure time.

The nature-lore student finally accumulates a wealth of nature experiences. These nature experiences are the materials out of which a structure is built. This structure is nature's laws. Out of the laws comes an appreciation of nature, and appreciation which may be termed a *local nature-patriotism*. And local nature-patriotism must precede *national nature conservation*. Without this appreciation in our people any attempt to develop the conservation of our natural resources is destined to failure. And with all this can we not sing "I love thy rocks and rills" with a new fervor.

IV. *The Camp Notebook*

A long and varied experience of taking notes in the field has led to the conclusion that a camp notebook must be:

(1). *Looseleaf.* a. As each camp and camper will have individual needs and desires;

b. So that only the pages needed for the trip scheduled are inserted;

c. So that the pages of the previous trip may be safely filed at camp;

d. In order that the pages may be arranged at the end of the season and neatly bound.

(2). *Handy size*, to be carried in a pocket or attached to a belt.

(3). *Standard size*, so that printed directions, identification charts, outline drawings, tables, etc. can be inserted or transferred to all field books.

(4). *Firm, smooth covers*, that they may be used for a drawing board or a writing desk in the field.

(5). *Durable*, to stand hard usage on the hike.

Camp Directors will appreciate the fact that the Comstock Publishing Company has found such a book and is at the same time placing on the market the pages asked for by each camp director. Such a cooperative policy bodes well for all concerned.

One of the most successful pages is the tree chart which shows in a compact space the outlines of leaves of our common trees. It is a page which is typical of other key sheets to be published. The bird outlines, already well known in schools, have been adapted to this "coat-pocket" edition.

Several camps have tried out the scheme and the following notes describe its use at Camp Chequesset. The idea was so successful that the directors of Camp Chequesset wish to have other camps share the benefits, not that they will do it the same way, but that they may start according to their environment and individuality. Our slogan is "The Nautical Camp for Girls" and all our doings whisper of the sea. Mountain camps will readily translate these breezes from the shore into the airy vernacular of the mountaineer.

Page One, The Topographical Map.

The government topographical map is the best map for use in the field. It may be obtained from the Director of the U. S. Geological Survey, Washington, D. C., at $6.00 per hundred. The Survey publishes a key map that shows just which areas have been mapped. It has also selected a set of 25 maps that illustrate an interesting variety of geographic features. Some of this set have descriptive texts printed on their backs. The set can be purchased from the Survey by any one for $1.00. In ordering, use a money order or certified check.

The map is most conveniently used when cut into sections and mounted on cheese cloth. The section most often used would be the area about camp. If the camp is located near the center of a three mile radius one may cut out a rectangle measuring $7\frac{1}{8}$ inches by $6\frac{1}{8}$ inches, and have a folding map that fits into the notebook. Using the lines of longitude and latitude as a starting place, and mark off the map into inch squares. Each inch represents approximately a mile. Number the lines and thereafter places may be located by their latitude and longitude.

We now have ready one of the most useful pages of the notebook. Safe pioneering, the compass and trail, and real scouting are based upon it. The description of two ways of using the map at Camp Chequesset may suffice to indicate the possibilities.

(1). *The King's Highway:* The old King's Highway was once the route of the saddle horse and the stage coach. This was before the railroad and the state road ran down the Cape. The roadway is then pointed out on the map. The campers are now told that ye old highway has been abandoned to trees and bushes which are rapidly claiming their right of domain. It takes a good scout to follow this hidden trail. For a short distance it is seen in old wagon ruts,—then keener searching for hub bruises or blazes on tree trunks,—or just sheer luck with the compass along the valley they must have "follered." The frontiersman never had a grander opportunity for a battle of wits. You will note that this byway is crossed by longitude 70 degrees and by latitude 42 degrees and 55 minutes north. Between these two points is an advance guard. Squads will be sent at twenty minute intervals with messages for the Captain of the guard. As a parting word of advice remember that haste makes waste, the fox is cunning, and sheep follow a leader without thought.

The King's Highway. At the Forking of the Roads

(2). *Bellamy's Kettle.* Long long ago Bellamy's Pirate ship was wrecked at South Wellfleet. The inhabitants of the Chequesset country harvested the plunder of old coins, flintlocks, and kettles. Each year one of these copper kettles is hidden near camp. One time the kettle was concealed at the lowest point on the camp grounds. This proved to be at the bottom of the pond by the garden. The discoverers found something worth while jingling in the kettle. They were coins found on Billingsgate Island and given to the camp by

Mr. Nye as medals for the winners of this game. This year the kettle has been buried with a "big secret" on the highest summit northeast of camp. Find the kettle and bring it back to camp without being captured. You may study your contour maps and work out the best method of locating and moving the trophy. For each person disturb-

CHEQUESSET CAMPERS SKETCHING AN OLD SEA CLIFF. Scouting is not limited to the footprints of today. Note the Indian shell heap (A) in the right of the bank. The dark line (B) in the cliff shows the contour of the original hill top. Where once was an ancient valley (C) is now a hill or sand dune (D). The original hill (E) was deposited by the glacier. The exposed roots of the bay-berry (F) and beach grass (G) show that the cliff is retreating. When written in story form this makes an interesting chapter in the past geography of the shore line of Wellfleet Bay.

ing the camp routine, such as lateness to a meal, a point will be taken off the final score of that team. The captain of each team will appoint a place for you to meet for council and maneuver. The success of the expedition depends upon strategy. Compasses, field glasses, pedometers, etc., may be obtained at the camp library.

The Page of Tree Leaves.

(A few suggestions for use.) On these trips the contour map and the tree page are carried in the notebook.

(1). *Tree Spying.* Stop at a tree, such as the wild black cherry. Each one identifying the tree by use of the leaf chart within three minutes time is given a point. For each mistake a point is subtracted. At the end of the trip add the scores and announce the winners.

(2). *Tree Scouting.* Appoint leaders to choose teams. Tell them to study the oak leaves on the chart and then at a given signal give them three minutes to obtain a white oak leaf. The tree given should be known to be nearby. At the end of three minutes blow a whistle. Those back in their places with a white oak leaf (no more, no less) score a point. Next send them scouting for a red oak acorn, a balm-of-Gilead bud, and so on. The team scoring the greatest number of points represents the group of best tree scouts.

(3). *Tree Trailing.* Hide messages "en route" and send out companies 30 minutes apart. The messages may read as follows: Take the valley trail to the east until you see a large yellow willow. In an abandoned flickers home is a note. Read it carefully. This note may read,—Within sight of this spot is a silver poplar. As far from the tree as it is high and in the direction of its noonday shadow is buried a message on birch bark. Please leave this scroll as you find it. The group following the directions farthest and in the quickest time wins the honor of the trail.

(4). *Tree Cribbage.* This may be played for a time when on a long hike. Assign a numerical value to certain trees. One group may take one side of the road and the other the opposite or the points may go to the side recognizing the tree first. In this case it leads to a long range recognition by form. (Trees "en masse" and silhouettes suggest interesting rainy day projects for the notebook.)

Taking the Notebook on Special Trips. As an example of this sort of use of the notebook we will describe our trip called,—"To the Hermitage:" Every bailiwick has its hermit. Ours is a grizzly sea-dog who has taken to land some two miles from the coast. He is a Thoreau-like individual reminding one considerably of that famous naturalist who walked the length of the Cape some three-quarters of a century ago. The present recluse has squatted on the site of his great grandsire's claim and his tract reaches unto the shores of the same pond. From the cedar swamp in back he has lugged, dragged, and rolled in turn the logs for the framework of his hut. The adz, an heirloom, has again played its part and the timbers have been slowly hewn into shape for the sills and rafters. A clump of lilacs, he will tell you all this, marks the east bedroom of the old homestead—long since tumbled and gone. Of the old days—naught remains to suggest ancestral fortitude or thrift but scraggly apple trees, decrepit, gnarled, and windblown, with a few belichened fence rails.

Our equipment for this trip is the notebook with a map, two pages of drawing paper, two pages for notes, colored crayons, a reverence for the crudities of pioneer days, an eagerness to hear and understand a backwoods language,—a woodsy speech which has all but disappeared, and a desire to express the experience in writing and in sketch with an understanding heart. Following a perparation in spirit it was suggested that the campers might like to make a list of the evidences of an early homestead, the methods of a pioneer, the reasons for believing that he has a love for nature, and quaint expressions. What greater wealth of material could one wish for a future school essay (if one can forgive such trespassing) or, if the spirit of the letter has not been killed, for the very joy of writing literature (note spelling with a small l).

Note, then, that every camp region has its hermit. Hermits vary. It ought not to be necessary to say "know your hermit." Hermitages vary as also does the region there-abouts. Therefore, every camp has an experience unto itself with respect to hermits. Not only that but every individual camper has a different reaction to the experience. The stories and drawings therefore show an individuality of response. A few of these are selected at random from the notebooks. These quotations have not been changed. They are borrowed from the personal property of the owner for a definite cause and not for criticism. And let it be emphasized that nature councillors are not to trespass on this private property with a red eye or a red pencil for spelling, split infinitives, or vertical twists to the penmanship. The number of poets and writers killed off by this method will never be revealed but let us not kill the spirit in camps. And as Mrs. Comstock says in her *Handbook of Nature-Study* "These books, of whatever quality, are precious beyond price to their owners. And why not? For they represent what cannot be bought or sold, personal experience in the happy world of out-of-doors."

Mr. Dyer (The Hermit). From the field book of "Bumps," age eleven years, the youngest girl at Chequesset.

"His grandfather settled here years ago. The pond was named for him. The lilac bushes and the fruit trees indicate the great age of the place. The Hermit has planted boughs on the north side of his corn to protect it from the cold. He also made a wheel-barrow with much patience and care. He has made a little bird house on the top of a stick driven into an old stump which has been there for many years. He has some timbers left from those used to build his house. Back of his house he has made a chicken coop of pine boughs. He has placed boards on either side to weigh the boughs down and keep them together. He shows his interest in flowers and trees by planting and taking care of both. Around his garden is a fence to prevent the deer's (they are seen frequently in this region) from eating all the beans over night as they did one year. Years ago the house used to be almost up to the water's edge but the new house which Mr. Dyer made himself is back much farther. Mr. Dyer seems to take an interest in camp girls and is not a bit timid about answering questions."

Photo Mount Sheets: These are special sheets for snapshots. These little "pleasantries" like the principal on compound interest, double in value in a few years. Is it any wonder that this book, probably Volume I of the author, becomes so highly prized as years roll on?

The Bulletin Board is an important adjunct to the notebook. At the end of a trip a few of the best and most original sketches, stories, poems, snapshots, and collections are exhibited on the bulletin board. This bulletin board is not a place for drying or airing ideas. It is a live place because it is newsy. Yesterday's hike differs from today's story and tomorrow's game. The blackfish are playing in the offing,—It is a kind of whale,—Watch for any proof? The bluefish are running down the Bay,—Ten fishermen are wanted on the wharf at 3 p. m. Beach plums are ripe,—Follow Barbara right after

EVERY CAMP SHOULD HAVE A TREE CHAMPION: The Camp Chequesset tree is the PITCH PINE (Pinus rigida). It stands for ruggedness and individuality. It overcomes difficulties. It never displays primness. It is picturesque. It sends out a healthy aroma and lulls campers to sleep with its murmuring branches. To those far away this spray from "The Pines" may revive pleas- ant memories of camping days Down on Cape Cod. (Post cards were made of this photograph and on a rainy day the campers colored them in with Jap- anese water colors and sent them to old camp friends. Such a spray never fades.) It takes three seasons for a pine cone to ripen. The three stages are shown in this picture. How do the pine needles of this spring's growth differ from older leaves? The staminate or "pollen blossoms" are also shown. Find a few. Find some scars of the needle clusters. How long do the leaves stay on the pitch pine? How can you tell that the larger cone is ripe?

breakfast. The sand pipers are hatching,—'Nuf sed!' Poultry men sign here for turn to collect the eggs. There is always something going on. It is refreshing. It is a medium for the exchange of ideas and more than one good resolution is made to an ideal (not idol) at this shrine of activity (not sanctum sanctorum).

Nature-study Forms (Bird, animal, and flower plates).

A resolution was made at Chequesset that no study of plant or animal life should be made except it be alive and in its native haunt. It is surprising how many of these opportunities are waiting for the one who looks for them. When ordering our supplies for the camp notebook we limit the order to the living things at hand.

Last June we knew that the barn swallow was feeding her young on the bungalow porch; that the chickaree had rented the mailbox in the pines; that a chippy, a robin, and a pine warbler were within 30 feet of the dining room door; that Bufo the toad was beneath the steps; that the blue-birds were rearing their second families in the boxes; that red perch would be caught at Gull Pond; that the squid, and the skate, and the hermit crab would be seen on the shore; that the swamp azalea would scent the ponds and arethusa dot the meadows. We knew many, many friends that would be at camp and because we knew these friends, we knew just which outlines to order from the publishers.

Then when summer came with a rainy day there were those who were anxious to color in the outline of the barn swallows as they fed the young under the eaves of the piazza. And just because someone was interested in doing this, others became interested. Then so many caught the spirit that it was quite the fashion. And just because it was not a school there were others who did not care or shall we say did not have to "join in." They had found something more interesting,—perhaps they were reading Cap'n Eri, or were making fudge, or a pine needle basket, or were just gazing at the colors of a drift-wood fire. What ever they were doing it was genuine. And we believe that it was all making for a broader intelligence.

The pine warbler arranged her class room for a sunny day. She had built a nest in a small pitch pine about six feet from the ground. In this case a small table was moved to the side of the tree. Here a camper could sojourn with water colors or merely pause for a peep at the young on the way to mess. She who studied the swallow might not care for the pine warbler or might have some other interest— and at the camp the interests are varied and many. Several passer-bys so gained the confidence of the mother bird that they were able to stroke her—not a mean accomplishment in itself. And to that nervous little somebody from so far away this touching of a wild bird was a contact with something far greater—it was a momentary connection with steadiness of motion—better than notebooks or rubies—for therein was the foreglow of self-composure and calmness. To all this the nature councillor must subscribe in vision and sympathy. What one should do on a particular occasion cannot be forecast. It is not preparing prescriptions nor is it dealing out patent medicines. It is furnishing opportunities for the love of the beautiful and timely suggestions for companionship with out door life. If in this the councillor be sincere her companions will be many.

Chequesset Nature-cards. Plate number VI shows another source of instruction. It is a story told by four pictures. These pictures are taken from the camp environment and arranged to tell an important truth or serial event in nature. The cards are given out as a surprise after having studied the phenomena or they may serve as a suggestive basis for the written description. In either case they form a bright spot in the complexity of future living when the writer has a minute to pause for nature reminiscences of camping days.

CHEQUESSET NATURE CARDS. No. 1—*From Ponds to Reclaimed Land.* Questions: Cause of ponds? How do long ponds become round ponds? How are round ponds conquered by vegetation? Uses of meadows?

Garden Competition

This was a novel competition in establishing boat gardens in old seventeen foot dories sunk in the ground. The flower guide sheet of the Camp and Field Notebook is a great aid in the game. Wellfleet is an ideal gleaning ground for such a fête. When the villagers turned their trade from whaling in the deep sea to quahauging down the bay the houses were gradually moved from the ocean side of the Cape to the harbor shore. The pretty posies of the old fashion gardens were abandoned to fate and now come camp girls scouting for the choicest bloomers to ornament their boat plots. Tansy, iris, spurge, house leeks, yuccas, and dusty miller were delegated to their proper places alongside the Hudsonias and stonecrops from the wilds. The opposing teams looked on with admiration as finishing touches were placed here and there before the arrival of the judges.

And then came the judges with their notebooks—the art councillor, the nature councillor, and a senior camper. The gardens were viewed most critically—with a fine tooth rake as it were. First the judges took notes as to color, then as to choice and variety of material, the arrangement and symmetry of form, how the material was planted, and then the originality. One crew had made a sundial while the other had purchased a sailor windmill and this was against them as they had spent money rather than ingenuity. And the idea of permanency,—one garden showed a tendency to run to plants that were in blossom for the time being. Even the trimness of the ground around the boats did not escape their scrutiny. Solomon's garden in all its glory could not have been surveyed as were both of these. *A Final Word as to Notebooks.*

We have been dilating about the opportunities and some of the uses of note books at Camp Chequesset. If you are to use a similar scheme you must get on deck and look at the beauties of your own harbor. There are no more beautiful skies, no more sweeter song birds, no greater history, no stauncher patriotism in any harbor than at Wellfleet and these things are not better in Wellfleet than in your town. Some day we hope to come over the mountain and hear about the notebook on your side.

V. Nature-Study Equipment

Aquarium
Half-barrels are good for water plants and fish.
Battery jars for glass aquaria—Whitall, Tatum Co., New York City.
Fruit jars make good small aquaria.
Home-made glass aquaria may be cemented together. See Hodges "Nature-study and Life."

Bee Hives
Observation Hive with protection when not observing. The A. I. Root Company, Medina, Ohio. "How to Keep Bees," Comstock, same Company.

Blackboard
Home-made blackboard—use linoleum in wooden frame. Paint black.

Bulletin Board
Use natural color art burlap. Andrew Dutton, Canal St., Boston, Mass. Twenty-five to thirty cents per yard. Note: All prices are liable to vary.

Charts
Use unbleached muslin. Camp Kehonka uses curtains on rollers.

Birds
Audubon Society, 1974 Broadway, New York City.
Miniature Bird Pictures, 80 subjects, 3 x 4 inches, 1c. each. Game of 52 Wild Birds; Game of 35 Wild Flowers, 35c. each. Post cards in natural colors of Wild Birds, Animals, etc.
Seventy-five leaflets, 43 colored plates, 41 outline drawings, 97 half tone illustrations, $1.75 postpaid. One hundred fifty colored lantern slides, 80c. each. Bird Charts No. 1 and No. 2 especially recommended.
International Harvester Company, Extension Department, Harvester Building, Chicago. Mimeographs Copy of Helps in Bird Study sent free. Working drawings for making things, 1c. each: 1. Wren House. 2. Bird Feeding Station. 3. Nesting Box for Robin. 4. Plant Protector.
Mumford, A. W., 160 Adams Street, Chicago, Illinois.
Birds and Nature Pictures, colored, 25c. each.
National Geographic Society, 16th and M Streets, Washington, D. C., "200 pages illuminated with 250 matchless subjects in full colors, 45 illustrations

in black and white, and thirteen striking charts and maps." $3.00 post paid in United States.

Perry Picture Company, Malden, Mass., 300 bird pictures in color. size 6″ x 8″, 2c. each. Special prices on large orders.

State of New York, State University, Albany. Birds of N. Y. $1 set.

Winnetaska Bird Charts, Dr. John B. May, Cohasset, Mass. Cards 5½″ x 3½″ with outline drawings, legends, and notes on habits. Eighteen land birds, 4 each of wading, swimming, and birds of prey. 30c. per set post paid. Recommended for field work.

Clouds

Charts of cloud forms: (a) Blueprints, 16″ x 23″, 5c.; (b) Classification of clouds, colored, 20″ x 24″, 25c. Free through Congressman or U. S. Department of Agriculture, Weather Bureau, Washington.

Geography (Pictorial)

Land, water and air, 48 sheets, $1.00; United States (Prelim.) (48 sheets) $1.00. Pictorial Geography, Dept. B, 16 and M St., Washington.

Insect Breeding Cages

Simple cages covered with mosquito netting. Have sods, soil, or potted plants inside. Adapted to studying the life histories.

Leaf Printing

Need the following material: printers roller, tube of printer's ink, photo-mount roller, piece of glass or old slate, and paper. See Comstock's Handbook of Nature-study for directions.

Maps

Most valuable is the Topographic Map published by the U. S. Geol. Survey, Washington, D. C. Obtain Monograph 60 by Atwood, Salisbury.

Outline maps, made by means of mimeograph, are most valuable. This is especially true of the immediate locality.

Minerals

A collection of the minerals of the region and exhibits of the local mining industries should be in the Camp Museum. Twenty minerals and 20 rocks in compartments, $4.50, Chicago Apparatus Company, 32 S. Clinton St., Chicago. Also Ward Nat. Hist. Est., Rochester, N. Y.

Museum

Making a case is a good manual training project. Use vials for seeds and soil. Labels. Exhibits of local industries.

Riker Mounts for insects. Kny-Scheerer Co., 225 4th Ave., New York.

Notebook

The *Comstock* Publishing Co., Ithaca, New York is putting on the market a loose leaf notebook especially adapted to nature-study in the field. It also has pages for accounts, music, and photos. The name of the camp or school will be printed on the outside. Every camper should have one of these notebooks. Price about $1.25, at least as near cost as market conditions will allow.

Pictures

Perry Picture Co., Boston, Massachusetts. 7″ x 9″ pictures in natural colors of birds, animals, minerals, fruits, etc., 2c. each, assorted as desired. No order for less than 25c.

Thompson Blue Prints, Syracuse, New York. Subjects general.

Stereopticon Lantern

Acetylene burner and prestolite tank for camps without electricity.

Tree Survey

Send for Horticultural and Pomological Investigations, B. P. I., Form number 219, U. S. Dept. Agriculture, Washington, D. C. Instructions for taking a Bird Census, U. S. Dept. Agric., Bureau Biological Survey, Henry W. Henshaw, Chief, were published in 1914 and 1915.

Trees

Trees in Silhouette, by Henry Turner Bailey. Published by Atkinson, Mentzer and Company, Boston, Mass., and New York City.

Forest trees, each picture 9″ x 12″, consists of 3 pictures. Price 1 set, 8 sheets, 24 pictures, 40c. Three sets $1.00.

CHAPTER III

THE NATURE WAY AT CAMP CHEQUESSET

"There is a pleasure in the pathless woods;
There is a rapture on the lonely shore;
There is society, where none intrudes,
By the deep sea and music in its roar:
I love not man the less, but nature more."
—*Byron: "Childe Harold's Pilgrimage"*

This chapter is written to illustrate how the Nature Lore method has been carried out in one camp. No two camps could furnish the same experiences. It must always be different. All camps do have, however, similiar possibilities. A good leader will discover the nature resources peculiar to the particular camp in which he serves. The method will be the same. The "daily dozen" which follow will be suggestive as to mining the nature possibilities in other camps.

Would you like to know how we study nature at Chequesset. We don't study it. They provide nature experiences for us. I want to tell you about a few of these experiences.

1. One day last summer we went on a sailing *Trip to the Heronry*. Professor E. H. Forbush, the State Ornithologist for Massachusetts, is a frequent visitor at camp. His coming usually means a trip to the heron rookery on Great Island, and the Tern Colony on Jeremy Point. We band these birds and have learned that some of the black-crowned night herons on the island go on a hunting trip to Quebec, in the fall, instead of going directly south as was expected. Mr. Forbush always tells interesting experiences around the drift wood fire. On the way home we made up a song to the tune of Father Time. It is one of our camp songs now.

2. *The camp flower* is the beach pea. This was decided after an enthusiastic campaign between the devotees of the beach pea and the backers of the sea lavender. (Fig. 1.)

3. *A visit to the Hermit* is full of nature interests; we are always welcomed, on this, our annual visit. The adz and the hewn rafters, his rustic bird houses, the white lily which came ashore in the British ship Jason, stories about the taming of the Canadian geese and "Gandy," the nest of the wild duck, and the feeding of his bob whites stir our imagination and broaden our nature conception. (Figs. 2–3.)

4. You would have enjoyed *The Crow Debate*. Last summer two families of crows started their career in the pines by the lodges. They were rather indiscreet in becoming garrulous early in the morning. They were brought to trial. There were lawyers for the defense and others for the prosecution. The jury found the crows guilty but because of their youthfulness and future use to the farmers of the Cape they were sentenced to live the remainder of their young life in an onion crate in the top of a pine tree some distance from the lodges. (Figs. 4–5.)

Fig. 2

Fig. 4

Fig. 1

Fig. 3

5. On *Foraging expeditions* we gather blue berries or strawberries for pan cakes, harvest succulent shell fish on the shore, or bake perch from Gull pond, any of these occupations is enough to whet one's appetite without the addition of the salty air for sauce. Getting clay, bayberries, quahaug wampum, pine needles, and cat-tails for arts and crafts makes that industry doubly interesting. (Fig. 6.)

6. *A Whale of a Story.* Wellfleet or Whalefleet is said to be the second whaling port in the country. On the old barns you will often see whale weathercocks. At the base of the cliffs, toward Herring River, or up by Blackfish Creek one often discovers the bleached bones of the leviathan partly hidden by the beach grass. Then again, on the back shore a keen eyed camper is often rewarded by glimpses of passing whales. This all goes to show that Wellfleet is a wonderful place for a whale of a story. (Fig. 7.)

Every story must have a heroine. Look now, as to how this port was settled by girl campers. In olden times the Chequesset tribe of red men held clam bakes on these shores and the shells heaps still testify to their ancient festivities. In 1914 girl campers resolved to use the same name and established themselves as Camp Chequesset. What wonder, then, that Chequesset Campers dig clams, push out in canoes, and as they get more experience explore Duck Creek in sailboats. Wellfleet Bay is the summer home of 50 sea-hermits. Camp Chequesset, the nautical camp for girls.

It was Tuesday, July 28 that the whole camp turned out to go "Down to the sea in ships" at the village movie house. We sat on the edge of our seats in suspense as the mighty whales lifted the boats high out of the water. Then we saw, "Belay!" "Where away?" "Two points off the starboard bow, Sir!" The Chequesset girls broke out with a hearty clap. It was such a sudden burst of pleasure that it had to be explained to inquiring villagers, who visited the camp the next day, that that very lingo was the Camp way of saying "Hello."

Now we Chequesset girls have "old salts" come in to tell sea yarns. Cap'n Stull, the ambergris king, and an old camp friend, came over to camp the evening of July 29th to tell us girls stories.

Ambergris, you know, comes from the intestines of a whale and gives the lasting qualities to perfumery. Well, sir, Cap'n Stull brought over his sea trinkets. In some old log books he showed us drawings of whales made by sailors, ambergris, high grade watch oil from the head of the blackfish, the claw of a fifteen pound lobster, and the sketch of a whale graven by a fisherman on a sperm whale's tooth. He had a whole sea-bag full of interesting stories and it was with great difficulty that Mother V. dragged us away to bed. Cap'n Stull is the hero of our yarn.

Imagine this! Just a few days after Cap'n Stull's visit to camp— August 3d to be exact—we were all in swimming. Cap'n Bill came running down to the wharf and announced that 105 blackfish were ashore in East Brewster. We were all aglow with excitement. Mrs.

Fig. 6

Fig. 8

Fig. 5

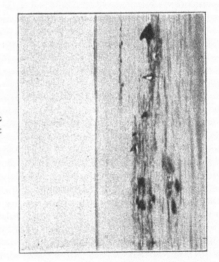

Fig. 7

Cram hurried her famous dinner of swordfish, mashed potato, peas, and all the fixings—topped off with ice cream. Cap'n Taylor was rounded up and by 1.05 P. M. all the mermaids and landsmen were aboard the camp launch, Mouette. We were outbound on a whaling trip, and right over the same waters as the whaling boats of yore. On the leeward shore was Eastham. We could see the white sand beach where the Pilgrims beheld their first blackfish in 1620. Chugging and puffing the Mouette bore us down past Billingsgate. Quahaug boats were on their moorings. The old rum runner with her weather stained hulk made a black shadow against the island. These things did not interest us today. We were bound whaling. All eyes were scanning Cape Cod Bay.

"Blackfish!" "Blackfish," "Right ahead!" "Off the starboard quarter!" Every Chequesseter was startled by the cry. In the distance black specks were bobbing out of the calm blue. Cap'n Bill grabbed the mooring hook and stood on the bowsprit. Cap'n Taylor grinned from the cabin window as he headed his boat toward the school. It was school in two senses of the word as you shall see. Cameras were held in readiness as the Mouette swiftly bore down on the prey. We gazed at the spouting blackfish. They would arise and dive, their glossy backs glistening in the sun. (Figs. 8-9.)

"Woopie!" "Hee-hee!" cried the campers up forward. And the wild screams were taken up by others aft. The boat churned on through the foaming mass. Thousands of questions were asked. A veritable school room, yet different, for the questions were being asked by the pupils. We were close enough to the blackfish to see their spout holes and to hear their puffing. Yes, these blackfish are a species of whale. No, not fish, for they breath by lungs. They will not tip the Mouette over. The fishermen drive them ashore with boats. It was at least half an hour before we got through asking questions and if there is such a thing as an examination I guess it's when we tell this story and show the pictures to our friends. Examination or no, we will never forget this whale of a story.

Cap'n Taylor now headed toward the Brewster shore. While the ship's company were gazing along the beach Chubbie cried out in her sonorous voice, "There they are! There they are!" The Mouette swung to. "Lower away!" came across the deck. The campers sprang over the rail into the dory. "Give way!" the Cap'n shouted. "Ay, ay, Sir!" cheerily cried the Pirate Crew, as the oarsmen pulled for shore. All eyes were riveted upon the black forms on the beach. "Give way there! Pull way together!" The loaded dory grated on the sand. All hands o'er the gunnel and the Cap'n shoved off for another load.

It was a strange sight to see these huge bodies boiling in the hot sun. And there was Cap'n Stull with his men. The heads had all been removed by his crew to get the famous watch oil. The sight was wonderful and awesome. Look at this one, piped up Speedie as she leaped upon the foaming prey. All hands lined up for a picture, singing the

Fig. 10

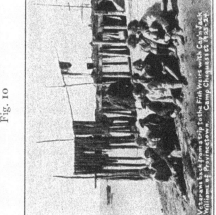

Fig. 12

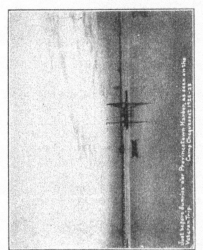

Fig. 9

Fig. 11

good old Chequesset chantey, "The Chequesset girls don't wear any combs, for they comb their hair with a cod fish bone," etc. "Mike" was so tickled that she danced a hornpipe right on the whale's deck. The steel blue of the sky and the sea were becoming one as the Mouette pulled up anchor to return to camp.

A note by the directors: This is a true story told by one of our campers. It answers two questions. One question is—what is your daily program and we always have to answer that we never have two days alike. The tide changes every day and we know not what the next tide will bring. The second question, is, how do you teach nature-study? Well, this is an example of a nature lesson at Chequesset. It is one of the many that a camper lives and experiences. How different from learning a list of 50 fish?

7. *A Chequesset Trip to the Fish Weirs.*

Dear Folks at Home:

The other night Dan Harvey and Cap'n Jack Williams came over from Provincetown. Dan Harvey sang sea chanteys and had us sing with him. It was simply great. Then Cap'n Jack told us fishing stories. He had caught a horse mackerel that morning and we seemed so interested that he told Cap'n Bill that we could go out with him some morning. So last Thursday we went. (Figs. 10-11.)

Imagine it! At 2:45 A. M. we were creeping around the pines trying not to wake up the whole camp. It seemed funny to be eating breakfast. We were soon speeding through the darkness toward Provincetown. Some were admiring the sunrise, but when we got along a little ways we discovered that it was Highland Light.

There was a grand rush for the wharf only to find Cap', Williams was going by old time. Edna sketched the sunrise and the boat silhouettes in the harbor. I am sending a snap shot of the sunrise. It does not half tell the story for it was just gorgeous as it came up out of the Atlantic. I never saw the sun rise before.

Cap'n Williams was surely good to us. He not only took us way out to his fish traps but when we got ashore he showed us around a cold storage plant. I am inclosing a picture of him and the girls that went out. It was great fun to watch the artists. (Fig. 12.)

We drove back to Highland light where a delicious luncheon was enjoyed by all. Then we visited the bayberry dipping place. Serat tried to buy a bayberry mint.

It is most time for taps and I am awfully sleepy. They call it putting out the anchor lights and piping down. We have a treasure hunt tomorrow.

Your loving daughter,
ANN OLE CAMPER.

8. *The Old Tree Stump* is beneath the tide line. The other day one of the camp daddies took us over to see the old stump. He dug around it and cut off some of the wood. We had a great time trying to figure out how this tree use to grow in that spot. When we got home

Fig. 13

Fig. 14

Fig. 15

Fig. 16

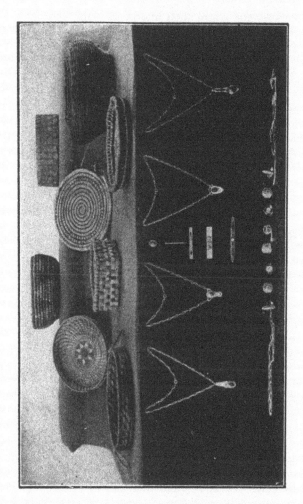

Fig. 17

Cap'n Bill told the story about Egg Island that use to be a little way out in the bay and can be seen at low tide. The Indians use to chase the white men onto this island. (Fig. 13.)

9. *The Buried Shell Heap.* We took our notebooks and went over to the cliff by Herring River. We had colored pencils and made a sketch of the cliff. The green beach grass, the brown cliff, the ashes from the old fire, the white shell line in the cliff, the light dune on top of the shells, and the contour of the former glacial hill made a wonderful sketch. Then we wrote the story. The glacier deposited the hill, the sea made the cliff, then the Indians had several feasts as shown by the heap of shells. After the generations of feasting the wind covered up the shells with sand. Then came the beach grass to anchor the dunes. The picture shows us trying our hand at the story. It was great fun. (Fig. 14.)

10. *Scene on the Gypsy Hike.* We learn how to make camp fires and beach beds, to pitch tents and to blaze trails, to use a compass and topographic map, and to understand signals; all these essentially belong to camp life. (Figs. 15–16.)

> To construct one's bed in the forest;
> To go to school in a sailboat;
> To say one's lessons to an old scout
> Is the pleasant lot of an outdoor girl.

11. *Arts and Crafts.* The quahaug pearls, wampum, cat-tails, pine needles, bay-berries and sweet grass lend a Cape Cod atmosphere. Weaving old fashioned rugs and tooling leather are other attractions. This novel industry is carried on by one of the skippers at the Bungalow. The Chequesset idea is to gather the same raw materials from the environment that the Pilgrims gathered and know some of the Pilgrim arts. (Fig. 17.)

12. *Shadowgraph.* Just as the fear of an eclipse has been removed from modern life by the aid of science, so has the dark shadow of conventionality been removed by the summer camp. Primitive man fled in fear as the moon blotted out the sun's light. Today we can go about with intelligence and assurance. A Chequesset Girl loves the Green Grass Moon as she is ushered into camp and the light of the Harvest Moon as she bids farewell. She experiences the tides as regulated by the moon and the cool breezes born of the sun. She knows them because she lives them.

Just as every one of the 15 objects in the above sketch is dependent upon the light-giving and life-giving qualities of the Sun, so is your health dependent upon simple outdoor living. Old campers will enjoy seeing how many of the 15 shadows they can name. New campers will enjoy an imaginative trip of the same kind. After you have listed these shadows write a sentence after each one telling how it is dependent upon the Sun. Then write a paragraph on how the sun can help you this summer, and send it to Cap'n Bill.

CHAPTER IV

The Camp Museum

"The beauty of the world has never been of great pith or moment to mankind. Its admirers are few, its destroyers are many. . . Will he never learn that *happiness is not a matter of possessions*, and that mental content, joy of heart, a love of loveliness, are more potent factors in human well-being than naval power or commercial gain."—*John C. Van Dyke.*

One should first of all recognize that the camp ground is a museum in itself, but it is often more fun to set aside some nook as a *Nature Den.* This area should include a pond and brook for water life, a rockery for certain ferns and lichens, a shady corner for native flowers like the mocassin and orchids, and an open place for field plants. The nature den may be provisionally marked off with gray birch rails but an arbor vitae or red cedar hedge is more permanent and gives it a secluded atmosphere. Two evergreens may mark the entrance with a rustic sign overhead or two rock piles covered with creeping vines. When one enters he sees markers indicating the Fern walk, the Rhododendron Path, To the Turtle Pond, The Wild Flower Garden, The Spring, follow this trail to the Woodpecker's Hotel, the Beaver Houses are this way—no loud talking.

If the visitor takes the Rhododendron Path he may see small wooden signs along the side telling the common names of shrubs, trees, and flowers. Beneath the name is written an interesting fact about that particular plant. Some that we see may be: Shadbush, Blossoms when the Shad go up the rivers to spawn; Trembling Aspen, A "Weed Tree" used with Spruce for paper pulp; White Ash, Useful for basketry; Sensitive Fern, Young shoots a substitute for Asparagus. Attention may be called to the Oven Bird's Nest, Do not step on it; or Red-eyed Vireo's Nest overhead, young hatching, do not disturb; Dog Toothed Violets, It took them seven years to produce seed, do not pull up; Can you find the four kinds of ferns on this ledge? Help us to protect them; Mountain Laurel, two thousand people visited this spot last year and did not break this shrub; this is a white oak stump. How old was the tree when cut? Which way did it fall?

We may see some young people at work with a trowel. Upon getting nearer we see them setting out ferns in the shade of some hemlocks. We investigate and find that they are carefully transplanting maiden hair ferns. This will be a new fern in the sanctuary and they tell us that whenever a new plant is brought in that it is put as nearly as possibly in the same kind of a haunt as it's home.

Back of this clump of Rhododendrons we see a sign that says, *Campcraft.* Walking out into an open area we see various cubbyholes in the shrubbery. The first one has a lean-to. We walk on and cross a corduroy bridge to the fire-places. Each is labeled as to name and purpose. In another opening are various camping shelters. Bird houses and feeding stations seem to be scattered all about.

43

Now we visit the *Nature Garden*. Our young guide explains that the hoarhound is used in making candy, that the peppermint flavors the ginger ale, and that the milkweed is used for a pot herb. Yes, we have future plans and have already contracted to raise Ginseng for a certain commercial firm. That row of cuttings is of the basket willow. By another year we are going to cut the year old shoots

A ROCKERY IN THE GARDEN of the Bridgewater, Massachusetts, State Normal School.

for basketry. We believe that it will be cheaper and more fun than using raffia. The golden rod garden is a marvel for color and variety of form. It hardly seems possible to get together so many kinds. The sunflowers and wheat are for seeds for feeding our bird visitors in the winter.

Each camp, of course, must make a survey of its grounds and see what it has that can best be made into a Nature Den. I once saw an amphitheatre of fire places made in an abandoned gravel pit. One of the most unique Nature Gardens I ever saw was in an old quarry. It was like a visit to fairy land. The seams of the rocks made natural stepping stones. The stone stairway led past seeping pockets of hanging ferns and polypody. The last descent was by ladder which took one onto the quarry floor. The whole had the effect of a room. The draperies were all sorts of vines, clematis, frost grapes, woodbine and bitter sweet, I remember, which had been trained along the walls. Here and there were white quartz with nature quotations carefully lettered in moss-green paint. A seam of talc was featured. From the mineral exhibit it was but a step to the pool which was growing native water plants, each being identified by a little map held flat on a rock shelf under a plate glass weight. Here and there were rock-chairs and fire places, perfectly safe, even with flying sparks. The speakers pulpit was nearly hidden with greenery. The whole setting was the work of a master and planned to the minutest detail but the little schemes did not catch one's eye for some time. What at first had been considered as a very undesirable feature of the grounds, perhaps by force of circumstances, had become a gold mine instead of a lime quarry. It was like a great painting. The campers were getting as much out of the Outdoor Museum as they were putting into it. And we must say that they had put their heart and soul into it.

The *Indoor Nature Museum* has been as little understood as the Nature Den. It is not a "canned goods" establishment where things are shelved. It is not a set of curios preserved with mothballs. It is not a collection of delicate wares marked "Don't touch." It is not a set of mummies pickled in preserve jars. The modern museum has outgrown the idea of being a repository..

There are three requisites of the Indoor Museum. It must be interesting, active, and useful. There can be no compromise. If this test were applied to some indoor museums that we have seen there would be some radical changes. We are going to keep these three things in mind before we allow anything to cross the threshhold and when anything ceases to be interesting, active, and useful it must leave the museum.

The size is not so important as the lay out. Any small room tent, or lean-to will suffice. Around the border will be benches or shelves somewhat below the eye level of the children. The walls will have art burlap. Beneath the benches are air-tight chests to store things away from mice and insects. Suppose we look around.

On the wall we notice a set of fire pictures. If a group is assigned a "trench fire," and they do not know how to do it, they do not ask questions but consult the picture. We also note a map of the region. We constantly refer to that when planning trips, or when we wish to know the name of this lake or that mountain. The weather map always appears in the frame. It has been found more useful and dependable than the words of weather prophets. And there are pictures of last year's pageant. It reminds us that we have to work hard to even live up to their standard. It was certainly beautiful. The knot board, paintings and sketches on trips, last week's photographic contest, the baseball team, posters of the athletic girl, the swimming chart, and all kinds of leaf prints, cat-tail baskets, curtain pulls, freshly painted door stops, and a tow net are some of the things we note on the wall.

Now we will glance along the back of the benches. We note a model of a bubbling spring that was made after the last mountain trip to show how a spring works. Someone says that the material in that box is being arranged for a camp fire talk. The camp master is going to tell us about the way a glacier carries and deposits gravel. These are models of log cabins. One group is going to make a cabin out of drift wood. The bird houses are being made to take home and will be on exhibition at the camp meet. We have to stain them yet.

We have tadpoles in these jars. They were egg masses last week. We have several caterpillars on these different potted plants. That milkweed chrysalis formed yesterday. That sign on the toad cage says, "Give me a fly." He ate 55 caterpillars for dinner. We have seen some one-celled animals through the compound microscope and some baby oysters caught in the tow net. That box of sphagnum moss? That has been here several years. It is interesting because we used to collect sphagnum for the Red Cross during the war. Some kinds are useless. They used it in place of absorbent cotton. There use to be an observation bee hive at that window. This section is reserved as a *"What is it Shelf?"* If any one does not know the name of a flower etc., he puts it on this shelf. Credit is given to the one that names it first. Yes, this is a storm window. It is cheaper than plate glass. We keep it over these Indian relics as people were always moving them and they got lost. The front of the benches we reserve for working space.

The chests contain loan material. There are geology hammers, cyanide jars, field glasses, minerals, samples of soil, a series of models to show how camp rocks formed, a good collection of bird's nests (We have a talk once a year about those), home-made butterfly nets, fruits of trees, woods of trees that we got at the saw-mill (We used those last when we heard the story about the Thousand Year Pine, by Mills) and many objects that we use for Sunday Night Talks.

The model in the centre of the room is of this valley and the two bordering mountains. It took us a month, in spare time and rainy days, to make it. First of all we made an enlarged contour map of the region from the government topographic map. We lined the government map in squares and then made larger squares on a large piece of brown paper. This made it possible to transfer the map "by eye." We then traced the 20 foot contour lines on the enlarged map using transparent paper. This tracing was placed on a 1-inch pine board and traced with a hard pencil. The line on the board was then sawed out with a jig saw. The same steps were repeated for the 40 foot contour and so on until all were sawed. Each pattern was then placed on top of the preceding and in proper place. Others had been preparing some papier mache. They tore up old newspapers into bits, and soaked them in water for three days. This was stirred into a dough. This dough was now smoothed over the model with a putty knife. When the model resembled the region it was left to dry. It was then varnished, colored, and labeled to show the trails, camp grounds, springs, interesting rock formations, etc., that we visit. We are always referring to this model and often have little talks around it before going on a trip.

Our flower show is a little different than using a botany press and mounting the specimens on a sheet of paper. One time we arranged wild flowers at the public library. We had about 8 steps (about 4 inches wide and 6 feet long) made out of boards from discarded grocery boxes. Over this we tacked manila cloth that we got by cutting up an old peanut bag. Then we got a supply of bottles and filled them with water. The wild flowers were arranged and labeled and attracted a great deal of attention. We took turns standing by the show and answering questions. It was great fun and of course instructive. Then we sometimes decorate our tables with flowers.

The camp museum presents those things which give a vivid and lasting impression in order to realize better the life and processes of the local environment. The geology of the locality is the foundation. The formation of the mountains, valleys, and lakes that surround the camp is told, with specimens and models, as a story so as to be readily understood by the children. The old idea of going to a museum and merely gazing is of little value. Visitors at camp usually look about the museum with idle curiosity. A camper visits the museum with a deep interest in one thing. It may be with the nature counsellor who uses the models to make clear just how the rocks of the quarry were put down. This may be followed by a field trip, which in its turn, must have a single, definite aim in view.

The museum is also an outlet for the collecting instinct. The collecting age leads to exploration. This is an opportunity for lessons in conservation. The camper may collect minerals but not rare flowers, leaves but not birds eggs. Collecting excursions are opportunities to establish right habits. They may start as recreational and become educational.

The camp museum, then, is simply a device to bring the child and nature together. Those things that the child has opportunity to see in nature do not usually need to be brought inside. The nature leader must use his judgment. If it is observing a butterfly visiting a flower it had better be in the fields. If the object is to see the butterfly suck up the nectar it can best be done in the museum by putting a little honey or sugar solution on a glass plate. When the interest leads to that, and the butterfly has answered the question, it is released and the museum is ready for the next thing that may be called upon the stage of nature experiences.

AN OUTDOOR THEATRE, Camp Hoffman, R. I. Girl Scout Camp, (made of Arbor Vitae)

FRIENDS

May I be friends to all the trees;
  To birds, and blossoms and the bees;
To things that creep, and things that hide
  Through all the teeming countryside;
On terms with all the stars at night,
  With all their playful beams of light;
In love with leafy dales and hills,
  And with the laughing mountain rills;
With summer skies, and winter snows;
  With every kind of breeze that blows;
The wide sea, and the stretching plain,
  The tempest, and the falling rain—
If I were thus what need had I
  To fear Death's solemn mystery
That takes me from the world's alarms
  And lays me in earth's loving arms?
           —John Kendrick Bangs.
      —(Connecticut School Document No. 2, 1923.)

CHAPTER V

ETIQUETTE OF THE WOODS

Camping more than any other life develops thoughtfulness, courtesy, generosity, and comradeship. One may be well mannered in his own home and yet wear out his welcome on the trail. Ignorance of the laws of the trail is the means by which the old campaigner spots the "greenhorn," and the "tenderfoot." The code of the woods is simple. It is strict. One is on trial until he learns it. It is mostly knowing what not to do.

The Trail Hog: One rule of the road is to leave it at least as beautiful as you find it. I know of trails one could follow by means of tin cans, string, egg shells, banana peelings, and paper plates which are strewn along the way. These trails are blazed by the trail hog. Perhaps he came in a limousine. He may be a society man in town but not in the woods. The path of the trail hog is most offending to the woodsman. One of the golden rules of camping is to leave no trace. All refuse is either burned or buried.

The Fire Criminals: The greenest of "greenhorns" is the fire criminal. He builds a big roaring, smoking fire that he cannot get near. He fails to clean up around and burn the leaves around the fire and does not put the fire out when leaving it. Different states have different fire laws and many do not allow the building of a fire without a permit. Campers disregarding these laws are subject to arrest. Firebuilding is an art and the quickest way to size up a woodsman is to look at his fire.

The Ignoramus on the Farm: It is not in good form to romp in a hayfield, to cut up fence rails for fire, to leave the gate open or bars down, to throw stones in the grass, to pull hay from a stack to make a bed, to steal apples that fall by the roadway, to troop across the lawn for a drink, to steal the farmers' blueberries, to take a short cut across the plowed ground. The farmer's latch string is always out. He is proud to have you drink from his well. He may be glad to give you some apples. He does not enjoy having you let his herd of cattle loose to run through his cornfield. Let us adopt the old law of neighborliness. Give a word of kind greeting when you meet

49

him on the road. Pay a compliment to the water from his well and enjoy the view from his porch.

The Garrulous Member: Every group usually has some one who talks much and says little. He sees nothing in the surroundings and hears less. He may be full of garrulity and giggles but the remainder of the group may wish to see the lake or hear the woodthrush. A woodsman sees and hears and is a man of a few words. "A wise old owl lived in an oak, the more he heard the less he spoke; the more he spoke the less he heard, why aren't we all more like that bird."

The Night Hawk: Most animals take great care in making their beds. Tenderlings raised in a green house do not. A "tenderfoot" is sure to show off all night. He discovers a sharp stone early in the night; he does not make a hip-hole; and does not know that for warmth he needs more blanket under than over him. He shifts and rolls, tosses and moans, and spends most of the night before the fire. If he has not awakened everyone about camp he surely will. About dawn the greenhorn, sleepy and haggard, routs the camp with his "gentle" whispers or his "soft" tread. There is no better test of the fine art of camping than the overnight hike.

Borrowers: Never borrow a tooth-brush, drinking cup, knife, axe, or gun. The first two may spread disease. The last three are the back-woodsman's chief tools. A novice is sure to appropriate the axe and try it out on the cherry tree or strike it into the roots of a tree. The chances are 99 to 1 that he does not know how to sharpen it. If you are a tenderfoot do not let the fact out by borrowing these personal belongings of the forester.

Hands Off Cached Property: A cache is an outdoor storage of food or supplies. Arctic explorers cache their supplies for the return trip. To disturb these goods might cause death. If you find a canoe hidden by a lake leave it there. A thoroughbred woodsman never disturbs cached property. A greenhorn usually does,—not so much that he is dishonest but that he does not know the ways of the woods.

Buck Ague: This is a disease of "sissies." It is sometimes known as "Fraid cat" or one is said to be "Hugging the ground", or "Looking for mother's apron strings." A boy once thought there was an owl in a tree shooting nuts at him and began to cry. He had an attack of Buck Ague. Some people think that it is fashionable to squirm if someone eats frogs' legs; to squeal when a toad hops; and to screech at the sight of a snake. This is plain ignorance in nature-lore. A real scout thinks that you are playing to the gallery. He knows that you are an amateur and nothing that you can do thereafter will square accounts.

The Sandwich Tosser: The sandwich tosser is one who does not exercise an earning power equal to the value of the sandwich. He is usually not weaned from the home larder. He never prepares the sandwich for his own lunch. He eats all but the crust. He has no fear of the future because his food has always been furnished him.

"The wolf is at the door of the world" but why should he care. He is too lazy to carry or repack the food and therefore throws it away.

Camp Chores: In camp no one is exempt from camp chores. There are no bosses. Everyone shares in the tasks. I know of a party of campers who went away and left their dirty dishes for the "Guide" to wash. This breach of wood etiquette was unpardonable. It meant a new guide for the next trip.

The Dawdler: Known as a "wall-flower." Does the "heavy looking on." Prefers to knit. Doesn't get into the game. A woodsman is always a good sportsman. He is always ready. He is not a dawdler. He plays the game because he likes it. He plays fair. He plays to win, but he is a good loser. When the crowd sings he sings heartily. He cheers lustily. When asked to tell a story he does it cheerfully. He has firm convictions. He is true to his principles. He is willing to die for what he believes is right.

Courtesies En Route: It is wood etiquette to pass the word back: as,—"Low bridge" when going under something; "All to the left" when an auto is coming; "Look out for poison ivy"; "See the Turk's Cap Lily"; "Drinking water on the right."

Water Wailers: Greenhorns on the trail are continually wailing for a drink of water. They have always had water at their demand. To wait for a drink is a lesson in self-denial which they have never learned. An old scout may be as thirsty as you, but he never complains. You might never guess that he is thirsty. He is a patient waiter.

Tree Butchers: Inexperienced naturalists, boys with new knives, curio gatherers, and tree butchers cut bark from the birch trees and leave the scars for all the future visitors to see. It is considered better form to cut down a white birch and use the bark from its trunk rather than to ring many trees, leaving them to mar the landscape.

The American holly, mountain laurel, pussy-willow, and black alder are subject to wood vandalism. Many people break off the branches and leave an ugly shaped shrub. Use the knife rather than mutilate the tree.

Then there is the custom of carving one's initials in a beech tree. One woodsman said that he would have one beech tree in camp for everyone to carve their names upon rather than have the whole forest a series of totem poles.

"Freeze": "Pipe down" on shipboard; "Ten-shun" in the army; "Silence" in the court room. One of the first laws in tenderfooting is to freeze immediately when the signal is given.

Protection of Native Plants: Woodsmen know that there is great danger of extermination of many of our most interesting plants. This is because they are torn up by the roots. Arbutus, ground pine, sabbatia, gentian, columbine, orchids, ferns, and Jack-in-the-pulpit need protection. These plants should be cut sparingly with a knife. It is better taste to cut a few sprays than to gather a bunch. Protect rather than destroy. Leave them for others to enjoy. Do as you would be done by.

"A mist on the far horizon,
The infinite tender sky
The ripe rich tints of the cornfield
And the wild geese sailing high.
And all over upland and lowland
The charm of the goldenrod,
Some of us call it Autumn,
And others call it God."
—*Carruth.*

WHITE-THROAT SINGIN' IN THE RAIN

Never had much shoolin', got no use fer church;
'D ruther go a-strollin' 'mongst maple, oak, an' birch,
Scramble up a hillside, or saunter down a lane—
D'j'ever hear a white-throat singin' in the rain?

Take some April mornin', never mind the showers,
Git up aroun' sunrise time, look fer early flowers.
Listin!—"*Ol' Sam Peabody-y-y-y,*" jest ez plain ez plain—
Bless his heart!—a white-throat, singin' in the rain!

You kin have yer robins an' yer song-sparrers, too,
An' all the rest but one of the hull dern springtime crew;
They ain't much more musical than a whoopin'-crane—
Lemme hear a white-throat singin' in the rain.

When my turn comes round to go, needn't say no prayers;
Go away an' leave me by myself, upstairs.
I'll fergit my sorrers an' my troubles an' my pain,
Thinkin' of a white-throat singin' in the rain.
—*George W. Dorland.*

From the last letter of Captain R. F. Scott to his wife as he lay watching the approach of death in the Antarctic Cold. "Make our boy interested in natural history if you can. It is better than games. Keep him in the open air. Above all, you must guard him against indolence. Make him a strenuous man. The great God has called me. Take comfort in that I die in peace with the world and myself,—and not afraid."

Subdue the earth.
Have dominion over every living thing.
Make the earth a garden.—*Bible.*

CHAPTER VI

"One impulse from a vernal wood may teach you more of man, of moral evil and of good, than all the sages can."—*Wordsworth*.

"These are the things I prize
And hold of dearest worth:
Light of the sapphire skies,
Peace of the silent hills,
Shelter of forests, comfort of the grass,
Music of birds, murmur of little rills,
Shadows of cloud that swiftly pass,
And, after showers,
The smell of flowers
And of the good brown earth,—
And best of all, along the way, friendship and mirth."
—*Henry Van Dyke*

Religion should not be unrelated to the child's surroundings. It stands for simplicity and naturalness. It is pathetic when one thinks how indirect and how foreign to our lives much of our religious education has been. It often begins with a language that means nothing to a child and amidst descriptions he will never meet. The real lessons come from familiar objects and scenes that the child can understand.

There is no better place to begin than in camp. There he is not fettered by the artificialities of theology, or the molds that make for set form, nor passing styles. After all is said and done—God made the out doors but not the church. One does not get soul-enrichment by memorizing words. He does not learn morality by being lectured. He does not learn service by merely quoting the sayings of wise men. He must experience soul-enrichment, live acts of morality, and meet opportunities for service.

One method of teaching the way of a rich life is through a story. All of the parables of Jesus are out-of-doors. The sayings of the poets abound with the out-of-doors. It is the way of Kipling in the Jungle Book. One of the best talks for a Sunday night may be obtained from Burr's "Around the Fire Stories." They are best heard about the glowing camp fire under the spell of night with a wind through the trees.

53

A Fire-mist and a planet,
A crystal and a cell,
A Jelly-fish and a saurian,
And caves where cavemen dwell:
Then a sense of Love and Beauty.
And a face turned from the clod.
Some call it evolution, and others
Call it God.   —*W. H. Carruth*

Another source of spiritual assets comes from a discussion of things found in the immediate environment. This method is adapted to a Sunday night service. It may be the sea or the mountain, the sand or the rock, the sunset or the flower. Whatever the object it should be near at hand. Study it together by group discussion and try to find its relation to the development of the spirit and life's ideals. The method of introducing and leading in such a discussion will be given in a few instances.

I. Sand

Camp Chequesset is located in a region of sand—namely Cape Cod. The camp museum has small glass vials with samples of sand from the Sahara, the Isle of Thanet, Florida,—in fact from all over the world. Each camper is given a vial and a "pinch" of Cape Cod sand.

The following may be read by the leader for his own thought. "The soil is the sepulcher and the resurrection of all life in the past. The greater the sepulcher the greater the resurrection. The greater the resurrection the greater the growth. The life of yesterday seeks the earth today that new life may come from it tomorrow. The soil is composed of stone flour and organic matter (humus) mixed; the greater the store of organic matter the greater the fertility."—John Walton Spencer.

The campers are asked to examine the sand closely and to feel of it and tell something that is characteristic of it. The list, as given, is written on the blackboard. They may observe that it is fine, uniform in size, angular, hard, insoluble, not easily fusible, clean, silica, different colors. When a lull comes ask for the uses of sand, and each time that a use is given have them select the characteristics which adapt it to that particular use. They will find that they need to add to the former list, but let that grow as it may. The final list will look as follows:

Use	*Characteristics fitting it for that use*
Sandsoap	fine, uniform, angular, hard
Sand paper	fine, uniform, angular, hard
Hour glass	fine, uniform, durable, not adhesive
Glass	hard, insoluble, not easily fused
Cement	fine, uniform, angular, hard, insoluble
Filter	fine, porous, insoluble, clean
Grindstone	fine, hard, angular
Blot Ink	absorbent, fine, clean, insoluble
Sand Foors	fine, hard, clean, angular
Icy sidewalks	fine, angular, hard
Cuttings	fine, clean, good aeration and drainage
To pack vegetables	fine, clean, good aeration, dry

Tell them some anecdotes about sand and Cape Codders. The inhabitants are said to have a certain gait which they have acquired by walking through the sand. The Cape girls have become very dexterous and have the reputation of being able to throw the sand out of their shoes at every step. Often times the windows of the houses near the shore become sand blown. The Cape Cod Coast Guard are noted for the sand in their character. They have acquired great endurance walking sandy stretches often in the face of sand storms. The sand bars make it one of the most dangerous winter coasts in the world. It was a sand bar that turned the Pilgrims from the New Jersey coast and headed them northward to Provincetown Harbor. Sand has always been an important factor in the history of Cape Cod.

Now that the campers have a better acquaintance with sand they are invited to think of it as a basis of working out a code of morals for the camp program. It is surprising to find how many ways of living can be thought of that are taught by sand. It takes several grains or individuals to accomplish anything. "In unity there is strength." Each grain has its own character yet works together for the common whole. Little things count in the long run. Endurance is an asset. The use of a thing depends upon its characteristics, and it can be no more useful than it has characteristics which adapt it to that use. Out of sand comes the beach pea with all the colors of the rainbow.

It is of further interest to recall the allusions to sand in the Bible. The campers usually recall the story of the man who built his house upon the sand. They may think that the wilderness of Judea is dry because of sand. This is not true. It is because of a lack of rainfall. Sandy soils are however, usually dry. Deserts mean a sparse vegetation. The animals have to wander in flocks from oasis to oasis. Sheep are easily adaptable to a desert life. They require a shepherd for protection. He leads them but does not drive them as would be the case with cattle. He comes to know his flock and they him. After this discussion they have a better understanding when they say together,—

> The Lord is my Shepherd, I shall not want,
> He maketh me to lie down in green pastures,
> He leadeth me beside the still waters.

> "Every clod feels a stir of might.....
> And climbs to a soul in grass and flowers."
> *—Lowell*

> "Here is a problem, a wonder for all to see.
> Look at this marvelous thing I hold in my hand!
> This is a magic surprising, a mystery
> Strange as a miracle, harder to understand.

> What is it? Only a handful of dust: to your touch
> A dry, rough powder you trample beneath your feet,
> Dark and lifeless; but think for a moment, how much
> It hides and holds that is beautiful, bitter, or sweet.

"The vision is compact of blue and gold,
  Of sky and water, and the drift of foam,
  And thrill of brine-washed breezes from the west:
  Wide space is in it, and the unexpressed
Great heart of Nature, and the magic old
  Of legend, and the white ships coming home."
                                —*Richard Burton.*

Think of the glory of color! The red of the rose,
  Green of the myriad leaves and the fields of grass,
Yellow as bright as the sun where the daffodil blows,
  Purple where violets nod as the breezes pass.

Strange, that this lifeless thing gives vine, flower, tree,
  Color and shape and character, fragrance too;
That the timber that builds the house, the ship for the sea.
  Out of this powder its strength and its toughness drew!"
                          —*From "Dust" by Celia Thaxter*

II. Mountains

The greatest sermon ever preached was not in a temple but on a mountain. If a camp is located in the mountains it affords an opportunity for a better appreciation of the Holy Land and its geographical influences. A map of Palestine will be of great assistance in getting a clear vision of the geographical features.

Since the time of Jesus, Palestine has been called the Holy Land. It is smaller than the state of New Jersey. Palestine is divided into three parts—Galilee on the north, Judea on the south and Samaria between. The highest peak, Mt. Hermon (9200 ft.), is snow clad, except for the last of August. It is at the headwaters of the Jordan which passes through the Sea of Galilee and the Dead Sea.

Judea is much like the Allegheny Plateau. If one stands on a hill top he will see other hills—in all directions—of about the same height with narrow valleys between. The region is inaccessible and consequently the customs do not change. The Judeans still make butter in goat skins and crush olives for their oil with a rock. (See pictures in National Geographic Magazine.) The mountain whites make butter in a ferkin and grind their meal by the old mill stream. "Jews have no dealings with Samaritans" and the mountaineers cling to their cabins. The highlanders are less worldly wise. In the secluded hills there is time to gather ideals and the prophets wail over the wickedness of the lowlands.

The lowlands of Galilee and Samaria, like the Mohawk Trail, was crossed by the main lines of traffic. Most of the caravan traffic between Egypt on the south and Phoenecia on the north passed through Samaria. It was the route of the armies of the Romans, Crusaders, Napolean, and the Turks. The Samarians had no opportunity to preserve their individuality. Trade and culture developed in the lower lands—rich cities in contrast to mountain villages.

In the mountains the folks come to rely mainly on their flocks for sustenance. Ragged shepherds, sturdy and alert, and the barter of peasants. The strength of the hills is theirs. Permanent villages are few and there is, of course, the danger of robbers.

III. A Lake or Pond

Almost every camp is either on or is accessible to a lake. Every park has a lake. If a man is rich enough to have an estate he also has a lake. Lakes are found in nearly all "tourist lands." William Wordsworth lived near a lake. The Old Oaken Bucket and the pond that stood nigh it have been a source of enjoyment to thousands. Thoreau had his Walden Pond. Jesus lived near a lake. The gospel gets much of its beauty from fisher folk and the Sea of Galilee. A lake surely can mean more to a camper than a place to swim and row. If the out door chapel is not already on a lake this particular Sunday Night should be held on the lake shore or on a hill that overlooks the lake.

There are several ways of introducing the subject. The leader may say, "What is a lake?" If some one says a large body of water the leader may say, "So is the Ocean;" or if a small body of water, "So is a mud puddle." It is impossible to define a lake. In fact, there is no need of a definition any more than there is need for defining a chair or a horse.

The leader may say, close your eyes and think of the most beautiful lake of which you know. After they have had time to think ask a few questions. How many thought of a lake with a definite name? A name is not necessary for us to enjoy it. What is on the shore of the lake? How many saw trees? Hills? Farms? How many could see across the lake? Why are these the characteristics of an ideal lake? Those who thought of a lake with a name,—was it Lucerne, Tahoe, or Lady of the Lake?

Or the leader may ask the campers to look at the lake or at some special cove. Think to yourself,—what do I see? After a short time hear from different ones. Some see the color, others the ripples, some the muddy bottom and others the reflection of the blue sky, or possibly the trees standing upside down. Why is it that we think and see so differently.

Tell them how the lake was formed and how it is being filled with soil and vegetation. Have them estimate the long time that is involved in the history of a lake. A lake is like a person. It is born. It is fed. It is active when young. It writes a diary. It breathes. It has moods. It gets angry. It may be very silent. It never babbles like the brook nor does it roar like the ocean. It paints beautiful pictures. It has children. It dies. Before undertaking this the leader should be well versed in the stages of a lake. This may be obtained from a good physical geography. The resemblance to a person should be thought out by the campers.

IV. Sunsets

"Wilt thou not ope thy heart to know
What rainbows teach, and sunsets show?"
—*R. W. Emerson*

How many saw the recent eclipse? Who went the farthest to see it? Why did you go so far to see it? Which is more beautiful—this sunset across the lake or the eclipse? Why do people not take a special train to see the sunset? What impresses you about this sunset? Why is the sun more beautiful when it is near the horizon?

(We are looking at it through a layer of dust just as you looked at the sun at the time of the eclipse through a smoked glass or film). How long does a sunset keep the same colors? (The colors are constantly changing). To whom does this sunset belong? (To all who can enjoy it). It is also interesting to know that each one of you has a sunset of your own. I do not know what your sunset is. I do not know how many colors you see. You do not know how many I can see. I wish that I could see the sunsets that the artists in Provincetown see. In what way is a sunset a miracle? A sunset is done according to law. Its variation depends upon the amount and kind of dust, the moisture, the wind, reflection, refraction, radiation, absorption, temperature, etc. The sun sets at a different time and angle every night.

Sing: Day is Dying in the West.

Softly now the Light of Day.

End by singing taps and going quietly to bed, perhaps going to sleep, with a violin solo.

Taps: Day is done | Gone the sun | From the Lake | From the Hills | From the Sky | Safely rest | All is well | God is love.

SERMONETTES

"'Tis Eden everywhere to hearts that listen
and watch the woods and meadows grow."
—*Theron Brown*

The term sermonette is used in contrast to the Sunday Night Talk. As the name suggests—it is brief. It also comes at any time the spirit moves and the object is at hand. The following are suggestive for sermonettes. The most effective ones come from a questioning by the child.

Birds: No one knows how or why birds migrate except those who have never studied them. No one can account for the blue-tinged, brown speckled eggs of the Chipping Sparrow hatching into a bird that will grow a brown cap, a gray streak over the eye and a black one through it; that will line its nest with horse hair; that will build its nest 10–15 feet from the ground; that will migrate south in the fall except by the blind word *inheritance*. Yet there is something in the egg of the chipping sparrow that causes it to develop these precise characteristics. Such facts make one humble.

Butterfly:

"And what's a butterfly? At best
He's but a caterpillar drest."
—*John Gay*

The butterfly is nothing but a glorified caterpillar.

Chambered Nautilus: An emblem of immortality.

"Year after year beheld the silent toil
That spread his lustrous coil;
Still, as the spiral grew
He left the past year's dwelling for the new,
Stole with soft step its shining archway through,
Stretched in his last-found home and knew the old no more."
—*Oliver W. Holmes*

Ferns:

"God made the fern for the most perfect leaf just to see how well he could do in that line."—*H. D. Thoreau*

Stand a fern spiral in a vase and watch it unfold.

"But the first glance at one of these little wooly spirals gives us but small conception of its marvelous enfolding, all so systematic and perfect that it seems another evidence of the divine origin of mathematics."—*Anna Botsford Comstock*

"If we were required to know the position of the fruit-dots or the character of the indusium, nothing could be easier than to ascertain it; but if it is required that you be affected by ferns, that they amount to anything, signify anything to you, that they be another sacred scripture and revelation to you, help to redeem your life, this end is not so easily accomplished."—*Henry D. Thoreau*

Fish:

THE ORACLE OF THE GOLDFISHES

"I have my world and so have you,
A tiny universe for two,
A bubble by the artist blown,
Scarcely more fragile than our own,
Where you have all a whale could wish,
. Happy as Eden's primal fish.
Manna is dropt you thrice a day
From some kind heaven not far away,
And still you snatch its softening crumbs,
Nor, more than we, think whence it comes.
No toil seems yours but to explore
Your cloistered realm from shore to shore;
Sometimes you trace its limits round,
Sometimes its limpid depths you sound,
Or hover motionless midway,
Like gold-red clouds at set of day;
Erelong you whirl with sudden whim
Off to your globe's most distant rim,
Where, greatened by the watery lens,
Methinks no dragon of the fens
Flashed hugher scales against the sky,
Roused by Sir Bevis or Sir Guy;
And the one eye that meets my view,
Lidless and strangely largening, too,
Like that of conscience in the dark,
Seems to make me its single mark.
What a benignant lot is yours
That have an own All-out-of-doors,
No words to spell, no sums to do,
No Nepos and no parlyvoo!
How happy you, without a thought
Of such things as Must and Ought—
I too the happiest of boys
To see and share your golder joys!"
                                        —*Lowell*

Flowers:

"Little flower; but if I could understand
What you are, root and all, and all in all,
I should know what God and man is."
                                        —*Tennyson*

"Beside the moist clods the slender flags arise filled with the sweetness of the earth. Out of the darkness—under that darkness which knows no day save when

the ploughshare opens its chinks—they have come to the light. To the light they have brought a colour which will attract the sunbeams from now till harvest."
—Richard Jefferies

A Sunday Call in the Village

ENJOYING A POSY GARDEN IN THE VILLAGE: On Sundays the camp girls dress in white and visit friends in the village. They sing to shut ins and never miss an opportunity to pay a friendly visit to old fashioned posy gardens.

"And toward the sun, which kindlier burns,
The earth awaking, looks and yearns,
And still, as in all other Aprils,
The annual miracle returns."   *—Elizabeth Akers*

THE DANDELION

"'Tis Spring's largest, which she scatters now
To rich and poor alike, with lavish hand;
Though most hearts never understand
To take it at God's value, and pass by
The offered wealth with unrewarded eye."   *—Lowell*

Consider the lilies of the field how they grow, they toil not, neither do they spin: and yet I say unto you, that even Solomon in all his glory was not arrayed like one of these.—*Matt. 6: 28, 29*

This passage in the Bible is probably thought of more than any other when referring to flowers. The species which is common in Palestine, however, is Anemone Coronaria which is bright scarlet in color. It is not a lily. At Easter all Christians think of the cultivated "Easter Lily" as a symbol of resurrection. There are white anemones in certain places in Galilee. What lesson did Jesus intend to teach when he referred to the "lily of the field?" Why did he select a common wild flower of the wayside?

The cyclamen, which is sold as a potted plant in early spring, grows abundantly in the mountains about Jerusalem.

The true function of a flower is to produce seed.

Lo, the winter is past, the rain is over and gone; the flowers appear on the earth; the time of the singing of birds is come and the voice of the turtle is heard in our land.—*Songs of Solomon 2: 11, 12*

The word turtle refers to the turtle dove.

If ye have faith as a grain of mustard seed, ye shall say unto this mountain, Remove hence to yonder place; and it shall remove; and nothing shall be impossible unto you.—*Matt. 17: 20*

In the beginning a mustard seed is small but it grows into a large herb with yellow blossoms. In this parable Jesus pictured the growth of the spirit and faith. The mustard referred to is probably Sinapis arvensis. We have several yellow mustards that are quite similar.

And God said, let the earth bring forth grass, the herb yielding seed, and the fruit tree yielding fruit after its kind, whose seed is in itself, upon the earth: and it was so.—*Gen. 1: 2*

Grass is the first plant mentioned in the Bible. It is the most beautiful plant and the most common. If we should remove all the grass it would be a sorry looking spectacle. Few people would name it, however, as the most beautiful plant. Why?

All flesh is as grass, and all the glory of man as the flower of the grass. The grass withereth, the flower thereof falleth away; but the word of the Lord endureth forever.—*Peter 1: 24*

Grass is usually mentioned in the Bible as a symbol of decay. The reason for this is evident. The rainy season in Palestine is a matter of a few weeks each year. When the rains cease the grass withers and turns brown just as in our winter season. Grass is referred to in the Bible more than any other plant.

What did Moses mean by describing Palestine as a land "flowing with milk and honey?"

The Gall Fly: When a fresh oak gall is discovered on a field trip remove the larva from its "sponge cake." What parts are missing? (Eyes, feelers, legs, wings). How account for these parts not developing? What does this teach us? This is an excellent example of "sponging a living." Good honest toil is the cause of development. It does not pay to be a hanger-on.

Insects and Flowers: Tell the story of the dependence of the clover on the bumble bee as an example of cooperation.

Leaf Miners:

"And there's never a leaf nor a blade too mean
    To be some happy creature's palace."
                                        —*Lowell*

Leaves: Most plants are very careful with their leaves. They have them arranged just so—at a definite angle so that they can get sun-

light. They are very good to each other just like some families.
They know how to make hay (starch) when the sun shines. Sit down
by a primrose or mullein or burdock some day and call attention to
the leaf mosaics. Point out some yellow insignificant plant whose
sunlight has been shut out by its neighbor. Grass on the other hand
is a plant that gets along well with its neighbors.

Snow: Let a few flakes fall on the coat sleeve and examine them
with a hand lens. A snow crystal is a six pointed star. The design
of one ray is always repeated in the other five, yet Mr. W. A. Bentley
has found a thousand types.

Thoreau says:

"A divinity must have stirred within them before the crystals did thus shoot
and set. Wheels of storm chariots. The same law that shapes the earth-star
shapes the snow-stars. As surely as the petals of a flower are fixed, each of these
countless snow-stars comes whirling to earth, pronouncing thus, with emphasis,
the number six."

Stones:

"Sermons in stones, and good in everything."—*Shakespeare*

Pick up a glacial pebble, a piece of granite, pudding stone, or coal—
in fact any rock and tell its life history.

Trail: Tell the parable of the Good Samaritan. Read Sam Walter
Foss's poem:

"Let me live in a house by the side of the road
and be a friend to man."

Some day when on the open road see who can find the most ways
in which it is a friendly road. People are friendly on the trail. They
tell you the way or gladly give you a drink from their well. Write a
song of the Open Road.

Trees:

"The Groves were God's first Temples."—*W. C. Bryant*

Hymn: O Beautiful for Spacious Skies.

There is a legend that the cross was made of Aspen Wood, and
that is the reason that the leaves shiver and tremble in the breeze.

There are several parables that refer to trees.

"Even so every good tree bringeth forth good fruit; but a corrupt tree bringeth
forth evil fruit. . . . Wherefore by their fruits ye shall know them."

"Behold the fig tree, and all the trees; when they now shoot forth ye see and
know of your own selves that summer is now nigh at hand."

"For if they do these things in a green tree, what shall be done in the dry?"

"A certain man had a fig tree planted in his vineyard; and he came and sought
fruit thereon, and found none. Then he said unto the dresser of his vineyard,
Behold, these three years I come seeking fruit on this fig tree, and find none: cut
it down; why cumbereth it the ground?" . . . And Jesus answered and said unto
it, No man eat fruit of thee hereafter forever. . . . And Peter calling to remem-
brance saith unto him, Master, behold, the fig tree which thou cursedst is withered
away. And Jesus answering saith unto them, Have faith in God."

"I think that I shall never see
A poem lovely as a tree.

A tree whose hungry mouth is prest
Against the earth's sweet flowing breast;

A tree that looks at God all day,
And lifts her leafy arms to pray;

A tree that may in summer wear
A nest of robins in her hair;

Upon whose bosom snow has lain;
Who intimately lives with rain.

Poems are made by fools like me,
But only God can make a tree.
—*Joyce Kilmer*

"Earth's crammed with heaven
And every common bush afire with God."

The assignment method is interesting for a Sunday Night and is liked best by campers. The campers may use Bibles with a concordance. Give such topics as the following for presentation about the camp fire.

Ant—wise instincts, provident, industrious
Apple—venerable, beautiful, law of gravity
Bee—protects home, bee-line, industrious, social
Bramble—beautiful but has thorns
Cedar—majestic, vigorous, long lived
Clouds—

"I am the daughter of Earth and Water,
And the nursling of the Sky;
I pass through the pores of the ocean and shores;
I change, but I cannot die."
—*Shelley*

Dove—peace
Dragon—mythical
Eagle—strong, swift, keen vision, cares for young
Fish—symbolic
Grass—numerous, beautiful, good to neighbors
Grasshopper—locust—ravages vegetation
Iron—strong, hard
Lily—resurrection, fair, tender
Lion—strong, courageous, fierce, vigilant
Mountain—strong, great, enduring
Oak—sturdy, stalwart, great oaks from little acorns grow.
Palm—symbolic
Salt—essential to life
Snow—purity, so common place that the miraculous part escapes
Sparrow—humble, common, numerous
Sun—light

CHAPTER VII

Nature Games

A Nature Game is not a substitute for Nature-study but is a part of it. Nature play is instinctive and has the power of developing the play habit. Childhood is the time for fixing this play habit in the out of doors. If neglected the individual will usually be deficient in that particular training. Man is the only animal that ever neglects or trains away from the games of nature.

Many of these games have been adapted from old games that have been handed down from generation to generation. It will take but a little ingenuity to modify them for new games. All the rainy day games may be played out of doors. They are classified this way as a matter of convenience in using them.

Most of these games have already been published in various magazines such as the Nature-study Review, The Trailmaker, The School News and Practical Educator. A few were used in C. F. Smith's Recreational Methods. This is the first time that they have been assembled in one booklet.

Test of a Game.—Is it used spontaneously afterwards? What does it teach? What physical, mental, or moral traits does it develop? What social traits does it involve?

Suggestions.—Know your game. Play with spirit. Everyone take part—no porch or lounge lizards. When possible play outdoors. Old games by request. Always teach a new game. Change to fit conditions.

1. Rainy Day Games

A. BIRDS

1. Bird Identification.

Give each one a stuffed specimen or bird skin, a card or ticket, and a good bird key as Walter's or Wilcox's. When the student gets the right name from the key the ticket is punched, and a new bird is presented for identification. The one who identifies the greatest number of birds in a given time wins. The cards are gathered at the end of the time limit and a curve is plotted on the board to show the number identifying 1, 2, 3, etc., birds. Each one then knows how he ranks with the average in ability to identify birds.

2. Bird Description.

Have cards on which is written the description of birds. Read slowly. The one guessing the name first is given the card. Anyone making a wrong guess has to give back a card.

3. Swat the Blindman.

Have Audubon Charts hung on the wall. Have a player stand back to the chart and describe a given bird. This always furnishes a great deal of amusement to the audience.

4. Hawkeye.

Cover up a bird quickly and ask questions. What color is the robin's breast? His head? His bill? How many white spots are there? Where is there black? What color is his throat? How many noticed white around the eye? This is fine training for observation.

5. Gyp.

The group thinks of some bird. They are given several books in order to become well informed about the bird. "It" then comes in and asks questions which may be answered by "yes" or "no."

6. Bird Silhouettes.

Either the outline of the bird is cut out of black paper or the form of the bird is thrown on a cloth by a light. The one identifying the greatest number wins.

7. Bird Picture Contest.

The colored pictures of birds are cut into four parts,—head, body, tail, and legs. The pictures of the legs, and in the case of the seed eating birds the beaks, are scattered on the following tables: ducks, other swimming birds, wading birds, tree trunk climbers, insect feeders of the air, birds of prey, and seed eaters. When the whistle blows the players select a picture of some bird's leg and from the characteristics of the legs tries to find the other parts of that bird. As soon as one picture is completed another leg is taken and so the game progresses until the supply is exhausted.

8. Bird Logomachy.

Use cardboard letters printed on one side. Place face down on the table. Players take turns drawing letters and placing face up on the table. When a player can make a bird name from these letters he takes the letters and spells the word in front of him. The person getting ten words first wins.

9. Bird Rogues' Gallery.

Slips of papers are passed out to players and they are given two minutes to draw a picture of a bird for the Rogues' Gallery. The exhibit is then set up and the judges walk by the exhibit. Recognizable birds are given honorable mention (1 point). Birds represented in action may be given red ribbons or red pencil marks (2 points). The best sketches may be given blue ribbons or blue marks (3 points). The judging may be made very funny. The team getting the greatest score wins.

B. TREES

10. Twig Matching.

Obtain several kinds of twigs 8 to 12 inches long. Cut into two parts. Mount the lower half on a board. Scatter the other halves on a table. At a given signal the players observe closely one of the twigs and then run to the unmounted group to get the other half. If the wrong half is brought back he tries again. This game requires close observation. Leaves may be used in the same way, or flowers with short stems may be fitted to longer stems, or leaves to leaf scars.

11. Jack in the Box.

A branch or flower is held up quickly from the back of a box. The players write down the names. See which team gets the highest average.

12. Getting a Clew.

Have a sheet of paper or cloth with a hole in the middle. Show the edge of a leaf, a little more at a time. Whoever gives the name correctly first is given the leaf. The one who gets the largest collection wins. Pictures of birds may be shown in the same way, the beak being the first to appear.

13. Indoor Twig Relay.

Have a group of winter twigs scattered at one end of the room. Have as many of each kind as there are players. Show a twig, as the white ash. The players may look as long as they wish. Samples are then passed back and they are given 30 seconds to get a white ash twig. Everyone back in his seat with a white ash twig at the end of 30 seconds gets one point.

14. Spot the Tree.

Give each player a sample twig of a tree that may be seen from the window. The players go to the window and mark on a map, if it be an elm for example, the elm trees spotted from the window. Maple twigs, horsechestnut, pine, spruce, etc., are well adapted. This is a very interesting game to train in the sense of long distance observation.

15. Tree Silhouettes.

Cut from black paper the silhouettes of trees. The trees particularly well adapted to this game are spruce, pine, elm, red cedar, weeping willow, palm, sugar maple, lombardy poplar, and white oak. Hold up the silhouettes for naming.

C. GENERAL

16. Nature Alphabet.

The leader names a letter of the alphabet. Each player in order names a bird, flower, or tree (decided upon before starting) which begins with that letter. Anyone who cannot do so in less than five seconds is out. No one is to name an object which has already been named. The patrol having the greatest number at the end of a certain time is the winner or the last group to name an object commencing with that letter wins a point for his team.

17. Authors.

A home made set of bird or tree authors may be made which is similar to the well known game of authors. In authors there are 20 books of 5 cards each. The cards are dealt evenly to the players. Each in turn asks for cards until a player does not have the card asked for. When a book is obtained the cards are put down as a book. The greatest number of books wins.

18. Game of Touch.

The players are blindfolded and a natural object is placed in their hands. They have 30 seconds to feel of it. The name is then written down. Some objects particularly suited to this game are: various seeds, leaves, fruits, evergreens, flowers, barks of trees, nuts, feathers, shells, vegetables and soils.

19. Game of Smell.

As a variation a team may elect its "best smeller." The players are blindfolded and allowed to smell of common objects. The name is then written down. Objects with a distinct odor are: mints, black birds, wintergreen or checkerberry, balsam, pennyroyal, skunk cabbage, onion, parsnip, tomato, tansy, rose, sassafras, sarsaparilla, spice bush, turnip, cedar, kelp, apple, orange, mold, strawberry, and cucumber. Strong odors such as onion or skunk cabbage should not be given first.

20. Game of Taste.

Many of the objects suitable for smell are adapted to the game of taste. Others are rhubarb, sorrel, licorice, sugar, salt, clove, cinnamon, radish, catnip, peach, cabbage.

21. Color in Flower Outlines.

This contest may extend over a long period of time, as a week. Divide into teams and no outline to be colored in unless the flower is found and brought in by someone in the team. All coloring to be done from specimens.

22. Acornicle.

A game for four or less. Each player names his kind of acorn and gathers fifty. The ground or a piece of paper is marked off into 200 or more squares. The object of the game is to get five acorns of one kind in a row (In five successive squares). One player puts an acorn in any square. Each player takes a turn placing one at a time in an empty square. Opponents may be blocked by placing an acorn at the end of his row. The first one to get five in a row wins. A good game for learning the differences between oak acorns. May be played with twigs or leaves.

23. Getting Partners.

If in the beginning it is desirable to get partners it may be done as follows: (1). Cut leaves in two, crosswise and lengthwise. (2). Give leaves to boys and fruit of same kind of trees to girls. (3). Give tree questions to boys and answers to girls. (4). Use bird characters, Mr. and Mrs. Robin, Mr. and Mrs. Downy Woodpecker, etc. (5) Give duplicate slips to boys and girls, having names of birds. Each boy imitates the bird by voice, walk, or flight. If the girl recognizes it as the one named on her slip, she claims him as her partner.

24. Flower Favor Game.

Use reeds for stems; colored plastocine for receptacles; pine cone scales, feathers, shaving, or shells for petals; Insert petals into plastocine. Sealing wax, dyes, paint, and enamel will be useful. Have a real flower as a guide and give a definite time limit. Judges choose the best.

25. Social-Quiz.

Questions which give fun and knowledge. Good social mixers. Audience shows approval by saying "How, How" or disapproval by grunting. Each person given a question in turn. Why can't a hen smile? What bird has eyes in front of its head as you do? What animal has a nose most like your own? What animal does your right hand neighbor's face most resemble? What animal has been your most favorite companion? What bird does your left hand neighbor walk like? What animal do you drink like? Who drinks like a chicken? Who has a figure like a wasp? What animal does A's dancing remind you of? Who has a backbone like a jelly fish? Who has a complexion like a baldwin apple? What animal does your teacher call you? Who walks like a duck? Like a pigeon? What sound in nature does A's eating soup remind you of? What kind of an animal does B's ears remind you of? What animal do you prefer to converse with? What sounds in the hen language do you understand? By what animal name would you prefer to be called? Who eats like a pig? Who laughs like a hyena? Who is a scare crow? Who has the complexion of a lobster. This list may be elaborated.

26. Shadowgraph.

Attach a white sheet as a curtain with a bright light behind it. Pass several objects behind the sheet so that a shadow is cast. Use such objects as a pine tree, horseshoe crab, cabbage, dog, turnip, rhubarb leaf, pear, hen, goat, stuffed birds, pine cone, wooly bear, cattail, lily, lemon. Audience have a sheet of paper and write names.

27. Stunts.

(1). Leaf throw. Throw a leaf as far as possible without folding or breaking. (2) Imitate the call of a cow or duck. Nearest wins. May decide by pointing to competitors and the one that the audience claps loudest for wins. (3) Who can make the best poem using the words rabbit, habit, white, fight. (4) Hop like grasshopper. (5) Make a speech on fishing.

II. Outdoor Games

A. BIRDS

28. Birds' Nest Jackstraw.

The catbird's nest may be taken to illustrate this game. It should only be played with an abandoned nest. Before dissecting it, each player lays claim to the kind of material he thinks most abundant in the nest. It may be a hemlock twig. He has first claim on hemlock twigs but on no other. Take out a stick for example. The same is true for all the sticks, leaves, weeds, grasses, fine roots and strips of bark. Remove the parts one by one. In case someone recognizes a part as grass and no one else knows the kind, he is entitled to the specimens of that kind. Ten seconds or a count to ten is allowed for each to claim their kind of twig. In case it is not claimed by the one entitled to it the object is given to the next one recognizing it. Have the players count the number of each kind of material. Add five points for the naming of any one kind. The second part of this game consists of finding the source of material. The director of the game holds up a strip of bark which probably is that of the grape vine. After everyone has examined it carefully the leader blows a whistle and the first one discovering the grape vine gets a point. This is repeated for each kind of material.

29. Birds' Nest Tag Day.

This contest lasts all day and the best results are obtained when a good library is available. The camp is divided into teams. A tree having a nest may be tagged with a conspicuous card which must not be so placed as to frighten the bird. The tag must have the name of the bird proprietor of the nest observed.

If the member of any other team discovers that the nest is wrongly named it may be re-tagged and the first tag cannot be counted in the final score. Otherwise the nest is only to have one tag. A team discovering a nest and not knowing what kind it is may watch for the builder or investigate in the library. Other ingenious methods will develop.

30. Spot Spy.

This game is great fun when resting on a hike or when loitering along the way. The leader says: "I can see 5 white oaks." The group are given one or two minutes to spot the white oaks. All those who see them may indicate it by sitting down, taking off their hat, or by some other agreed signal. All those who see the object get a point.

31. Sign Language.

This game is similar to pantomimes. A player does the whole thing by the sign language. He may come out and point to himself. This means "I." He then flutters his hands like a bird. This means "I will fly like a bird." He may then imitate the flight of a swallow. It may be a hawk. He shows that it is not a swallow by measuring. He holds his hands apart the length of a swallow and shakes his head. He then holds his hands apart the length of a hawk and points into the air and with a sweeping motion of the hand indicates the spiral soaring of the hawk. Perhaps he now imitates a frog. He again comes back to the hawk and by signals shows that the hawk is looking for the frog and when he sees the victim pounces upon it and eats it. The hand held over the eyes means "Look." Pat the abdomen and smile means, "That tasted good." Or the food might have given a stomach ache. This game gives an unusual opportunity for ingenuity. The one guessing the name of the bird gives the next pantomime.

32. Snipe Hunting.

This is an old game with fifty-three variations. It is usually carried on at night. The "victim" usually holds the bag with a lantern to attract the snipe. The skirmishers make a large circle each having a tin can full of pebbles to frighten the snipe into the center of the circle where the bag is held. Sometimes everything is quiet, at a given signal, and the hunters creep off leaving the innocent one with the bag. At other times a hen is smuggled into the bag and the "Struggling snipe" is taken home in triumph.

B. TREES

33. Tree Tag Day.

Give each player 10 tags with the names of ten trees common to the tag area. Give them 20 minutes to pin the tags five feet from the ground on the north side of the trees named. No tree is to have more than one tag. A great number of tags may be given for an all day tagging. The one tagging the greatest number correctly wins. A second game of "Calling in the Tags" may then be played. A player may bring in any tag except his own. If he finds a tree incorrectly tagged he leaves it and on a later tour of inspection obtains two points if he can correct the mistake.

34. Tree Spying.

Stop at a tree, such as the wild black cherry. Each one identifying the tree by use of the leaf chart within three minutes is given a point. For each mistake a point is subtracted. At the end of the trip add the scores and announce the winners.

35. Tree Scouting.

Appoint leaders to choose teams. Tell them to study the oak leaves on the chart and then at a given signal give two minutes to obtain a while oak leaf. The tree given should be known to be nearby. At the end of two minutes blow a whistle. Those back in their places with a white oak leaf (no more, no less) score a point. Next send them scouting for a red oak acorn, a balm-of-Gilead bud, and so on. The team scoring the greatest average represents the group of best tree scouts.

36. Forest Good Turn.

The Scout law, "Do a good turn daily" is well illustrated by this game. Give each group a few minutes to discuss the subject "Forest Good Turn." When the whistle is blown they are given 5 minutes to do a good turn. Each good turn is worth one point. If no one else did that particular good turn it is worth two points. This encourages originality. The team performing the greatest score wins. The reports are not the least important in this game. Some good turns are: Labeling poison ivy, destroying a tent caterpillar's nest, neatly cutting a broken branch, removing a tree fungus, hiding a rare flower, cleaning away fire bait, picking up rubbish, burying broken glass, hanging out a piece of suet for birds, and planting the seed of a desirable plant. If each scout troop played this game once a year the amount of good accumulated would be inestimable.

37. A Forest Census.

This game is well adapted to a permanent or temporary camping site. Mark off a forest area as the "Out door museum." Have a large number of sale tags such as used in a department store. Divide into groups called foresters, miners, florists, birders, etc. If there are a large variety of trees and a small display of minerals give each tree the value of one and each mineral the value of three. The naturalists are then given fifteen minutes to label and list the natural history objects under their department. In the case of birds it would be the evidence of the bird rather than the bird itself. The reports around the council ring are instructive and often amusing.

38. Tree Trailing.

Hide messages "en route" and send out companies 30 minutes apart. The messages may read as follows: Take the valley trail to the east until you see a large yellow willow. In an abandoned flicker's home there is a note. Read it carefully. This note may read: Within sight of this spot is a silver poplar. As far from that tree as it is high in the direction of the noonday shadow is buried a message on birch bark. Please leave this scroll as you find it. Before sending out the trailers it should be announced that skill and not speed is the essential thing. The group following the farthest wins. The trail, therefore, should become more difficult as it goes on. Such a trail may be made more interesting and exciting if it follows a story. Possibly some pirates landed and hid some booty the night before.

39. Hare and Hound.

Instead of paper use leaves such as the chestnut leaf. This does not litter the country and is much more instructive.

40. Tree Cribbage.

This may be played for a time when on a hike. One group may take one side of the road and the other the opposite or the points may go to the side recognizing the tree first. It should be limited to trees on the road side of the fences. Counting the number of legs on the right and left is fun and usually ends up when coming to a poultry yard or cemetery. Sometimes it is stated beforehand, as a joker, that a rabbit or a white horse, seen first will count as five points.

41. Tree Pantomime.

Epoch 1. Have a person with smoked glasses, cotton batten in ear, gloves on hands, adhesive over lips, and clothes pin on nose. Nothing in the environment makes any impression. Epoch 2. Meets a scout or woodsman. The woodsman teaches the greenhorn. He gets his eyes open, the cotton batten out of his ears, etc. He is now able to recognize trees by feeling, taste and smell. Being in the woods becomes a delight.

42. Pitch Pine Tag.

Something like Puss in the corner except player cannot be tagged while touching a pitch pine, or some other tree agreed upon.

43. Leaf Passing.

Choose a broad leaf. Players stand in rows. At a given signal the one in front passes the leaf overhead to the one behind who passes it between his legs to the next and so on, alternating over and under. The one at the back of the line runs to the front and the leaf is passed back again. This is repeated until the one who started in front is back again. If the leaf is torn or injured in any way the game is lost.

44. Prove it.
Players sit in a circle. The one starting the game says,—"From where I am I can see a gray birch." The next one says,—"From where I stand I can see a gray birch and a black cherry." The next player repeats all that the previous players have said, in exactly the same order, and adds another tree or bird. It may be limited to what is seen on one gray birch tree. If anyone doubts the statement he may challenge the speaker. Anyone caught drops out of the game.

45. Opposite and Alternate.
The players are divided into two lines, one called "opposites" (refers to buds and leaves) and the other "alternates." They face each other. When the leader calls "opposite" all the players in that line run and tag a tree that has opposite buds or leaves, and the "alternates" try to tag them. Anyone tagged before he finds a tree of his own becomes an "alternate." The side having the greatest number of players at the end of a given time wins. The same may be played with annuals and perennials.

46. Twig Diaries.
Suited to late fall. A tree is named, as tulip. The players must run and get a tulip twig that grew in the preceding summer. He counts the number of leaf scars and searches for that number of leaves that grew on the tulip tree.

47. Tree Jerusalem.
On a hike through the forest holler the name of a tree and give 30 seconds to claim a black birch. All those getting a tree in an allotted time get a grain of corn. The one getting the greatest number wins. Can add beaver by way of interest. Anyone seeing a man with whiskers hollering "Beaver" first gets 5 grains of corn or "double beaver" (whiskers plus moustache) gets 8 grains.

48. Leaf Relay.
Line up in groups. Give each one at the head of the line a list of trees. At a starting signal the first player hands the list to the second player and runs and gets a leaf of the first tree on the list. When he returns with the correct leaf the second player passes the list to the third in line and runs for the leaf of the tree second on the list. The group getting the greatest number of leaves in a given time or which finishes the list first wins.

49. Grand Change.
Players are divided into four groups, as black oaks, red oaks, chestnuts and elms. Players stand by their tree—no two at any one tree. "It" stands in the centre and calls elms. All elms must change trees while the centre player tries to get a vacant tree. If the centre player calls forest, everyone is required to change, but must keep to their particular tree.

C.   GENERAL

50. Curious Shaped Animals.
This game is well liked by children on Nature Guide Trips. Give them five minutes to get a curious shaped animal. The scaling bark of the yellow pine is particularly well adapted to this use of the imagination. Drift wood and washed roots are suggestive. Pine cones as the bodies of filli-loo birds, knots as curious heads, and berries as eyes add to the fun. In camp it may be announced that there will be an exhibition that evening. It furnishes a good time for the day.

51. Blind as a Bat.
It is well known that a bat is not blind. Some people who are blind can see more than some people who are supposedly not blind. Blind the player with a neckerchief. Each player has a keeper who may have a string attached to the player to guide him. He is not allowed to carry on conversation. The keeper writes a list as given by the "blinded player." The one naming the most things correctly in ten minutes is the winner. The keepers may then take their turn.

52. Trailing.
Let a person walk through an untraveled region. Walk rather recklessly without taking care about footprints, snapping twigs, or breaking branches. At the end of ten minutes sit down in an inconspicuous place.

53. One Old Cat.

Have a home base, umpire, and any number of fielders. The game is played with a volley ball or an indoor baseball. The batter hits the ball with his hand. He is out if a fly is caught or when hit with the ball. When the ball is hit the umpire hollers the name of a plant, such as primrose. The runner has to get on one of these plants to be safe. A fielder getting on one of these plants first makes it unsafe. The same plant cannot be named in succession. A runner on a plant must change plants when the batter hits a fair ball. Fielders cannot hold the ball but must keep passing it. The umpire acts as pitcher and tosses up the ball for the batter to hit.

54. Camouflage.

The group hide their eyes while one person or a stuffed animal is hidden in some conspicuous place, but not entirely out of sight. A confederate may assist in the camouflage by using green boughs, grasses, etc. He may make misleading sounds as breaking of limbs to suggest climbing a tree, etc. When all have seen through the camouflage the first discoverer is entitled to be camouflaged.

55. Aggressive Resemblance.

This game is similar to the one given above but is more exciting. Post a lookout. Players render themselves indistinguishable and try to creep up on the Lookout. If the Lookout names and points to a person, that person is out of the game. The one who gets nearest to the Lookout in a given time is the winner.

56. Outdoor Smelling.

This is a continuation of the Indoor Smelling Game. To count, the object must be scented before being seen. It may be new plowed ground, a pine wood, new mown hay, a salt marsh, a dinner cooking, a forest fire, a stove coal fire, an underground fire, peat burning, a tar barrel, etc.

57. Nature Sounds.

The group are given five minutes to see who can make the longest list of things heard in the woods during that time. It may be a raindrop, crow, cow, rooster in distance, rustling leaves of oak or the swish of the pine, tapping of the woodpecker, or song of the brook.

58. Dark Night Anatomy.

Players sit in a line or circle. Pass out the parts of a cat. Use oyster for the liver, grape without skin for the eye, wet rubber tube for intestines, etc.

59. Seed Dance.

The player who can keep a certain kind of seed in the air longest, without using the hands, wins. Milkweed seed is good for this amusement.

60. Exploring Game.

Parties are sent out to discover good blueberry picking, clay for pottery, frog's eggs for the aquarium, a good region for nutting, sphagnum moss, etc.

61. Sand Tracking.

Make puzzles on the beach, such as: Someone has a piggy back ride, someone falls down and is helped up, someone crawls on hands and knees to view a bird, etc.

62. Observation Game.

This is a game commonly given in scouting. It is allowed to take place in a store window. It is better to use common objects of the out-of-doors or even rare birds that should be known about. Allow to observe the coloring of the woodchuck for example. After it has been written, read a description and have the players check off each point mentioned.

63. All Round Scouting.

This game is a real test of scouting. Group is divided into patrols. Tell them that this game is to see which patrol is best adapted to go into the woods and shift for themselves. First of all if they do not have matches they need something to make a fire. Show them a piece of flint or quartz that is in the neighborhood. When all have had a good look at it give them one minute to get a piece. Everyone having a piece and back in place at the end of a minute is given one point. Next say that they need good tinder. If they bring back good, tinder, such as grey birch bark, shreds of red cedar, nest of a mouse, give them one point. Next they must have kindling wood. It should be fine, dry, and preferably pine. Then they are sent for edible plants, such as have been found in the neighborhood. The specimen must be shown each time. Milkweed for greens, pine knot for a candle, sumach berries for lemonade, maple sugar leaf to identify the sugar tree, poke weed leaf for locating starch food, nuts for fat, etc. They are given a minute for each. The patrol having the best average is best suited to roughing it.

64. Compass-Pacing-Nature Trailing.

This again is a game for real scouts. A meter is given which may read something like this:

LEFT HAND CHECKS	COMPASS AND PACES	RIGHT HAND CHECKS
	NE	
White Pine	2	Granite Boulder
Open field	16–30	White Oaks
	N2OW	
Brook	7	Alders
	N15E	
	10 Fence	
Fox den	13	White Ash
Poison Sumach	25	Ledge

65. The Plant Geography Game.

There should be a library of good botany books, which tell whether certain plants were introduced from Europe, or Asia and their geographical range. The players are taught how to find this information. A native plant counts one point, a plant from Europe one point, and a plant from Asia two points. The contestants are given a day or two to make their collection. All assemble for the final reckoning. The plants that come from Asia by the way of Europe will count two points if the side presenting the plant discovers that its original home was in Asia and only one point if they say Europe. They will discover that the majority of plants introduced from Europe are weeds. Some interesting "whys" result from this game.

66. Foraging Expedition.

This is a game of life which is played by all animals. Civilization has made it impossible for most people to play the game, successfully. Have edible plants count one point, medicinal plants one point, and plants useful in the crafts one point.

67. Spelling Bees.

Divide players into groups. Play the game with fall flowers, insects, or trees. Hold up a fall Flower. The first in line must name it and give an interesting fact about it. If he fails he drops out of line. The side having the greatest number remaining wins. It is better to commence with the most common and well known plants.

68. Intelligence Trail.

This is one of the few trail games where time as well as intelligence may be the important factor. For example: Trail starts at a granite boulder. If this granite has

mostly pink feldspar	go SSW	16 paces to a grey birch
considerable hornblende	go NNW	10 paces to a grey birch
no black mica	go N	20 paces to a red maple
over 50% fine grained quartz	go S	15 paces to a sassafras

The next station is the red maple. If this tree has

bark resembling beech	go E	12 paces to an elm
alternate leaves	go E	24 paces to an elm
leaves with same shade of green above and below	go W	18 paces to a red maple
twigs with strong odor when crushed	go W	30 paces to a smooth sumach

The next station is an elm. If this tree is
over 125 feet tall . . . . go NNE   20 paces to a woodchuck hole
less than 50 feet in crown
    spread . . . . go SSW   30 paces to an ant hill
4–5 feet in diameter . . . . go SE   38 paces to a coral mushroom
has leaves averaging over 6
    inches in length . . . . go NW   17 paces to a limestone boulder
(To get the height of a tree sight it with a pencil held vertically, at arm's length. Turn the pencil to a horizontal position and locate a corresponding distance from the trunk and on the ground. Pace off this horizontal distance.)

69. Best Curio Collector.

Played with a group walking through the forest. Name the curio and crowd scatters to find it. The one discovering it first gives a war-whoop and others gather around him. If he is successful that is the starting place for the next. Send for such things as the following: A hump back tree, a tree struck by lightning, a tree with last year's catkin, a tree with scale insects on it, a tree infected with galls, a tree with branches on one side, last year's fruit stem, a tree with moss on the north side only, a tree with lichens on the south side, a tree that has a stone in the centre of the fruit, a deciduous tree that has cones, an evergreen tree that does not have cones, a red maple that has had fruit, a red maple that has not had fruit, a twig that took 10 years to grow an inch, a twig that grew ten inches in a year, a twig that grew 36 inches or more in a year, see who can find the oldest twig five inches long, a sumach bush five years old, a rock with a quartz vein, a tree with a rock callous, where a wood pecker has been feeding, a woodpecker's home, the work of the sap-sucker, a feldspar crystal, pine pitch enough to fill a thimble, fruit of the ash tree, a mud dauber's nest, a leaf miner's home, nuts gnawed by a squirrel, an owl's pellet, evidence of a rabbit, a robin's nest, an animal foot print.

70. Sentinel and Marauders.

A game adapted to rest hour on hike. Seat one patrol in a circle with individuals at least ten feet apart. Cut shrubs three feet high and stick in ground five feet in front of each member of patrol. Blindfold with neckerchiefs. A judge stands in the middle of the circle. The second patrol is given 10 minutes to steal into the circle and get a shrub and walk away with it without being detected. The seated patrol, or sentinels, when hearing a footstep, twig snap, or unnatural rustling of leaves, points in that direction. Any marauder pointed at in this way is eliminated. The judge indicates this by a wave of the hand. The eliminated individual must go to the auditorium, some 60 feet away, decided upon before the game starts. The score equals the number of shrubs taken out of the circle for at least a distance of thirty feet in the ten minutes. The Sentinels and Marauders exchange positions. The team securing the greatest number of shrubs in 10 minutes wins. Dead twigs and dried leaves may be placed around the circle or it may be played amongst a thick growth of shrubs. This game develops great skill in stealthiness.

71. Sound Locator.

A good woodsman can locate a sound quickly both as to direction and distance. As a preliminary training have "it" stand back to 10 or 12 people. No person to be nearer another than ten feet. A leader points to some one who whistles. "It" turns around quickly and names the one who whistled. If correct the whistler takes his place or the best average for 10 trials. Out-of-doors this may be tried by rustling autumn leaves on the ground, by wading in water, by jumping in sand, by dropping a stone, or by throwing a stone a few feet away and having the person identify the stone, by snapping a twig, by taking three steps, etc.

72. Firefly Tag.

Played when it is dark. A fast runner is provided with a flash light. He must give occasional flashes. Pursuers must tag the runner. Whoever catches him becomes the firefly.

73. Shell Divers.

The diver scores the number he brings up at one dive or different values may be given to different kinds.

74. Pine Cone Baseball.

Use pine cones instead of baseball. The pitcher throws a cone at the batter. If the cone hits the batter a "strike" is made. If not, it is a "ball." Count strikes and balls as in baseball. The batter may hit the ball. If it is a fair hit he becomes a base runner. Runners may be put out between bases by being hit with the cone in play. Players cannot run with cones. The batter may catch the cone and throw it into the field.

75. Gall Test.

The players try to collect the greatest number of galls in a given time. One point for each kind obtained and one additional point for each one identified. Use Professor E. Laurence Palmer's booklet published by the Comstock Co.

76. Trail Observation.

A good scout can retrace a trail because he remembers certain objects along the trail. Walk for a certain distance over a trail and then ask ten questions or each one may write ten questions and then exchange. Arguments will follow and it will usually be necessary to go over the trail again. Various objects may be placed along a backyard walk. They then return to the house and are given a pencil and paper to make a list of what they saw.

77. The Balance of Nature.

The hunter traps the skunk, the skunk eats mice, the mice eat bumble bees, the bumble bees assist the clover, they may sting the cow, the cows eat clover, the hunter drinks milk. The skunk is safe in a hole in the ground. In the game it may range from an ant hole to a hole of the ground hog. The mouse is safe in a hollow tree or in an abandoned bird's nest. The bumble bee is safe under a stone (standing on stone). The cow is safe under a tree. The clover may receive a friendly visit from the bee or be nipped by the cow. The hunter is safe at home. At a given signal the struggle for existence begins. Anyone tagged by his particular enemy—when away from his safety, becomes that animal. The game may be ended by saying that after a given signal any one tagged becomes Hors de Combat and is out of the game.

78. Sense of Direction.

Greenhorns get lost going into the woods and bragnuts when they try to get out. Zigzag a group through a forest and then have them try to go straight back to camp. This is not a running race. Everyone is expected to walk. The one getting back first may have gone on the most direct route. If a fire tower is at hand the leader may check up or all scouts may point toward camp and one climb a tree and determine the winner.

79. Breaking Through the Line.

Outline the region in which the contest is to take place. May be between the lake and a certain road. Half of party guards the line and other half tries to break through. May use pine cones for ammunition. To be killed the scout must feel the bullet. Anyone killed is out of the game and must lie down until scalp (neckerchief) is taken or until rescued by own army. Cannot take active part in game after killed. May score on point basis. 5 points for a patrol leader, 2 for a scout, and 10 for the scoutmaster. Everyone getting through the line gets two points for his side.

80. Spying.

Have one group quarter of a mile away do something. The other groups are on a hill top and take notes on what they are doing. Each group on the lookout may have binoculars. The group having the most accurate description, according to the judgment of the party in action wins.

81. Compass Training and Testing Trail.

A testing trail becomes a training trail after once having used it. Anyone failing should take the test on a new trail. One may name the points of the compass and yet be unable to use it. A woodsman should be able to follow the trail in either direction. This trail is made out for Mausoleum Hill, Syracuse, New York, but can be adapted to any region. The notes in paranthesis are hints to leaders laying out trails and were omitted in the original directions. The one being ex-

amined should fill in blank spaces. Commencing at the N. W. corner of the area
fenced in on Mausoleum Hill, follow the fence east. (This gives the scout a
chance to test his compass used in laying the trail) for ————
paces (A pace is not a stride); thence ————for ————paces; thence————for
paces————; thence along fence line (By fence line is usually meant an abandoned
fence, or where it used to be) due————for paces————to a corner tree. The
trunk growth of a tree is upward outward (Cross out one of these words). Proof
————. Set course S 5 degrees west and proceed 30 paces (this gives an
opportunity to check up on the average pace of the one who laid the trail) to a
woodpecker hotel. The woodpecker got room and board for his services. This
structure is also a squirrel storehouse and feeding station. From here continue in
same course so as to pass a shagbark hickory in 9 paces, a hop hornbeam in 3 paces,
a red oak in 5 paces, a large-toothed aspen in 6 paces, and an American elm
in 21 paces. Observe the American elm closely and continue pacing in the same
direction until coming to an elm that is somewhat different from the American
elm. This is the slippery elm which is ————paces from the American elm. If
you have the correct tree you will find that the inner bark is fragrant and mucilaginous. Cut off a twig and test it. Set your compass at this tree and travel W
15 degrees N 160 paces to a rock. Is this rock a resident or a transient? How
know?————. What artificial fragments do you find at the western end of this
rock?————(Bits of broken glass. This is a means of checking up that they find
the right rock). Shift your course 20 degrees more to the north and go 31 paces to
an apple tree. It differs from the surrounding hawthornes in not having————and
in having————. Near this tree is a stake. Sight from the apple tree past stake
to some natural object. Take bearing and walk in a bee line until reaching a
cairn (Rocks piled up. Not rocks arranged for a fire place). Keep on in the same
direction to a second cairn. Check up directions now and then with the compass.

82. *Scenic Locator.*

Now proceed to summit of the hill. Orient your topographic map and locate by
compass the following hills (Number refers to elevation given on map). Fill in
compass readings to right of numbers.) 681————; 686————; 670————; 761
————; 845————; 647————; Onondaga County Home————; University
Farm————; Crouse College————; Water Tower————. A diagram of a compass might be made on a board and varnished to prevent wearing by the weather.
This could be fixed in a permanent position on a hill or at camp and then the
various points of interest, visible from the site, might be printed on the compass
lines. The compass is then known as a *scenic locator.*

83. *Nature Locator.*

Set a board compass in a forest. Make out a tree table for 10 trees as follows:
Name of Tree; Compass Direction; Number of paces from compass, or make a bird's
nest table as follows: *Name of nest; Compass Direction; Kind of tree; Heighth from
ground.* Another form of nature locator is to have a foot of one inch pipe on a
swivel joint so that it can be rotated and raised or lowered. Have a list of nature
objects and direction, such as: NNE elevated 32 degrees is an oriole's nest. A
person looking through the pipe when in this position will see the oriole's nest.

CHAPTER VIII

I. *Blazes and Cairns*

Equipment: Four car checks, compass, one more sandwich than you think you will need, sharp pencil and eye, a longing for the out-of-doors, a goodly measure of "pep" and red blood.

Where: Take Woonsocket car from Union Station to Cobble Hill Road.

Co-operation: It is requested that those who follow this trail heed these simple rules: Put up bars after passing through. Do not disturb the signs of the trail. Dispose of refuse.

Leave the car at Cobble Hill Road. Divide the trailers into groups of 5–10 each. Follow car track past Olney pond, turning to the right at post 107. Each group should be given five minutes start on the one before it.

This game may be played while waiting for the various groups to get a start. See how many things made by man can be gathered on the track between Cobble Hill Road and post 107, such as buttons, matches, etc. Each article counts one point. This is car-track cribbage.

At post 107 begin at once to look for the boulder trail. This consists of large rocks which have recently been turned over. They lead to an object which looks something like a picket fence. At the end of the boulder trail—on the north end of this object—set your compass so as to take the range N20 degrees E. Walk in the direction of this range. You will soon pass an old abandoned fence row on the easterly end of a former field. The fence is not there now. Follow this range to the NE corner of the field, then along an abandoned cartway. This cartway is now a footpath. Follow the footpath most usually trodden until coming to some juniper trees. The juniper tree resembles the red cedar, except that it grows low near the ground.

At the juniper trees turn left, following this trail a few feet to a clearing. This clearing is rather recent, as shown by the chestnut stumps where the trees were cut in the winter of 1922.

Find the tote road where the logs were hauled from the cut area. Recent wheel marks on the rocks and the broken twigs of the young growth mark the trail. Follow this until coming to an auto road.

Follow the auto road north, looking on the right for two white oaks nearly 30 feet apart. The first white oak has three old blazes of an axe and the second has three new blazes just made with a knife. Just beyond these oaks turn to the right, following the trail. This part of the trail has knife blazes on trees, and when crossing a cleared area has a stick through a forked branch. The forked branch is placed in the ground and the stick points the way to the next forked branch or blaze.

You will soon come to a collection of boulders on top of a hill. There is a wonderful view from the hill. On the surface of these boulders may be found rock-tripe, a species of Umbilicaria. This plant has often saved the lives of arctic explorers and of hunters and trappers in Alaska. Knife blazes about six feet from the ground lead one from the boulders to a well-worn trail going to the right.

Follow this path a short way, looking on the right for a cairn—rocks placed on top of each other. The cairn marks the beginning of a new trail which leads across ledges. As these ledges remind one of the bare slopes of the northern peaks, Mount Washington, Adams, Madison, etc., the way is marked by a line of cairns such as used on these mountains. The top rock on each of these cairns is milky quartz so that it can be seen at a distance. The next cairn is always in view. Proceed cautiously so as not to lose the trail until coming to a red cedar on a cliff.

At the red cedar take a range S20 degrees east, to a large white oak over one foot in diameter. This oak is quite a little distance away, but in sight. This particular oak has three horizontal lines half encircling the tree and about 18 inches apart. Go to this oak.

At the white oak take a range 35 degrees east of south. Follow this line to a cairn.

At the cairn sight west 20 degrees south. In this direction, but not visible, are two glacial boulders over six feet high resting side by side on a glaciated rock, or ledge. Keeping the course stated, go to these boulders. If you have followed the trail correctly you are now at Twin Rocks.

From the top peak of the left-hand rock, as you approach, you may set your compass. The largest rock visible between S and E is our destination. How many degrees east of south is it? Go to this rock. It is at Rockside Point, the fireplace where we will cook our dinner.

II. *Picture Trailing*

Winter is not a closed season at Quinsnicket. Every school citizen has the right to experience and enjoy our snow-clad forests. In Figure number 1 may be seen a group of young folks who are out in Lincoln Woods on a skiing trip. They have just finished an outdoor meal of broiled steak, fried onions, hot rolls, steaming cocoa, and toasted marshmallows. Did you ever whet up your appetite for just such a hot dinner with the crisp February air for frosting? No wonder that they are forever asking, "When may we go again?" Yet there are people in Rhode Island who think it necessary to go to the northern country—North Conway, Jackson, East Jaffrey or Laconia. But be it north or be it west, our slogan is *"See Quinsnicket First."* We are going to suggest a day's Program for those who wish to pioneer in this way. Although the schedule is written particularly for winter, it can be used at any time of the year. Let us make Quinsnicket an all the year park.

We plan to take the 9:15 a. m. car for Woonsocket from the west incline in front of railroad station, Providence, to Cobble Hill road. The fare is exactly 10 cents if we use the new metal tickets. We appoint captains and organize teams of about ten each. Each team will follow Cobble Hill road easterly. All try to follow these directions.

Fig. 1. A Skiing Party at Quinsnicket

The rustling leaves on the large pasture oaks tell us that it is winter. In olden days the white oaks were considered of great value for ship building, and when other trees were cut they often escaped the colonist's axe that they might become timber-size. This may be the reason for the white oaks being the veterans of Quinsnicket. We shall meet several of these stately monarchs. Count them. The teams reporting the correct number will be credited with one point.

In order to get into the quaint atmosphere of these oldest inhabitants let us read the description of a similar scene from Virgil.

> "Jove's own tree,
> That holds the woods in awful sovereignty;
> Nor length of ages lasts his happy reign.
> And lives of mortal men contend in vain.
> Full in the midst of his own strength he stands,
> Stretching his brawny arms and leafy hands,
> His shade protects the plants; his heart the hills commands."

Curiously enough, along with these thrice-octogenarians are the gray birches—perhaps the shortest lived of forest trees. Young though they are, many are round shouldered from carrying loads of winter snow and ice. Shake one of the gray birch catkins into your hands. The smaller objects are winged seeds. These seeds are protected by the larger scales, which resemble soaring birds. These bird-like scales make interesting silhouettes on the snow. Obtain one of these scales as proof that you have made a visit to the gray birch settlement.

Some of the less flexible black oaks—across the way—had their heads snapped down by the ice storm of 1921. How are these black oaks trying to "carry on?"

Fig. 3. Winter in the Clearing

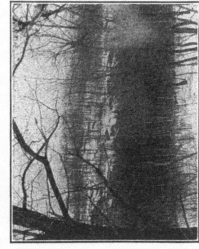

Fig. 5. Gray Birches

Fig. 2. A Natural Totem Pole

Fig. 4. Black and White Oaks

We now pass the "3 H" Elm keeping vigil for the grotesque figure caught by the camera man in Figure 2. How do you account for its weird shape?

We may now have the fun of "running out" an old boundary line. This post is but one of several that mark an old fence line. Surveying scouts often have to find the limits of a woodlot. George Washington and Henry D. Thoreau were very clever at this kind of scouting. Those who are successful will discover the beautiful view in Figure number 3. Observe that this picture was taken on the edge of a clearing.

The two Japanese-like trees in the foreground are noted for their many names,—Sour gum, black gum, pepperidge and tupelo. You

Fig. 6. Table Rock

may take your choice. Our course is now along the southern margin of Olney pond, toward Stump Hill on the east. As we pass through the grove of mixed hardwoods, look for other tupelo trees. Take a small twig from one of these trees and cut a lengthwise section through the centre. Keep the twig to show the cross partitions in the pith.

This grove is mostly chestnut, black oak and white oak. The chestnut are dying from the blight and this is probably the reason for the clear-cut area over which we have just tramped. Figure 4 shows the difference between the trunks of the white and black oaks. The lighter colored trunk is that of the white oak. The white oak sheds its bark. Pieces of its bark are shown on the white snow. We may collect the following as evidence of having identified these trees. A piece of bark from the white oak; a leaf with sharp pointed lobes (instead of the rounded lobes of the white oak) for the black oak; and a twig of a chestnut sprout, with its inclined buds, for the chestnut.

Fig. 7. The Sycamore Tree

Fig. 8. Gray and Black Birches

Fig. 9. Shagbarks

Our route is now along Stump Hill road past the site of the old thread mill. We can name the evergreens from the silhouette page in this book. Which way do the tips of the evergreens point? This is an indication of the prevailing wind.

We shall not need to put on our spectacles to find the sycamore tree. This tree is at its best in winter when there are no leaves to hide it. The white blotches shown in Figure 7 are characteristic of its appearance. We find the tree by a stream—as the poet, Bryant described it:

> "Clear are the depths where its eddies play,
> And dimples deepen and whirl away;
> And the plane tree's speckled arms o'er shoot
> The swifter current that mines its roots."

Emerson refers to the sycamore as "the largest, grandest and loftiest deciduous tree in America." There are three possible proofs of having discovered this "plane tree of the west" (Platanus occidentalis). We may find a large piece of bark such as it has the habit of pushing off; or a fruit ball (sometimes called buttonwood tree); or a leaf and twig, to show how the leaf stem fits snugly over the bud like a skull-cap.

We now continue north, crossing Lincoln road and entering a foot trail. Let us watch out for a pure stand of gray birch in a ravine as portrayed in Figure 5. This valley contains the headwaters of Lily Pond brook, which flows past Tablerock. We follow the trail which parallels this valley until we reach fireplace number eight. Figure number 6 is taken at Tablerock. This fireplace does not show in this picture. One of the most interesting trees in this scene is a black birch growing from a niche on the side of the ledge. We have no trouble in recognizing the white and the black oak. Between the two oaks and beyond is a red cedar. The wood of this tree is handled more than any other wood as it is used in making lead pencils. It is also used in making cedar chests, because its fragrance is said to keep away moths. We are particularly interested in it at this time, as the bark is dry and its fine fibres may furnish a means of kindling our fire. It is here that we planned to cook and eat our dinner. (To reserve this fireplace telephone to the office of the Metropolitan Park Commission, at the State House at least a week before the trip.)

After dinner we may count the scores of the teams and then continue the competition. Figure number 8 was taken within a three minutes walk of Table rock. The first team finding it receives a point. This picture is of a gray and black birch forest. One of each kind of tree is standing in front, as if they had stepped out of the crowd to have their photographs taken.

Another tree that can be easily recognized by its trunk is the American Beech. All squirrels and blue jays know this tree. The clean gray bole of the tree cannot be mistaken. The smooth bark of this beech looks like a totem pole. It was fashionable a few years ago to carve one's initials on a beech tree. This custom prevails no longer, and is even condemned because it defaces the beauty of the clean, smooth trunk. What team can find the nearest beech?

Fig. 10.   The Shelter at Quinsnicket

Our next search will be for the nearest shagbark or shell bark hickory. This too, is a friend of the squirrel. James Russell Lowell describes a squirrel that—

> "On the shingly shagbarks bough,
> Now saws, now lists with downward eye and ear,
> Then drops his nut."

The shagbark shown in Figure 9 is near the pump by the Glen. From the silhouette page we can find the name of the tree back of the shagbark.

Figure 10 is of the shelter at Quinsnicket lake. There is another shelter at the recreation field. In case of storm a party may take refuge at one of these places.

The long shadows of the trees tell us that it is getting late. We must gather up our belongings and "leave no trace." This means that we must burn all paper and rubbish, and put the tin cans and bottles in the container. And, what is still more important, we must make sure that every spark of fire is out. He who loves the out-of-doors must not risk ruining the beauty of Quinsnicket—the very thing that attracted him to it.

Now that everything is ship shape and the clans are ready to depart for their several homes it will be in keeping with the trip to sing taps.

> "Day is done!   From the Lake.
> Gone the Sun!   All is well,
> From the sky,   Rest in Peace,
> From the hill,   God is nigh."

III. *The Druid Circle Trail*

In ancient Celtic countries are found rude structures of stone which are thought to be the altars of the Druids. Often times a hundred youths would gather about these altars for instruction. All their communications were oral. The Druids held their meetings in the deep forest, especially in groves of oak. Every day was tree day with them. The trees, the stars, the earth, and the rocks were their favorite studies. The Druids also set forth and preserved the laws of the land. Their emblem was the egg of the serpent. The Druids were not only a learned class, but they had a code of ethics and were brave. It is said that before battle they would often throw themselves between two armies to bring about peace.

There is a Druid Circle in Rhode Island. It is older than the Druid altars of ancient Europe. The stones of the altar are larger than those ever moved by the hands of man. They were brought from the north by the great glacier and placed where one of our beautiful forests now grows. In this circle one may learn about the trees, the stars, the earth and the rocks. We know of no better place for Rhode Island youth to revive the ancient custom of tree worship.

These notes are suggestions for the gathering of modern clans. The clan may be a school, a scout troop, a neighborhood party, a family or just one person. The gathering braves should be divided into squads of 5 to 10 in each group. The group then elects its chief. The chief calls his clan to order and may arrange his tribe in a circle. When any member of the circle wishes to address the chief, he arises and stands silent. The chief recognizes him with a nod. The brave then says: "O Chief, I suggest the following name for our clan." This continues until agreed upon. Someone else then rises and when recognized says: "Oh Chief! I suggest that our tribe have the following emblem." If the tribesmen agree they say, "How! How! How!" in true Narragansett fashion. If they do not approve they show displeasure. Songs, and warwhoops are organized in the same way. Then comes the code of wood ethics. Someone says: "O Chief! I move that this tribe go on record as opposed to the scattering of rubbish, such as paper, tin cans, etc., in our forests." Someone else says: "O Chief! I would like to hear from the other warriors as to their opinions concerning the 'Grouch' who goes on a hike." Before disbanding the chief appoints the time and the place: "Tomorrow morning, five hours after the sun has passed the horizon, I want my braves to meet the Grand Sachem (ordinarily known as teacher) in front of the big council house. Bring food to cook over the fire. We will take the white man's big caboose marked Woonsocket. And this takes us along the Louiquisset Pike to the Quinsnicket station."

The Grand Sachem, upon reaching Quinsnicket, calls for organization of the tribesmen. The tribes then march in order, trying a few warwhoops on the way, until they come to a sign which says:"Break

Fig. 12. Boulder on the left has recently lost cap

Fig. 14. Three White Oaks

Fig. 11. Gray Birches

Fig. 13. Chestnut Blight

Neck Hill Road." Enter the woods at this point. The tribe should leave the starting point five minutes apart and follow the trail, as blazed in this book.

Starting at the sign "Break Neck Hill Road" take the left-hand trail into the woods until you find a sign which reads, "To Quinsnicket Hill and Lake." Follow the arrow until you come to the view shown in Figure 11. This a clump of gray birches. Observe the black

Fig. 15.   Chestnut Stumps

patches below each limb. Note the limbs broken by the heavy ice storm of 1921. Why did the birches not break? Take the extreme left-hand patch which leads past a glacial boulder to a clearing. On the north is a winding white road. It goes toward Lime Rock and the whiteness is due to the limestone soil. On the east you may look over the aspens and see Quinsnicket Lake. Look about for some evidence to prove to the Grand Sachem that you have seen Quinsnicket Lake. This will count as one honor for your tribe.

Now return to the gray birches. One trail leads across a glaciated ledge. Follow this until reaching Figure 12. Collect evidence of being here for another honor. Note the white quartz vein and the cap which has slid from place on one of the rocks. Your evidence must not be either of these two observations. The snow was still on the north of the rock when the picture was taken. Why?

Your route from here is due south along a fire line. Make three observations about the fire line. If they are correct you win one honor point. Follow the fire line until it is crossed by a foot trail and then take this path east. Continue slowly and be on the lookout for Figures 13 and 14. Figure 13 shows a group of chestnuts which have been attacked by the chestnut blight. This disease will probably exterminate our chestnut trees. Figure 5 is of three white oaks. The oak was the favorite tree of the Druids. Whatever they found growing on that tree they thought was a special gift from heaven. This is why they worshipped the mistletoe. You will find something growing on some of the oaks at Quinsnicket. See how many kinds of things you can find growing on the oaks from this place to Druid Circle. One honor will be given for every five specimens collected

Fig. 16. Quinsnicket Woods

from an oak tree. Estimate, by pacing, how far apart the white oaks are from the diseased chestnut. One honor will be awarded for discovering these trees and giving a reasonable estimate of their distance apart. Hereafter, the number of honors for each certain thing will be indicated by a number in parenthesis.

Continue on this trail until coming to a cartway. Send several scouts to observe the chestnut stumps, Figure 15. These chestnuts were cut down last winter. Why? (1). How old were these chestnuts? (1). How do you account for the peculiar shape of the stump? (1). In what year did this event take place? (1). Examine several stumps and determine from what direction the enemy came. Steal along the roadway in this direction.

If your scouts made a correct report of their last skirmish you will soon come upon the view shown in Figure 16. One great rule of woods ethics has been violated here. It is a common violation in the white birch region of New England. Report this in the council ring with a just reason as to its being poor ethics in the woods. (1). Estimate the amount of wood in terms of cords in the pile on the right. (1).

Continue on the same course until reaching a much-traveled road. This leads you directly to Druid Circle. You will first see this magic group as viewed in Figure 17. You will now sit down for a rest and reckon up accounts. Add the honors and determine which tribe has proved to have the best woodsmen. You may now wish to continue to the Glen and cook in the fireplace or if you have an hour you may wish to have the leading tribe defend its title. If the challenge is made and accepted you may deposit your dunnage and prepare for the fray.

The Grand Sachem takes the middle of the council ring, and around the circle are the chiefs with their tribesmen in back. The Grand Sachem speaks:

"Fellow Narragansetts! The fact that you are here is due to your cleverness as scouts. The Wampanoags have shown the greatest

Fig. 17. Druid Circle

skill. The Pequots doubt their superior ability and wish to continue the contest. In the old days such disputes were settled by the bow and arrow. To-day we are to settle this by prowess of brains and the knowledge of woodcraft. No one is to leave the circle until the sound of the Gourd.

"You may now look at Figure 18. This is a glacial boulder. When the ice lowered it into place it came down onto smaller rocks. It was set there more than 10,000 years ago. The first team finding this rock and having all its members stand near the place where the camera man took the picture will have one honor. Your war cry will signify that you are there. You may start when I tap the Gourd. We will re-assemble here after discovering the particular spot mentioned."

Figure 19 is another member of Druid Circle. It is particularly interesting because the large even surface shown in the front of the rock shows where another rock has fallen away from it. This second rock is in this vicinity. The tribe surrounding this second rock first wins an honor. Three proofs must be given that it came from the rock in the picture.

Figure 20 is of a ledge which was smoothed and worn by the heavy ice sheet. It extends in the direction in which the glacier travelled. At the southern end of the continuation of this ledge is a red cedar which has had a misfortune. What was the mishap? Your proof of

Fig. 18. A Glacier Boulder set down on other Rocks

finding this tree will be a correct report as to what limb has become the new leader (one honor). Your report must be in within 10 minutes for the honor. Trees, like birds or people, upon losing a leader, develop a new one.

Figure 21. The first person standing where the camera was when this picture was snapped wins an honor for his tribe. The picture

Fig. 19. The Counterpart of this Rock is nearby

must be carried in the mind. Any scout taking the printed picture along forfeits the honor that he may win. Do the same for Figure 22 and for Figure 14.

Figure 23. This boulder shows a quartz vein. A fire has been built near it and the intense heat has chipped off some of the rock.

The tree in the picture is a black locust. The scout finding it first wins an honor for the tribe.

Figure 24. This is an illustration of what the lumberman calls coppice reproduction, i. e., growth from the stump. When the white man came to America the majority of our trees had grown from seed. To-day the majority of our hard woods have come as sprouts from a stump. The tribe first finding this cluster of chestnuts and bringing a fairly accurate report of the size of the original chestnut receives one point.

Fig. 20.   Ledge Worn by Glacial Ice

Beyond this chestnut is an abandoned road on the right. Follow this road until it comes to an old fence. This will require more ingenuity than the other explorations. What kind of a fence is it? (1). Take one leaf of the vine that grows on this fence. What kind of a leaf is it? (1). In what way do trees tell where this road used to be? (1). How do the rocks guide you? (1). Carefulness rather than quickness is the keynote to success in this instance. You may report back to the circle in twenty minutes. Any team not in place when the whistle is blown will not receive credit.

Rhode Island has many beautiful views. Figure 25 is one of them. It is at the end of a formation similar to Figure 20. You may wish to recall what you were told about Figure 20. Across the valley from this viewpoint is the northern slope of a hill. We know that it is not a southern exposure because the white snow may be seen through the trees. In order to take this view the camera stood beside the white oak which is shown in the picture. Less than ten feet away is the first "thing" to have been seen on top of this ledge

Fig. 22.   Broken by the "Teeth of Time"

Fig. 24.   Coppice Reproduction

Fig. 21.   Trees vs. Rocks

Fig. 23.   Glacial Boulder Showing Quartz Vein

after the glacier retreated. The scout first laying claim to this "thing" wins an honor for his tribe.

We may now take our second rest. While we sit here and take in the beautiful view we can have a tribal song contest. Or perhaps the Grand Sachem will tell a story about the early Narragansetts. And as the twilight and dusk comes up the valley we will pick our be-

Fig. 25. Where the Glacier slid over

ongings and leave Druid Circle as we found it—enshrouded in the mysteries of the past—and dawning on the morrow with new problems for other scouts.

IV. *The Trees and Rocks: Sunset Point*

(Take a group to the region in which the pictures were snapped. Tell the story as given below. At the word "go" have them hunt for Figure 26, etc. Give points to group finding it first. Then tell about Figure 27 and have them find that.)

A long time ago this part of the country was very mountainous. Frost and snow broke up the coarse rock materials, and spring torrents washed the soil down the steep slopes. The high elevations were gradually worn down, age after age, until nothing remained but rock cores and low hills. Then came the great glacier from the north. It rounded off the hills, planed the rocks, and pushed up hills of gravel. Huge boulders were torn from rock cliffs and carried to new regions. It is difficult to realize that the glacier came some ten thousand years ago, as it is a little over 425 years since Columbus discovered America, and 2000 years have not passed since the birth of Christ. Yet the results of the invasion of this ancient ice sheet are but as yesterday.

One picture shows an ancient "bed-rock" that has always resided at Quinsnicket, and "drift-rocks" that were brought by the glacier. What is the character of the surface of the older inhabitant? What

Fig. 26.   Parted Company during Glacial Times

Fig. 27.   A Tree Callous

Fig. 28.   The Black Birch is a Foremost
Rock Breaker

caused this? What is the weight of glacial boulder? (Give a rough estimate.) The glacier is the only known force in nature that can transport objects of this weight. What other pictures show permanent rock residents? Transported rocks?

The ledges often have a gentle, smooth slope on the north, and are steep and ragged on the south. How do you account for this fact? Point out these two kinds of slopes in the pictures.

Fig. 29. A Black Birch Prying off a Rock Lid

The frost and heat are slowly breaking up the Quinsnicket rocks. Some have been split apart and others have merely cracked. As a result the crevices are becoming filled with rock waste, and a mantle of soil is beginning to cover the bare rocks. Various plants have come to the aid of the weathering agents, and the rocks have been further broken by the roots of trees.

What picture shows rock crevices? What caused the crevices? What was needed in the crevices for the plants to grow? The plants growing in the seams of one of these rocks is the false lily-of-the-valley. How do they aid in rock decay?

What picture shows that a tree is mightier than a rock? What picture shows a rock split by heat and cold? Wedges of ice pry rocks apart with the same force that is exerted when ice breaks water pipes in winter. Trees growing against the edge of a rock become calloused. What picture illustrates this fact?

Fig. 30.   False Lily-of-the-Valley Growing in a Rock Seam

Fig. 31.   Polypody Fern Subduing a Rock Surface

Fig. 32.   Twin Rocks left on a Firm Foundation by the Glacier

If it were not for Quinsnicket being so rocky it might not now be a park. A region of this kind is not suited to farming. Consequently it was not cleared of its forests. When Providence and its suburbs began to get crowded a park commission was formed to pick out breathing spaces. The Lincoln Woods Reservation, thanks to its rocks, was one place that had not been robbed of its natural beauty. Glacier, rocks, trees, reservation, therefore, are the titles to the four chapters in the story of Lincoln Woods.

You may already have guessed that this writing is designed to be suggestive as to an interesting study that may be made when one goes to Quinsnicket. The following is a list of things to do at Quinsnicket:

1. Point out bed rocks and glacial boulders.

2. Find a rock, or portion of a rock, recently exposed to the weather. How did you recognize it? What caused the new exposure?

3. What changes have taken place in long-weathered rocks?

4. Point out ridges, pits, and furrows caused by weathering.

5. Find a rock that you think was split by heat and cold.

6. Find a rock from which the surface shell has broken loose. This was probably due to unequal heating and cooling, as a glass cracks when plunged into hot water.

7. The Quinsnicket rocks are mostly covered with rootless plants called lichens, which slowly aid in destroying the rocks. Collect six kinds of lichens.

8. Find a rock with the false lily-of-the-valley growing in its crevices. What made this possible?

9. Find a ledge with the rock fern (Polypodium) growing upon it. How much soil has collected beneath the ferns? How did it get there?

10. Locate a tree that is slowly wedging a rock apart.

11. There is one tree in the park that is lifting a large boulder. Be on the watch for it.

12. Trees that grow against the edge of a rock become calloused. Find one.

13. Observe little trees growing from a high niche in a rock. How did they get there?

14. What is the character of tree roots which grow along the surface of the rocks?

15. What is the ultimate goal of tree roots which grow along ledges?

16. Where have bends in the roads been determined by rocks? By trees?

17. Find rocks that are being sculptured by brooks.

18. Find localities where brooks have worn the soil away from the roots of trees.

THE WONDERFUL "ONE-HOSS SHAY."

So the Deacon inquired of the village folk
Where he could find the strongest oak,
That couldn't be split, nor bent, nor broke—
That was for spokes, and floor, and sills;
He sent for lancewood* to make the thills;
The cross-bars were ash, from the straightest trees;
The panels of white-wood,† that cuts like cheese,
But lasts like iron for things like these;
The hubs from logs from the "settler's ellum"—
Last of its timber—they couldn't sell 'em—
Never an axe had seen their chips,
And the wedges flew from between their lips,
Their blunt ends frizzled like celery-tips;
Step and prop-iron, bolt and screw,
Spring, tire, axle, and linchpin too,
Steel of the finest, bright and blue;
Thorough-brace bison-skin, thick and wide;
Boot, top, dasher, from tough old hide,
Found in the pit where the tanner died.
That was the way he "put her through."
"There!" said the Deacon, "naow she'll dew!"

Oliver Wendell Holmes.

"The love of rural life, the habit of finding enjoyment in familiar things, that susceptibility to Nature which keeps the nerve gently thrilled in her homliest nooks and by her commonest sounds, is worth a thousand fortunes of money, or its equivalents."—*Henry Ward Beecher.*

"Animals learn solely by direct observation and first hand experience with nature."

"Every animal, except some men, which educates its young ones at all—educates them on the basis that 'Life is the response to the order of nature'."

"A chick and duckling from the same nest soon part company, as their differing interests dictate."

"The teacher of a brood of ducklings, if she happen to be a hen, has a trying task."

"The circus draws the crowd, and fads attract attention; but fads pass by, and the circus is only for the hour. The nature-study that is true to life must be like wholesome daily bread."

—Quotations from Clifton F. Hodges.

"I come upon it suddenly, alone—
A little pathway winding in the weeds
That fringe the roadside; and with dreams my own,
I wander as it leads."

—James Whitcombe Riley.

CHAPTER IX

A Nature Guides' Dictionary

It is not many generations ago that everyone used native materials for food, medicine, clothing, and protection from storm. Materials for weaving baskets and mats, dyeing, and designs, and even the folk-lore came out of the environment. To present day citizens it is almost a lost art and acquaintance, yet many plants and animal products might be useful to people of the trail, and to the members of camp communities in their arts and crafts, if made known to them. A Nature Guide should have the ingenuity to supply many of the necessities of life from the natural environment. To these ends this little dictionary has been begun.

The information herein arranged, has been collected for years. A great deal of it represents the experiences of the writer as a country boy, and other items have been obtained through conversation with elderly persons and by knocking about as a guide. The information is arranged alphabetically so that if a person knows the plant but not its use or if he should have need for a certain thing and not know its source he can still find it. The materials herein noted represent only a small portion of the native materials that can be employed. This dictionary is therefore merely presented as a preliminary to a more extended research.

Abalone—dried meat is powdered for soup by Chinese and Japanese.
Absorbant cotton—sphagnum moss, wool of cotton wood or cinnamon fern, frayed cedar bark.
Acorns—gather in autumn, dry, remove shell and grind to powder. Filter water through the flour for few hours to leach away the tannin. Cook and serve as breakfast mush. Mix with four parts corn meal for acorn bread. Fruit of white oak and chestnut oak are best. Farmers use acorns to fatten hogs. Doll furniture and dishes.
Adhesive—use the pitch from pine trees. Adhesive tape good for mending holes and tears in clothing; for preventing blisters; for sealing cans, as condensed milk.
Amaranthus (Pigweed)—very good boiled as greens.
Anise—seeds and roots, give Anise flavoring.
Anvil—drive axe into stump and use head.
Arrowhead—drive nail into arrow and flatten head. Use thorn or sharpened bone.
Arrowhead or Saggitaria—boil roots as potato substitute.
Arrowwood (viburnum dentatum)—shoots used for arrows.
Artichokes(Helianthus tuberosus), Jerusalem artichokes—tubers eaten raw or cooked in early spring, good pickled. For sale in most seed catalogs. Easily grown about camp.
Asparagus Substitutes—shoots of ferns, stalk of young cocklebur, young shoots of pokeweed (root poisonous).
Astringent—Oak bark, hemlock bark, acorns, root of geranium maculatum or cranesbill.
Awl—splintered bone, thorn from honey locust or hawthorne.
Axe handles—use sound, seasoned oak or hickory. Use an old handle as a pattern.
Bag—corn husk, rawhide, woven cedar bark. Hotwater bag from stomach of freshly killed animal.
Bait—white grubs in rotten logs, larvae of dragon fly or caddisfly, earthworms, grasshoppers, snails, fish eyes.

Balm of Gilead—buds have a fragrant resin, often soaked in alcohol for a liniment. May be reproduced by cuttings.

Balsam—obtain from blisters on trunk of fir balsam, used in medicine and to mount microscopic preparations. Soothing and refreshingly fragrant. Needles used for balsam pillows.

Barberry—acid of berries quench thirst. Barberryade good with sugar. Gives good flavor to preserves.

Basket—bark of birch, woven cattail leaves, two leaves of the palm clasped like hands and then braided. Basket willow sprouts, pound log of swamp ash or red cedar until falls into strips at annual rings. Cut into ribbons.

Basswood—buds and bark mucilaginous. Good for coughs. An emergency food. Inner bark a cordage.

Bayberry—berries put in pads for flat irons to give fragrance to handkerchiefs. Berries simmered in water yield a wax for candles.

Beach Grass—basket weaving.

Beads—Job's Tears, acorns, fruit of eucalyptus, clay baked and painted, various seeds, small shells.

Bedding—evergreen bows, moss, ferns, grass.

Beds—Mats of willow sticks, woven cattails, bows of fir balsam, dry seaweed, pile of hay.

Bee sting—use mud or vinegar.

Beech—3 cornered nuts, edible.

Bellows—use hollow elderberry twig to encourage fire.

Belt—snake skin or mammal hides, woven cattail leaves or bark of leatherwood.

Birch Beer—steep bark and twigs of black birch.

Birch, White—bark useful for camp utensils, model of teepes or canoes, writing paper, drinking cups, hanging baskets, and tinder. Do not remove from standing trees. Bind in bundles for torches. Sprouts used for brooms.

Black Haw (Viburnum prunifolium)—fruits.

Black Mussel—cheaper, more nourishing, and more easily digested than oyster. Cook in same way as clams.

Black Alder (Ilex verticillata)—steep leaves for "Chinese Tea."

Black Birch—twigs and bark edible. Black Birch beer made by fermenting sap.

Black Cherry (Prunus serotina)—half wilted leaves poisonous to cattle. Tea of bark a strong tonic.

Black mustard (Brassica nigra)—seeds crushed to flour make mustard and paste useful as mustard plasters.

Blankets—make of dog or goats hair, wool, feathers, heads of cattails.

Blisters—prevent, if chafing use adhesive. If blister has formed build around it with strips of adhesive.

Bloodroot (Sanguinaria Canadensis)—rootstock gives a crimson juice used by the Indians for war paint.

Blow pipe—elderberry twig with pith removed.

Bones—useful for pipe stems, awls, handles for tools. Vertebrae for neckerchief slides.

Boneset (Eupatorium perfoliatum)—"thorough wort tea" used as a tonic.

Bouncing Bet (Saponaria officinalis), or soap wort—Juice of roots is soapy. Introduced from Europe.

Bow—use shagbark hickory.

Brachen (Pteris aquilina), Eagle fern or Brake—young shoots boiled in salted water and served on toast with butter resemble asparagus.

Broom—birch broom, cattail leaves, splints of ash.

Brush—husk of cocoanut, grasses tied together, wing of bird, bristles of pig, needles of pine. Paint brush for fine work from feather and quill.

Buckthorn (Rhamnus cathartica)—berries used as a dye.

Bulrush (Scirpus)—stems for weaving.

Butter and Eggs—juice said to be a fly poison.

Butternut—steep inner bark for cleansing digestive tract. Green husks of fruit make yellow or orange dye.

Callus—if on foot, means improper shoe. Pare down until tender. Cut doughnut shaped adhesive. Build several layers around callus. Put little salve in center every night.

Candle (see torch)—bayberry wax, mutton tallow, pine knot, rolled birch bark, meats of some nuts, cannel coal, bee's wax. Trench candle is made by boiling tightly rolled newspaper in paraffine.

Candle stick—hole cut with knife in sod, potato, hollow bone, empty can or bottle shells.

Caraway—seeds used for flavoring on cookies.

Cattail—(Typha latifolia)—head dipped in fat used as a torch, mats of leaves used in winter homes of Fox Indians, cattail floss used to stuff pillows and cushions, root starch used for bread by the Iroquois, leaves used on Palm Sunday, leaves used to cane chairs and making hats. Down used as absorbant cotton and for baby padding by Pawnees.

Ceanothus Americanus or New Jersey Tea—dried leaves used for tea during the Revolutionary days.

Chafed spots—lather spot at night with a pure soap.

Chair bottoms—cattail leaves, splints of swamp ash, palm leaves, heavy twine.

Checkerberry—see Wintergreen.

Chickweed—similar to spinach.

Chinese Tea—steep leaves of the black alder.

Chokeberry (Aronia)—red and black, sweet apples size of a "berry."

Choke Cherry (Prunus Virginiana)—red cherries edible when cooked.

Clay—tiles, beads, leaf impressions, relief for animal and flower sculpture, marbles, chinking cabins, cover for baking potatoes, pottery.

Clothes moth—keep away from wool by packing in cedar chests or closets. Cedar shavings on woolen clothing is an aid.

Coat hanger—cut stick or roll newspaper and tie a strip around middle.

Cocklebur—peel shoots and eat like radishes. Take rind off stalks before it blossoms and boil like asparagus.

Cocoanut—make dippers from the shell. Elderberry wood fitted into lower part of the cup as a tube makes a good funnel. Stencil brushes from portions of natural husk. Cocoanut bottles.

Coffee substitutes—dried root of dandelion, chickory root, Job's Tears, bean of Kentucky Coffee Tree, parched corn.

Colds—steep leaves of sweet scented golden rod, steep sea moss and lemon juice, balsam, onion.

Colic—infusion of pennyroyal leaves.

Compass—point hour hand of watch toward sun. Halfway between the hour hand and XII is south.

Conch shell—horn to call to dinner.

Cordage—inner bark of linden or cypress.

Corks—burn for blacking, fit with tooth picks to imitate animals, large corks from fish nets for candle-sticks.

Corn cob—used in shelling corn, use with sand to scour kettles, corn cob pipes.

Corn husks—mats and bags, Figi dolls, hats.

Corn—parched corn a concentrated food for hikes. Parched corn for coffee. Tortillas or aboriginal Mexican corn cake—grind softened hulled corn and water. Roll thin and toast on flat stone over slow wood fire.

Corns—cover or build around with sphagnum moss soaked with iodine. Wear shoes that fit.

Crabs—blue crab and Lady crab are as good as lobster.

Cranberry—excellent acid sauce.

Cranberry, Bush (Viburnum Americanum)—fruit edible.

Cranesbill or wild geranium—rootstock an astringent.

Crayfish—boil and eat as lobster.

Cucumber root—Indian Cucumber Root. A good food but should only be used in emergency as it is disappearing.

Cup—cocoanut, shell, gourd shell, clam shell, grape leaf. Knob cups are made from knobs formed by tree over old branch stubs. Saw off knob and dig out old branch.

Cypress (Taxodium distichum), Southern Bald Cypress—inner bark used in braiding light cordage.

Dandelion (Taraxacum Taraxacum)—root used as a tonic and for coffee. Leaves used as a green.

Deep Sea Clam—meat edible, shell used as a skimmer by colonists, also as sugar spoon. Used as a hoe by Indians.

Diarrhoea—use an astringent.

Digger Pine (Pinus sabiniana)—an important food of the Indian.

Dock (Rumex)—boil leaves for a green.

Dog Tooth Violet—a good green, use only in emergency.

Drum—use tightened skin of an animal.

Dyes—Inner black oak bark for yellow, inner bark of alder for orange, butternut bark for black, butternut husks for yellow, pokeberry, black walnut bark, elderberry, iron rust, pollen tiger lily.

Egg shells—receptacles for planting tomatoes or lettuce seed for transplanting. Heads for dolls,—chinese, darkey or clown.

Elderberry—(Sambucus Canadensis)—berries used for elderberry pie and wine. Jellies and pies. Pith easily removed for tubes. Indians used for flutes. Wood makes good skewers. Source of material for pop guns by boys. Use hollow twig to encourage fire.

Elm—good wood for rubbing sticks. Sheets of elm bark used to cover framework of light poles to make summer house for the Fox Indians.

Elm, slippery (Ulmus fulva)—inner bark has pleasing flavor, also a good fiber for rope.

Equisetum (Horsetail)—"Indian sandpaper."

Eucalyptus—fruit suitable for necklace.

Evening Primrose—root a celery substitute.

Everlasting—used as component of permanent boquets. Sweet scented everlasting gave aromatic odor to grandmother's garret.

Feathers—useful for paint brush, head gear, ornaments, quill pen, sofa pillows toothpicks, arrows, darts, insignia.

Feet, care of—bathe often and dry thoroughly. Dust between toes. Use clean woolen socks.

Feet that "swet"—Use boric acid powder.

Fir Balsam—needles for pillows.

Fire brand—rushes bound together.

Fire building—focus sun's rays with camera lens or field glass; ıse flint and steel; use rubbing sticks.

Fire extinguisher—sand, beat with evergreen bough hitting toward burnt area, start backfire at a stream, lake or roadway.

Fish (see hook and bait)—to clean hold on a board by head and scrape off scales from tail toward head. Slit each side of back fin and lift out small bones. Slit belly and remove entrails. Cut off fins, head and tail and wash in cold water.

Fish hook—Indians used sharp-pointed bird bone. Also greenbrier thorn lashed by own tendril.

Fixative—cover white shellac with alcohol and shake frequently for a week. Use the clear liquid for fixing charcoal and mushroom sketches.

Flag root—Steeped in sugar makes an old fashioned candy. Young shoots and inner leaves edible.

Flint—strike with steel for a spark to make a fire (use back of closed knife blade in emergency).

Frogs—meat on legs similar to white meat of chicken.

Flour substitutes—sunflower seeds, root of skunk cabbage, ground acorn meats, wild rice, rootstocks of cow lily or saggitaria, artichokes, cooked root of poke weed, corm of the Jack in the Pulpit, corn meal.

Fungi—do not attempt to eat unless familiar with the common edible and poisonous varieties. Do not take fungi which have a cup at the base, nor those which are bright colored, those that have white gills at maturity, that have tubes or pores on the under side, or a milky juice. Some fungi are well suited for etching. Use sharp stick.

Funnels—use cocoanut shell fitted with the tube of an elderberry stem.

Fuzz stick—whittle a dry pine limb so that shavings remain on stick. Stand with shavings pointing downward to start a fire. Use golden rod blossoms when dried on stalk.

Ginger (Asarum Canadense), Wild Ginger—rootstock has flavor of ginger.

Glue—boil down hoofs or horn, being careful not to burn. Draw off fluid.

Goldenrod, Sweet scented (Solidago odoro)—anise fragrance, often called Mountain Tea.

Goldthread (Coptis trifolia)—long bright yellow root good for sore throat. Grows in damp woods.

Gooseberry—fruit used by pioneers for jelly.

Gourd—cup, ladle, rattle, wren house.

Greens—dandelions, marsh marigold, pigweed, purslane, milkweed, narrow leaf dock, chickweed, chickory.

Groundnut (Apios tuberosa)—starchy tubers are potato substitutes. A climbing perennial with milky juice, 5–7 leaflets, growing in marshes. Cultivated by Indians for its tuber (Indian potato).

Hairbrush—fruiting summit of the teasel, dried corncob.

Hats—corn husks, cattail leaves, paper bags.

Hawthorne (Crataegus)—Hawapples make a good jelly.

Hazelnut (Corylus Americana)—nuts used as basis for soup.

Hemlock (Tsuga Canadensis)—bark an astringent used in case of diarrhoea. Leaves used to make a favorite drink in lumber camps. Not the herb used by Socrates.

Hickory—good bow wood. Shoots make good rope. Tough wood for axe handles and baskets. Shagbark has edible nut.

Hoarhound (Marrubium vulgare)—old household remedy for colds. Laxative in large doses. Flavoring for hard candy.

Hog Peanut (Amphicarpa Monoica)—edible bulbs or nuts on root.

Hole or tear in clothing—turn inside out and cover with adhesive tape.

Honey Locust—thornes used for skewers. Slow burning. Lasting coals. Good side logs or back logs in cooking fire as scarcely burn when green.

Hornbeam—durable wood for implements.

Horn Spoon—horn boiled until soft—split—and pressed into a shallow hole in ground with a round stone. After shaped it is allowed to cool and then trimmed into shape.

Horse shoe crab—Candle holder, ink stand, food for hens.

Horsetail (Equisetum)—"Indian sandpaper."

Hot water bottle—wrap soapstone or other hot stone in newspaper; canteen with hot water; stomach of freshly killed animal.

Ink—use inkberry, pokeweed berries, or juice of onion for secret messages.

Inkberry (Ilex glabra)—leaves used to make Chinese Tea, berries for ink.

Indigestion—root of sweet flag.

Insect bite—use juice of raw onion, salt, mud, iodine or ammonia.

Iris—leaves for weaving mats and baskets.

Irish Moss—wash and dry. Use to make Blanc Mange.

Iron rust—basis for paint. Remove stains from white cloth by sour milk.

Ivy poison—use strong soap and then alcohol.

Jack-in-the-Pulpit (Arisaema triphyllum)—stings if eaten raw. Seneca Indians used for bread, should be used only in emergency as the plant is becoming rare. Purple in woods, pale in sunlight.

Jerusalem Artichoke (Helianthus tuberosus)—a yellow sunflower. The Indian's Potato. Good raw or cooked.

Jimson Weed—poisonous when eaten.

Job's Tears (Coix lachryma)—native of India. Used as a substitute for coffee.

Kentucky Coffee Tree—seed used for coffee by colonists.

Kindling—see tinder.

Kinnikinnick—Plains Indians rarely smoked pure tobacco, preferring a mixture of tobacco, sumac leaves and the bark of the red willow or dogwood. This mixture was called Kinnikinnick by the whites. Algonquin word meaning "mixed."

Labrador Tea (Ledum Groenlandicum)—Evergreen shrub of cold bogs. Revolution substitute for tea.

Larch—(Larix Americana)—steep bark for a laxative. Makes good rubbing stick.

Laxative—wild senna leaves, tamarac bark, hoarhound.

Leaf prints—use printers ink and rubber roller.

Leaf skeletons—boil leaves in quart of water containing 2 ounces of slaked lime and 4 ounces of washing soda. Remove pulp and dry on blotters.

Leatherwood (Dirca palustris)—wood and bark flexible and tough.

Lemon—basis for making imitation of chicken.

Lemonade—Indian lemonade made out of sumac berries. Barberries make an acid drink.

Lemonade powder—saturate lemon juice with sugar, dry and put in vials.

Lettuce seed (Lactuca sativa)—common garden lettuce—seeds yield excellent illuminating and cooking oil. Valued in all oriental countries.

Lighting—oil of sunflower seed, lettuce seed, and nuts. See candles.

Linden—inner bark useful for matting and cordage.

Lodge Pole Pine (Pinus Murrayana)—used as tipi poles by Western Indians.

Lotus (Nelumbo lutea), America Lotus, Water Chinquapin—cousin to oriental lotus. Boil seeds and bake tubers.

Mallow (Malva rotundifolia)—"cheeses" are good tidbits.

Maple-sugar—by boiling down sap of sugar or red maple. The Dakota Indians made sugar from the sap of the Box Elder Maple. Keys good imitation of wings of birds or dragon flies.

Marsh Mallow (Hibiscus moschentos)—fibre of stalk equal to jute and hemp. Root edible raw.

Marsh Marigold (Caltha palustris)—a pot herb.

Mat—corn husk, inner bark of linden, cattail flags.

Matches—dip in paraffine to waterproof. Dry by rolling briskly between palms.

May Apple (podophyllum peltatum), mandrake—fruit edible when ripe. Fruit makes a good jelly. Leaves and root poisonous.

Mildew—to remove stain, moisten, rub on it castile soap and chalk dust, wash off.

Milkweed—Indians prepared a crude sugar from flowers. Excellent pot herb.

Morel (Morchella)—eaten boiled.

Moths—keep away with red cedar shavings.

Mountain Ash—fruit used for marmalade in Scotland.

Mountain Laurel—spoon wood of Indian.

Mud—good for bee sting.

Mullein—Romans dipped stalk in suet for a funeral torch.

Mushroom prints—gather before ripe and lay on paper, gills down. Fix the print with fixative.

Muskrat—produced for fur and meat. (See Farmers Bulletin 869.) Soak meat over night in salt water to remove gamey flavor. Good for roast or stew.

Mustard—use black or white mustard seeds crushed to flour. Can flavor with sweet gale.

Nannyberry (Viburnum lentago)—edible black berry.

Necklace—use fruit of eucalyptus, acorns, Job's tears.

New Jersey Tea (Ceanothus Americanus)—leaves used as tea substitute in Revolution. Red root a dye.

Nosebleed—rub ice or snow on back of neck. Roll up a leaf and place under upper lip.

Nut Grass—(Cyperus esculentus)—has a sweet tuber.

Oak—bark an astringent. See acorns.

Onion—raw onion good for insect bites. Juice used as a secret ink. Quenches thirst when carried in mouth.

Onion, wild (Allium)—substitute for garden variety.
Opuntia (Prickly Pear)—fruit edible.
Oxalis—use acid leaves in salad.
Paint brush—use feathers.
Paint spots—remove from clothing with equal parts of turpentine and spirits of ammonia.
Palm leaves—woven like clasped hands to make an emergency basket. Palm leaf fans.
Parched corn—good as concentrated ration on hike, make a coffee.
Peanuts—basis for making miniature animals. Good food.
Pemmican—a condensed food of the Plains Indians. Pounded meat, berries and fat.
Pennyroyal—(Hedeoma pulegioides)—infusion of leaves a remedy for colic.
Periwinkle—roasted like chestnuts in England. Meat picked out with a pin.
Pigweed (Amaranthus)—young plants make a good pot herb. Substitute for spinach and beet greens.
Pine—dead limb from deep in trunk makes a good torch. Splinters good for kindling. Inner bark mucilaginous and suited to chewing. Gum made by burning wax and chewing part that drops into a kettle of water. Pitch makes a good adhesive. Steeped inner bark a good cough medicine. Long needled varieties used in basketry, pillows, and making mats. Cones adapted to making of imaginary birds, nuts edible.
Pipes—soapstone, bones, corncob.
Pitch—adhesive for woodsman, stop leaks in a canoe, use to start fire.
Plum, wild (Prunus)—fruit good raw or dried. Beach plum jelly and preserve a dainty dish.
Poison Ivy—white berries, 3 leaflets. Wash with soap and water if been in contact. Baking soda relieves inflamation. If blisters, open and apply alcohol.
Pokeweed—root poisonous, young shoots and leaves substitute for asparagus, berries make a good ink.
Pop Corn—good food. Eat with milk and salt. Make into corn balls with molasses.
Poplar—soft wood is easily carved.
Potato—rub hands with raw potato to remove vegetable stains.
Prickly Pear (Opuntia)—fruit edible. Do not handle leaves.
Primrose, Evening—root a celery substitute.
Puffballs (Lycoperdon)—use in young stage for food.
Purslane—a tender pot herb.
Quahaug shell—Indians used blue part for wampum.
Quartz—strike with steel for a spark to make a fire.
Razor Clam—edible. Cook in same way as common clam.
Red Cedar—Berries and leaves boiled by Dakota's for a cough decoction. Bark frayed for tinder. Long strips of bark used as skirts by Indians. Bark used as pads for Indian babies. Sheets of cedar bark, overlapping, used to cover framework of light poles for summer houses of Fox Indians. Wood used to line closets to keep moths away from woollen clothing.
Red Pine—bark tannic.
Red Seaweed—float in water onto paper. Its own gelatine makes it adhere. Good for decorating place cards, name cards and letter paper.
Rice, wild (Zizania aquatica)—seeds used in similar way to cultivated rice. See 19th Rep. Bur. Amer. Ethn. pt. 2, article by Dr. A. E. Jenks, for uses by Indians.
Rock tripe—(Umbilicaria)—a food used by Arctic explorers.
Root Beer—see Birch Beer.
Root foods—marshmallow and artichokes good raw, yellow pond lily roasted, arrowhead boiled.
Rose—petals for sachet bags.
Rubbing stick—larch, elm.
Rushes—material for basketry.
Sage (Artemisia)—steeped by Indians as remedy for indigestion.

Sagitarria variabilis, Arrowhead—root good boiled. Found in shallow water. Indians called it wappatoo.

Sand—scouring, hour-glass, modeling mountains, to show formation and changes along sea shore, modeling animals, studying animal tracks.

Sarsaparilla (aralia nudicaulis)—an aromatic root substituted for sarsaparilla.

Sassafras—root, leaves and bark edible. Steep root bark for sassafras tea or root beer. Leaves used to flavor soup.

Scouring Rush—used by colonists to scour cooking utensils.

Sea Clam—shell used as a hoe by Indians. Shell used as a sugar spoon and to skim milk by the early settlers.

Sea Moss—(Chondrus crispus)—a red alga known as "Irish Moss." Wash and dry and use for Blanc Mange.

Sea Weed Mounts—float red sea weed in water onto writing paper, arrange with small brush and dry.

Sedative—white pine or wild cherry bark.

Seeds for food—acorns, parched corn, beach pea, mustard, pumpkin.

Senna (Cassia Marylandica), Wild Senna—infusion of leaves valuable as a laxative.

Service Berry (Amalanchier), June Berry, Shadbush—dark purple berries ripen in June. Make excellent pies.

Sheep Sorrel (Rumex acetosella)—use acid leaves in salads.

Shells—to clean use dilute hydrochloric acid. To polish use emery powder or pummice stone and oil.

Shoe Preservative—½ beeswax, ½ Neet's foot oil, or 3 mutton tallow to 1 beeswax. Thin enough to penetrate and thick enough not to melt in sun.

Silver Poplar—difficult to burn.

Skins of animals—hotwater bags, drums, mocassins, belts, rawhide.

Skunk cabbage (Symplocarpus foetidus)—leaves good pot herb when water is changed several times. Root a good bread flour.

Skewer—a pin to fasten meat for roasting. Use splints with good taste, as black birch or sassafras; thorns from hawthorne or honey locust, sharp bones, split elderberry twigs.

Slate—useful for temporary or permanent records.

Slippery Elm (Ulmus fulva)—inner bark edible. Gives a mucilage. Good for coughs. Injures tree to remove bark.

Snakes—rarely strike higher than knee, therefore wear leggings in snake regions. If bitten by a poisonous snake make small cuts, suck out wound (if no sores in mouth) and spit out poison. Twist handkerchief or shoe lace above bite. Loosen tourniquet if longer than an hour.

Soap—root of Bouncing Bet or Spanish Bayonet.

Soapstone—easily carved into pipes, bowls, slate pencils.

Solomon's Seal (Polygonatum biflorum)—root a potato substitute.

Sore throat—root of the gold thread, leaves of pipsissewa, leaves of raspberry. Gargle with strong solution of salt water. Bind woolen cloth or sock, saturated with turpentine and lard, about throat at night.

Sorrel (Rumex acetosella)—use acid leaves in salad.

Spearmint—(Mentha piperita)—leaves used for flavoring sauces and drinks as Ginger ale. For growing see Farmers' Bulletin 694.

Sphagnum moss—a good antiseptic, substitute for absorbant cotton, made into ball and used as a hanging flower basket.

Spice Bush (Lindera Benzoin)—powdered berries for spice, leaves and twigs a tea substitute.

Spoon—deep sea clam shell, cocoanut shell, boiled horn of mammal. See horned spoon.

Spruce—cordage for general camp use. Boil green branches for spruce beer. Source of spruce green.

Squash—shell dried for cup, ladle, or rattle. Crooked necks basis for making geese.

Squid—eaten by Japanese. Equal to scallops in taste.

Star Fish—Pin arms out on a board and dry in shade until odor disappears.

Stings—apply wet salt, wet earth or ammonia.

Stone implements—knives, spearheads, arrows, mauls, hammers, pestles, sinkers, mortars, pencils.

String—twisted gut, swamp milkweed, fiber of stalk of Marsh Mallow equal to hemp, inner bark of Linden, and Cypress good for cordage. Twist by rolling in one direction with palm. Rootlets of spruce, hemlock or tamarack.

Sumac—steep crimson berry-like fruit in water to make Indian lemonade. Stems give it an undesirable woody taste. Acid taste of berries alleviates thirst when held in mouth. Poison sumac has white berries and grows in swamps. Wash with strong soap and then use alcohol.

Sunflower (Helianthus annus)—seed makes good bread flour, and oil for cooking or illuminant.

Sunstroke—get patient into shade. Use cold water.

Swamp Milkweed (Asclepias incarnata)—Indians twisted silk fibre by twirling a stone. Make a very strong twine.

Sweet Bay (Magnolia Virginiana)—cover petals with alcohol for a perfume. Leaves give flavor to roasts and gravy.

Sweet Cicely (Asmorrhiza longistylis)—root has a pleasant flavor.

Sweet Clover (Melilotus Alba)—bunches hung up in camp give a pleasing fragrance.

Sweet Flag (Acorus calamus)—root is a remedy for dyspepsia. Root steeped in sugar make an old fashioned candy. Inner leaves edible. Favorite food of Muskrat. Leaves have a pleasant odor.

Sweet Gale (Myrica)—gives pleasant flavor to mustard and roasts.

Sweet Grass—mats and baskets.

Sweet Scented Goldenrod (Solidago odora)—leaves used as a stimulant for colds and colic. Pleasant odor, resembles licorice.

Tamarac (Larix Americana)—, Larch—steep bark for a laxative.

Tea—leaves of black alder, catnip tea, thoroughwort, spice bush, See coffee and Indian lemonade.

Teasel (Dipsacus sylvestris)—fruiting summit frequently used as a hair brush.

Thirst—carry raw onion, sumac berries, barberries, green tendrils of grape vine, sorrel leaves or even a small pebble in the mouth.

Tinder—grey or white birch bark, fuzz from cedar bark, pine knots, gun powder, linen, nest of the mouse, dry leaves from a shelter, dry wood cut from middle of a dead limb.

Tobacco—see Kinnikinnic.

Tonic—root of dandelion, leaves of boneset, yarrow leaves.

Torches—pitch cast in large leaf, cattail dipped in oil, dead pine limb from deep in the trunk.

Trillium—good greens. Use only in emergency.

Tubers—or potato substitutes—American lotus, artichokes, arrowhead root, nut grass, ground nut.

Turtle—meat of snapping turtle a good soup stock.

Viburnum—shoots used for arrows.

Vomiting—leaves or berries of the holly, soap suds, tickle inner throat with feather, run finger down mouth.

Walnut—shell basis for imitation of turtle shell, cloves for legs.

Wappatoo—Indian name for arrow head plant.

War Paint—blood root or celandine. Powder from limonite or hematite.

Watch for Compass—point hour hand at sun. Half way between that and XII o'clock is south.

Water arum (Calla palustris)—rootstock edible when cooked.

Water bag—stomach of freshly killed animal good for a few days.

Water cress—(Radicula Nasturtium)—leaves and stems good in salads.

Waterproof—dip in paraffine.

Whelk—meat removed by boiling. Basis for good chowder.

Whistle—alder with pith removed, willow or apple.

White Ash—burns when green if split fine.

White Pine—bark a sedative. Inner bark good for chewing. Pitch a good adhesive. Use pitch to repair leaks in canoe. Inner dead limb good torch.

White Sweet Clover (Melilotus alba), white melilot—flowers serve as a flavoring.

Wild Cherry (Prunus serotina)—infusion of the dried bark in cold water a mild sedative.

Wild Ginger (Asarum canadense)—root has ginger flavor.

Willow—use young stems for mats, beds, baskets.

Winter Bouquet—a permanent bouquet may be made of everlasting, rabbits foot clover, bayberry, bush clover, grasses, cattails, sea-lavendar.

Wintergreen (Gaultheria procumbens), checkerberry, Mountain tea-leaves aromatic. Substitute for tea. Berries edible.

Witch Hazel (Hamamelis Virginiana)—distilled branches give 15% ethyl alcohol.

Withes for Baskets—ash, white oak, hickory, yellow birch.

Wood ashes—use to scour cooking utensils. Do not scratch.

Wormwood (Artemesia)—steeped in boiling water by colonists as a remedy for indigestion.

Yarrow (Achillea millefolium)—pour boiling water on leaves for infusion. Bitter tonic. Yarrow Tea.

Yellow Dye—Celandine juice, or boil inner bark of black oak.

Yellow Pond Lily—Cow lily refers to cow Moose which eats roots. Roots good roasted. Parch seeds and eat like pop corn.

Yucca—dried stem a good base board for making fire with rubbing sticks. Root used as soap by Indians of Arizona. Fiber used by Indians for rope and cord.

BIBLIOGRAPHY

Gilmore, Melvin R. Uses of Plants by the Indians of the Missouri River Region. 33d Annual Report Bureau American Ethnology.

Robbins, Harrington and Freire-Marreco. Ethnobotany of the Tewa Indians. Bulletin 44, Bureau American Ethnology.

Stevenson, Matilda C. Ethnobotany of the Zuni Indians. 13th Annual Report of Bureau of American Ethnology, 1915.

Horace Mann complained "that, as a child, he had never enjoyed the free intercourse with nature that his ardent mind craved. Speaking of himself and of the children with whom he mingled, he says that although their faculties were growing and receptive, they were taught very little; on the other hand, much obstruction was thrown between them and nature's teachings. Their eyes were never trained to distinguish forms and colors.—*Hinsdale, B. A.*

Horace Mann and the Common School Revival in the United States, p. 80.

A LEAN-TO MADE OF PINE BOWS. (Camp Hoffman, Rhode Island.) It faces south to catch the warm sun, is back by evergreen trees to break the cold winds, and the stems point upward so that the needles will drain the water.

A SEASHORE LEAN-TO. It is in the making and illustrates a simple truth, Woodcraft and Nature-lore must be determined by the materials of the environment.

THRUSH SONG

I.

"Knee deep, knee deep, knee deep,
  Cherry du, cherry du, cherry du, cherry,
White hat, white hat,
  Pretty Joey, pretty Joey, pretty Joey."

Song 2.

Swank! swank! swank! swank! swank!
Get yer beak clipped! Get yer beak clipped! Get yer beak clipped!
Tut! Tut! Tut! Tut! Tut! Tut!
Silly fool! silly fool! silly fool!
Cheese it do! cheese it do! cheese it do!
Naughty! Naughty! Naughty! Naughty!
Pip, pip! Pip, pip! Pip, pip!
Swelled head and empty too! Swelled head and empty too! Swelled head and
  empty too!
She's a peach, peach, peach, peach, peach, peach, PEACH!
For you to eat? For you to eat? For you to eat?
I *don't* think! I *don't* think! I *don't* think!
Cool cheek! Cool cheek! Cool cheek!
I fill the bill—I'm IT! I fill the bill—I'm IT! I fill the bill—I'm IT!
(Pause to take breath and passing fly.)

—*From "Girl Guide Book."*

Cooking by Crews on beach beyond Chequesset Inn.
Camp Chequesset 1925-26

CHAPTER X

OUTDOOR COOKING

THE THREE DEGREES OF HIKING

Cooking is one of the most interesting and available of outdoor activities. Most enthusiasts do not realize, however, the obligations which go with it. In order to make it definite the subject has been covered by the "question-answer" method. The answers appear separately so that the leader may drill himself. Furthermore, the series is arranged in three degrees,—the *Hiker*, the *Fire Builder*, and the *Over Night Hiker*. Before venturing on a hike it is absolutely essential that the leader know certain "rules of the road." The large amount of damage caused by camp fires makes that step one of serious responsibility and the time is near at hand when every guide will be required to have a license to build fires outside of certain designated spots. When it comes to staying over night and cooking several meals the health and life of the campers requires all this knowledge plus other safe-guards. The importance of a grading system, or camp catechism, similar to the one that follows cannot be too stringently emphasized.

I   The Hiker

1. What announcements should be made in arranging for a hike?
2. What permission should a hiker have?
3. Should a hiker be allowed along who has a cold?
4. What should a hiker put on his feet?
5. What care should be taken with the socks?
6. What care should be taken with the shoes?
7. What care should be taken with the shoe laces?
8. What should be in the first aid kit?
9. Who is responsible for the first aid kit?

10. Who checks up the equipment?
11. When should the equipment be inspected?
12. The time set for assembly for a hike is 9 a.m.
    What time should the hikers start?
13. It is a duty of hikers to prevent accidents.
    To whom is this a duty?
14. Name a sensible "dare" that might be accepted.
15. Give an example of a foolish "dare" that should be refused.
16. The hikers "shack" is just outside the city limits.
    Where should the hikers start to hike for the "Shack"?
17. Which is better for hiking:—cross country or the state road?
18. On which side of a road—when necessary to hike in the road—should hikers march?
19. What "command" should be given when an auto approaches?
20. How does the scout law apply to hikers?
21. What should the leader do about a hiker who does not obey?
22. What is a good hiking pace?
23. What is a piedometer?
24. What speed record should be tried for on a hike?
25. What distance should be tried for on a hike?
26. What is the best manner to rest?
27. What animals should hikers kill?
28. What snakes are harmless?
29. What three snakes are injurious?
30. How would you recognize poison ivy?
31. How would you recognize poison sumach?
32. What flowers should you take for a bouquet?
33. What flowers should not be taken for a bouquet?
34. A good hiker will stop at how many houses for water?
35. Where else may one drink water?
36. How often should a hiker drink water?
37. What land areas should not be "crossed"?
38. What care should be taken of fences?
39. What attitude should be taken in regard to fruit?
40. Where may one throw stones?
41. What relation is there between hikers and farm animals?
42. What should be done with waste from a lunch box?
43. How soon after lunch should a hike be continued?
44. What activities are good on a hike?
45. How far should a hiker straggle from the party?
46. What should a hiker do if his feet ache?
47. How would you treat a blister?
48. How would you prevent a blister?
49. Where should dismissal take place?
50. Where should a hiker go after dismissal?

II. The Fire Builder

1. How much food is necessary for eight people?
   (Potatoes, bread, cocoa, pan-cake flour, frankfurters)

2. Why is canned food not so good for hiking?
3. What is essential for a cooking kit?
4. How should a hiker carry matches?
5. Where should a hiker build a fire in relation to: rocks; lakes; brooks; pine trees; logs; leaves.
6. What would you do on a windy day?
7. How would you prepare an area for a camp fire?
8. How is an underground fire started?
9. What is the law in regard to forest fires?
10. What is the fine or term of imprisonment if a fire results from carelessness?
11. What is the best size for a cooking fire?
12. How should you lay material for building a fire?
13. When is a fire permit necessary in this state?
14. What is courteous to do before building a fire upon any property?
15. What is courteous to do before collecting wood upon any property?
16. Who should gather the wood?
17. What kind of wood should be gathered?
18. What should be gathered for tinder?
19. How much large wood should be gathered?
20. How would you break large pieces?
21. Do you put the large pieces on before or after you start to cook? Why?
22. Should you cook over a flame or coals? Why?
23. How many living trees should be cut down? Hacked? Barked?
24. How many trees should be blazed?
25. How does a bonfire differ from a camp fire?
26. Why should a hiker never build a bonfire?
27. How can you clean dishes in the field?
28. What should you do with food left over?
29. What should you do with paper, peels, and string?
30. Wht should you do with tin cans and bottles?
31. What evidence should there be of your having camped there?
32. How would you extinguish the fire? When?

III. *The Over Night Hiker*

1. How much food is necessary for eight people for three meals?
2. What equipment is necessary for an overnight hike?
3. What should the hiker do first upon arrival at destination?
4. When should the hikers use a camp guard?
5. How would you prevent getting wet?
6. What precautions would you take in regard to insects?
7. How near or far apart should the beds be grouped?
8. How would you make a camp bed?
9. How would you arrange a camp light?

10. How would you build a camp latrine?
11. How would you make a camp stove?
12. What should be done with wet clothing?
13. What should be placed beneath the camper?
14. How many hikers should have "duties"?
15. What "night rules" would it be best to observe?
16. When should the blankets be rolled?

I. The Hiker:

1. Time and place of meeting; car fare, if necessary; material needed, such as knife, compass, map, cup, spoon, or axe; food arrangements; distance to walk; time and route planned for return. Place and time of dismissal.

2. A young hiker should have the permission of his parents. Fire and trespass permits may be necessary.

3. Leader will have to use judgment. If cold is in an infectious stage and the group has the habit of passing about a canteen and drinking from a common cup I would send such a "liability" home.

4. Low-heeled tramping shoes that will keep out moisture.

5. Woolen socks are best. Do not have wrinkles. Carry extra pair for long hike. During rest period bathe feet, dry thoroughly, powder toes, and change right to left. If any part of foot chafed cover with adhesive.

6. Shoes should fit, be water-proofed, and somewhat flexible (new shoes should not be broken in on a hike).

7. Shoe laces should be strong, snug but not too tight.

8. Adhesive, a clean bandage, absorbent cotton, iodine. First aid is first aid. Nothing more.

9. The leader.

10. The leader, or someone assigned by the leader.

11. Before starting.

12. 9 a.m. If you start at 9:15 they will always come 15 or more minutes late.

13. It is a duty to self, to the group, to the leader. If a scout it is a duty to scouting.

14. To pick up a garter snake.

15. To slide across unsafe ice until it breaks.

16. Energy should not be used up hiking through the city. Get to the shack as soon as possible and start to hike from there.

17. In hiking avoid the highway as much as possible.

18. Hikers should walk on left hand side of road. They will then see autos coming and can step one side.

19. "All left". "Graveyard" has been used in some camps.

20. The scout law applies to hiking. Illustrate by concrete examples.

21. Should not send home in the middle of a hike but should refuse to take him next time.

22. Do not have a foot race or an endurance test. Three miles an hour with rest periods is a fair average.

23. Instrument to measure distances walked by person who carries it.

24. Should not try for speed records on a hike except in case of an emergency.

25. Fifteen miles should be the day's limit.

26. Lie down and relax. Feet up in air relieves blood pressure.

27. Should not kill any needlessly. Some think that they should kill a snake on sight.

28. Most snakes are harmless.

29. The three injurious snakes are the rattle snake, the copperhead, and the water mocassin.

30. White berries and three leaflets. Woodbine has blue berries and five leaflets.

31. Poison sumach has white berries and grows in swamps.

> Berries red,
> Have no dread!
> Berries white,
> Poisonous sight!
> Leaves three
> Quickly flee!—*Hamilton Gibson*

32. Those in the open fields, as daisies, buttercups, Queen Ann's Lace, goldenrod, everlasting, dandelions, asters, milkweed, and cattails.

33. Laurel, flowering dogwood, arbutus, Indian Mocassin, Adder's Tongue, columbine, anemones, maiden hair ferns, hepaticas.

34. Greenhorns are forever stopping to get a drink. This may become a nuisance to house dwellers and has an element of exposure to disease. Seasoned hikers prefer not to stop to get a drink of water.

35. One may carry a canteen or drink from a spring. Sumach berries or barberries will help quench the thirst.

36. Hikers tend to drink too often, and too much water. They should train themselves to go without water. Those who stay in the house do not drink enough. We should drink about 12 glasses a day.

37. We should not cross lawns, gardens, hayfields, orchards, or areas containing domestic animals.

38. Leave them the way you find them. If you knock down a stonewall replace it. Do not leave bars down or gates open.

39. Fruit belongs to the owner. Often times he is willing to give it to hikers. It should never be disturbed without permission, even if it appears to be going to waste.

40. Do not throw stones into hayfields, at domestic animals, into private ponds, or toward buildings.

41. Hikers should respect farm animals. Do not frighten them. Do not leave bars down for them to get out of enclosures.

42. Waste food may be left for birds. Papers, string, and refuse should be buried. Do not throw away sandwiches.

43. Should rest for half or three-quarters of an hour after eating.

44. Nature Games and woodcraft, scouting and trailing, story telling, and photography.

45. No one should straggle from the party. It always causes alarm and inconvenience.

46. If temporary bathe, change socks, and rest. Have feet examined for fallen arches.

47. Insert sterilized needle under skin near base and drain. Wash with antiseptic. Dry with powder. Build up around with adhesive or use a corn cushion. Remove cause.

48. Use adhesive on threatened part.

49. At some place agreed upon, as starting place or headquarters.

50. In the case of young hikers they should go home.

II. *The Fire Builder:*

1. See *Outdoor Cooking* recipes.

2. Water in canned goods is an added weight.

3. No dishes are essential. A pan, tin cup, and spoon are handy.

4. In a water-proof container.

5. Should clean away duff. If there is danger from sparks should make a stone fire place or trench the fire. A beach with the wind blowing toward the water is best.

6. Build a fire in a gravel or sand pit. I have gone without a fire rather than take a chance of having the camp fire escape.

7. Rake away all forest litter and build a fire place. Be sure that there is a fire extinguisher handy such as loose soil or water.

8. By fire catching in the humus. Such a fire may smolder for days before the blaze appears. It is better not to build a fire where there is thick humus.

9. In many states there is a law that the person who sets a forest fire is financially reliable for the amount of damage done.

10. Should look this up for the community in which you are leading.

11. Beginners usually build a bon fire, in contrast to a cooking fire. It should have hot coals and be small enough so as to get near to it to cook.

12. The tepee or log cabin is a good way to lay material to start a fire. These methods give a draft.

13. Should know dates for your particular community.

14. To get permission of owner.

15. To get permission of owner.

16. Everyone except those assigned other duties, as preparing the food or getting water.

17. Dry lower limbs rather than damp wood from ground.

18. Birch bark, cones, dry leaves, cedar bark, chestnut frays, standing stalks.

19. Usual mistake is not to gather enough. Should not be much larger than 2 inches in diameter.

20. Break large pieces by hitting over a sharp corner of a boulder or better by prying between two trees growing from same stump.

21. Put large pieces on before cooking so as to burn down to coals.

22. Cook over coals. They give more heat and less soot to blacken kettle.

23. Living trees should not be cut. Bark should not be cut from live birches.

24. Trees near cities should not be blazed. Other means may be used for laying a trail.

25. A bonfire is larger than a camp fire.

26. Bonfires are wasteful and not adapted to cooking.

27. Use a damp sod for first cleaning. Wood ashes and water are good. If followed with good hot water they may be air dried.

28. Leave scraps for wild animals. Do not throw away good food.

29. Burn or bury all refuse.

30. Bury all tin cans and bottles.

31. There should be no evidence that anyone has been there.

32. Extinguish the fire when through with it and sometime before leaving. Use considerable water. Keep stirring ashes and using water until there is no smoke or sizzling. Then cover with soil and weight material down with stones from fire place.

III. The Over Night Hiker:

1. See *Outdoor Cooking Recipes.*

2. Food, blankets, poncho, matches, flash-light, latrine.

3. Choose site for bed and make it.

4. When playing a game or for the experience.

5. Poncho under and another over. Make trench each side of bed. Camp near a shelter.

6. Go when or where insects are not pests. Carry mosquito netting. Camp where the breeze keeps insects away.

7. Depends upon group discipline and how many green horns are present. Far enough apart so that one group will not disturb their neighbors. Not more than two in a group.

8. Dig a hip-hole. Fill with leaves or spruce boughs. Make comfortable. Put poncho on top. Fold blankets and pin with horse blanket pins. Put poncho on top.

9. Each group should have a flashlight. If possible have a lantern to show way to latrine. Campers often get bewildered and lost at night.

10. Dig a trench and use a small log for seat. An old nail keg is handy.

11. Make a stone or clay stove.

12. Wet clothing should be removed and dried.

13. See answer to No. 8.

14. Everyone should be assigned duties. No loafers.

15. Time of retiring, talking, and getting up.

16. Sun and air upon getting up. Roll up after breakfast.

In Camp Cooking

Greenhorns usually do this	instead of	these things which are earmarks of old timers.
1. Stand around and watch the wood gatherers		"joining in" on getting wood
2. When they realize their duty, get green and wet wood from the ground		lower dead branches from evergreen trees.
3. Leave the duff		cleaning away leaves and vegetable matter.
4. Have a small wood-pile just enough to start the fire		reserve supply to maintain fire.
5. Put large pieces on bottom or put them on too soon		having goodly supply of twigs for kindling.
6. Pile wood horizontally and airtight		loosely in a tepee.
7. Light fire on top and to leeward		on bottom and to windward.
8. Cook over a flame		over live coals.
9. Leave papers, cans, banana peels, and string		no trace.
10. Leave smoldering coals		being absolutely sure that fire is out.

Recipes

The object of publishing this little collection of recipes is to put them into convenient and accessible form for outdoor planning. Most of the recipes have been used by campers in varied forms, in the preparation of outdoor meals. They have been planned for a group of eight or a patrol of scouts. The quantities given should vary according to whether fruit and candy are added. Hints for the use of wild materials may be found in the Nature Guide's Dictionary.

Goulash (Cap'n Bill's Chowder):

1 qt. can tomatoes, 1 qt. (2 cans) corn, 4 good sized boiled potatoes, 6 slices of bacon, 2 medium sized onions, salt & pepper to taste. Serve on pilot biscuit or in tin cup. May put in any left overs in vegetables or meat. Quickly prepared, filling, and good on a cold day. Leave out one can of corn if all of party does not come. It is well to cut up onions and fry a bit with bacon before adding other materials.

Slum Gully.

More expensive than Goulash. 2 cans Tuna Fish (1 lb. tins), 2 cans peas, 1 pt. cream sauce seasoned, 1 can pimento, 4 hard boiled eggs. Serve on pilot crackers. To make cream sauce use two tablespoons of butter (or bacon fat), two tablespoons of flour, and two cups of milk.

Hunter's Stew.

Use an enamel pail. Cut up 6 slices of bacon on bottom of pail. Put pail over thin layer of coals (not flaming fire or deep coals). Cut up and brown 6 onions. Braise 1 lb. top round steak cut into 1 inch squares. Stir or "jiggle" often. Put in carrots sliced very thin. Pour in boiling water. Cut up 4 large potatoes and 2 parsnips. 6 green peppers may be added. Wild carrot root, horseradish, or wild onion may be used for seasoning instead of cultivated material. Add salt and pepper. Let simmer for an hour or two. Thicken with flour or oatmeal. Serve in tin cup. Excellent on cold day.

Rum Tum Tiddy.

1½ lb. cheese (Young America), 1 good sized pepper (green), spoonful butter, 1 small can tomato soup, 1 small can evaporated milk, 1 teaspoonful flour to thicken, pinch of salt. Melt butter in pan. Cut up cheese and melt in pan. Stir *constantly* to prevent sticking. Pour in can of soup. Mix flour with milk gradually. Pour into mixture. Add salt and red pepper. Stir until smooth. Serve hot on toasted bread.

Baked Stuffed Peppers.

Cut out the core of ten peppers and cook in boiling water about ten minutes. Fry 1 lb. hamburger steak flavored with onion. When done mix with cracker crumbs or cold potato or left over vegetables or can of corn. Season and moisten with milk. Stuff peppers and bake in reflector oven or clay oven.

Clam Bake.

Line a hole in the ground (2.5 feet deep and diameter) with stones as large as the two fists. Keep a big fire in it for a couple of hours until the stones are very hot. Remove unburned chunks or wood. Quickly cover embers with layers as follows: moist rockweed (or seaweed); potatoes (small); corn in husks; thin layer seaweed; clams; thick layer of seaweed. Let steam about 45 minutes or until the clams gape open. Outside husk of corn should be removed leaving just one layer of husks. Tie loose ends with a piece of husk.

A Bean Hole.

Prepare the fire as for a clam bake. Have hole plenty large enough. Have pot of beans completely surrounded with hot coals. Cover with burlap and thick layer of soil. Bake for about 12 hours. Beans should be soaked over night and parboiled before being put in the hole.

Dogs in a Blanket.

Skewer thin slices of bacon wrapped spirally around frankforts. Roast over hot coals. Turn frequently to save drippings. They baste and sear it on all sides.

Dog-on-the-Rind.

Slit a frankfort lengthwise and press a string of pork into it, using a skewer to hold it in place. Roast over a fire turning frequently.

Fish Cakes.

Remove bones and mince fish mixing with mashed potato. Season with salt and pepper. Mix in thoroughly a well beaten egg to hold cakes together. Fry in deep fat.

Gypsy Bake.

Clean fish leaving outside skin and scales. Cover with a thin layer of clay. Place in a bed of coals from ¾ to 1 hour. Clay oven may be made. Make form out of rocks or green sticks. Plaster with clay. Dry with fire, filling in cracks. Bake on a flat rock closing door and chimney.

Reflector Oven Biscuits.

Reflector oven may be made from galvanized iron. Use prepared flour, as Reliable. Pour evaporated milk into package and make small patties with fingers. Have pan well greased. Build a hot fire in front of a rock. Stand oven in front of fire, moving it nearer or away according to the amount of heat. In season make strawberry short cake. i.e. Split biscuit and add berries crushed with sugar.

Twisters.

Get sassafras or black birch club about three feet long. Wind ribbon of dough around it spirally. Place over hot coals. Turn slowly.

Biscuits in "10c Store Oven."

Get two pie tins with straight sides and about 1 inch deep. Grease pans. Mix flour and put in one pan using other for a cover. Can invert and bake on both sides or may cover completely with embers.

Baking in a "Spider."
Grease pan. Mix up prepared flour and flatten in pan. Hold over coals to get a bottom crust. Prop pan up before the fire to brown the top or turn bread to brown other side.

Frog's Legs.
Remove skin and soak over night in salt and water. Wipe dry, roll in flour, and fry.

Kabob.
Get green stick as large as small finger. Impale 1½ in. slice of steak, 1½ in. square of bacon, slice of apple, and ½ an onion layer obtained by cutting the onion vertically and making a hole in the center with the curved blade of the scout knife. Repeat these in order as many times as needed for a meal. Roast over coals.

Cottage Cheese Sandwich.
Spread cheese on two slices of bread. Moisten with chili sauce. Put two slices of crisp bacon between and press bread together. Toast over coals. Tasty and filling.

Rock Fry.
Heat a flat rock in hot ashes and fry steak or ham on it. Eggs may be held on by a triangle, bacon fence. Pan cakes from a package of prepared flour may be browned and turned when the cake has distinctly bubbled. Grease rock before pouring on batter and when turning it. The fire may be built under thin rocks.

Barbecue.
Dig a pit three feet wide and three feet deep, and as long as necessary. Put in small rocks to a depth of about 20 inches. Fill with dry wood and burn until get coals. Place iron bars over pit of coals to hold meat. Make a sauce using water, butter, vinegar, salt, pepper, tomato sauce, celery sauce, and cayenne to taste. Tie rag on stick and use as a swab. Baste meat turning often. The slower the meat cooks the better. Time depends upon size of meat.

Glazed Apples.
Make a syrup of Karo and water. Keep warm over fire. Spike a tart red apple on a stick and dip apple into the syrup. Drain and let cool until able to eat it.

Apple Fritters.
Make a batter out of one cup of prepared flour, pinch of salt, 1 egg, and enough milk to make thin. Pare and cut into batter tart apples. Stir and drop by spoonfuls into hot fat. Fry brown and coat with powdered sugar.

Cap'ns Toast.
Chop a small onion and cook slowly in butter. Add flour mixed with salt and cayenne, and stir until smooth. Add small can of tomato soup, one tablespoon mustard, ½ pound sliced cheese. Cook in double boiler until smooth. Add well beaten egg, stirring until thickened. Serve on toasted bread.

Creole Spaghetti.
Put ½ pint spaghetti in water. Salt and boil for 30 minutes. Drain. Add one quart canned tomato. Brown a chopped onion in frying pan and add with flour stirring until smooth. Brown 1 lb. Hamburg steak in frying pan and turn into mixture. Add one quarter pound of thinly sliced cheese. Stir until evenly mixed. Prepare in double boiler to avoid burning.

Cottage Cheese.
Heat skim milk and buttermilk at 90°F. until firmly coagulated. Pour onto clean white cotton cloth. The "whey" drains off and leaves a "curd". Salt the "curd".

Camp Cocoa For Ten.

1½ cups powdered milk. 10 teasp. cocoa. 10 teasp. sugar. Pinch of salt. Mix and carry in a covered can. Add two quarts of water and boil 5–10 minutes.

Hike Ration.

If going on a long hike or mountain climbing, can carry a hike ration. Put dried fruit-prunes, apricots, raisins, nut meats, in pocket and a cake of chocolate with a few crackers.

Berry Flap Jacks.

1 small package prepared flour, 1 egg, salt, and can evaporated milk. Add milk to make above consistency of batter. Add blueberries, wild strawberries, or cut pieces of fruit, as apple. Melted cheese, thinned with milk, makes a good sauce for pan cakes.

Tuna Omelet or Salmon Omelet.

2 cans of fish, 1 lb. cracker crumbs, salt, pepper, and butter to taste. Mix well. Add one egg and enough milk to moisten. Try out three slices of bacon. Turn mixture into hot fat and cook as an omelet.

A Sandwich Meal.

Fry three slices of bacon. Drop two eggs directly from shell into fat. Put bacon, lettuce leaf, and two eggs between two slices of bread.

Corn Bread.

2 cups corn meal, 2 cups flour, 4 eggs, 2 tablespoons melted butter, 4 teasp. baking powder. Add enough evaporated milk to make paste and bake brown.

Bran Bread.

2 cups of bran, 1 cup flour, 1 tablesp. melted butter or bacon fat, 3 teasp. baking power, 3 eggs, 1 cup raisins, ½ cup nut meats. Mix dry and then add milk to make a thick batter. Make into patties and bake in reflector.

Mint Sauce.

To 2 cups of water add juice of half a lemon and three tablesp. sugar. Add 2 cups of mint leaves. Bring mixture to a boil.

Flag Root Candy.

Dice the root of Sweet Flag. Boil in slightly salted water to get out strong taste. Drain and drop into a syrup (equal parts of water and sugar). Allow to simmer.

"Rabbit."

Melt 1 tablespoon butter. Add 1 heaping tablespoon flour. Level 1 teaspoonful mustard. ½ teaspoon salt. Cut up ½ lb. mild cheese and add one can tomato soup. Melt all together and thin with milk. Eat on Uneeda biscuits.

Club Sandwich.

Toasted bread, lettuce, tomato sliced, bacon or ham, salt, pepper, mustard or mayonnaise.

Sweet Chocolate Sandwich.

Put between two slices of buttered bread and toast.

REFERENCES:

Camp Cookery Hints for Leaders — *Agathe Deming*
945 West End Ave., New York City

Camp Fires and Camp Cookery — *E. Lawrence Palmer*
Professor of Nature Study, Cornell University

"What cheer is there that is half so good,
  In the snowy waste of a winter night,
As a dancing fire of hickory wood,
  And an easy chair in its mellow light,
And a Pearmain apple, ruddy and sleek,
  Or a Janetting with a freckled cheek?"
Close by the jolly fire I sit
  To warm my frozen bones a bit.
             —Robert Louis Stevenson.

'Tis born with all; the love of Nature's works
Is an ingredient in the compound man,
Infused at the creation of the kind.
             —William Cowper.

"Lord, what music hast Thou provided for the saints in heaven when Thou affordest bad men such music on earth."—*Izaak Walton referring to Bird Songs.*

"I think heroic deeds were all conceived in the open air, and all free poems also.
I think I could stop here myself and do miracles,
I think whatever I shall meet on the road I shall like, and whoever beholds me
    shall like me,
I think whoever I see must be happy.
    *   *   *   *   *   *
Now I see the secret of the making of the best persons,
It is to grow in the open air and to eat and sleep with the earth."—*Walt Whitman*

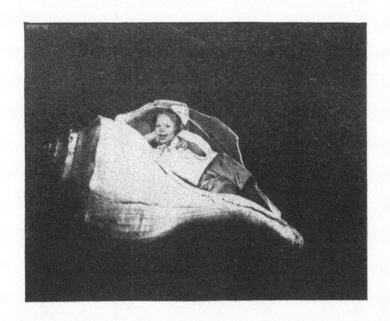

CHAPTER XI

CRAFT OUT OF THE ENVIRONMENT

The most campy camps do not repeat the activities of the school room. They play nature games instead of basketball, enjoy nature music instead of practicing the scale, and use local raw materials instead of raffia, plastocine, and glass beads. Every camper should have a favorable opportunity to collect craft supplies out of the environment, to come in contact with a skilled leader, and to create craft material. The camper should gain an understanding of why the old industries of the homestead were so excellent in technique and training. He should get the feeling of how these old fashioned gifts were expressed in craftsmanship. He should live in the atmosphere of expressing himself and his ideals in his work. The following are suggestive as to a few ways of striving toward that end.

1. *Place Cards.* Have a contest in making place cards. The "shell animals" shown in Fig. 1 give an idea of the possibilities in this contest. Pine cone animals are wonderful creations.

2. *Camp Equipment.* See Fig. 2. Camp furniture such as stools and benches of the lumber camp, a birch broom, a rustic toaster, hat racks, coat hangers, pot hooks, gourd cups, and corn husk mats make the cabin attractive.

Fig. 1

Fig. 2

Fig. 3

Fig. 4

3. *Indian Craft.* Fig. 3. Read about the Indians of the locality. What did they use to make baskets, mats, ornaments, and houses. What did they use for dyes, designs, and canoes? This picture shows the wigwam of the Yosemite Indian which is made out of the bark of the Incense Cedar. The structure in the foreground is a "Chucka" or granary. Here the Indian stored acorns for food.

Fig. 5. An Outdoor Stone Stove, Camp Andree Clark, A National Girl Scout Training Camp, Briarcliffe, New York. Note the Log Cabin Chimney and the hot water boiler.

The branches were woven tightly to keep out squirrels and mice. Try to find out what the Indian expressed in his crafts.

4. *Colonial Craft.* Fig. 4. Visit old homesteads and discover how they spun, wove, and dyed cloth. Have them demonstrate how they made candles, chests, tables. What did they use and how did they make baskets? Get them to teach you how to cane a chair. Ask to see their samplers, hooked and braided rugs, and corn husk mats.

5. *Fire-making.* Learn about the various kinds of fires. Make stoves out of clay or stone. The stone stove in the picture was made at Camp Andree. The crane in the other fire picture was tied on with a grape vine. (Figs 5, 6, 7)

6. *Woodcraft.* Fig. 8 shows the weaving of a bed. The second, Fig. 9, an outdoor bath tub and a tin pail over head with nail holes to be used as a shower bath.

7. *Musical Instruments.* Make a corn fiddle, an elderberry fife, or a kettle drum. The xylophone, Fig. 10, could be heard a mile away on a still night. Dead pieces of cedar were sounded with a wooden

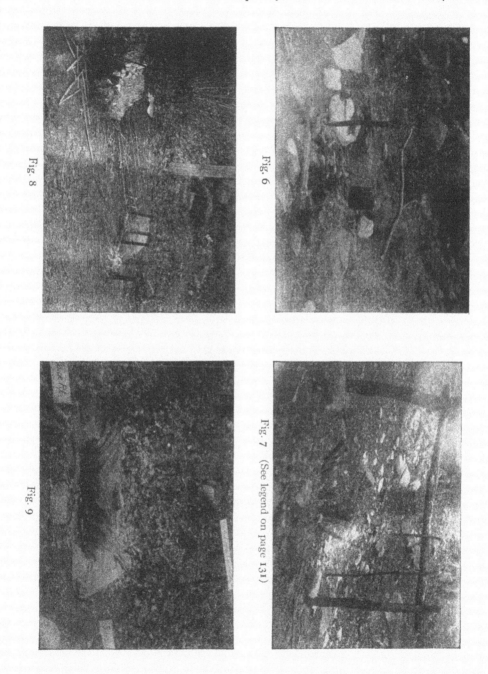

Fig. 8

Fig. 6

Fig. 9

Fig. 7 (See legend on page 131)

Fig. 10

Fig. 11

mallet until a complete range of notes were acquired. If the piece sounded too high the end was chipped off, if too low it was slashed off on the side. Quaker oat boxes, and others of various sizes, were placed under the key boards to give resonance. The instrument was very successful.

8. *A Picture Frame.* Fig. 11 shows a tree that has been burned by fire and then exposed to the weather for many years. The weathering has given a rustic effect of quarter sawing. Enos Mills has used such pieces for picture frames.

Fig. 7. Automatic Stew Fire which will burn safely while the builder does camp chores or goes fishing. The sticks settle downward and feed the fire.

IMAGINATION GUIDES OUR RACE

During the long centuries between cave and cottage our good ancestors traveled Nature's inspiring pictured scenes. With interest and with awe they watched the silent movements of the clouds across the sky, they listened with speechless wonder to the mysterious, unseen echo that lived and mimicked in the air, they puzzled over the strange, invisible wind that shook the excited trees and whispered in the rustling grass. They heard the echoing crash of thunder, saw lightning's golden rivers in the cloud mountains and looked with childish joy upon the silken rainbow. They marveled at the wondrous sunrise, the light of day, the fireflies in the forest, and the lonely, changing moon. The mysterious darkness was never understood but the silent, faithful stars they named and watched with nightly wonder. By trail and campfire these thought-filled wonders took life and color, became poetic stories. Through the changing seasons and the passing years Nature built the brain and kindled the illuminating imagination—the immortal torch that guided our advancing race and which triumphantly leads us on.—*Enos A. Mills.*

CAMP EDITH MACEY

(A Girl Scout Leaders Camp) Briar Cliffe Manor, New York. Photos by Mattie Edward Hewitt. Courtesy Girl Scouts, Inc. A Wonderful example of a Camp which fits its surroundings. Designed and carried out by Mr. James Rippin

Photo by M. E. Hewitt, N. Y. City

Fig. 12. "THE GREAT HALL" is a memorial to Edith Carpenter Macey. The "Singing Steps" are made out of native stones which harmonize with the rugged ledges of Westchester County. It is here that city ills are vanquished by the sun, air, and simple life.

Photo by M. E. Hewitt, N. Y. City

Fig. 13 (See legend on next page)

Photo by M. E. Hewitt, N. Y. City

Fig. 14 (See legend on next page)

Fig. 15

Fig. 13. INTERIOR OF THE "GREAT HALL." Camp Edith Macey. The hewn rafters, straight-backed settles before the fire, benches, rustic chairs, and mutton-tallow candles are in keeping with the mansion. Harking back to primitive times is essential to the outdoor scout.

Fig. 14. A UNIT ASSEMBLY LODGE. Where a troop encampment of 30 scouts may gather for a "social bee" as did the old-fashioned families two generations ago. The fire place is just the incentive for folk songs, country dances, and Woodland stories, and really essential for the "back to the country" movement.

Fig. 15. OUTDOOR COOKING SHELTER AND CAMP DINING TABLE. Comfort in the open is an Art. Here girls work with the environment instead of fighting it. Note that the framework of the shelter is chestnut timber which was killed by the chestnut blight and was otherwise doomed to decay. The fireplace is of field stone, and the roof is shingled with slate slabs. The bench seats are from chestnut logs. The woodwork has a "weathered oak" stain.

CHAPTER XII

Common Mistakes in Natural History

Man has roamed over the surface of this earth for at least a half million years. His progress has been incredibly slow. Most of his superstitions are the superstitions of the cave man. Many of his errors have a direct lineage to folk-lore and nursery rhyme. Is it not remarkable how little we change? A man-made war in Europe may kill thousands and man-made traditions live. A great step in mental conservation would be taken by eliminating our load of errors and mysticism.

The following notes are based on a recent examination which was given to test the kind and extent of mistakes pertaining to natural history. The total number examined was 281. This included 14 different classes ranging from the Junior High School through the College. The results indicate how little the average student thinks and how little the most of them see. Without doubt thousands more will roam the same road until our school authorities open the avenues of opportunity to think and discover.

If there were any way of knowing many people would be glad to learn, sub rosa or otherwise, whether they should be classified as traditionalists or progressives. This article has been written so that the reader may test himself as to his belief in Natural History ideas which are ill-born, and of a crude age. The questions introduce a usual misconception and immediately following is told the origin of the erroneous idea and the correction. The percentages are given so that the reader will know where he ranges with others in respect to the right answer. Once realizing the value of live interests it is hoped the reader will help smother the system which glorifies mute memorials of a chaotic past.

Old Sayings Versus Economic Facts

Has your training been one of tradition or of realities? Try the following test around the camp fire and then read about the results of the test in various schools and the interpretation.

These are common "sayings." Fill in the missing words.

Name one economic fact about each animal mentioned.

a. ....as a bee;	Bees................................
b. ....as an owl;	Owls................................
c. ....as a bat;	Bats................................
d. ....as a loon;	Loons...............................
e. ....as an adder;	Adders..............................
f. ....as a peacock;	Peacocks............................
g. ....as a mouse;	Mice................................
h. ....as a crow;	Crows...............................
i. ....like a hawk;	Hawks...............................
j. ....as a hornet;	Hornets.............................

Many of our "old sayings" are anything but true. "Busy as a Bee" is the most familiar. 90%–100% of the pupils in the various schools examined remember the old adage. Yet a bumble bee is the antithesis of thrift and does not store enough honey to keep the colony over winter. Only the queens survive the rigors of winter. If by the "busy bee" is meant the honey bee we have but to recall that the community is noted for its drones. The queen and drones do none of the work of the hive, have no pollen baskets, cannot sting, and cannot secrete wax. All the members of the hive loaf or hibernate during the winter. It would be far more appropriate to say as "Idle as a Bee."

A naturalist does not need to be reminded that bats are not blind; owls are not wise; loons are not crazy; adders are not deaf; peacocks are not proud; crows are not black; and hornets are not mad. The results show, however that people are more apt to remember an old saying which is an untruth (prosaic ignorance) than even one economic fact about an animal. Is this because of or in spite of our education systems? The results of this test are as follows:

The average number of sayings remembered in proportion to one economic fact:

Bees	1.06	Peacock	1.4
Owl	1.8	Mouse	1.33
Bat	2.6	Crow	1.24
Loon	8.0	Hawk	.47
Adder	.57	Hornet	1.9

It is rather astonishing to find that the Adder and the Hawk are the only two animals in the list that are more apt to be remembered in connection with some economic fact than in an old saying. It is also surprising to find that with the exception of the seniors in an Agricultural College the pupils of Junior High School are the only ones to know more about the economics of these common animals than about their myths.

In the case of the Bat five girls out of fourteen in a Junior High School mentioned the possibility of Bats getting into ones hair. Only one boy out of fourteen mentioned that this might happen. This belief is a feminine trait,—perhaps because they are more concerned. Tradition is long-lived. The tax that progress has to pay superstition is the worst kind of taxation because along with it vanishes the powers of reasoning.

Old sayings are passed down for the most part by "word of mouth." They receive a great deal of encouragement in first grade readers and early-grade literature. These sayings pertain to our commonest animals. At the same time our current literature and government publications furnish a host of facts about these same animals. Then there is the opportunity of gaining information by observation. Hearsay, reading, and observation are the three vehicles of information and if judged by present day results the effectiveness diminishes in the order named, hearsay being far more potent in our present working conception than observation. One is passive absorption, the other active production. One is parasitism. the other kinetic energy. One is nursery rhyme mysticism, the other is work bench service. One is charity entertainment, the other has everyone in the play. One is predigested food, the other food for digestion. One is senile sport, the other youthful enjoyment. The function of the school is to make production more interesting than elegant consumption.

Ten ideas like "Blind as a Bat" remain ten ideas. There is nothing to provoke new thought. The discovery of one idea about the Bat, as—it eats insects—demands more ideas. One can be intellectually as well as physically blind. The first method is one of conspicuous wastefulness whereas the second tends toward further service. Shall we spend our wealth of education in self-indulgence or in production for the community? If educational courses mean self-indulgence they stop there. If they mean "I have come to make life more abundant" the curriculum needs revision.

Some of the economic facts for which credit was given were due to reputation rather than fact. The ratios have thereby been greatly softened and should we double the sayings as they now stand in proportion to the economic facts the results would be nearer scientific. Take crows for example: fifty per cent of the economic facts given were that crows eat corn. Only twenty-two per cent mention anything of credit to the crow yet authorities say that the crows' credit account with the farmer far outweighs the debit side. A small minority—yet a serious number—mention such things as "pick out your eyes." One is reminded of the Old Danish Proverb—"A Crow is never the whiter for often washing," or the Chinese saying that "Crows are black all the world over." The axiom of today as regards pupils reared by the tribal-folk-lore method might be "They are never the wiser for often observing."

The Hawk record parallels that of the Crow. Fifty per cent of those examined mention the reputation which has been established in the human mind that "Hawks steal Chickens." Twenty-one per cent speak of the Hawk as dangerous to small birds. The unclassified answers were such as: "It steals;" "Dangerous;" "Eats people;" "Take away children at times;" "Injurious to small children." There is a man-raised notion that anything a Mouse, Crow, or Hawk assumes for its own use is stolen property. Stealing has been over-emphasized. If we pick a blueberry by the roadside or cut a tree from our woodlot is it stealing? Hens lay eggs for reproducing their kind. Are we stealing when we take them to boil? The potato plant stores starch in tubers for reproduction. Is man purloining when he uses them for sustenance? There is the underlying principle of the rights of others. When studying Hawks we should get their point of view and when studying Chickens our focus must be on Chickens. The world needs training in these fundamentals which must first begin in a study of nature's laws at home. It might help him understand the other fellow's point of view and possibly temper his sentiments as regards the rights of a starving Europe.

What Color is the Robin's Breast?

The Robin Redbreast (Erythacus rubecula) of Europe is a warbler and is less than half the size of the Robin (Planesticus migratorius) of North America. The Robin Redbreast of the Old World has been described as having a yellowish-red breast. It has become endeared to the English by coming near to their homes in winter and has won a distinguished place in English rhyme and lore. A few of the colors mentioned by standard writers are given:

> "Art thou the bird whom Man loves best,
> The pious bird with the *scarlet* breast,
>   Our little English Robin?"
>   —*Wordsworth, "Redbreast Chasing the Butterfly"*

> "In the Spring a fuller *crimson* comes upon
>   the Robin's breast."
>   —*Tennyson, "Locksley Hall"*

> "Robin, Sir Robin, gay, *red-vested* knight,
> Now you have come to us, summer's in sight."
>   —*Lucy Larcom, "Sir Robin"*

When the early settlers came to America they cherished in their minds the Robin Redbreast. When they espied the American Thrush with his home-loving spirit and chestnut hue it was an easy matter to transfer the title "Robin Redbreast" to our native species. In the same homelike fashion the bluebird was called the Blue Robin. The English colonists to India and to Australia performed a similar feat. In each place of adventure a new species received the name "Robin Redbreast." It is not surprising to find that this idea of Robin Redbreast permeates the minds of American folks with an impelling force. It has a traditional partisanship with the English race; an endurance of three centuries on a new continent.

How do students of today answer the question,—what is the color of the Robin's breast? 42% say red; 17%, reddish brown; 15%, orange red; and only four out of 281 mentioned that there is a difference between the male and female. Other colors mentioned were orange, grey, pink, crimson, scarlet, yellow magenta, maroon, brick red, rustic red, and chestnut. The results show as much versatility as one could desire from a student career of nursery and rhythmic feeding.

The next step consisted of placing a male robin (The male is brighter colored) in a paper bag and cutting a small hole to show a small area of the breast. Eighteen graduates of various high schools were asked to write the name of the color. The answers were as follows: brown, 8; yellow brown, 4; grey brown, 2; red brown, 2; golden brown, 1; orange, 1. When observing the Robin's breast without the blinders of nursery rhyme no one called it red. The concealment of the Robin in the bag was unnecessary as no one recognized it when it was removed. When the group were told that it was a Robin one immediately said: "But the Robin has a red breast?" This little episode is typical. From time immemorial—today as of yore—we are led to view facts through the rims and spokes of tradition and hearsay.

Sentimentally none of us would abolish the phrase "Robin Redbreast." It has a home-spun attractiveness. Educationally, however it is important to know that red is red. We must see, not through the smoked glasses of nursery rhymes, but with a clear vision. Not that we will appreciate folk lore and Robins less but Robin Redbreast more.

That the Robin Redbreast does not have a redbreast is representative of a group of contradictions in the language of natural history. The wing of the Red-winged Blackbird is not red; the shell of the Soft-shelled Clam is not soft; the Black Mussel is blue; the Starfish is not a fish; the Potato Bug is not a bug; Nuthatches do not hatch nuts; Flying Squirrels cannot fly; Darning Needles are unable to darn; wormy apples do not contain worms; and waves are never, no never "mountain high."

How do Squirrels Open Nuts

The answers were as follows: with their teeth (60%); crack them open (18%); with their mouth (7%); gnaw them open (4%).

Contrary to prevailing opinion squirrels cannot crack nuts. They gnaw them open. It is interesting to note that amongst a hundred college freshmen (50 men and 50 women) this knowledge was limited to 12% of the men. The answers might indicate that the men get more of the outdoor experience. That 60% of those examined say teeth indicates that the question was: What is the principal organ used by the squirrel in opening nuts? This reply is typical of the many indefinite answers. Pupils should be trained to answer questions intelligently.

The origin of the idea that squirrels crack nuts is ambiguous. The fallacy is being perpetuated by first grade readers. There are evidences in literature which suggest that the terms teeth, crack, and nut are associated. Lamb in a letter to Wordsworth (August 9, 1915) humorously uses the term cracker in place of teeth,—"I conjecture my full-happiness'd friend is picking his crackers." Wilberforce in his life of S. Wilberforce (1868, p. 380) in describing the nose and chin said,—"She is a toothless, nut-cracker jawed old woman, but quite upright and active." The Penny Encyclopedia refers to nut-cracking Squirrels and Grove Matthew in his poetical work "The Most Famous and Tragical Historie of Pelops and Hippodamia" (1587) writes of "The little crack-nut Squirrels." The writer has been unable to find any such species mentioned in any scientific treatise.

When do Buds Form?

Even if one never observed that buds begin to form early in the summer they could easily reason this out as it is in summer that plants do their growing and form their various structures. In winter the buds are in a resting stage. In spring the bud may develop into a leafy shoot. Sometimes it develops into a flower or a flower cluster or it may produce both leaves and flowers.

The average per cent of the answers classified by seasons shows that fact and belief are very remote. Fall, 23.5%; winter, 24%; spring, 60%; summer, 8%. The percentages, in this case, add to more than a hundred per cent as several pupils mentioned two seasons in which the buds could form. The majority of pupils believe that buds form in the spring. This is true until we get to the seniors in an Agricultural College and teachers just out of Normal School. The larger per cent of these students believe that buds form in the fall. That one-quarter of the pupils think that buds form in winter may be due to the old proxy of things happening instead of growing. We must place more emphasis upon the differences between these two phenomena.

What do Buds do in the Spring?

Buds begin to grow very early in the spring. Many people glibly say that they begin to "swell." This growth keeps up for several months. The scales then spread apart and fall off. Twigs set in water in the house show how slowly this process takes place.

Out of 191 answers 100 said that they *open*. 14 others said that they *burst open*. The use of the term burst was used almost exclusively by the female sex. 22 others said that buds *form* in the spring. The remaining 55 used such terms as *grow, bloom, unfold,* and *develop*. After looking over nearly 200 answers to this question one draws the conclusion that the majority of people have a mental picture of the bud and its "spring opening" as a sort of spontaneous generation. They believe that they burst open as mushrooms which are said to grow over night.

What is the Advantage of Horsechestnut Buds Having Varnish? Of Having Wool?

The varnish not only keeps too much water out which is rarely necessary but serves the more important function of preventing the tender leaves from drying out. The wool does not serve to keep out the cold since the buds freeze. The covering is a non-conductor and prevents *sudden* freezing and thawing. This "slow process" applies in the same way to the successful thawing out of frozen ears. The majority fall easy prey to the idea that varnish is to keep out the moisture and that wool is to keep the buds warm.

What is the Difference Between an Insect and a Bug?

In a popular way the term insect includes bacteria, spiders, and wood lice. A coral-polyp has wrongly been called a coral-insect. A bug is thought to be synonymous with the word insect. Strictly speaking a bug is an insect which possesses a sucking-beak and belongs to the order Hemiptera. All bugs are insects but all insects are not bugs.

In no class were there less than 10% and in three classes at least 35% gave as a difference that *insects are smaller*. Probably the word connected up with a confusion that exists as to the meaning of the words insect, germ, and bacteria. However, in following the idea up with a request for a list of bugs and a list of insects the list of bugs consisted of erroneous examples, as: potato-bug, rose-bug, etc., along with the true bugs, such as: squash-bug and bed-bug. A less common notion ranging up to 20% of the class was that *insects fly* and bugs do not. A third difference, less pronounced, is that *bugs are hard-shelled*. The best answers were limited to seven pupils who recognized that a bug is a kind of insect. No one gave a more specific difference.

What is the Difference Between Daylight and Sunlight?

People are not apt to credit the light of day to the sun. Sunshine is the direct rays of the sun whereas sunlight and daylight may be direct, indirect or both.

The following are typical answers:
1. Daylight is light of day; sunlight, light of sun.
2. It might be daylight but the sun might not shine.
3. Daylight is a natural daily occurrence; sunlight only when the sun shines.
4. Sunlight, the sun shines; daylight, ordinary light.
5. Daylight is light before sunrise and after sunset.
6. Daylight comes after darkness but sunlight from the sun.

What is the Difference Between Shrubs and Small Trees?

Shrubs consist of several woody stems from the same root. Shrubby and scrubby in early use were often applied to trees having a stunted growth. A small tree has a single main stem.

The following types of answers form an interesting basis for analyzing the various degrees of mental reactions to the question.
1. Shrubs are sometimes plants and bushes and not trees.
2. Shrubs is a cluster of small trees.
3. All shrubs are not small trees.
4. Shrubs are a group of plants; a tree is one plant.
5. Shrubs have many limbs.
6. Shrubs are bushier than many small trees.
7. Shrubs never grow into trees.
8. Shrubs are bushes and do not grow very tall at any time.
9. Small trees will grow taller; bushes will not.
10. Shrubs are short chunky bushes; small trees are tall and slender after the shape of a large tree.
11. Shrubs contain more leaves and are larger in circumference.
12. Shrubs are thicker and branches close to the ground.
13. Shrubs are many trunked plants and small trees have one trunk.

What is the Difference Between the Lung Capacity and the Capacity of the Lungs?

Unfortunately there has developed in physiology a difference in these two expressions. By capacity of the lungs is meant the total amount of air that the lungs are capable of holding at one time but lung capacity means the amount that one can exhale at one time. There is always a certain amount of air left in the remote parts (not only the bottom part), of the lungs.

If any of the Following Statements are Incorrect Give the Correct Form

1. *Remove the shell of the oyster and look at the animal.* The shell is the skeleton of the oyster. The internal or soft parts are only a part of the animal. 81 substituted the word fish for animal. The oyster has no back-bone and is not a fish. Amongst the 281 examined only 12 recognized that the whole thing is the animal. Amongst the university freshmen five men and two women noted the difference.

2. *Earthworms rain down*—It would be nearer correct to say that earthworms rain up, for, as 153 stated earthworms come out of the ground when it rains. Five individuals changed the word rain to reign. This may be a result of the tendency of some schools to deal with words rather than ideas.

3. *Night air is "unhealthy"*—Nearly three-fifths of those examined recognized that night air is as "healthy" as day air.

4. *Flies carry typhoid fever*—76 gave the correction that flies carry typhoid fever germs which cause the disease. They do not carry the disease. One answered,—"Right—let's kill 'em."

5. *Mosquitoes bite*—It is rather a difficult thing to prove to some people that the mosquito does not bite. One man wrote "I know it." 73 wrote the fact of the case that they pierce and do not bite. It is also interesting to note that only "she bites," "he is a gentleman."

6. *Little flies grow into large flies*—An adult fly, when it comes from the pupal case is as large as it will ever be. Few people know that the fly does its growing when in the maggot stage. 37 of the people examined knew about the growth of flies.

7. *Man and animals eat food*—"Man and animals" is a common mistake. It should be "Man and other animals" for man is an animal. A total of 20 noticed the error. 12% of the men in proportion to 6% of the women of college freshmen recognized this.

8. *Oxygen purifies the blood*—The red corpuscles of the blood simply act as vehicles of transportation. They carry the oxygen from the lungs to the cells. Only five indicated that oxygen does not purify the blood. The purest blood in the body is the blue venous-blood as it leaves the kidneys.

9. *Oxygen builds up the body*—Oxygen tears down the tissues and thereby we obtain energy to do work. This knowledge was limited to twenty.

10. *Plants breathe*—Many believe that animals take in oxygen and give off CO_2 and that plants take in CO_2 and give off oxygen. Both respire and in the same way. Plants have the additional power of taking in CO_2, assimilating the carbon and giving off oxygen. Plants do not breathe. They do not have a respiratory movement.

About What is the Size of the Largest Animal Cells?

The majority of students believe that cells are always microscopic. Birds eggs are cells. The ostrich egg is the largest living cell. Nerve cells are often $2\frac{1}{2}$ feet in length.

Where do Dragons Exist?

The answers may be tabulated as follows:

Students	Number	% saying don't exist	% saying in Mythology	Number saying in Woods	Number saying China, Africa
College	176	22%	27%	8	15
High School	102	43%	28%	3	2

The per cent saying that dragons do not exist, tabulated according to sex is as follows:

	Male	Female
College Freshmen	72%	58%
Junior High Schools	91%	56%
(ninth grade)	98%	96%

These figures are insufficient to warrant an infallible conclusion but they suggest tendencies. To the young child, hobgoblins, cock-horses and dragon-like beings are quite real. In high school the dragon of Tanglewood Tales is explained as a myth. In college the earlier ideas have persisted, possibly subconsciously, but sufficiently indeed to mystify his mental activities. What an indictment against nursery rhymes should this prove true. On the other hand there is the possibility that students of college age reasoned out that the question as it stands is rather foolish. They concluded, perhaps, that what is really meant is where are dragons supposed to exist? However as the figures now stand they indicate that a far greater per cent of college freshmen than of grammar school seniors believe in the existence of dragons.

Is it possible that this is the result of a premature classical career? Is it due to a lack of ordinary scientific training as a basis? Literature is the artistic, aesthetic expression or belles-lettres of the evening and science is the working foundation of everyday life. Educators have for the most part placed literary courses at the beginning and science courses at the end of the curriculum. As a result students view facts through the fancies of literature. The situation needs investigation in order to discover the modus operandi by which ideas arise. Cure will follow the discovery of the mode of infection.

What Shape are Stars?

Stars are spherical and not star-shaped as one might suppose from representations on flags and Christmas decorations. The idea that stars are star-shaped decreases and the idea that they are spherical increases with higher education. The word round was often used for spherical.

What do Bees Gather from Flowers?

The idea that insects gather nectar and pollen increases and the mistaken notion that bees gather honey persists into adult life although not to so marked a degree as in youth.

What is the Use of the Hollow Stem to the Flower of the Dandelion?

The hollow stem is stronger than the solid stem.

The answers indicate that the majority think of the hollow stem as a passage way. Some think of it as a storage place yet they never see anything stored in the hollow part. Perhaps they think that the food mysteriously disappears upon cutting the stem.

Underline the Parts Which are Present in the Pussy Willow Plant

(The parts of a plant are root, stem, leaf, bud, flower, fruit).

If one did not already know he might guess that all these parts are present in the pussy willow plant.

What Relation is There Between Toads and Warts?

There is no relation. Only twelve of those examined knew friends who had had warts as a result of picking up toads. It is probably not a coincidence that these twelve were not limited to one such acquaintance. If they knew one they knew several. It is easy enough to obtain witnesses who have picked up toads and never had warts and also the testimony of persons who have had warts and never handled toads.

Where do Hair Snakes Originate?

Although there is no such animal twenty-five students indicated the method by which hair snakes originate. Out of 50 men and 50 women in a college freshman class the idea was limited to 26% of the men. A state agricultural college and a class in high school that had had biology, contributed more means by which these animals originate than the other classes. Chamber's Encyclopedia (1753) speaks of animated horse-hairs as horse-hair worms.

The places where these fictitious animals were supposed to originate were such as: in dirt; in the body; in water; eggs; intestines of mammals; swamps; crossing of reptiles; among rocks. Dragons were also supposed to come from swamps and water. A swamp seems to be a mysterious place. Anything that is mysterious, ipso facto, occurs there. Such reasoning would give-way in a good course in nature-study.

How do Weeds Get Into Gardens?

The majority of weeds are annuals and come from seed.

52 think that weeds just grow in the garden a la Topsy. 24.2% (Average) say that they are due to seed. In a college freshman class 32% of the women to 10% of the men believe that weeds just grow. This wide spread belief reminds one of the parable of the seed growing unobserved. (Mark 4: 26–27). "This is what the Kingdom of God is like—like a man who has scattered seed on the ground and then sleeps by night and rises by day, while the seed is shooting up and growing—he knows not how." The following sentences show how the same parable is expressed today. The disconcerting thing about these statements is that the authors think that they are giving real reasons. We can no longer be content with such statements, as—"Come there." "Grow there by nature." "Weeds have a natural tendency to grow." "The seed was probably near a stone or bad soil." "Unhealthy land." "Just grow from poor part of soil." "Grow over night." "Grow from fertilizer." "Naturally grow from the ground." "Natural growth of earth." "From lack of attention." "By lazy people not hoeing their gardens." "Weeds being wild grow anywhere." "Seeds of weeds are always present." Such statements show that we may have spontaneous generations in the mind as well as physical spontaneous generation. The latter has been disproved but belief in it, though often subconscious, is still rampant. It is of paramount importance that education do away with the superstition that permeates our intellectual processes. Nature-study, gardening, and general science contribute a great offset to this mental obstruction.

What Shape is the Heart?

The heart is not heart-shaped. Heart-shaped refers to the conventional form seen on valentines, jewelry, and playing cards. In the same way the botanist calls certain leaves heart-shaped or cordate. The heart is sometimes spoken of as pear-shaped.

Other early notions referring to the heart have been passed on by certain words and expressions, such as—sweetheart, heart-breaker, hard-hearted, hearty, heartless, and heart-strings.

It may be surprising to know that many people think of the heart as cordate. This is true of 21% of those examined in an agricultural college; junior high school, 21%; high school freshmen, 39%; high school seniors, 35%; normal school freshmen 45%. The latter are our future teachers. They no doubt will learn much about the shape of the heart in physiology and it is to be hoped that they at least will not pass on the mistaken ideas.

Is the Fear of Snakes "Inborn" or due to Education?

Snakes have played an important part in the delirium tremens of literature. With a scriptural foundation aided by the old time immunity of Ireland the fear of snakes is exhibited at a very early age.

Amongst college and normal school students 38% believe this fear is due to being inborn; 33% believe it is the result of education or attitude of adults; and 6% say that it is due to both. There exists an undoubted gullibility as to the evil doings of snakes. From this belief it is very easy to manufacture a fable as the hoop snake which is supposed to take its tail in its mouth and roll like a hoop. Superstition and fear is a tax on intelligence. We should in no way more especially by that of awe—subjugate the reasoning powers. Dickens once said, "What a beautiful thing human nature may be made to be." He might have said in this connection: What a fearful thing human nature may be made to be.

A little investigation indicates that the fear of snakes is due to the attitude of parents, teachers, and other associates of the child. Professor John B. Watson of Johns Hopkins University has carried on some experiments to test out the truth of the older statements which maintain that violent emotions appear at the child's first sight of animals. His results are published in "Kindergarten and First Grade" for January, 1920. He concludes that babies have fear but that there are few positive results in the reaction of children to their first sight of animals. I once took a baby rat and a garter snake into the first grade. The children were told that one of the animals was warm and the other cold. They were then asked if any one would like to pat them to see which was warm and which cold. There was a stampede to the front. I then realized that I had not had the courage of my convictions. The children were then asked to take their seats and told that each one would be given an opportunity to pat the rat and the snake. They all did this without the least sign of fear. It may be claimed that this too was due to education i. e., the expression of the teacher's face, manner of handling the reptile, etc. If this be true, which kind of education do we want? The kind that handicaps clear thought or the kind that takes things on their face value? If there are no legitimate reasons for fearing snakes why make the assumption.

What Danger is There in Picking up an Adder?

There is a legend that once the Python was the only poisonous snake. It could sting a footprint and the poison would kill the man. One day a Crow told the Python that a man had not been killed. The Python then climbed a tree and spat out all its poison which was swallowed by the smaller snakes. Possibly on no less a foundation rests the assumption that all snakes are poisonous.

The word snake does not occur in the scriptures. The term adder is given to several venomous serpents and sometimes to the Horned Viper (Cerastes). Ps. lviii 4. "They are like the deaf adder that stoppeth her ears." This may have given origin to the old saying "Deaf as an Adder." Pro. 23: 32. "At the last it . . . stingeth like an adder." From this may have arisen the idea that adders sting. Adders are not deaf and neither do they sting. Behold their wondrous colors. Their colors are beautiful and blend and the lilies of the field do not have patterns like one of these. Yet man in all his studies of harmony does not admire them.

The danger in picking up an adder, as expressed in school, are hereby given:

	Poison	Sting	Bite	None
College and Normal School	16%	12%	12%	8%
High School	28%	14%	31%	2%

The data indicates that these erroneous ideas are later corrected, to a slight extent, but not as much as one might wish. The mistake comes in allowing them to originate.

All adders of North America are harmless. (The only poisonous British reptile is the common adder of Europe, Vipera berus.) The Puffing Adder is protected by his "puffing." The Milk Adder or Milk Snake searches around barns and old cellars for rats and mice thereby performing friendly acts for the farmer. The name of this snake comes from a reputation of stealing milk. Ditmars writes that he cannot be induced to drink milk unless suffering from great thirst and goes on to mention that if the snake should drink its full it could not consume more than two teaspoonfuls. Snake facts and snake fancies are therefore quite remote.

Scientists believe that man decended from . . .? (Supply the missing word).

No scientist believes that man descended from Monkey. 58% of the students in college and normal school nevertheless think that such is the case. What many scientists do think is that Man and Monkey probably came from a common ancestry. Zoologically they belong to the same group. From a structural point of view man differs less from the apes than they do from other monkeys.

Scientists believe that animals originated from . . .?

Scientists do not believe that animals descended from plants. They are supposed to have come from a common ancestry. There are organisms which cannot be classified as either plants or animals.

Rabbits should be lifted by the . . .?

There is no more logic in thinking that a rabbit should be lifted by the ears than that a cat should be lifted by the tail or a baby by one leg. The tail of a cat is a convenient handle for lifting her but she has taught us that it is not good form. The rabbit has suffered by not having such an effective means of communication.

The majority of people believe that rabbits should be lifted by the ears. (The function of the ears is to hear.) This belief is more prevalent amongst boys and men than with girls and women. It is exceptional to think of lifting the rabbit by the body. This, however, is the most humane method of picking up the animal.

What is the Name of the Fourlegged Animal with Bright Spots Which is Found Under Logs and Stones in Damp Woods?

This animal is the Salamander and not the Lizard which it resembles only in external form. The Salamander is an amphibian and the Lizard is a reptile.

The results indicate the marked existence of the erroneous idea that the animal is a Lizard. The term Salamander is more widely known amongst the men.

The name Salamander is a synonym for fire-proof. The ancient naturalists (or fabulists) Pliny and Aristotle maintained that the salamander was incombustible. This fire-proof fame has come down through the ages being applied to various utensils used around the fire.

Give a Sentence Using the Correct Pronoun (it, she, or he) in Referring to Nature

The expressions mother earth, mother nature, mother west-wind and the like have led to the misconception that "nature" is "she." Some have come to think of nature always as "mother nature." This personification is liable to lead to what "mother nature" tells her children and then to talking-ducks and weeping Lady-Bird Beetles, and so on, ad infinitum. "Mother nature" is not a mother, as we wish to know mothers. The oyster is said to have $1/1,145,000$ of a chance of living. This is not the act of a mother but a "cold blooded" means of perpetuation. The vase-shaped egg-sac of one of the spiders contains several hundred eggs. These eggs hatch and most of the little spiders are eaten by the stronger members of the family. Only a few young spiders emerge in the spring. We must either change our views of the spirit of a mother or acknowledge that "mother nature" exists entirely in the imagination.

What Causes Autumn Colors?

The autumn colors are due to the breaking up of the chlorophyll which gives different species of trees their peculiar colors. The brightest colors are produced in swamps where it is cool and moist before the frost occurs. A frost turns leaves brown and not into bright colors.

	Nature	Frost	Weather	Sun	Season	Sap
College and Normal School	2%	17%	4%	2%	1%	3%
High School	8%	29%	11%	6%	1%	2%

The majority believe that autumn colors are due to frost. Weather and season answers are probably hazy notions of the same idea. To answer this question by saying it is nature is begging the question. All questions pertaining to nature could be answered this way.

Why Are Leaves Placed Around Shrubs for the Winter?

This common practice amongst horticulturists is to keep the frost *in* and not *out* as is commonly thought. The mulch prevents a sudden freezing and thawing. The reasons given with average % were: Warmth, 21%; frost, 37%; protection, 15%; fertilizer, 8%.

Would you Eat Butter or Drink Milk that had Chemicals in it?

	No	Depends	Yes
College and Normal School	21%	9.0%	30%
High School	49%	0.5%	17%

In thinking of chemicals we are too apt to have a vision of laboratories, bottles, and labels. If our perspective stops there we do not have the broad outlook that chemistry should give us. We see symbols and not ideas. The symbol is only a means to an end. Our educational courses should not stop with the means of seeing, these tools should be made to reproduce ideas.

Have Some Friend Give You the Following Spelling Test

Potato, radish, pole beans, larva (of an insect), harelip, plantain, dandelion, mullein, caterpillar, development, definite, occurrence, pistil (of a flower), develop, squirrel, nasturtium, moss, diptheria, dessicate, paraffin, inoculate, tonsillitis, singular of species, leaves, seedsmen, and bacteria.

This list of biological terms is a collection of words which are most frequently misspelled. Although many well-educated people are poor spellers the correct spelling of these words may indicate the amount of Biology one has had. In the case of radish, however, the lowest per cent of correct spelling was 45% made by high school graduates who had had Botany. The misspelling in this instance may have been an index to the kind of Botany they had had in high school.

The singular of species is most often misspelled. It should be species. The highest average was 26% made by the freshmen of a woman's college. The word specie was the incorrect form most commonly used. One scientist has said that specie is something that a biologist never has. The difficulty may be easily removed in other cases by knowing the meaning of the word as in the case of harelip, pronunciation as in develop, or the origin as dandelion from dent-delion. Such words as occurrence may be made indelible by calling attention to the two r's. The second most difficult word is mullein. Six per cent of the men in the freshmen class of a noted university were able to spell the word correctly. The Winston Simplified Dictionary gives mullen as a possibility but this book was not published until 1919 and it is doubtful if any of these men obtained their modern method of spelling from this source.

The relations according to sex are of interest. The boys from a ninth grade headed the list in spelling *development* correctly with an average of 84%. The girls from the same school had an average of 63% but when it came to the word occurrence the girls headed the whole list with 84% correct and the boys of the school had but 56% correct. The girls from two junior high schools obtained the higher per cent in 11 cases and the boys in 10 cases. The difference is more marked when we come to the freshmen in college where the girls win in 15 cases and the boys in 6 cases. Three of these words, where the men averaged higher were in the list won by the boys in the grades. In general, as expected, the spelling efficiency increases with age and grade.

This list of words is not intended for a school spelling lesson. They should be added to the pupils' vocabulary when occasion arises for their use. As they are words which are misspelled over and over again the teacher should make sure that they are mastered. Since most of these words are commonly used by children in the grades they should be effectively dealt with at that time.

In passing, a story may not be amiss. A tailor wished to order 12 smoothing irons or gosses. He did not know whether to write 12 tailor's gooses or geese. He finally ordered one tailor's goose and eleven more like the first order. The moral is clear. In times of doubt use "ready-made" words to suit the occasion.

Pronounce the Following Words:

abdomen	ab-do′-men	not ab′-do-men
acorn (Anglo-Saxon meaning field or acre)		ā-kern
alimentary	al′ and not el′	
almond	ä′mŭnd	
ant	ánt	not änt
asparagus	ăs-păr′-ā-gŭs	not sparrow-grass
biology	bĭ-ŏl′ō-ji	not bĭ
cerebrum	sĕr′ē	not sē-rē′
chestnut	ches′	not chest′ nut
cranberry	kran′ beri	not kram
fungi	fun′ ji	not fung′-gi
hoof	hŏŏf	not hŏŏf
horseradish	hôrs-râdĭsh	not rĕd-dĭsh
insect	in′-sēkt	not in′-sĕk
intestine	in-tes-tĭn	not tin
isolated	either i′ or ĭs′	
laboratory	lăb′-ō-rā-tō-rĭ	not lăb′-rā-tō-tĭ
lettuce	lĕt′-ĭs	not lĕt-ŭs
loam	lōm	not lōŏm
mackerel	mak′-er-el	not mak′-rel
muskrat	musk′ rat	not mush′ rat
poplar	pop′ lar	not pop′ u lar
porpoise	por′-pus	not por′ poise
pumpkin	pump′ kin	not punī kin
root	rŏŏt	not rŏŏt
spiracle	spĭr′ or spī	
squirrel	skwŭr′ ĕl	
stomach	stum′ uk	
sumac also h	sŭ′ mak or shŏŏ′ mak	
teat	tēt	
tomato	mā to or mä to	not to-măt-o
turnip	tûr′ nĭp	not turnup
zoology	zō-öl	not zŏŏ-öl

A Providence lady once asked her gardener to plant salivias near the walk. The gardener replied: Wouldn't spetunias look good over there?

What is a Germ?

The etymology of the word germ is doubtful. The Latin word *germen* means to sprout. In botany it refers to the rudiment of a new organism. Huxley referred to the budding of corals, as multiplying "by means of germs." In Linnaean nomenclature it is the ovary or the seed. Muir and Ritchie's Bacteriology says "germ, microbe, and micro-organism are often used as synonyms with bacteria, though, strictly, they include the smallest organisms of the animal kingdom." Evidently there is ambiguity in the use of the term amongst biologists.

The answers of pupils may be listed as follows; the figure representing the highest per cent in any one class thinking that that word is the meaning: Organism, 6; micro-organism, 36; micro-animal, 14; micro-insect, 28; and bacteria, 70. The term *organism* is not used as often as a generation ago yet it ought not to be confused with micro-organism which refers to a microscopic animal or plant. Organism has come to be an abbreviated form of micro-organism.

What is a Meadow?

Originally a meadow meant a hay field. Later it included a pasture and in some places was extended to include a low well-watered ground. In North America, according to the Oxford Dictionary, it is "a low level tract of uncultivated grass land, especially along a river or in marshy regions near the sea." If we condense this statement it might read,—a low, level, moist grass land. The word meadow has been prefixed to the names of animals and plants which occupy the meadow land, as meadow hen (name applied to various herons); meadow-lark; meadow-sweet (spiraea); meadow beauty (rhexia or deer-grass). The word meadow occurs frequently in literature but without uniformity as to meaning.

Amongst students of today there are 80% who recognize that it is a hay-field. It is a decidedly masculine trait to know that it is in a low area, the knowledge being held in a ration of 4 males to one female. Not more than 12% of any class, and only a total of 18 recognize that it is a low hayfield.

The geographical distribution of the various meanings of the term would make an interesting study. "Up from the meadows all filled with corn" suggests another departure.

What is Dirt?

Primarily dirt means excrement, secondarily filth, and only colloquially soil. The answers were in the following proportions: soil, 141; filth, 64; soil and filth, 5; matter out of place, 7. The majority have the false impression that soil is dirt. Soil may be perfectly clean, as sand, which is used for filtering. In preparing the garden we are tilling the soil.

What is a Biennial?

Biennial (bi-two, annua, year), existing for two years. In botany a biennial plant vegetates during the growing season of the first year and dies after producing fruit the next year. Amongst the freshman class entering college or normal schools many answered that a biennial means twice a year (33%) as answered that it means two years.

How Long Does an Annual Plant Live?

An average of 59% said one year and an average of 16% said less than a year. The highest number of correct answers were made by boys in a junior high school (28%), and freshmen in a normal school that have had botany (27%); high school freshmen (29%); high school seniors (28%). The figures of the higher institutions would indicate that as education continues the concept of the length of life of an annual plant diminishes. It still remains a masculine trait, 20% of the men in the freshman class to 12% of the women saying that an annual plant lives less than a year.

What is a Flower?

The following are typical answers:
1. A stationary living creature.
2. A thing growing out of the earth to beautify it.
3. A plant with colored variations.
4. A plant grown to admire and not to eat.
5. The petaled flourish of a plant.
6. The blossom of a plant.
7. That which later bears fruit.

Over one-third of the pupils in our schools when using the term "flower" are thinking of a "flowering plant." Such plants are ordinarily grown for their blossoms. The first four answers show such a use. Early English writers made a similar use. Shakespeare, 1593, in Lucerne (p. 870) mentioned that "unwholesome weeds take root with precious flowers" and Milton in Paradise Lost (XI, p. 273) wrote, "O flours that never will in other climate grow." In a popular way when we mention flowers we think of the colored (not green) parts and do not say flower when the petals are absent as in the pussy willow. A corolla and calyx are not necessary. Botanically we should think of the flower as a means for reproduction. The children should be taught that garden vegetables, weeds, and grass are flowering plants. They should think of the flower as a "seed-maker" and not as an ornament just meant for us.

Where do Plants Occur?

Plants do not merely sprout from the ground. They may be found in the ocean breakers, on the backs of shell-fish, in clear ponds and on rocky cliffs, on the roofs of all houses and in every room below, on the leaves of trees and on their roots in the ground.

Use the Words Sitting, Setting, Laying and Lying in Reference to a Hen

Many will no doubt sympathize with the farmer who told the Boston School teacher that he didn't give a darn whether the hen was sitting or setting but when she cackled he wanted to know whether she was laying or lying. To set means to cause to sit. It is correct to say that he is setting the sitting hen on a sitting of eggs. Setting hen or a setting of eggs is incorrect. There is such an animal as a setting dog. To lay means to cause to lie. Hens lay eggs.

What is an Animal?

The meaning of the word animal has had rather a checkered career. Bishop Gawin Gouglas in his translation of Virgil's Aeneid (1513) wrote "Undyr animal beyn contenyt all mankynd, beist, byrd, fowel, fisch, serpent, and all other sik thingis." The word itself originates from the Latin meaning breath of life or anything living. It was hardly in the English before the end of the 16th Century. It was not used in the Bible in 1611. In 1875, Helps, Animals and Masters (iii, 53) says that "When I use the word 'animals' I mean all living creatures except men and women." In New York, October 18, 1911 a sea captain was brought into court for piercing the flippers of large green turtles in order to tie them. The magistrate held the captain in $500 bail on the basis that within the law the turtle is an animal although without the law it is a reptile. The same kind of a case had failed in 1867. On the 1920 calendar of the Rhode Island Humane Education Society are the following sentences: "The object of the Bands of Mercy is to teach children to be kind to each other and to all who need help and protection, as well as to animals." "The child who respects the rights of animals will also respect the rights of human beings." Amongst the uneducated the fur bearing quadrupeds only are considered as animals. These various meanings of the word have led to a mixture of ideas in the schools.

The opinions of college freshmen as to various animals are hereby summarized:

The woodchuck is a mammal. 1. *Woodchuck.*—40% females to 10% males think it is not an animal. 34% females to 12% males think it is a bird. This is probably a confusion with the woodcock which is a bird. This also indicates that girls think that a bird is not an animal.

2. *Man*—Man is an animal, belonging to the class mammals. 32% of the females to 2% of the males think that man is not an animal. This is further proved by the fact that 12% of the males to 6% of the females correct the statement "Man and animals eat food" to "Man and other animals eat food." 18 out of 281 examined recognized that man is an animal. It is also used in this sense in statute books in reference to the prevention of the cruelty to animals as "man and animals."

3. *Whale*—The whale is not a fish. It is a mammal. 36% females to 68% males think that the whale is an animal. 62% females to 38% males think that the whale is a fish. The males in every case tested seem to have a more accurate and broader view as to what is an animal.

4. The *perch* is an animal belonging to the class fish. 16% females to 42% males think that the perch is an animal. 78% females to 94% males think that the perch is a fish.

5. The *bat* is a mammal and not a bird. 38% females to 62% males think that the bat is an animal. 36% females to 42% males think it is a bird. Every thing that flies is not a bird. Fossils show that there used to be flying reptiles. There are so called flying fish, flying squirrels, and a flying lemur. Angels are also credited with the art of flying. The latter are the only animals, not birds, represented as having feathers.

6. The *heron* is a bird. 26% females to 50% males recognize this. 30% females to 16% males believe it is a fish. This probably comes from a confusion with the herring which is a fish.

7. The *turtle* is a reptile. It is a relative of snakes and lizards. 28% females to 48% males recognize that it is an animal. Only 42% males to 10% females recognize that it is a reptile.

What is Meant by Ugly?

A study of the idea ugly in literature will give a basis for understanding the present day feelings.

1300–1400, Cursus, Mediaeval and Vulgar Latin (Gott), "Fell dragons and tadis bath . . . ful laithsum on to here and se. . . .

1666, Bunyan, Grace, Abounding to the Chief of Sinners (p. 84), "I was more loathsome in my own Eyes than was a Toad."

1667, Milton, Paradise Lost (XII, 178), "Frogs, Lice, and Flies, must all his palace fill with loath'd intrusion."

1600, Shakespeare, As You Like It (ii I. 12–14),

"Sweet are the uses of adversity,
Which, like the toad, ugly and venomous,
Wears yet a precious jewel in his head."

The fable of the jewel in the toad's head is probably based on the glistening cartilage which represents an unossified basioccipital.

1611, The Bible, Exod. 8: 2, "I will smite all thy borders with frogs" Rev. 16: 13. "Frogs (come) out of the mouth of dragons."

The following terms are used in the Bible in reference to the serpent: subtile, beguiled, fiery, brasen, crooked, poison, biteth piercing, and wise.

1748, Thomson, Castle of Indolence (543), "In chamber brooding like a loathly toad."

1886, Besant, Children of Gibson, II, vi, "A knight was sent forth to kill a dragon or a loathly worm."

Animals have borne the burden of false misrepresentations down through the ages. It is not surprising that the toad gets the highest vote as to ugliness, 60% giving him that characteristic. The other ugly animals were as follows: Bat (58%), earthworm (54%), fly (51%), snake (51%), ant (47%), and eggs of clothes moth (42%). 16% of the females in a college freshman class think that the cow is ugly. It is a male trait to think that the dragonfly and the silkworm is ugly. Ugliness seems to be a matter of enchantment or prejudice. All these traditions have been interwoven with modern progress. The heavy tax levied by superstition on intelligence is well known to scientists.

What Animals are Injurious?

56% of all the pupils examined believe that the snake is injurious; 27% the dragonfly; 26% the bat; 14% the earthworm; and 28% the crow. It is a female trait to think the bat and eggs of the clothes moth are injurious. Of course the eggs are harmless. The larvae that hatch from the eggs do the damage. It is a male characteristic to think that the dragonfly and crow are injurious. The value of these animals as insect destroyers is now becoming generally recognized. It must be admitted however that the process is slow and the only remedy is progressive work in the grades to prevent the confounding of fancy with fact.

What is Cross-pollination?

Cross-pollination is the transfer of pollen such as by the wind, or by insects, from one flower to the stigma of another. Cross-pollination is not cross-fertilization. Cross-fertilization is the union of the male and female elements. Cross-pollination makes cross-fertilization possible.

Epilogue: In the beginning of this article I suggested that too much seems to centre on the reputation of the buried past and not on direct observation of the living present. Antiques from tradition-shops are still coddled by sentimentalists. These musty relics are handed down from generation to generation thereby cheating would-be workers with their false value. My task is finished if I have been able to show, at least to some extent, that whatsoever is most exact and regular concerning nature is also most useful and excellent.

COMMON MISTAKES IN GEOGRAPHY

Since ideas gained in childhood are firmly fixed in the mind, it is essential that they be correct ones. Colloquial expressions learned in the earlier days of life, become a great source of annoyance at a later period. There are certain persistent errors in geography which are in such common usage that they are frequently passed unrecognized. It is the purpose of this article to call attention to such mistakes in order that the present day pupils may not be handicapped in later years by having acquired erroneous opinions and phrases from the elementary studies.

What is the form of the earth? The form of the earth is an oblate spheroid, but for all practical purposes it may be considered a sphere. The fact that approaching ships first show their sails does not prove the earth to be a sphere. A curved surface would give the same effect and consequently, so far as this reason is concerned, the earth might be considered as shaped like a pie. The fact that men have traveled around the earth does not prove it to be a sphere. Men could travel around the earth if it were a cube.

What is the "crust" of the earth? It was formerly thought that the interior of the earth was a molten mass and that the exterior was a crust. This condition is impossible since the tremendous pressure of the soil and rocks would not allow substances to exist in any state except that of a solid. If the earth were not a solid it would behave differently toward other planets, and there would be internal tides. Another evidence that the earth is a solid is shown by earthquake waves. The shocks from an earthquake travel at a velocity that would be expected in a solid and not in a liquid. There is no such thing as the "crust" of the earth.

What is the sky? The blue sky seems to rest upon the earth, and as children we have often thought of walking to it. The sky is no more a real object than a rainbow, but is the appearance produced by the reflection of blue light from the air.

Where is the sky? The sky, or "heavens," is below our feet as well as above our heads. Stars shine on the opposite side of the earth as well as on this.

What part of North America is in the same latitude as England? Many think that England is opposite New York City. England being in the latitude of Labrador, is as far north of New York as the Bahama Islands are south.

Distinguish between the following: British Isles, Great Britain, the United Kingdom, and the British Empire. Many people use these terms synonymously. Great Britain includes England and Scotland; the United Kingdom includes all the British Isles; the British Empire includes the United Kimgdom and all its possessions.

Why does the North lead the South in manufacturing? The North does not have better advantages than the South for manufacturing. The South has the raw material, the coal, and fine water power. The North has led in manufacturing because it got the head start.

If a place is directly below on the map, is it directly south? It is a common error to think that what is directly below on the map is directly south. Only places on the same meridian are north or south of each other. For the same reason a straight line on a map does not represent the shortest distance between two places.

What direction does the top of the map indicate? Some students think that the top of the map must always be north and the right hand east. These conditions usually exist but a map of the United States, although Texas were at the top, would still be a map of the United States.

Do Arctic explorers go "up north?" The expressions "up north" and "down south" are incorrect. Down is toward the center of the earth and up is in the opposite direction. This notion probably arose from the fact that the top of the map is generally the northern part.

Do all rivers, in a general way, flow south? The expression "down south" has led some to think that rivers must necessarily flow south. The Nile, Mackenzie, and St. Lawrence are examples of rivers which flow toward the north.

Point in the direction of the north pole. Ask a pupil to point at the north pole and he either points to the northern horizon or to the North Star. If a cannon ball were shot in such a direction it would travel out into space. One would have to point down into the earth at an angle, depending upon the number of degrees the observer is from the pole.

How correct is the scale of maps? The globe is the most correct means of representing the distribution of land and water, but it is not large enough to exhibit the features of small areas. It is impossible to represent correctly the curved surface of the earth upon a plane surface. The central part of our hemisphere maps is in true proportions, but at the edges the distance between two places is doubled. On a common Mercator map, measurements at the equator are correct but the distances on the sixtieth parallel are doubled. Christiana for illustration is represented twice as far from Cape Farewell as it actually is.

What is a river divide? A river divide is not necessarily a mountain. It might be a plain.

What is the cause of irregular coast lines? It is generally due to the sinking of the coast. In Norway however the fiords are due to glacial erosion.

Is mineral water pure water? The fact that the water contains minerals shows that it is not pure. The rain which falls after it has washed the impurities out of the air, is as pure water as we find naturally. People who go to mineral springs to be cured of nervousness are probably cured by the amount of water they drink rather than the kind. The greater per cent of nervous tissue is water and the nerves naturally need water as food. As a race, we are starving our nerves.

Where is the Great American Desert? The Great American Desert is almost a thing of the past. It has been effaced from the continent by irrigation. Land which formerly bore sage and cactus is now producing from fifty to thousands of dollars worth of alfalfa, peaches, apples and grapes.

What is the fertility of desert soil? It is commonly thought that desert soil is always poor soil. It would not pay to irrigate soil which is unproductive. Desert soil is usually rich soil. It needs moisture and not fertility.

What enables rivers to traverse deserts? The Nile and the Colorado Rivers are examples of rivers which maintain their course across a desert. These rivers do not get the bulk of their water supply from the desert. Their water supply is regulated and kept constant by the melting of snow in the mountains.

What is the climate of deserts? The desert is not limited to warm regions but may be a cold expanse in Northern Siberia. Deserts are not limited to absolutely dry regions, but have some rainfall.

What is meant by "dry land?" The expression "dry land" probably originated amongst seafarers and meant comparatively dry. No land is absolutely dry, not even the desert. A film of water adheres to the particles of soil even in the longest period of drought. The same is true when speaking of dry air; no air is absolutely dry.

Which is heavier, moist air or nearly dry air? Since vapor goes up it is lighter in weight than air. The vapor in the air displaces some of the nitrogen and oxygen and therefore makes the moist air lighter.

Explain the motions of heated air. We should avoid the expression "warm air rises." Heat causes air to expand in all directions. The heavier cold air pushes up the lighter warm air the same as mercury pushes up water when mercury is poured into a test tube holding water. Warm air does not rise, it is pushed up.

What is the origin of dew? Dew does not fall but collects upon objects cooler than the air. A poetical license may give a wrong idea. A description may be both scientific and beautiful. The two styles are illustrated as follows:—

> "The dew was falling fast,
> The stars began to blink."—*Wordsworth*

> "Came as silent as the dew comes,
> From the empty air appearing,
> Into empty air returning.
> Taking shape when earth it touches;
> But invisible to all men
> In its coming and its going."
> —*Longfellow "The Song of Hiawatha"*

What is Snow? Frost? Snow is not frozen rain and frost is not frozen dew. Snow is formed by the condensation of water vapor below freezing point, and rain is the condensation of vapor above the freezing point. Frost and dew are formed the same way except that they are formed on objects instead of in the air.

Why is a rainy day a "good day?" It is customary to call a sunshiny day a "good day" and a rainy day a "bad day." The rainy day is just as good a day as the sunshiny day, and is just as essential to our well being.

Do forests cause rainfall? The popular opinion is that forests cause rainfall, but in reality the rain is the cause of the forest. The want of rain prevents the growth of trees.

Why do forests grow on the hills and not in the valleys? Trees do not grow on the hills because they do not like the valleys, but because the valleys have been cleared away for agriculture and other industries.

How far north do corals grow? Corals grow as far north as the Bermuda Islands which are in the same latitude as Cape Hatteras.

Why can we look at the sun late in the afternoon and not at mid-day? This is not because the sun is farther away nor because it is less bright. The rays have to pass through more dust at sunset than at noon and it is consequently like looking through smoked glass.

Where does the sun rise? It is a common saying that the sun "rises in the east and sets in the west." Sunrise and sunset are convenient terms for the apparent motion of the sun. The real motion is the rotation of the earth into the sunlight. The sun practically "stands still" as Joshua commanded it to do. The sun does not rise directly in the east nor set directly in the west but twice a year.

Where does the day commence? Some have thought that the day commences at Greenwich, possibly because longitude is reckoned from that place. It has been decided by arbitration that the International Date Line, which is near the 180th meridian, should mark the beginning of day.

Where is the sun at noon? The sun is not directly overhead at noon except for persons on one parallel. That parallel must be in the Torrid Zone, and the sun is never over one parallel but twice in one year.

How long is a day? It is usually thought that a day is twenty-four hours long. It is, for any one locality, but the total length of each day is forth-eight hours. The new day commences at midnight. It takes midnight twenty-four hours to pass around the earth, so that when the new day commences just east of the International Date Line, that day has already existed twenty-four hours; yet it will be twenty-four more hours before the day will end at that place. Christmas, therefore, is celebrated for forty-eight hours on the surface of the earth.

When is the sun nearer to us, in the summer or in the winter? The sun is nearer to us in the winter but the rays are more direct in the summer. It is the direct rays that produce the warm season. Since the distance of the earth from the sun has an effect on temperature the summers of the southern hemisphere should be warmer than ours.

Which name is more appropriate, Winter Solstice or December Solstice? What is called the Winter Solstice by us is really the Summer Solstice for the southern hemisphere. It would be better to call it the December Solstice.

Where is the line of greatest heat? When speaking of the equator the geographical equator is usually meant. The geographical equator is a fixed regular line but never represents the line of greatest heat. The heat equator is an irregular line which moves from about 23.5 degrees north to the region of 23.5 degrees south of the geographical equator.

"And she climbs at last to a berg set free,
    That drifteth slow;
And she sails to the edge of the world we see:
And waits till the wings of the north wind lean
Like an eagle's wings o'er a lochan of green,
    And the pale stars glow
    On berg and flow * * * *
Then down on our world with a wild laugh of glee
She empties her lap full of shimmer and sheen.
And that is the way in a dream I have seen
    The Weaver of Snow.
        —*Fiona Macleod in "From the Hills of Dream."*

"Welcome are both their voices,
    And I know not which is best—
The laughter that slips from the Ocean's lips,
    Or the comfortless Wind's unrest.
There's a pang in all rejoicing,
    A joy in the heart of pain,
And the Wind that saddens, the Sea that gladdens,
    Are singing the self-same strain."
        —*Bayard Taylor.*

"We walk silent among disputes and assertions but reject not the disputers, nor
        any thing that is asserted,
We hear the bawling and din—we are reached at by divisions, jealousies, re-
        criminations on every side,
They close preemptorily upon us, to surround us, my comrade,
Yet we walk unheld, free, the whole earth over, journeying up and down, till
        we make our ineffaceable mark upon time and the diverse eras,
Till we saturate time and eras, that the men and women of races, ages to come,
        may prove brethren and lovers, as we are."—*Walt Whitman.*

Down swept the chill wind from the mountain peak,
        *   *   *   *   *
It carried a shiver everywhere
From the unleafed boughs and pastures bare;
The little brook heard it and built a roof
'Neath which he could house him, winter-proof.
        *   *   *   *   *
No mortal builder's most rare device
Could match this winter-palace of ice;
'Twas as if every image that mirrored lay
In his depths serene through the summer day,
Each fleeting shadow of earth and sky;
Lest the happy model should be lost,
Had been mimicked in fairy masonry
By the elfin builders of the frost.
        —*Lowell.*

CHAPTER XIII

"Timber Lines"

People, like trees, are peculiarly the product of their environment. Both are gregarious. First come the early settlers in the valleys. In a few years there is a stand of seedlings and they in turn breed more offspring, and thus in half a century or more there are the old patriarchs, almost hidden in the crowd, the second generation, and the grandchildren. The valleys have become crowded and both people and trees march upward to the hills after more and more sunshine.

Behold, the people maketh the earth empty; and maketh it waste, and turneth it upside down, and the inhabitants thereof scattereth abroad. And it has come to pass that Nature's only reservation, that she may call her own, has been pushed unto the "timber-line." Here steel has not stirred, hoofs have not trampled and hands have not hacked. It is the only extensive and purely wild place not guarded by keepers. These areas too will pass out of their natural existence—will be gone forever and a day—like the prairies and the deserts—unless a love of untrampled nature is awakened within the hearts of American youth.

The outermost edge of tree world is favored above all wild gardens. In the fullness of beauty it opens a panorama across mountains or into the open ocean. Seas of water or cloud make its border and in the middle of it, over damp bottoms or in low hollows, are mossy tundras. For many a century the wind has visited its shrubby growths,— working like a dependable gardener; bending and pruning; bringing clouds to water them; sifting pollen onto flowers; and scattering resulting seed. Wind-branching pine or spruce in endless variety, irrepressible and undaunted, stand sentinel along the coast of the Atlantic and over the stony crests and folds of Appalachia.

Venture out some day beyond "timber-line," let night come on, or a storm—and see how quickly you turn back to the shelter of the forest. Perhaps this is the primal instinct of the race. Anyone who has not returned to the forest as he does to the welcome of his own hearthstone is oddly insensitive but he who has not been into unspoiled nature still lives in dusty history.

The importance of "timber-line" recreation is not merely a dream· There is the Appalachian Mountain Club with over 3,000 members, 257 miles of trails, and its huts. The New England Trail Conference—its membership consisting of mountain clubs, hotel companies, athletic associations, landscape architects, and boy scout organizations,—plans a through trail route from Mt. Katahdin in Maine to the Jersey Highlands and thence to North Carolina. And under the Weeks Act, the United States Government plans to purchase about a million acres in the White Mountain region, not for a National Park but for a National Forest. Nearly half of this amount has already been purchased.

No less significant, and one of the most recent features in our educational progress, is the rôle of the summer camp. There are at present about 500 private camps for boys and girls. This does not take into account the scout and Y. M. C. A. camps nor the many types of camps for adults. And it is not by mere chance that at least 90% of these camps are within a days hike of "timberline." Their programs tend toward simple health, simple interests, simple

Fig. 1. At "Timber Line" in a Cape Cod Forest. Note the clump of scrub oaks sheltered from the fury of the winds by waving beach grass.

happiness, and simple beauty. Their leaders realize that besides school lessons there are nature's lessons, and that nature is at her best near elfin forests. We hope that the time is near when all the children of all the people will have the opportunity of being trained in the ways of unspoiled nature.

The most truly "timber-line" school is the Trail School of Enos Mills which is on Long's Peak in Colorada. The country has not yet awakened to how great a son it has lost in Enos Mills. In his preface to the *Adventures of a Nature-guide* he wishes that "every park had a nature-guide and that every wild place might early become a park." The magnitude of his visions were so large as to require longevity and his sudden disappearance leaves a chasm broken in a trail which no one else can blaze. His soul was made for the noblest of nature guides; he had in a short life laid bare the possibilities of "helping people to become happily acquainted with the life and wonders of wild nature." Wherever there is peace and good will· wherever there is vastness, patience, and beauty, he is at home.

Yet the work of Enos Mills must go on. No government or association can save a natural reservation without an intelligent public to enjoy it. Safe camp fires, camp sanitation, sane hunting, conservative fishing, etiquette of the trail, emergencies in the woods, outdoor cooking, are the names of some of the courses that must be given

Fig. 2. Resting in the crater hole of a Cape Cod scrub oak forest. Such crater holes are evidences of air raids from the Atlantic.

along side of, if not in the place of, split infinitives, compound interest, trick spelling, and endurance tests in geographic names. A woods

Fig. 3. "No tree land" on Cape Cod. Along this bleak expanse lies the patrol-line of the Coast Guard. This is one of the most dangerous winter coasts in the world.

can be no better than the people want it to be. This can only come through education in the personality of the trail.

"Timber-lines" have always been misunderstood. In the first place the word is untruthful. Timber does not cease at a line. Trees get smaller gradually toward an upper limit of growth. Like some other lines it is becoming less dogmatic with the increase of knowledge concerning it. As the "snow-line" varies with the season. often

Fig. 4. Cape Cod Pitch Pines often lead a low life on high ground: These girls are lifting the edge of a tree-lawn—a large limb which has avoided the winds by crawling.

times day by day, so does the "timber-line" vary from year to year. It also varies in different mountains ranging from 6,400 feet in the Alps up to 12,000 feet in some parts of the Rockies. Then again, a "timber-line" is not limited to high mountains or to frigid latitudes. It may be along the coast as shown by the photographs of the Cape Cod "timber-line." Furthermore, "Timber-lines" are not caused by a cold temperature but have to do with the amount of available moisture. Excessive evaporation due to high winds stunts tree growth on the seashore or in the mountains. In both places the trees are dwarfed, bent and crouching, with leaves thick and leathery to prevent the loss of moisture. Contrary to expectation the flowers of the dwarfed herbaceous plants of alpine countries assume great brilliancy. It were as though the flowers were making up for the insignificance of the leaves. Switzerland is nearly as popular for its flowers as for its mountains. The most inaccessible craigs of the Tyrolese mountains make possible the Swiss Edelweiss, which signifies noble purity.

We rejoice that "timber-line" knowledge is becoming more general and that "timber-line" folks are becoming more numerous. It is still a "high-brow" subject in that all the people have not "arisen" to its possibilities. To shake the dust of the valleys is almost a necessity. By "timber-line" contact one develops a personality totally different from the one who struggles along in the crowd. The man of the trail has an individuality of sunshine and air. He is physically fit and has endurance. He enjoys camping and has a natural history background which is contagious. He is a keen observer—a good thinker and because of these things is versatile in his environment. He appreciates nature in darkness as well as at noon day, in rain or shine,

Fig. 5. And a little tree shall lead them: Falling out of the tree-tops is a common occurrence when striding over the elfin forests of the Cape Cod "Timber Line."

in winter or summer. The most hopeful thing in "timber-line" education is that all who are initiated into its lore live out the spirit of "not to be ministered unto but to minister" the knowledge of the trail.

But we started to tell about people and "timber-lines." Here on rugged land have grown wind blown trees, not wanted by man, yet emulating his love of ruggedness. Like him, they are migrating to this alpine region making their way amongst stones and ledge. Both, perchance, belong not to the aboriginal race of mountaineers, but have strayed up from the tender valley stock. None have more difficulties to meet. None meet them with greater sturdiness. These are the ones we wish to describe.

Near the summit of the White Mountains we notice little islands of spruce springing up amongst the boulders—their very birthplace for a time defending them against the demon wind. The next year, when they appear above the bulwark, the wind cuts them down. They do not despair but put forth two branches where one grew before. The new twigs lie low or creep between the rocks growing

more scrubby each year. If perchance, some raise their head they are
clipped each year by the wind, like an estate by a lawn-mower, until
they form tree lawns, from one to three feet high, and quite even, as if
trimmed by a gardener. Just before sunset these tree lawns appear
like large carpets laid along the low places of the mountains. In

Fig. 6.   A Tree Lawn at Long's Peak, Colorado: This tree lawn is as regularly clipped by the wind as the well kept lawns on Cape Cod. It is about
three feet high and as even as though trimmed by a gardener.

time some of the densest and most impenetrable forests have been
woven, easier to walk on than through.

Many of these elfin forests are old forests, if you reckon from the
time they started, but infants still when you consider their growth
and prospects. The wind continues to keep them down for years,
until at last they become their own keepers. Some leeward shoot
which the wind cannot reach starts upward in triumph. No longer is
it suppressed and it now starts with new energy towards an upright
life and lives a second span, as it were. Thus these pigmy forests
create their own protection, a united we stand and a divided we fall
sort of policy.

And thus it is with man. It is only here and there—in the rugged
places—that a Lincoln or a John Muir or an Enos Mills springs up.
Small clans of boys spring up among the boulders, their very birth
places for a time defending them against the conventionalities of life.
Then when they appear along the privet hedge or in the yard of a
tenement house the landlord cuts them down. They do not despair

but put forth into the street and grow two evils where one grew before. Society continues to keep them down until at last they become their own keepers, and the most persistent prevails and maintains his rights to wild nature. He gets away from the ungrateful host of men. Such is always the pursuit of freedom. "Timber-lines," the haven of which they dream and aspire to reach, have been pushed farther and farther away and is ever guarded by the hundred tentacled octopus of conventionalities, so that it is an herculean job to escape them.

The era of the barefoot boy with cheeks of tan will soon be past. He is an animal which will probably become extinct in southern New England. He may still wander through vacant lots and throw tin cans into a dump heap. If fortunate he may even have access to a park where the chestnut trees and the apple trees have been cut down for fear they should break the dignity of his training. Ah, poor fellow, there are many pleasures he will never know!

"Hear this, ye old men, and give ear, all ye inhabitants of the land. Hath this been in your days, or even in the days of your fathers? . . .

"That which the palmer-worm hath left hath the locust eaten; and that which the locust hath left hath the canker-worm eaten and that which the canker-worm hath left hath the caterpillar eaten. . . .

"He hath laid my vine waste, and barked my fig-tree; he hath made it clean bare, and cast it away; the branches thereof are made white. . . .

"Be ye ashamed, O ye husbandmen! howl, O ye vine-dressers! . . .

"The vine is dried up, and the fig-tree languisheth; the pomegranate-tree, the palm-tree also, and the apple-tree, even all the trees of the field, are withered: because joy is withered away from the sons of men."

This much I have to say unto you for the sake of your children and their children's children. The "timber-line" is their refuge. May we say to all the children of all the valleys that they shall not want. May we still lead them in green pastures. May they walk in the paths which it is righteous that they inherit that they may restoreth their souls. May their cup runneth over. Surely goodness shall follow them all the days of their life; and they will dwell in the house of the Lord forever.

SUGGESTED NATURE ACTIVITIES FOR EIGHT SUCCESSIVE WEEKS
(Including Sundays, rainy days, sunny days, and hike days.)

	First Week	Second Week	Third Week	Fourth Week	Fifth Week	Sixth Week	Seventh Week	Eighth Week
Sunday Nat. Hike	Flowers for Tables	Wonderful View	To sit and hear birds	To beautiful flower nook	To a Mt. Top (Sermon on Mt)	To an old Churchyard	To an old fash-ioned flower Garden	Visit with Sheep—Read about in Bible
Nat. Poems	Songs of Outdoors,	Van Dyke;	Outdoor Services,	Mattoon;	High Tide,	Richards.		
Sunday Night Talk	Sand or Landscape	Mountain or River	Sea, stars or trees	Rainbow or Wind	Birds or Lakes	Sunset or Flowers	Mists or Sky	Storm or Sun
Rainy Day Nature Craft	Bird Houses Mt. Map	Aquarium Leaf Prints	Fungi Etchings Fungi Print	Color Flower Outlines	Bayberry Candles	Seaweed Mts. Paint Shells	Making curious Animals	Blueprints Leaf Skeletons
Nature Talk	Wood Etiquette	Insect & Flow.	Dev. Fruit	Adaptations	Snakes	Poison Plants	Bird Songs	Forestry
Nature Song	Alouette	Old MacDonald	Violets	Cricket	Pussywillow	Row Boat	Hike	Taps
Rainy Day Nat. Games	Bird Chart Games	Twig Matching	Twig Relay	Tree Silhouette	Bird Authors	Smell	Observation	Seed Dance
Nature Talk	Fire Spirit	How and Why	1000 yr. Pine	Grasshopper	Just So	Stickeen	Stalk Flowers	Animals
Nature Story	Museum	Geology	Fire	Weather	Lakes	Rivers	Soil, Mts.	Heroes
Nature Hike	Map Hike	Photography	Bird Trip	Tree Trip	Insect Census	Blueberries	Hare Hound	Trailing
Vil. Friend	Garden	Hermit	Collection	Rose Garden	Duck Farm	Hollyhocks	Bee Hives	Landscaping
Forage	Greens	Ind. Lemonade	Sassafras Tea	Flag Candy	Mushrooms	Potato Subs.	Asparagus Subs.	Herbs
Games	Birds Nest	Tree Trailing	Tree Cribbage	Compass	Spot Spy	Exploring	Tree Spy	Trees
Hike for Nature Craft	Clay	Bayberries	Sweet-Grass	Willow Baskets	Beads	Tinder	Relics	Decorations
To Get Somewhere	Birds Nest	Story When Resting	Something Apro pro	Attention to Vista	To Collect Minerals	How to Read Blazes	Glacial Evidences	Folk Tales
Sunny Day In Camp	Birds Nest Tag Day	Forest Good Turn	Forest Census	Blind as Bat	One Old Cat	Camouflage	Aggressive Resemblance	Sand or Snow Puzzle
Games	Plant Geog.	All Round	Trailing	Pitch Pine Tag	Tree Tag Day	Snipe Hunt		
Night Stunt	Stars Sounds	Scouting	Sing Lang.	Tree Panto-mime	Dark Night Anatomy		Feeling Game	Silhouettes

Nature-Lore should be on the schedule. There should be opportunity for trips, games, songs, stories, and Sunday Nights. It is not fair to expect it to compete with horseback riding and baseball. It should contribute to the arts and crafts and pantomimes. It should be prominent on Rainy Days. It should be voluntary. It is not necessary to have the same activity twice in a season. There should be a nature library. There should be Audubon Charts for Rainy Day Games. There should be Government Topographic Maps for outdoor trailing and games. There should be material for making leaf prints on Rainy Days. There should be bird outlines and colored pencils for coloring them in on a rainy day.

PART II
NATURE-STUDY FOR SCHOOLS

NATURE-STUDY AND TEACHER

We ought to study Nature just from books, is what I say;
It does not do for Teacher dear in any other way.
Because when once I found a spider, brown and very fat,
And brought him carefully to her in my best sailor-hat,
My teacher cried aloud in fright, and squealed, and took on so,
I had to hurry to the door, and let my spider go.

One time I found the finest kind of long, soft, fresh green worm;
But, my! you ought to see the way it made my teacher squirm!
Then on her desk I put a snail, a harmless little thing
That would not hurt a bit, because it could not bite nor sting;
But when it came half-way from out its shell, and tried to crawl,
The noise my teacher made they say they heard across the hall.

Another time a baby mouse I brought her in a box;
She gave a look, and then a scream that folks could hear for blocks.
I thought she'd like to see a snake, and brought one in a pail;
But Teacher yelled a lot, and would not even touch its tail!
So Nature-study in a book is all that she can stand,
For when it comes to *samples*, Teacher hasn't any sand!

—By Blanche Elizabeth Wade, in "St. Nicholas."

CHAPTER XIV

TEACHER TRAINING IN NATURE-STUDY

I. *The Present Situation*

Nature education has too frequently failed and has been nearly exterminated by being presented under wrong concepts. It is now showing healthy signs of budding out in the reorganization of Normal Schools into Teachers' Colleges. The Recreation Congress has also given a decided impulse to the work of training teachers in nature-study. It is inevitable that it will vary greatly in its new organization and also in the specific work offered. While standardization of teacher training should particularly be avoided in a field of study that is built upon the immediate environment, there are certain salient points which would increase the value of the course.

The analysis of the situation consisted in making a study of the catalogs of Teachers Colleges offering work in the natural sciences. Requests for the catalogs were sent to the members of the American Association of Teachers Colleges, which consists of 143 members.

There were found to be 275 teachers in the natural sciences for the 117 Teachers' Colleges responding to the request for a catalog, or an average of 2.55 teachers per school.

The following conclusions are drawn from a study of the titles of Professors.

1. The large majority of professors in Teachers' Colleges have titles that pertain to technical subjects which are taught in secondary schools rather than to the elementary subjects of the grades and junior high schools. This is a custom which is probably taken over from the academic college. If nature-study, or school gardening, or elementary science is taught at all, it is usually given by a man trained in technical subjects and not prepared in the subject from a professional or teacher viewpoint.

2. There is not much agreement as to what should be taught nor upon the place of emphasis. There are five classes of professors of natural sciences according to their titles:

   a. *Elementary grade subjects* such as nature-study, elementary science, and school gardening. This group is in the minority and is given the value of 1 for purposes of comparison.

The number of teachers of natural sciences is an indication of the enrollment rather than the number of courses in a given institution.

   b. *General titles* such as professor of science or professor of science and athletics. The professor of all natural sciences is becoming extinct. There is still, however, a goodly number in the smaller teachers' colleges. They average 1.2 as many as teacher of elementary subjects. The science and biology teacher in addition to laboratory work are commonly expected to do extra work in other departments. Five of these professors as indicated by their titles, teach athletics. It may be that they are usually better adapted to such work than the professor cf music or pedagogy.

c. *Special titles* such as professor of beekeeping, professor of veterinary science, or the professor of horticultural landscape gardening. These titles sometimes depend upon conditions peculiar to the institution, the professor often being in the vicinity is taken on as an annex. This is no reflection on the ability of the professors with the special title. The number of these professors is 1.2 more than the professors of elementary subjects.

d. *Economic subjects* such as professor of agriculture and allied subjects. This phase is gaining rapidly and now numbers three times as many professors as do the elementary subjects.

e. *Technical subjects* such as biology, botany or zoology. Although there are still five times as many professors of technical subjects as there are of the elementary grade subjects this phase is giving away to the economic and professional courses. The number of professors teaching technical subjects as compared with the number of professors teaching elementary school subjects is in inverse ratio to the number of pupil teachers preparing to teach these subjects.

A few Teachers' Colleges have made liberal provisions for the Natural Sciences—notably the George Peabody College for Teachers, Nashville, Tennessee, which offers over 3000 hours and the Illinois State Normal University which offers over 2000 hours. The number of hours given in the natural sciences ranges from slightly over 100 hours per year up to 3300 hours—rather of a decided contrast when comparing the extremes. There is no agreement as to the phase to emphasize or to the amount of time to be given. Some omit one part and others another part. This is irrespective of what the environment has to offer. Most Teachers Colleges give but one or two courses in Nature-study. None offer a complete opportunity for specializing in Nature-study. Since Normal Schools are becoming Teachers' Colleges we can rightfully expect teachers of Nature-study.

Agassiz revolutionized teaching from books to the laboratory. We now need to get from the laboratory to the out-of-doors. Spermatophytes can be taught in any region—Maine or Florida, but not so with Alpine Plants. What teachers see of flowering plants is mostly in the class room. The titles of the courses indicate a purely academic treatment where the subject matter is regarded in its systematic development. There is nothing given to bring out a love for the subject. (The academic ear will scoff at this.) There is nothing given to meet the specifications of teaching the dandelion or the violet or the care of geraniums to little children. There is no provision for the personal touch with the objects of the environment. It is not surprising that most college-trained Biologists fail with Nature-study.

Courses in Entomology have a decided economic trend as indicated by their titles. There appears to be a tendency to emphasize the economic treatment by giving original and catchy titles. Even in these courses it is not always a knowledge of insects in the home garden, on the shade tree in the dooryard, or pests of the house. Most teachers are pedagogically reared to believe that the only place to study an insect is under a microscope, in formalin, or from a specimen mounted in a box. They do not know that every boy and girl has a laboratory at home. They do not know that living insects are more interesting than insects pinned to a board. They do not know that a child can have a happy time with insects. They do not know how to provide occasions for children to have interest and delight with insects. They have not been specifically trained for that kind of a job. Our Teachers' Colleges do not provide sufficiently for teaching Nature-study in the Grades and Junior High Schools. They are not meeting the needs of the times.

It hardly seems conceivable that some Teachers' Colleges would give more time to training their pupils in the science of Farm Meats, or the technique of swine than in the skill of cultivating the out-of-doors in the heart of every child. The enjoyment and prosperity of hogs cannot be instilled into the makeup of our teachers at the expense of the enjoyment and prosperity of children. The children who will later be the raisers of hogs will enjoy hog culture better and be more prosperous if they have a genuine interest in the out-of-doors.

II. *Suggestions for Teacher Training in Nature-Study*

To the teacher of Nature-study falls the most difficult job of teaching. He must know the out-of-doors in general and the nature study of the community where he is teaching in particular. He must be a capable teacher in order to adapt his methods—probably to originate new ones—which fit into the particular needs of that community. He must know the working principles of all the sciences —Chemistry, Physics, Geology, Botany, Zoology and Hygiene in order to answer the *whys*. He must have the *ing* ability on the farm. He must have the confidence of the boys and girls and their parents. He must be able to lead them on field trips. He must be able and cheerfully willing to render community services not expected of other teachers. He must work constructively and co-operatively in developing various organizations such as Scouts, Campfire Groups, Natural History Societies, Mountaineering Clubs, Audubon Societies, and Conservation. He must keep abreast of the general educational field. Ability to write for Teachers' Magazines and in public speaking will also contribute to his success.

There are two classes of students in Teachers' Colleges. The first, and in the majority, includes those preparing to teach in general. They cannot afford time to train and master all the methods in the particular subjects. Others are to become special teachers in history, arithmetic, botany, agriculture, and nature-study. The beginning courses in these subjects are required of all. Those chosen for training in Nature-study must be selected from the many available who show a desire and adaptability. In fact the admission to specialization should be restricted to those of such nature experience, personality, and interest as give promise of success in the teaching of Nature-study. It would be better if the prospective teacher be born in the country where he acquired a nature consciousness in his early days. He then has a background of experiences from which he may draw upon at will. He recognizes an old wood road, the nearly overgrown corduroy bridge, the fallen grist mill; he knows how to cross the meadows dry shod, jumping from hummock to hummock; he can go bushwhacking without tearing his clothes or getting lost; and a thousand other country experiences. He has a country vocabulary and understanding heart. He is in sympathy with the simple life and the country way. He enjoys the countrified and goes to the country for every vacation. He is successful with chickens and gardening, has had orchard experience, knows how to milk a cow, take off the cream, and churn butter. He enjoys camping and the trail. To be nature-minded is more important than to be nature-wise.

Furthermore, the Nature-study teacher is not a walking encyclopedia yet he should possess information much in advance of those whom he instructs. It is not practical nor advisable to have a technical preparation for the basis of teacher preparation. The subject and method course must go together. If I were to sacrifice either of the three,—nature-minded, nature-wise, or nature-pedagogy—I would sacrifice the knowledge. If one is interested and knows how to teach, the information will come with the teaching, and richness of experience.

Nature-study teachers should be trained in a separate department. It should be separate from the technical work of Biology, the economic work of Agriculture, and the pedagogical work of the education department yet with the privileges of co-operation with all of these departments. There should be close association with the science departments wherein materials and equipment for presenting subject matter are essential and available on the one hand, and with the training schools for the training of teachers on the other hand. There should be no duplication of work within the institution. The nature-teacher training in the professional phases of his job would undoubtedly render service in the improvement of methods in the subject departments.

The curricula for teacher training in Nature-study must differ according to the differences in the organization and teaching force of the Teachers' College offering the training. It is evident, however, that there is a need of a common viewpoint. All students should pursue the same studies in their Freshman year. In the first year it has been my experience that it is necessary to spend more time in acquiring facts. They are merely enabled to get a general view of Nature-study in the large, and methods of study in general. Incidentally, they are enabled to decide with more certainty whether they shall enter that special field. Each institution should prepare its nature teachers according to the definite needs of the schools wherein they are to serve.

We can also agree that there must be active participation in the lines of activity in which the teacher is to act. Not enough attention is now given to practice teaching in Nature-study, to leadership in the field, and to the functions of a nature teacher in the community. Experience as a Nature Counsellor in a summer camp brings the pupil to an appreciation level and a doing ability. The nature teacher in training should organize and conduct a complete project in the practice school under definite critics. He should take over a phase of an enterprise in a museum and conduct it on a museum basis. He should organize and direct home gardens. He should have the experience of being a scout leader. He should be versed in outdoor cooking, camp fires, nature songs, and nature stories. We may think of these requirements as professional activities of a nature study teacher or supervisor.

This professional training follows the introductory—perhaps we should say orientation—courses in Nature-study. The Freshman year and possibly the first semester Sophomore are required. The professional courses are then offered as electives for each of the remaining terms rather than many nature courses in the Senior year. This gives the pupil opportunity over a period of time to become nature minded, and to make a more permanent decision. By the Senior year he should be nearer competent and therefore not as wasteful of time in practicing on boys and girls. The student teacher keeps a diary of the various activities, holds conferences with the Professor of Nature-study and the various critics, and visits others in the same line of activity whenever possible. He becomes an efficient teacher of nature, has a background of sympathetic understanding of nature obtained through a love and contact with the out-doors, is a student of nature problems, and is fairly well grounded in the other sciences. He is ready to face the nature activities of the Junior High School and the Grades.

The nature room should also give an atmosphere. It should be a combination recitation and work room, have a department library, bulletins in pamphlet cases, illustrative material, a stereopticon lantern, work table, window boxes, aquaria, dark room, and a connecting office for conferences.

There is evidence that educators generally are awakening to the necessity of an increased emphasis on teacher training in Nature-study. Many Universities are introducing Nature-study Departments. It is a live topic amongst leaders. These introductory remarks are written to give a survey of the situation and to serve as a guide in mapping out a Nature-study program for the training of Nature-study teachers and supervisors.

In matters of great weight go to school to the animals.—*Democritus.*

Mother of marvels, mysterious and tender Nature,
Why do we not live more in thee?—*Amiel.*

The poetry of earth is never dead.
                              —*Keats.*

Come forth into the light of things,
Let Nature be your teacher.
                              —*Wordsworth.*

CHAPTER XV

THE NATURE-STUDY MOVEMENT IN AMERICA

Possibly no other subject has so frequently shifted its point of view. With each style of presentation it has had its palmy days followed by a decline. It is quite essential that a prospective Nature-study teacher know the pitfalls and, what is more important, know the background on which he must build and the goal toward which he must aim. Notwithstanding the many fads and periodic "bad repute" it is significant that nature-study still lives. Nature is slow to change. On the other hand, educational methods are constantly changing. This has made it difficult for the majority of teachers to keep up with the pace but through all this chaos there have been patient, quiet, sympathetic teachers carrying on Nature-study in a simple, interesting, and absorbing way.

If you seek the first nature students in America, you will inevitably go back to the settlers. The pioneer was a born naturalist. He located his home in the wilderness. He selected his pine and hewed it straight with a broad ax for the log cabin. He then sought out clay to chink the logs, and to make bricks for the chimney. The fire was started by flint and tinder. Aromatic herbs were gathered from the fields and hung in the garret as a source of medical supplies. Wild fruits formed an important part of his diet. Meats were preserved by salting in barrels or smoked for three weeks by smouldering hard wood chips. He acquired the knowledge of an anatomist at hog killing time in late fall, carefully dressing the animal into hams, shoulders, and ribs. He became an animal psychologist and acquired a reputation for judging horses far above anything described in David Harum. The sheep were shorn and homespun clothes were made in the winter. His nature-study was seasonal, on his immediate premises, and met his particular needs. Such aims might well guide the nature-teacher of today.

Then followed the urbanizing of America. At the end of the first quarter of the nineteenth century 7% of the population lived in the city and 93% in the country. The country boys and girls were being trained in colonial nature-study. They were not only hard at work in it but in their spare time played it, by going nutting, berrying, fishing, or hunting. At the end of the century 40% of the people were in the city with their children entirely divorced from the training that had been obtained from the school of nature. The sewing machine, kerosene lamp, stoves, matches and the great work of steam superseded the Colonial methods. The gathering of peoples into large centers was a decided check in the understanding of nature.

Natural history lessons, to teach morals and religion, were introduced into the common schools about 1830. This was the period of the three R's, which were supposedly the tools to all learning. The following is taken from Lovell's, "Young People's Second Book"

(1836) as an example of a nature story written to teach the wonders of the creation.

The Hen. Of all the feathered animals, there is none more useful than the common hen. Her eggs supply us with food during her life, and her flesh affords us delicate meat after her death.

What a motherly care does she take of her young! How closely and tenderly does she watch over them and cover them with her wings; and how bravely does she defend them from every enemy, from which she would fly away in terror, if she had not them to protect. While this sight reminds you of the wisdom and goodness of the Creator, let it also remind you of the care which your mother took of you during your helpless years, and of the gratitude and duty which you owe to her for all her kindness.

These stories soon became fanciful and untruthful. Many of them concerned foreign objects rather than local material. Darwin's Origin of Species (1859) upset the special creation motive of these stories. This fact plus the loss of contact with nature brought the period of sacred natural history to a close.

I. The Contribution To Nature-Study From Education

The roots of American nature-study go back to the great educational movements of Europe. Rousseau blazed the way for Pestalozzi who, in turn, became a teacher of Froebel. The torch was brought to America by their pupils. The most successful teachers in Nature-study have been decidedly influenced by these educational reformers and the story of the continuity in the development of Nature-study is the rich heritage of the student of today.

One of the first to preach the doctrine of modern nature-study was Comenius (1592–1670). He said that "as far as possible men are to be taught to become wise, not by books, but by the heavens, the earth, oaks and beeches, that is, they must learn to know and examine things themselves and not the testimony and observation of others about the things."

Rousseau (1712–1778) carried out the idea of Comenius by educating the boy Emile "according to nature." Rousseau's philosophy was to return to simplicity, reality, and personal experience rather than to be led by authority. His ideas were revolutionary and often extravagant. Many statements were made in "Emile" (published 1762) that are useful to us today.[1]

"I would have him (refering to tutor) a child, so that he might become a companion to his pupil and secure his confidence by taking part in his amusements." p. 19.

A summer camp Director in selecting someone as a nature counsellor always wants a leader who can play and enjoy nature with the camper. A teacher who is able to say, let *us* see if we can find out together becomes a companion. Nature-study, more than any other subject, offers this opportunity for companionship.

"Children brought up . . . where spiders are not tolerated, are afraid of spiders. . . . I have never seen peasants who were afraid of spiders. . . . If during his infancy,

---

[1]These quotations are taken from William H. Payne's (1893) translation and the pages are given as a matter of convenience should the student wish to follow up the selection.

he has seen toads, snakes, and crabs, without being frightened, he will see without horror, when grown, any animal whatever." p. 27. Modern psychologists have proven that the fear of animals is acquired and not inborn. The fear of snakes, for example, is due to the actions of adults before children.

"We no longer know how to be simple in anything . . . toys of all kinds and prices—what useless and pernicious furniture! Nothing of all this. . . . Little branches with their fruits and flowers, a poppy head in which the seeds are heard to rattle . . . will amuse him . . . and will not have this disadvantage of accustoming him to luxury from the day of his birth." p. 35.

"I say that a child does not understand the fables that he is made to learn . . . the instruction which we wish to draw from them necessarily brings into them ideas which he cannot comprehend, and the poetical form, while making them easier to retain, itself makes them more difficult for him to understand; so that entertainment is purchased at the expense of clearness. . . . In the fable of the crow and the fox, children despise the crow, but they all form a liking for the fox."

"You wish to teach this child geography, and you go in search of globes, spheres, and maps. What machines! Why all these representations? Why not begin by showing him the object itself, so that he may know, at least, what you are talking about! . . . The child who reads does not think—he merely reads; he is not receiving instruction, but is learning words. . . . He is not to learn science, but to discover it." p. 137.

"In your search for the laws of Nature, always begin with the most common and the most obvious phenomena, and accustom your pupil not to take these phenomena for reason but for facts." p. 153.

"Emile will never have dissected insects, will never have counted the spots on the sun, and will not know what a microscope or a telescope is. Your wise pupils will ridicule his ignorance, and they will not be wrong, for, before using these instruments, I intend that he shall invent them." p. 188. . . . "Once more, my purpose is not at all to give him knowledge, but to teach him how to acquire it when necessary, to make him love truth above everything else." p. 188. Nature-study is not dissecting. The microscope is occasional. Nature-study is not to acquire knowledge.

Rousseau has a clear idea as to the aim of a field trip. "Whatever is done through reason ought to have its rules: Travels, considered as a part of education, ought to have theirs. To travel for the sake of traveling, is to be a wanderer, a vagabond; to travel for the sake of instruction, is still too vague an object, for instruction which has no determined end amounts to nothing. I would give to the young man an obvious interest in being instructed; and this interest, if well chosen, will go to determine the nature of the instruction." p. 308.

"Men were not made to be massed together in herds, but to be scattered over the earth which they are to cultivate. The more they herd together the more they corrupt one another . . . cities are the graves of the human species. After a few generations, races perish or degenerate, they must be renewed, and this regeneration is always supplied by the country. Send your children away, therefore, so that they may renew themselves, so to speak, and regain, amid the fields, the vigor they have lost in the unwholesome air of places too thickly peopled." p. 24.

Pestalozzi (1746–1826) tried to educate his own child according to the plans for "Emile" but failed. In 1774 he started a school on his farm to train 50 poor children in the three R's and in gardening. This school failed financially at the end of two years. Pestalozzi believed that all could be educated to an intellectually free and morally independent life. He substituted discussion for reciting, group instruction for individual hearing, and thinking in place of catechism. He maintained that "Observation is the absolute basis of all knowledge. The first object then, in education, must be to lead a child to observe with accuracy; the second, to express with correctness the results of his observation." This is why Pestalozzi is called the "Father of Object Teaching."

Froebel (1782–1852) from 1808–1810 was a student and teacher under Pestalozzi and opened a private school at Yverdon, in 1816, along Pestalozzian lines. This also was unsuccessful financially. Froebel so directed self-activity as to develop inborn moral, social, and intellectual capacities. Nature-study and school gardening, therefore, became prominent. He considered the child as a social animal rather than independent as Rousseau had trained Emile. This same play and game spirit has been best developed in the summer camp to its real moral, social, and educational value. The recent camp development of arts and crafts out of the environment—clay for pottery, willow basketry, bayberry candle dipping, and cat-nine-tail mats is a continuation of the Froebel idea. This is not teaching a trade, as Rousseau would have it, but a development of the creative power.

Froebel's advice to parents is to "Take your little children by the hand; go with them into nature as into the house of God, allow the wee one to stroke the good cow's forehead, and to run about among the fowl, and play at the edge of the wood. Make companions for your boys and girls of the trees and the banks and the pasture land." Froebel also urged that "if the boy cannot have the care of a little garden of his own, he should at least have a few plants in boxes or pots." He would encourage the child to have pets and to observe wild life. Froebel's idea was to develop the spirit rather than the "faculties." The spirit of nature-study is the key to successful nature-study. Much of this spirit has been inherited through Froebel's kindergarten. The first English speaking kindergarten was organized in Boston (1860) and the first public school kindergarten under Superintendent W. T. Harris in St. Louis (1873).

II. The Contribution to Nature-Study from Science

Another great influence on nature-study came down through the halls of science. It dates back to Aristotle (384–322 B. C.), the greatest pupil of Plato. Aristotle is the founder of Natural History. His use of the inductive method strikes the key note of the teaching lesson in nature-study of today. He believed in the use of the senses followed by reasoning—the observation of facts followed by explanations. His reasoning led to definitions and principles. His knowledge led to a classification of plants and animals. His classification evolved like that of the child naturalist. It grew gradually out of his personal observation and thinking. Classification may be the result, but is never the aim, of Nature-study.

The first half of the 19th Century was one of species hunting. Linnaeus (1707–1778) published his Systema Natura in 1735. Twelve editions followed. His classification was artificial. If there is such a thing as the youth recapitulating the history of science this is the age of collecting specimens. Provision should be made for the child who wishes to collect but all children should not be made to do this as was the custom in some schools a few years ago when the sole work of botany was to collect and mount so many plants or in zoology to collect and pin onto a board so many insects.

The next period was that of morphology and anatomy. Cuvier (1769–1832) is known as the founder of Comparative Anatomy. The improvement of the microscope led to the study of minute anatomy and Schultze (1825–1874), who established that the protoplasm of the animal and the vegetable cell is the same material (1861), became the "Father of Modern Biology." The laboratory has developed what Dr. Charles Adams so aptly calls the "closet naturalist." Huxley, the pioneer in laboratory teaching, attempted some primers of science, but he had the principles of biology in mind and the child was not ready. A large number of nature teachers were recruited from the laboratory and had had no experience or interest with living plants and animals. Although the laboratory period was most important in the advance of Biology it was undoubtedly a check in the teaching of Nature-study in the public school.

The first scientist to take his pupils into the field was Louis Agassiz (1807–1873). His motto, "Study nature, not books," has been quoted in all attempts at Nature-study. Huxley says,—"Agassiz is a backwoodsman in Natural History. He clears the forest, cutting down all errors, theories, without regard to persons or established reputation. What a pioneer!" Agassiz, like Froebel and his countryman Pestalozzi started a school for children but failed financially. Agassiz had great influence on the masses and made nature more popular wherever he went. The Saturday Club had the great American nature writers—Lowell, Holmes, and Emerson. They all loved the out-of-doors and camped in the Adirondacks, the beginning of a very popular movement.

Agassiz's greatest contribution to American Nature-study was made by his summer school for teachers on the Island of Penikese, in Buzzard's Bay. David Starr Jordan, in his Science Sketches, says it is the "School of all schools in America which has had the greatest influence on American scientific teaching." In May, 1872, John Anderson, a rich New York merchant, gave $50,000 endowment and the Island of Penikese which had a house and barn. Agassiz announced the opening of the school for July 8 and when the boat was coming into the landing the carpenters were driving the last nails. The school lasted but three months but included most of the leading science teachers of the country. They studied sea life in its environment. The death of Agassiz brought the career of the school to a close, but its methods were deemed to be far reaching.

The next reaction against the "closet naturalist" was brought on by the "ecologist." Although Ecology is an old subject and received some attention by Linnaeus, Humboldt, and others it did not assert itself until about 1884. This was about the time that Nature-study was coming into its own. The first course in animal ecology was given in the University of Chicago, in 1902, by Dr. Charles C. Adams.[2] The Ecological Society of America was organized in 1914. In both subject matter and method, Ecology is nothing more than advanced Nature-study. Ecology has supplemented the nature-study field with rich material and the sympathy of its leaders has meant much to the movement.

III. Geography and Nature-Study

Geography minus political geography is good nature-study.

The natural history of the Colonial period was the kind found in the common school fact-geography. The plants and animals of Asia, Africa, and South America were pictured and described. These same objects became the source of education in building blocks and alphabet books. The scholars were made to memorize the lists because of the supposed disciplinary value. That Asa Gray was ahead of his time is seen in his protest: "I do not suppose that the mere treasuring up of facts will affect the object of education..... I venture the assertion that, if the truth were known, the child acquires a greater number of useful ideas, more real development and strength of mind, during his play hours with his rabbits, his kites, from his story books, than from the lessons assigned him during his hours of study; he is really educated more out of school than in school."

Home Geography arose in connection with the object teaching of Pestalozzi. Carl Ritter, a German, developed this idea beginning about 1817. Arnold Guyot, a pupil of Ritter, was agent of the Massachusetts State Board of Education from 1848–1854. Guyot, like Pestalozzi and Agassiz, was a Swiss. He came to the United States, as did Agassiz, and lectured to thousands of teachers. Home Geography is good Nature-study. The hill, the valley, the brook, and the pond are the homes of our plants and animals. Several home geography lessons are included in this book and when these subjects are not taken care of in the geography department they should receive attention by the Nature teacher.

IV. The Centers of Development of Nature-Study in the United States

1. *The Early Development of Nature-study in Illinois.*

The Natural Science Section of the Illinois State Teachers' Association was organized in December, 1888 and held its first meeting December 27, 1889. Professor S. A. Forbes was elected president. At the first session Professor Forbes read a paper on the History and Status of Public School Science Work in Illinois. The following notes, on the early development of Nature-study in Illinois, are taken from this important paper.

[2]The New Natural History—Ecology. The American Museum Journal, Vol. XVII, No. 7, pp. 491–494. 1917.

"Beginning in 1851, we find a superintendent of Stark County, (Illinois) saying hopelessly that, desirable as it is, he sees little prospect of a study of science in his schools, and, indeed, that some of his people still object to geography, even as contrary to the Bible, because it teaches that the world is round instead of having four corners."

Professor Forbes goes on to say,—"That reaching upward of the masses for more power and more light, which, spreading from Illinois eastward, gave us later the long line of land-grant colleges, and gives us now the State Experiment Stations, gave us also, as a sort of second growth from the seed first sown, the recognized acceptance of the natural sciences as a necessary part of the course of study in a true people's school. *That this fruitful movement arose earlier and went further here than elsewhere,* I attribute to the fact that it had here an able and devoted leader, who, himself an educated man, had those great human qualities which no learning can overlay, and which gave him access to all classes and power with all." He refers to Professor Jonathan B. Turner.

Professor Turner, in 1851, called a convention of farmers in Putnam County to consider education toward the farm. This not only led to the national land grant act but started a movement toward accepting "natural sciences as a necessary part of the course of study in a true people's school."

Three powerful allies furthered the movement. Ninian W.Edwards the first State Superintendent, in his first report (1854) stated that the teachers should have a "practical education, in which should be included not only what is commonly embraced in the common school course, but a practical knowledge of the sciences in their application to the ordinary pursuits of life."

The first State Normal (1857) by 1860 attempted, in the words of Principal Hovey, "to put our pupil teachers in possession of the leading facts of these sciences, (Physiology, Chemistry, Botany, and Geology) and the method of teaching the facts to children. . . ."

As early as 1868 the State Natural History Society assumed "the duty of supplying natural history materials to schools prepared to use them."

"In the Aurora (Illinois) Schools, Principal Jones has introduced in 1868 an elaborate course in natural science, beginning with the first year of the primary, and running through the high school."

These various movements culminated in 1872 in the introduction of four new sciences on the list required for a county teacher's certificate. This gave a great impulse to the work but by 1874 these requirements were limited to the holders of first grade certificates. As the country teachers usually held second grade certificates the study of nature underwent a decline.

It remained for Colonel Francis W. Parker (1837–1902) to revive Nature-study in the Cook County Normal School. Colonel Parker was of New England stock being born in Piscataquog, New Hampshire. His father died when he was six or seven years of age and young Parker went to work on a farm. He always said that altho he went to school in the winter that his real education was on the farm. Colonel Parker started to teach at the age of sixteen.

The Cook County Normal School, January 1st, 1896, became the Chicago Normal School. In June, 1899, Colonel Parker went to the Chicago Institute of Education which was later merged with the University of Chicago. Colonel Parker wrote an account of the work of the Cook County and Chicago Normal School in 1902.[3] His statements are forceful and clear cut.

"It was our good fortune to take the initial steps in subjects that have since become of general application. The great book of nature, God's infinite volume of everlasting, inexhaustible truth, had had scarcely a place in the courses of study in American Schools... The question was: How may nature be adapted to growing minds, to hearts that have loved and lived in nature until, indeed, they entered school?.... Things must be learned thoroughly; and a natural object was taken, examined, dissected, painted, drawn—exhausted, and the interest of the children exhausted at the same time. Another way must be found. Professor H. H. Straight, a pupil of the great scientist and educator, Agassiz, entered upon the work, in 1883, with boundless enthusiasm.... It was found that mere laboratory work was not close enough to the children; nature refuses to be viewed in bits and rags.... Field excursions, with their wealth of observation, were early introduced. Failures in nature-study, failures that were, however, prophecies, were the rule, until Wilbur S. Jackman, in 1889, undertook to grapple with the problem. The idea of thorough exhaustive work was abandoned. The phenomena of the "rolling year" were taken as the general guides; the child was brought into loving contact with nature; the subjects were adapted to different stages of child growth; art and nature were correlated. We have taken a step, and only a step, in the inexhaustible book."

Wilbur S. Jackman (1855–1907) was also a farmer boy and teacher. He graduated from Harvard in 1884 and taught natural science in the high school at Pittsburgh. Colonel Parker took him from here to the Cook County Normal School and then to the Chicago Institute of Education. Jackman became Dean of the School of Education during the last two years of his life. In 1891 Jackman published his book Nature-study. The book brought the child into direct contact with nature. It contained questions, but not answers, and the teachers found difficulty in using the book as they had not been trained in observation. Nathaniel Butler said at a memorial service at the University of Chicago. "To him more than to any one else is due the position of nature-study in the Elementary Schools.[4] Jackman said

[3]Elementary School Teacher, Vol. II, No. 10, pp. 765–766, June 1902. An Account of the Work of the Cook County and Chicago Normal School from 1883 to 1899, by Francis W. Parker.
[4]Elementary School Teacher, April 1907, Vol. VII, No. 8, p. 439.

that "The spirit of nature-study requires that the pupils be intelligently directed in the study of their immediate environment and its relation to themselves; that there shall be, under the natural stimulus of the desire to know, a constant effort at a rational interpretation of the common things observed." Ira B. Meyers followed Jackman

Courtesy Oswego Normal School

DR. E. A. SHELDON

and Elliot R. Downing has had charge of Natural Science at the School of Education since 1911.

2. Nature-Study in New York State had Two Centers of Development

A. *The Oswego Normal School.* Cubberley states that "in 1848 object teaching was introduced into the state normal school at Westfield."[5]

---

[5]Cubberley, Public Education in United States, p. 295.

It remained for the Oswego Normal School to become the influential center in the distribution of the Pestalozzian principles and practice. Dr. Edward A. Sheldon (1832–1897) was superintendent of schools in Oswego. He first began with Saturday classes which led to a city training school (1861) and later to the state normal school (1866). Sheldon introduced object teaching in 1859. In December 1861 Dr. Sheldon invited prominent educators to observe the Oswego work. Professor William F. Phelps, principal of the State Normal School at Trenton, New Jersey was appointed chairman to prepare a report. This report helped in the spread of the Oswego idea and in conclusion—"*Resolved*, That this system of primary instruction, which substitutes in great measure the *teachers for the book*, demands in its instructors varied knowledge and thorough culture, and that attempts to introduce it by those who do not clearly comprehend its principles, and who have not been trained in its methods, can only result in failure."[6] "Notwithstanding the diffusion of the principles of object teaching in the country during that period, its practice died out through the want of teachers trained in the system and its methods."[7] It was the lack of teachers that caused Physiography to be dropped from the curriculum and has led to a decline in the teaching of General Science. Nature-study has managed to survive altho it has been seriously handicapped by the scarcity of trained teachers.

Object teaching was the emancipation from the words of the textbook. It was a training in observing, reasoning, and expression. Professor S. S. Green, of Providence, R. I., in his report for a committee, appointed by the National Teachers' Association in 1864, asks, "Would you really know whether a candidate for the teacher's office is a good teacher or not? You need not examine him with difficult questions in Arithmetic, in Algebra, in Geography, or in History. You need not examine him at all. But put him into the school room, take from it every printed page for the use of the teacher or pupil. Give him blackboards,—give them slates. Let him have ears of corn, pine cones, shells, and as many other objects as he chooses to collect, and then require him to give lessons in reading, spelling, arithmetic, geography, and the English language. If the children come home full of curious questions,—if they love to talk of what they do at school,—deeply interested,—intent upon their school exercises,—then employ him,—employ him at any price, though he may not have graduated at the University, the Academy, or even the Normal School. Whenever needed, allow him or the children books. You are sure of a good school."[8] Professor Greene's test would apply to good teachers today and but a small per cent

---

[6]Barnard's American Journal of Education, Vol. XII (1862), p. 605.
[7]History of Object Teaching, an Address delivered by N. A. Calkins in 1861. Published in Barnard's American Journal of Education, Vol. XII (1862), p. 639.
[8]Object Teaching. Its General Principles, and the Oswego System. Professor S. S. Greene, in Barnard's American Journal of Education, Vol. XVI (1866), p. 258.

would be able to qualify. The majority of teachers are still carrying on recitations.

The method of the object lesson was to present the material to the class. The pupils felt, weighed, measured, tasted, and observed the color, composition, solubility, tenacity, transparency, ductility, brilliancy, and other qualities. They used all the senses. The various qualities were brought out by conversation. New words were introduced as they were needed. The object was then described in

L. H. BAILEY

writing. Natural history objects were occasionally brought in for a lesson but it was more usually a laboratory type of work on lifeless objects.

In 1878, H. H. Straight (1846–1886) came to Oswego Normal to teach natural science. He had studied at Harvard and had been at Agassiz's summer school on Penikese. Professor Straight started Nature-study in New York State in Oswego Normal School (1878). He changed object teaching to a study of living plants and animals. In place of the laboratory-schoolroom lesson he took his classes on field trips. Colonel Francis Parker took Professor Straight to the Cook County Normal School in 1883.

B. Cornell University. The following notes on the development of Nature-study at Cornell University are taken from the preface of Professor Anna Botsford Comstock's Handbook of Nature-study. "It was inaugurated as a direct aid to better methods of agriculture in New York State. During the years of agricultural depression 1891–1893, the Charities of New York City found it necessary to help many people who had come from the rural districts—a condition

ANNA BOTSFORD COMSTOCK

hitherto unknown.... A conference was called to consider the situation.... Mr. George T. Powell, who had been a most efficient Director of Farmers' Institutes of New York State was invited to the conference as an expert.... He made a strong plea for interesting the children of the country in farming as a remedial measure, and maintained that the first step toward agriculture was nature-study.... In 1894 ... eight thousand dollars was added to the Cornell University fund, for Extension Teaching and inaugurating this work. The work was begun under Professor I. P. Roberts; after one year Professor Roberts placed it under the supervision of Professor L. H. Bailey, who for the fifteen years since has been the inspiring leader of the movement, as well as the official head.

In 1896, Mr. John W. Spencer, a fruit grower in Chautauqua County, became identified with the enterprise; he had lived in rural communities and he knew their needs. He originated the great plan of organizing the children in the schools of the State into Junior Naturalists Clubs.... Some years, 30,000 children were thus brought into direct communication.... A monthly leaflet for Junior Naturalists followed; and it was to help in this enterprise that Miss Alice G. McCloskey, the able Editor of the present Rural School Leaflet, was brought into the work."

The "Handbook of Nature-study" by Professor Anna Botsford Comstock, 1914 is the best textbook that has ever been written upon

Anna Botsford Comstock,
Dean of American Nature-study.

Nature-study. It is based upon the Cornell Leaflets which were published from 1903–1911. The "Teacher's story" is to help the untrained teacher and this is followed by the subject lesson for the children. Its informal but interesting style has made it equally useful in the home. The popular demand for the book has made it necessary to get out a new edition each year. Mrs. Comstock is today the "Dean of American Nature-study." Upon becoming Professor-emeritus her place was taken by Dr. E. Laurence Palmer who has been getting out the very efficient and popular Rural School Leaflets.

3. *The Rise of Nature-study in St. Louis.*

William T. Harris (1835–1908), superintendent of schools in St. Louis (1868–1880) and later United States Commissioner of Education (1889–1906) wrote a syllabus on How to Teach Natural Science in the Public Schools (1871). In his introduction he says "The course is arranged with reference to method rather than quantity and exhaustiveness." He had a spiral arrangement of topics—plant life in grades 1st and 4th, animal life in 2d and 5th, and physical science in the 3d and 6th. His work had a widespread influence in the spread of Nature-study.

4. Bridgewater Normal School, the Early Center of Nature-study in New England.

Albert G. Boyden (1827–1915) was principal of the Bridgewater Normal School from 1860–1905. He was born on a farm in Walpole, Massachusetts and like other leaders in the movement was a teacher. "He was a leader in the sports of his fellows, and knew the products of all the fields, woods and streams in the neighborhood of his native village." (p. 18)[9] Principal Boyden in his report of 1885 said that

A. C. BOYDEN

"It is of great importance that the teachers of our country schools should have the familiar acquaintance with nature which will enable them to be guides and interpreters to the opening minds of the children." The writer use to sit in the Assembly Hall during what was called general exercises, and marvel at Mr. Boyden's powers of observation. His questions were quite apt to be on nature. What is the first tree on the right as you go down Summer Street or what color is the foliage of the ash this morning? In 1907, Mr. Boyden gave nearly two acres of land to the school for a natural science garden.

[9]Albert G. Boyden and the Bridgewater Normal School. By A. C. Boyden, (1919).

Mr. A. C. Boyden, who succeeded his father as principal, inherited his father's interest in the outdoors. One educational periodical in telling about the World's Fair in St. Louis (1904) said "while it is true that the West is more strongly represented then the East, the exhibit from the State Normal Model School at Bridgewater is pre-eminently the fullest and richest and the most carefully prepared nature-work in the whole educational exhibit. It is arranged under the personal supervision of Mr. Arthur C. Boyden, who for twenty-five years has been an enthusiast on Nature work and is today the leading exponent of the subject in Massachusetts."

V. The Contribution of the Museum to Nature-study

A REVIEW OF THE WORK DONE BY THE BUFFALO SOCIETY OF NATURAL SCIENCES IN CO-OPERATION WITH THE PUBLIC SCHOOLS.[10]

"From the day of its inception, one fundamental principle has controlled the policy of the Buffalo Society of Natural Sciences. Realizing the important place which a great museum must eventually take in the educational system of its home town, we have always endeavored to get into the closest and most effective relationship with the public schools of our city. Every facility which we had to offer to the student has been freely and continuously placed at his command. For years the science teachers of the city have been in the habit of bringing their classes to our building, and we have supplied them with room and materials for their work. In the study of geology, thousands of high school pupils have received great benefit form our collections of rocks and minerals, and our display of native birds and animals has been of the greatest help to the classes in zoology and natural history.

At the beginning, it was the custom for the teachers to accompany the classes and take charge of their work while here, the museum offering simply its collections and rooms, no attempt being made to provide lectures or instruction beyond what was displayed. In time, however, it was found that certain topics were of such universal interest that they would warrant special attention, and so the plan of special lectures for the schools came into being. A series of talks on "Bees," "Birds" and "Insects" was arranged for Saturday afternoons, and were open to such of the grades as cared to come, and met with much success. The attendance at these talks was entirely optional with the classes, the Department of Education simply recommending that as many schools as possible take advantage of them.

As a part of the work, it was expected that the classes which came to the talks would take the opportunity of visiting all the rooms of the museum, and in this way become acquainted with the fact that there were here displayed for their benefit many interesting and valuable specimens illustrating topics which would be an important part of their later studies along scientific lines. As a further inducement to

---

[10]Published in Bulletin No. IV, Vol. VIII of the Buffalo Society of Natural Sciences, Buffalo, 1906.

the schools to come to our museum, the large lecture room was equipped with a suitable apparatus for the projection of slides, and this was announced as being available for any school or class that cared to come and bring slides to illustrate a talk by the principal or teacher. For some months this room was in nearly constant use by the schools, the teachers coming with the classes and making whatever explanation was necessary as the pictures were thrown on the screen by the operator. In one series given to the third and fourth grades, on geography, over 14,000 children attended. Shortly, however, after this plan had been inaugurated, the Department of Education considered it advisable to equip nearly all the schools with

Courtesy Mrs. Enos A. Mills

THE FATHER OF NATURE GUIDING, ENOS A. MILLS, LONG'S PEAK, COLORADO. Nature Guiding, to quote Mr. Mills, "Creates more permanent interest in the biography of a single tree than in the naming of many trees."

lanterns, and therefore at this time only a few schools are under the necessity of coming to our rooms when they wish to display slides to their classes.

In the spring of 1905, through the efforts of Hon. T. Guilford Smith, the President of the Society, a plan was arranged with the Department of Education by which our Museum became an important factor in the work as required from the pupils of the grammar grades. A suitable collection to illustrate the weapons and utensils of the colonial days, used by Indians and whites, was arranged, and notice was sent to all classes studying American History that these things were on exhibit for the schools, and could be seen by applying at the Museum for a suitable hour to be assigned. It was expected that the teachers would bring their classes and explain the utensils and other interesting specimens to the classes from the cases. This plan did not meet with great enthusiasm. The teachers, already having as much special work as they could be reasonably expected to carry, preferred having some one else give the talks, and Dr. Carlos E. Cummings, the Secretary of the Society, was asked to take this matter in charge. This being done, the visits of the schools took the nature of regular lectures, and met with great and immediate success. Although attendance was not required, so many teachers applied for dates that in this course, as can be seen in the appendix, thirty-five separate lectures were given, to an attendance of nearly 5,000 children. These talks were followed by a series on birds and bees to the seventh grade, thirty lectures in all, to an attendance of 6,700."

VI. The Nature-Guide Movement

The happy idea of NATURE GUIDING, both in practice and in the interpretation of nature, was originated by Enos A. Mills, (1870–1922). Mr. Mill's guiding dated from about 1888. For 15 or 18 years he conducted parties up Long's Peak interesting them in the forests, the timber-line, the glacial chasms, and the life stories of the individuals which were met on the trail. His first published article is probably "Guides wanted" in the Saturday Evening Post ————— 1917. "The Children of my Trail School" appeared in the Saturday Evening Post, March 1919, and his book, The Adventures of a Nature Guide (Doubleday Page and Co., 1920), gives concrete information on the subject.

"As an experiment in internationalizing such recreational culture, the California Nature-study League[11] undertook to offer Californians the results of these investigations from Nordic Europe. The work commenced with a series of bulletins, utilizing the California County Library System.

Out of their circulation came several concepts. One was having a high power scientist act as Nature Guide at a string of adjacent summer resorts. The first test was in 1918 at three widely scattered California resort areas. These were made by the State Fish and Game Commission as a part of their conservation work. Having proven satisfactory, the Commission, co-operating with the League, decided on a wider experiment at Lake Tahoe. During 1919, Dr. H. C. Bryant of the University of California acted as Nature Leader. The work caught the attention of Superintendent of National Parks Mather of Washington, D. C."

In 1920 the Federal and State Governments commenced the Nature Guide Service in the Yosemite National Park. It was estimated that 27,000 visitors to the park made use of the service. The work proved tremendously popular and in 1921 it served about 50,000 tourists. In 1921 a similiar service was furnished in Yellowstone Park and in 1922 was installed in Glacier National Park. The movement in Yosemite consists of trips for adults and children, lectures, a

---

[11]Bulletin No. 57A, California Nature-study League.

wild flower show, and a museum where local things are exhibited and hundreds of tourist questions are answered by a naturalist.

By 1925, 10 Naturalists were on the staff in Yosemite, whereas 6 were employed in the past, branch museums were established, a field school of natural history organized, and 113,875 attended lectures and field trips. "Yosemite Nature Notes" is published weekly in summer and monthly in winter.

VII. The Nature-Lore School

In the same year (1920) that the Nature Guide Service began in the Yosemite a Nature-lore School was organized on the eastern coast by Dr. W. G. Vinal under the auspices of the National Asso-

Staff at Nature-Lore School in 1923

Left to Right: ANNA GALLUP, Curator Brooklyn Childrens' Museum; E. H. FORBUSH, State Ornithologist for Commonwealth of Massachusetts; MARIE STILLMAN RUSSELL, Artist; W. G. VINAL, Director of the Nature-Lore School; ANNA BOTSFORD COMSTOCK, Professor of Nature-Study Emeritus, Cornell University; MARY STILLMAN, Writer; SCHUYLER MATTHEWS, Naturalist and author.

ciation of the Directors of Girls Camps. Although the main purpose of this school was to train Nature Counsellors for summer camps there was a good number of registrations by teachers, physical education leaders, art students, and scout leaders. This course is usually given for the last week in June just before the opening of camps.

VIII. Camp Directors' Course

A special course in camping has been offered by Columbia University since 1920. The students spend a week at Bear Mountain in April or May and during this practice week learn a great deal about the principles and methods of Camp Nature-study.

CHAPTER XVI

ORGANIZATION OF A NATURE CLUB

One way of carrying out Nature-study is by means of the Nature Club. The first step in the organization of a Nature Club is to select a practical theme by vote of the prospective members. Write a list of subjects suitable for investigation in the particular locality, such as: insects, weeds, trees, birds, city beautiful. Next choose a name for the club, as—The Lincoln School Bird Club, The Wampanoag Woodsmen, The Nature Guards, The Woodcrafters of Springfield, The Roosevelt Wild Life Protectors, The Burroughs Club, The Blackwolf Tribe, The Agassiz Guides or Junior Audubon. The officers may consist of a club counsellor (usually the teacher interested in the project), president, vice-president, and secretary-treasurer. The meetings should be held regularly at a stated place and time. If an entire class of pupils should form a club and the teacher were the leader it would be permissible to occupy the regularly assigned school period with the club work making it a part of the school program. The danger of this is that it does not function as a club but often becomes another class. The most successful clubs obtain their membership from the different classes of the school. In this case several clubs might meet at the same time—the attendance depending upon their individual interests. The school period would still be occupied and the benefits would be extended to all members of the school without discrimination. The time and place of the meetings having been decided upon the next order of procedure would be the appointment of a committee to draw up a constitution to be presented at the next meeting. Upon acceptance it may be signed by the charter members.

METHOD

The method of the Nature-Lore Club is the socialized-project. Each meeting of the club should have a business meeting with parliamentary procedure,—a short social meeting of songs, yells, and stunts, and a work period. Every member is to be a leader and should be able to demonstrate before the club and in the various school rooms. The counsellors differ from most teachers in that they are leaders, or advisors. They are members of the club working on the same project with the other members. The club is self-governing and democratic in every detail.

It is helpful to have a chart on the wall of the clubroom which shows the individual achievments of the members. If this is the first club experience of the members it may be well to have a suggestive list although the self-starting system is better than the cranking up method when used in a club organization. This list, which should pertain to the theme selected, may be somewhat as follows: 1. Lead in a successful trip; 2. Have a story well received; 3. Teach a new song; 4. Invent a related piece of construction, as a trap; 5. Report on a lecture heard at another club; 6. Make a life history chart; 7. Show resourcefulness; 8. Make an uninteresting topic interesting; 9. Perform an act of service; 10. Read a book for a report; 11. Write an original essay; 12. Relate an experience with the subject at hand; 13. Demonstrate before an audience; 14. Organize or maintain a clipping service. This list of achievements is merely suggestive. No one achievement is required and the number of achievements that one may acquire is unlimited. The thing that interests one member may not interest his neighbor. Every member of the club is considered as an individual with individual capacities and interests. Members having the experience of being put upon their own initiative for the first time will have a peculiar and possibly helpless sensation. They are use to being told what to do—if not by the Czar at least by a teacher. The chart is an incentive to achievement and should it be necessary to give school credit it should be stated that ten or some other number of achievements are necessary for credit. It will also be readily recognized that individual achievements are not of the same numerical value. Spirit rather than completeness is sought in the club plan.

AIM

The following quotation from "The Nature Study Idea," by Dr. L. H. Bailey, is the keynote to the Nature-Lore Club. "I like the man who has had an incomplete course. A partial view, if truthful, is worth more than a complete course, if lifeless. If the man has acquired a power for work, a capability for initiative and investigation, an enthusiasm for the daily life, his incompleteness is his strength. How much there is before him! How eager his eye! How enthusiastic his temper! He is a man with a point of view, not a man with mere facts. This man will see first the large and significant events; he will grasp relationships; he will correlate; later, he will consider details."

SUGGESTION FOR AN INSECT CLUB.

Insects are selected for a type of club work in September.

1. *Insect songs.*

Obtain "Nature Songs and Stories," Creighton, published by the Comstock Co., Ithaca, New York. (Cricket, honey bee, mud wasp, woolly bear caterpillar.) Have a group tell the old jingle about "Lady Bird, Lady Bird." Then tell about the introduction of the Lady Bird Beetle into California to save the orange crop. This story should be found in an insect book and told to the club by a pupil. The group having this project then sing an original parody on Lady Bird, Lady Bird which tells about the TRUE Lady Bird Beetle. Truth is stranger than fiction. Invent an insect chorus to the tune of Old MacDonald.

2. *Insect Stories.*

Some good insect stories true to nature are: "Grass Hopper Greens Garden," Swartz, Little Brown; "Hexapod Stories," Edith Patch, Atl. Mo. Press; "Interesting Neighbors," Jenkins, Blakiston; "Social Life in Insect World," Fabre, Century; "Insect Stories," V. L. Kellogg, Holt; "The Romance of Insect Life," Edm. Selous, Lippincott, Stories for the social period should always follow and not precede the work. The story of the Untidy Fly, for instance should not be told until club members know about the fly and the harm that it does. The stories should be told by members. Original stories have a greater value. The club leader may suggest something as follows: How many have heard "This is the house that Jack built?" The insect world has real "houses that Jack built" and if you will write some of those stories we would like to have them at our next meeting. They might be told something like this: This is the girl that caught the germ, brought by the fly, that came from the egg, that was laid in the barn, back of the house that Jack built, etc. Investigate the story about the Hessian Troops sent over to America by King George III. How much damage did they do? The Hessian Fly was introduced at the same time. In 1916 the Hessian Fly destroyed more than $1.00 for every person in the United States. Find all the figures that you can about the Hessian Troops and about the Hessian Fly. Bring in your decisions. What does this have to do with the introduction of new insects? The Gypsy Moth and the Corn Borer are recent arrivals. The fable about the Ant and the Grasshopper is an impossible event. Tell the story and show what is impossible. Find a true story that is more interesting, as, the Agricultural Ant that plants and harvests a crop, or how ants carry on war, or about bees in their home. The story of the life of Henri Fabre is an inspiration. He was poor. He was considered a dunce. He did not have costly books or apparatus. He had to study in a kitchen. His folks had no sympathy with his interests and threw away his collections. In spite of these obstacles he became the world's greatest entomologist. Tell the story of his life. Get "Freckles" and the "Girl of the Limberlost" by Gene Stratton Porter, for the Insect Bookshelf.

3. *Insect Plays.*

A play might be given on Achievement day. Have a competition by teams and the one presenting the best original play might appear on the special program. Give suggestions, such as—The House Fly Visits an Untidy Family. Characters: Untidy fly, mother, baby, and a little girl. In the second act the fly visits a tidy family. This may be acted out in pantomime. Plays should always come after subject matter.

4. *Construction Projects.*

Some pupils may wish to collect and display the economic insects of the neighborhood. These insects may be classified as friends and enemies. This may form a part of the school museum. This will involve many construction projects. The best box for mounting specimens is the Riker Mount which is sold by Kuy-Scheerer Co., New York City. The better thing to do would be to make these boxes. Other things to be made are: Net for catching insects, spreading board, rearing cages, aquaria for aquatic insects, killing jars, and fly traps. Directions for making these things may be found in any good book such as Comstock's Manual for the Study of Insects. The "Moth Number," *Nature Study Review* (15c) is very good. Life history boxes are very valuable, such as that of the potato beetle. This becomes very useful during garden time. When we realize that the insects devour a little less than one-half of the vegetable matter that grows we can readily see the economic importance of insect study.

The plasticine model is another form of construction. Some members of the club will enjoy making such models as the following: Fly's foot, an adult fly or mosquito, or a dragon fly using maple keys for wings.

5. *Demonstration Projects.*

The breeding of flies in stable manure, the exhibition of mosquitoes in a glass of water to show how the oil kills the wriggler, fly tracks on gelatine as described in Comstock; the killing of scale insects on a fern brought from home; the treatment of plant lice; how to spread butterflies on a mounting board; etc. Everyone in the club is expected to be able to demonstrate. This is an excellent opportunity for correlation in language, and in drawing.

6. *Lectures.*

Each club should have a reference library. Books may be obtained from the homes, the library, and the Museum. Magazine articles are up to date. The newspaper has something about insects in almost every issue. Ordinarily the lectures should be given by the club members. These lectures could be on individual insects such as the dragon fly, scale insects, etc. Insect architecture, the relation of insects to flowers, how insects are protected, the department of Agriculture, the Silkworm industry, the honey bee would be interesting. Have outside lecturers, as—a doctor who would tell about the housefly, malaria, yellow fever, typhoid, typhus, bubonic plague, leprosy, and sleeping sickness. A member of the mothers' club might be willing to tell what the housewife should know about insects: the meal worm, flour weevil, moths in the carpet and rugs, ants, cockroaches, fleas, flies, and mosquitoes. A local gardener could tell about the enemies in the garden. The plumber and the architect have to consider the insects. The inventor got the idea of the aeroplane, forceps, hypodermic needle, saw, file, auger, or at least could have from the insects. The useful products are: honey, wax, silk, ink, dyestuffs, medicines, and shellac. Invite some one from the Museum or neighboring college to speak. Borrow insect collections. Get bulletins and also pamphlets from the Department of Agriculture at the State House.

7. *Handwork.*

Another form of hand work is to make colored posters which show the fly nuisance, the introduction of injurious insects. Make a map to show how this insect has spread. Color in outline drawings of the beetles. Color in moth outlines published by the Comstock Publishing Company. Make a club room border of colored butterflies on the blackboard.

8. *Community Service.*

The opportunities for community service in a locality depend upon the needs of that community. The housefly may be the commonest and the most dangerous insect. The story is told of one city that offered a reward for the boy or girl who could bring in the greatest number of flies. One boy who had an unusual eye for business went to work breeding flies and obtained the reward. Whereas the slogan use to be to SWAT THE FLY it is now thought to be better to say PREVENT THE FLY. The housefly lays its eggs in horse manure. Map work, clean up days, newspaper articles, and posters would serve an important part. The mosquito furnishes a great field for a survey and campaign. The children might seek for specimens in the community looking in any stagnant pools, swamps, etc. and note the results upon a map. Specimens in a bottle placed in show windows with posters would tell the story. Models to show draining of swamps and the use of petroleum in one bottle would show another method of prevention. A home survey of the damage of the clothes moth for the past summer with an estimate of the damage in dollars and cents followed by a poster to illustrate the prevention. If everyone follows the directions for next summer—how many dollars has the club saved to the city for the coming year? This may be one argument for the support of the club. A child has as much right to study insects as to read about Napoleon and Hannibal and other destroyers of mankind. One enemy is here and the battle is going on. The latter is a matter of history yet how many are taught about past heroes and remain ignorant of enemies about home. A campaign against head lice or bedbugs or cockroaches might be timely. Again we emphasize that nature service varies with the community needs. Does the gypsy moth and the brown tail, the tent caterpillar or the corn borer need attention? Herein lies the individuality of the community.

9. *The National Point of View.*

It is essential that the club work does not stop within or is not limited to the horizon of the community. The price of cotton is determined by the cotton boll weevil which is said to wage a tax equal to one-half of the cotton crop. For every bale of cotton sent to the factory another goes to the weevil. How does this compare with the tax imposed upon the colonies by King George the Fourth? Compare the reaction to these taxes. What is the government doing to prevent the introduction of such enemies as the gypsy moth and the corn borer? How does this concern us? How many people should this concern? Compare the taxes demanded and paid to insects with the taxes paid for education. In what way would an increased tax for education decrease the insect tax? Which would you prefer to pay? How can this be brought about? What are you going to do?

"Children should be interested more in seeing things live and in studying their habits than in killing them. Yet I would not emphasize the injunction, 'Thou shalt not kill.' I should prefer to have the child become so much interested in living things that it would have no desire to kill them."
— *L. H. Bailey in "The Nature-Study Idea."*

"The sedges flaunt their harvest,
In every meadow nook,
And asters by the brook-side
Make asters in the brook."
— *Helen Hunt Jackson.*

Summer fading, winter comes,
Frosty mornings, tingling thumbs,
Window robins, winter rooks,
And the picture story books!
— *Robert Louis Stevenson.*

CHAPTER XVII

The Organization of Nature Instruction on the Psychological Basis.

Suggestions for the Presentation of Fall Flowers by this Method.

In the last chapter you were told how one might organize nature instruction as club work with a view to socialization and practical application. Insects were selected to illustrate that method of presentation. In this chapter the subject is treated from a psychological basis, i. e. the interest of the child. Suggestions will be given for the presentation of fall flowers by the psychological method. Any method must guarantee progressive inspiring appreciation to the child. It is the child's right and heritage. Teachers may choose their method but as to results there can be no question. In the psychological method we start with the child's interests. Up to the time that they enter school all that they learn of nature is because of their interest in it. The function of the school is to further this interest. We are developing children and not daisies or dandelions.

Some courses in nature-study include "everything." Each grade has its list of minerals, plants, animals, and physical phenomena. These subjects are usually unrelated and dealt out one at a time as if no other subject would do. This method indicates that the aim is encyclopedic information. The schemer or author of the course has mistaken the means of nature-study for the results of nature-study. There is no subject that is best. Any list of fall flowers is capable of educating the child. What a child is, is more important than what he knows. The particular flower with which the child deals is incidental, for "A man's a man for a' that and a' that."

The selection of material has also depended upon the interest of the teacher. This has been lately illustrated in general science textbooks. The physiographer makes his book 90% physiography, the chemist believes that Chemistry is the all important, the botanist would linger upon plant study, and the literary inclined are often controlled by the sentimental and the mythical. Is there to be no choice of subject by the teachers? Most certainly. That is one reason for the teacher. It is the aim of nature education to retain the child in right living in his natural environment. If the interest of the child does not put him into first hand relation with these problems it is the duty of the teacher to show the child his needs. The selection of nature material should be determined as far as possible by the interests and needs of the pupil, which are based on his environment as determined by the seasons.

1. *The Collecting Instinct:*

Children are especially interested in collecting fall flowers not only because of their collecting instinct but because of their interest in bright colors and their sense of beauty. Have teams for yellow flowers, "whites," "reds," etc. The "reds" might arrange their flowers on the "red" table, etc., each team having an alloted space. The "whites" might challenge the "yellows." When the children discover that there are more yellow than white flowers the teacher might assign values to the colors and renew the interest by keeping a score. The game spirit is added to the collecting spirit by posting the scores of the teams. For bringing in a new flower give one point, for correctly naming it one point, for writing an interesting fact on the label card give a point. Have a small library for identification purposes, such as Reid's Flower Guide, Mathews' Book of Wild Flowers for Young People, and the New York State Museum Flower book.

This is the time of year to collect winter or dried bouquets. Plants especially adapted for this purpose are: Rabbit's Foot Clover, everlasting, bush clover, grasses, sea lavender, bayberry, and pine cones. Arrange in vases without water.

The so-called "fern-dish" usually consists of a glass bowl with a glass top. In the bottom of the dish is placed sand with charcoal to keep it "sweet." Various kinds of moss are arranged as a mat over the sand and given a good sprinkling. The collection takes care of itself as the moisture evaporates to the glass cover and then falls back as "rain" as it would out-of-doors. Some plants that may be collected for the "fern dish" are: partridge berry, pipsisewa, checkerberry, prince's pine, gold thread, cranberry, sundews, and if large enough the pitcher plant. Various seedlings and ferns will germinate from the moss. The "fern dish" is not only an ornament but furnishes material for other occasions as the cranberry for geography and the pitcher plant with its method of catching insects is as interesting as any story in fiction.

2. *Social Approval:*

Those things which society approve give a dignified enjoyment. Four examples of how we may make use of this interest in teaching fall flowers will be given.

a. *Potted Plants:* This is the time of year to dig up wild flowers and pot them for window decoration during the winter. These plants are most successfully transplanted by taking up sod and all. Each child should have a potted plant of his own with a wooden pot-label. The spirit of ownership, the responsibility of care, the cultivation of the right feeling in regard to his neighbor's property are early steps in training for citizenship. The practical training, the economic feature of the project, and the results in decoration will call exclamations of approval from all visitors to the room.

The children in the primary grades do not appreciate economic values. The transplanting of fall plants would have a larger value from the activity involved and a love of the colors rather than from the commercial standpoint. In the upper grades the pupils have opportunity to use their artistic taste in improving the economic value of the school grounds and the home grounds by transplanting from the fields.

A Few Suggestions

Goldenrod: Over fifty varieties. Select when in bloom. Will flourish in rich soil. These and asters make a good hardy border out-of-doors. Transplant in the fall.

Violet: Transplant to border between tall plants. Requires same kind of soil in which it was found.

Ferns: Set out in a shady place. North side of buildings.

Evening Primrose: Passes winter in rosette form. Blossoms into October. Rosette form may be potted.

Hedge Bindweed: Transplant after first frost. It will start a new growth. Have a support for plant to climb. Becomes a pest out-of-doors.

Burdock: Vigorous taproot. Good illustration of leaf arrangement. Keep a plant of burdock indoors for several years and note its change in habits due to a change in environment.

Dandelion: Leaves will die but large root soon sends out new rosettes.

Mullein: A biennial. Obtain plant that has not gone to seed, getting as much of taproot as possible. Sold by florists in England as "American Velvet Plant."

Queen Anne's Lace. Wild Carrot. Biennial. Blossoms into November.

Jack-in-the-pulpit: This plant is a great favorite with young folks. Recognized in the fall by cluster of scarlet berries. If potted will send up a shoot which will unroll and blossom. Suggests a story of mystery. Children enjoy keeping a dairy. Hectograph outline drawings and have children color in as plant develops. Use pictures for a moving picture show to tell the story of the "Jack" to the Mothers' Club.

Yarrow:

> "I like the plants that you call weeds,
>   Sedge, hardhack, mullein, yarrow,
> Which knit their leaves and sift their seeds
> Where any grassy wheel-track leads
>   Through country by-ways narrow."
> —*Lucy Larcom.*

Has fern-like foliage. Transplant rosette to pot in fall. Its leaves have the odor of tansy which is very pleasant in the winter months.

Peppermint: Easily potted and gives the familiar odor of peppermint in the winter. Edible.

Bull Thistle: Transplant the rosette of spiny leaves. This will bear several heads of purplish flowers. National flower of Scotland. The story is that a Danish soldier stepped on a thistle and his cry gave warning to the Scots that they were being attacked. This happened over a thousand years ago. Good type for teaching biennials and adaptions to environment.

Common Plantain: Large leaves with rosette arrangement. By breaking leaf stalk one can see the tough fibrous bundles. These bundles are typical of the bundles in the higher forms of plants.

They are the food channels. Pot in the fall. Feed seeds to canary. Called "white man's foot" by Indians. Find the expression in "Hiawatha."

Purslane: or wild portulaca. Its fleshy leaves make it easy to pot. Forms a mat. This plant and the live-forever are easily slipped.

b. *Purposeful bouquets:* i. e. bouquets for the sick in the neighborhood, for the hospital or for the library stimulate the desire to arrange flowers beautifully. Arrange two similar bouquets, one in a dull colored, simple vase and the other in a bright colored, fancy vase. Which do the children prefer to send? Arrange tall stemmed goldenrods in a tall vase and in a broad shallow dish; have one vase of goldenrods with the same length of stem and another with varying lengths of stems; have a tall single chrysanthemum in a tall straight vase and a "bunch" of the same kind of flowers in the same kind of vase; use one kind of flower with its foliage and compare it with the same flower with another foliage. The children will have the interest and the desire to learn all that they can about the arrangement of a bouquet. Have the children make up bouquets and then vote for the most beautiful one to be sent to a friend. The flowers in the library might be arranged by colors with labels and interesting facts. Committees could keep the "Flower Show" fresh and up-to-date.

c. *A Weed Exhibition* for an Agricultural Fair, or for the Grange, or a Teachers' Institute to show How Weeds Win in their Struggle for Existence would win social approval. The teacher might write the following table on the board as a guide for the arrangement of the exhibition.

SEED

Tumble Weeds	Seeds Carried by Wind	Seeds Carried by Animals	Long Seeding Period
Pigweed—A	Dandelion—P	Cocklebur—A	Chickweed—A
Ragweed—A	Jimson—A	Wild Carrot—B	Purslane—A
Smartweed—A	Milkweed—P	Burdock—B	
	Canadian Thistle—P	Burmarigold—A	
	Common Thistle—B	Beggar's Ticks	
	Wild Carrot—B		
	Ragweed—A		

ROOT   STEM

Tap Root	Creeping Root	Runners	Creeping Under- Stem ground
Dandelion—P	Milkweed—P	Cinquefoil—P	Grass—P
Dock—P	Sorrel—P	Hawkweed—P	Mints—P
Wild Carrot—B	Sweet Flag—P	Poison Ivy—P	Canadian Thistle—P
Burdock—B	Wild Iris—P	St. Johns- wort—P	
Pokeweed—P	Purslane—A		
Mullein—B			

LEAVES

Fleshy	Reduced	Hairy	Spines
Chickweed—A	Shepherds	Mullein—B	Thistle—B
Purse—A	Purse—A	Nettle	Prickly Lettuce
		Everlasting	

d. *Christmas Presents:* November recalls to our minds the nature students of Pilgrim trails. I am not so sure but what their methods were more satisfactory in a pedagogical way than the majority of our nature lessons of to-day. They selected the subjects which pertained to their needs. Harvesting nature's crops was a very important time. It was both an occupation and a fascination. Why not revive some of these old fashioned good times? They gathered wild grapes for jelly and grape-juice; barberries give a delightful wild flavor to preserved pears; bayberries were gathered for candles; the fruit of haws for preserves; and wild plum preserve is still appreciated by very old-fashioned people. There were a great many aromatic roots and medicinal plants that were hung in the attic; corn husks were saved to braid into mats; fragrant grasses such as sweet grass, native sedges and cat-tail leaves for baskets; leaves of the balsam, sweet fern, and bayberry for pillows; they made cider vinegar, shelled corn, and gathered faggots. I am sure that the nature class, the cooking class, and the arts and crafts class can plan a "thousand things" for Christmas and this is just the time to start.

The making of bayberry candles will be described as typical of this form of nature activity. The bayberry picking season begins early, for, as frosty nights come on, the waxy coating begins to drop off. It is important to gather a very large quantity of the berries and we will carry two large gunny sacks for the purpose. The wax is obtained by placing the berries in a wash boiler of water and bringing it to a boil. When all the wax has melted off the berries it will come to the top and when the water cools will harden. Be very careful not to let the wax burn. If the mixture smokes that is what is happening. It may be necessary to boil again in a double boiler to get all the dirt and refuse out of the wax. The bayberry wax may now be melted with tallow in the proportion of 1 part of bayberry wax to 3 parts of the tallow. The wax will still keep its green color and give its particular aroma. The wax is now ready for the bayberry dips.

In preparation for the bayberry dipping party obtain candle wicking. Cut the loose white cord twice the desired length of the candle. Double it and twist lightly to hold the strands together. Bend a wire hook through the loop which is convenient to hold it by or to hang it on a line. Melt the wax in a kettle and place newspapers on the floor near the kettle. Have the children form in a circle with their wicks. As each one passes the kettle he dips the wick into the hot wax and then as it cools shapes the wick with his thumb and fingers so that it will hang straight. Each time that he passes the kettle he dips the wick and adds another layer of wax in the making of the candle   This process is kept up until the candles reach the desired size.

4. *Problems:*

The problem-puzzle interest rises into prominence in the grammar grades. The reason of things and the practical application of knowledge is a strong appeal. Those who have read these articles up to the present point will note that nothing has been said about a teaching lesson by the question answer method. This does not mean that the Socratic method should be left out altogether. It is most valuable. This brings us to a *lesson* that we may call *nature-study*. Herein nature-study differs from nature-lore. The topic, plant societies will be presented as a type of a problem lesson.

The local distribution of plants is determined chiefly by moisture. Select an area for a detailed study of its plant population. Collect the plants which grow in this area. What problem do these plants have to meet in order to exist in this environment? Study each species as to its special devices for meeting the problem. Tabulate results.

What characteristics are dominant?

1. *Pasture Plants:*

This area is described in detail to show the method of studying a plant society. A great many cattle are kept in a nearby pasture. More cattle are kept in this pasture than can find good feed, yet some plants are able to thrive. These plants are collected by the children and arranged in bottles of water. They are then labeled by the children and those which they cannot name are identified by the teacher who may use Dana's, "How to Know the Wild Flowers."

The children try to discover how each plant is able to grow in the pasture. At the time of recitation the teacher writes the record on the board as follows. The material for the table is obtained by questioning.

Society Name—Pasture Plants.

Problems—Not to be eaten. Not to be pulled up. Not to be trampled.

Inhabitants	Special Characteristics	Use to Plant
Thistle	Spines	Keep off cattle
Mullein	Hairs	Distasteful
Cinquefoil	Tough and hairy	Uneatable
Dandelion	Bitter, short stem	Not palatable; not easily taken hold of
Hawthorne	Thorn	Wards off animals
Goldenrod	Tough and fibrous	Not easy to chew
Everlasting	Cottony	Uneatable
Buttercup	Bad taste	Unpalatable
Daisy	Bad taste	Unpalatable
Yarrow	Bad taste	Unpalatable
Grass	Creeping root stock	Easily reproduced
	Soft and juicy	Eaten by cattle

General Statement: Pasture plants are able to overcome their dangers by having spines, or by being unpalatable or by being tough and fibrous.

2. *Water Plants.* Problem: To obtain air (oxygen.)

Example	Characteristic	Use
Bladderwort	Epidermis thin	General absorption
Algae	Root system reduced	Not needed
Pickerel Weed	Water conducting tissues undeveloped	Not needed
Duckweed	Mechanical tissues undeveloped	Held up by water
Cow-Lily	Air passages	Buoyancy and breathing
Rockweed	Bladder-like floats	Buoyancy
Kelp	Few roots	Anchorage
Floating Plants	Breathing pores on upper surface	To obtain air
Floating Plants	Waxed upper sorface	To shed water

3. *Drought Plants:* Problem to obtain and hold moisture.

Example	Characteristic	Use
Trees	Shedding leaves	Reduce leaf surface
Corn	Leaves rolling up	Reduce leaf surface
Compass plant	Leaves edgewise to sun	Reduce surface
Peppergrass	Small leaves	Reduce surface
Mullein	Hair covering	Prevents evaporation
Bayonet Plant	Thick epidermis	To hold moisture
Portulaca	Fleshy leaves	To store moisture
Dandelion	Long roots	To obtain moisture

Most lawns are in a drought condition as the soil is usually sand or gravel thrown out when digging the cellar.

Plants standing in stagnant water are in the same predicament as the Ancient Mariner with "Water, water, everywhere and not a drop to drink." These plants cannot use the water on account of the poisonous substances which it contains.

Plants in the frigid zones, and hereabouts in winter, are in a drought period on account of the low temperature not allowing them to use the moisture.

5. *Other Plant Societies:*

Hard-tramped door-yard, fence row, dry, open field, dusty roadside, meadow bog, barnyard, dripping rock cliff, dry hillside, oak forest, hemlock or pine forest, brookside, swamp, marine plants, a dry stream bed, a running brook, aerial plants, red sea weeds, railroad embankment, river bank, sea beach, pond waters, sand hills, edge of salt marsh, gravel pit, cornfield, sandpit, a small island, high hill-top, a thicket, lake or pond shore, submerged aquatics, floating plants, a sphagnum moor, reed swamp, rock plants and trees recently cut.

6. *Activity Interest:*

The interest of the child in activity is one of the earliest interests and probably the predominant. The action of animals is more pronounced and claims greater interest in the first grades. Plant growth is slower and claims a greater interest in the upper grades. The kind of interest that is sought here is not the kind that attracts to the circus because that is a passing show. It must be a wholesome interest that is real and permanent. The relation of insects and flowers is of great interest to children and is another nature-study lesson.

One way of introducing the subject, write this poem on the board.

"Roly-poly honey bee,   Why are you so busy, pray?
Humming in the clover,   Never still a minute,
Under you the tossing leaves,   Hovering now above a flower,
And the blue sky over,   Now half buried in it!"
                                     —*Julia C. R. Dorr*

Subject Matter. Pollination is the transference of pollen from the anther to the pistil. If the substance of the pollen grain unites with the substance of the ovule, the ovule is said to be fertilized and it grows into a seed. Most of the characteristics of flowers enable them to secure this end.

If the pollen meets the pistil of the same plant, the flower is said to be self-pollinated. Cross-pollination is made possible in the grasses and conifers by the wind, in a few aquatic plants by the water, and in the showy flowers by insects. Conspicuous flowers often have an odor and nectar, a sweet liquid found at the base of the corolla. Insects visit these flowers to gather pollen and nectar. Bees always visit during the day, the same kind of flower that they first collect from in the morning. They do not gather honey but nectar to make the honey. Cross-pollination produces the more vigorous seed and fruit. That is why it is necessary to keep a hive of bees in a cucumber house.

Some insects have special structures which facilitate pollination. The tomato-sphinx moth has a long sucking tongue for reaching the nectar in tubular flowers. The honey-bee not only has a long tongue but the hairs are so arranged on the hind-legs that they collect the pollen. The ants are so small and have such smooth coats that they are considered undesirable guests. The humming bird and some snails also aid in this important work.

Flowers aid cross-pollination by insects, as follows:

Being inconspicuous but sweet scented—blueberry, Dutchman's Pipe.

Being inconspicuous but carrion scented—carrion flower.

Coloring parts, as:

     Corolla—trillium, geranium, oxalis, etc.

     Calyx—hepatica, anemone, clematis.

     Bracts—flowering dogwood.

Pistil, matures first—plantain, figwort.

Stamen matures first—some mallows, gentian, fireweeds.

Sexes of flowers separate—willow, maples.

Stamens and pistils different lengths—bluets, primroses.

Protect pollen from rain.

Natural position—nodding trillium.

Change in position—daisy, clover.

Concealing nectar from ants, etc.—snapdragon, butter and eggs.

Nectar only reached by long-tongued insects—honeysuckle, clover, nasturtium, morning glory, jimson weed, thistle, sages, evening primrose.

Bending stamens so that they will snap against insects—barberry, mountain laurel.

Arrangement causing stamen to be pushed down against insect—salvia.

Giving off odor when insects which visit them are most active.

At night—petunia, tobacco.

In sunshine—pea family.

A pinch-trap which fastens onto the leg of the insect and causes it to carry away pollen masses—milkweed.

A box-trap arrangement, insects do not easily find way out—Jack-in-pulpit, skunk cabbagde.

Method of Procedure. Have the class collect as many kinds of the flowers mentioned as possible. When the pupil is on the collecting trip he should wait near a group of flowers for the coming of an insect visitor. What is the name of the visitor? What is the name of the host? Where does the insect alight upon the flower? How does the insect gain entrance to the flower? Have questions answered in class. Instead of this some might follow a bee for five minutes and tell exactly what it did.

Write terms on board that class will need to use or draw a typical flower, labeling parts.

Calyx, sepals. Corolla, petals. Stamens, anther (pollen) filament. Pistil, stigma, style, ovary (ovules).

Butter and Eggs: Why is this plant easily seen from afar? (Bright colored; in clusters). On what part of the flower would an insect land? Is there any special guide for landing there? (Orange patch on the lip). What would cause the flower to open? (Weight of the insect). Does the flower give away the nectar? Explain pollination to class. Would bees or ants be apt to gain entrance? Why is it good policy to conceal the nectar from ants? Why is the bee such a good guest? Show the class the pollen baskets under the compound microscope. Where would you expect the nectar to be stored? Is there any guide to the nectar? (Hairs with a groove between.) What is above the bee? (Anthers). As the bee shakes the flower in obtaining the nectar, what would be taking place overhead. (Pollen dust shakes on his back.) How would this pollen reach the pistil of another plant? Have the class observe adaptions in as many other flowers as time permits.

The Violet. Where is the nectar probably stored? (Spur of lower petal). Observe the doorway to the nectar, to the nectar cup. Against what would the visitor brush when obtaining the nectar? Of what advantage is the shape of the lower petal? Are there any guides to the nectary?

Daffodils: Where is the nectar? Cut a flower lengthwise and taste of the nectar. How far would an insect have to reach to obtain this nectar? Show class the sphinx moth with its tongue uncoiled. How would the insect become dusted with the pollen?

Nasturtium: Find the nectar. Satisfy yourself that it is nectar. What led you to suspect the location of the nectar. What might lead the bee to the nectar? What would be in the way of the ant? Examine several flowers and determine if the stamens and pistils mature at the same time? This is of what advantage?

Larkspur: What part of the flower is colored for attraction? (Sepals). What parts form the nectar-tube? (Two sepals). The petals are guides to the nectar. Do the stamens or pistils mature first.

Salvia: What is the color of the calyx? Of the corolla? Which forms a long tube? What protects the stamens and pistil? The teacher should draw a longitudinal section of the flower on the board as the class describes it. Where is the doorway for the bee? What will the head of the bee strike as it enters the flower. The stamens are T-shaped. The head hits one arm and that pushes the other arm, which bears the anther, onto the back of the bee. Have class cut open the corolla to see this mechanism. What is the position of the pistil in older flowers? This is of what advantge?

Pea Family: Locust, Garden Pea, Sweet Pea, Clover. Where would bees alight? What would result when a bee stands on this doorstep? What parts of the flower fall off? What part continues to grow? Open a pod. Were all the seeds fertilized? Small undeveloped seeds were not fertilized.

Squash: Pass out staminate and pistillate flowers orstudy these flowers in the school garden. Have class discover that so me flowers produce pollen and that others produce the squash. This is one way of inuring cross-pollination. The willows and the red maple also have the sexes separate.

7. *Expression:*

Children of all ages take pleasure in telling others what they have seen or done. Tell about the potting of the Jack-in-the-pulpit to another grade; tell the story of the Bull Thistle and the Scots; report about the visit to the Childrens' Ward in the hospital with the bouquets; explain about the weed collection to the garden club; demonstrate the dipping of bayberry candles to the Mothers' Club; investigate as to why Mr. Brown keeps a hive of bees in his green house where he is raising cucumbers; give exact directions as to finding the spot where the sweet grass grows.

Children enjoy purposeful writing. Have them write to the School Nature League, Public School Dep't., New York City for assignment to some definite school to which they may send a collection of wild fall flowers. Each child write a letter to his new city friend telling how they gathered the flowers and some of the uses that have been made of them in his school.

Wild flowers are better studied in the fall because there is less danger of exterminating many interesting and beautiful forms. Send for leaflets published by the Society for the Preservation of Native New England Plants, Horticulture Hall, Boston, Mass.

Organize a seed exchange. List the seeds of fall flowers in your vicinity giving the particular values of each plant and send to a school in the far south, or to the Canadian Northwest, the Pacific states, Bermuda, Australia, etc., asking for a list of seeds of plants in their community that they will exchange. Write to the Department of Agriculture to learn about the law in regard to introducing foreign plants. Interesting material related to plant geography will be gathered.

8. *Physical Activity:*

It is the duty of the teacher to furnish the opportunities and pleasure of physical exercise. This may be done by means of games, collecting trips, transplanting, etc. These activities have been described elsewhere. In fact, every interest is sure to involve other interests. When there is a combination of interests it is often difficult to unravel the complication into simple interests. Neither is it necessary for the chief object is to obtain the interest and those that appeal to several interests are more apt to be successful.

a. Organize a *weed brigade* to clean up a vacant lot or a river bank. Replace the rubbish and weeds by organized planting. Set out willows along the water's edge. A public spirit will soon appear in the school. It may become contagious and like the measles infect the whole community.

b. *Attract the birds* by transplanting to the school grounds their favorite food plants. The birds prefer the wild fruits. Poke weed, sunflowers, wild sarsaparilla, buckwheat, and wild rice prove a great attraction. Elders, mulberries, sumachs, barberry and mountain ash are the best trees and shrubs.

c. *Poison Ivy Campaign.* Getting rid of poison ivy is a dangerous occupation. The amount of suffering that the plant causes has led some to believe that every community should take measures to get rid of this pest. Only those who are immune should be selected for the work and they should wear gloves and wash often, using plenty of soap. It does not do any good to simply cut off the tops. The roots must be dug up. The smoke caused by burning this weed and the poison sumac is poisonous to many people.

d. *Sandbox project:* There are many ways of preventing weeds. To satisfy the constructive instinct this could be well demonstrated by a sandbox project. Plant weed seeds and then show how they may be prevented by tillage, crops, lawns, smother crops, as alfalfa, frequent cutting; and building paper.

e. *Blackboard frieze:* Have a colored frieze of fall flowers on the blackboard. Place at the top so that it will not interfere with daily work.

9. *Story Interest:*

There are not as many opportunities for the child to become interested in fall flowers through stories as in the case of animal study. Gene Stratton Porter's "Freckles" is perhaps the best for upper grades. Hero worship and stories of achievement appeal strongly to the upper grade pupils and they would get a great deal out of reading about Chinese Wilson in the World's Work for Nov. '13. Luther Burbank's life reads like a story and the work of the Department of

Agriculture in the introduction of foreign plants is fascinating.
Most flower stories for the younger grades are too sentimental. The
"Child's Own Book of Wild Flowers" and Gibson's Blossom Hosts
and Insect Guests may be suggestive to the teacher.

10. *The Dramatic or Play Instinct:*

Nature-study offers abundant material for creative imagination.
It supplies the great fund of well ordered sense material which is
the necessary start for the process of real education. It gives
opportunity to overcome the danger at this age of being overfed with
myths and fairy tales, because it starts with the real. The teacher
must remember that the chief value to the child is not in the presentation of the play but in the planning of the details and in the interpretation. The foundation for the play comes from his external environment and the expression of the play comes from his internal self.
These things are his heritage and right.

The play, therefore, cannot be produced until the child has the
necessary material. It is necessary to have ideas about fall flowers before one can excite the imagination about them. The drama or
pantomime or whatever the method of presentation is, must come
toward the latter part of the study of any particular unit. Otherwise the drama becomes like the written composition in school where
the title is assigned without regard to the pupil's experience in that
definite line.

It may be necessary for the teacher to give a start in the organization of the drama. The following is given by way of illustration: Tell
the class that you are thinking of a rich lady who is very poor. She is
rich in wealth but poor in the love of the out-of-doors. In the case
of fall flowers she is blind to their beauty, does not smell their fragrance, has never experienced their tastes, has never gathered them,
and although having ears has never heard the hum of their guests.
This is the content of episode 1.

A girl scout comes along and takes the poor rich-lady by the hand
and walks with her in paths that lead to the removal of the blinders
and the hood. The lady realizes fully that she too is a companion in
this out-of-door life and that she is a working unit with responsibilities
and pleasures.

To her apply the words of Longfellow to Agassiz:

"And he wandered away and away, with Nature the dear old nurse,
Who sang to him night and day, the rhymes of the universe.
And when the way seemed long, and his heart began to fail,
She sang a more wonderful song, or told a more wonderful tale."

Now Dame Fashion appears and asks: "Why are the ladies of the
land so much more natural? I see women without rouge and with
low-heeled shoes; and girls are flocking along country by-ways with
simple, inexpensive dress of bloomers and middies. Everyone goes
about joyously yet obeying nature's laws. Who persuaded the
young ladies not to be slaves to a vanity case? They appear to be
enjoying life. Have they discovered a new secret? And what is it?

And among these walks with the commonplace we find nature making itself the guide of fashion.

Divide the class into teams and have each work independently in developing the play. One group of people that worked out the story given above represented the blindness by smoked glasses, the deafness with large pieces of cotton batten in the ears, the inability to smell by a clothespin over the nose; the lack of delicate touch by mittens on the hands; the missing taste by adhesive over the mouth; and the general failure to grow by a brick on the head. This person goes about little concerned with her environment. She is absorbed in the use of cosmetics, chewing gum, patting her hair, and arranging her clothing. Her high heels make her tired and she sits down often.

A girl scout comes along. She is alert. Her step is elastic. She is interested in things about her. She has a keen eye and a sharp ear. Her face is tanned and robust. The bended form on the stone observes the scout. A cloud of darkness seems to fall away from her. She is greatly agitated and begins to weep. The scout understands and goes to comfort her.

In the next scene the two are walking through the woods in hiking costumes. The "spirit of sight" dances out from amongst the trees and restores that pleasure to the blind friend. And thus in turn all that oppresses and fetters her life are removed and she again feels the joy of a free life.

11. *The Song Interest:*

Some will be interested in collecting and some in the drama but all will enjoy group singing. Probably no subject has been written more about in song than that of the flowers. There is a flower song in every good collection of songs. In addition to these general collections there are nature song books such as Katherine Creighton's Nature Songs and Stories published by the Comstock Company.

12. *The Game Spirit:*

The game spirit is one of the greatest factors in education. More time should be given to it in the school room. Quick observation, sound judgment, and logical reasoning are used in games. Games should vary according to the age of the pupils. Those best adapted to the lower grades tend toward the physical. As an example of this sort of a game with the fall flowers I will describe the Game of the Senses (see chapter on games).

The Game of the Senses: The teacher has a collection of plants for the purpose hidden in a bag. Some of the plants best suited to this work are the following: Touch, mullein, leaf, water-lily leaf, pearly everlasting plant, the blue flag, a bracket fungus, and a cat-nine tail. Smell: tansy, peppermint, catnip, root of sweet flag, geranium, skunk cabbage, garlic. Taste: sorrel, grape (mashed so as not to be recognized by feeling), checkberry leaves, sassafras leaf, dandelion leaf, rhubarb. Sight: hold up the commonest plants such as: primrose, jewelweed, Queen Anne's lace, and burdock.

The class is then divided into teams. The teams hold a meeting and elect their best representatives for smelling, one for tasting, and so on. The "smellers" are blindfolded and given chairs in a row. The tester then crushes some tansy leaves and holds it to the nose of each representative who whispers the name. Those who get the correct name are awarded one point for their team. The team having the highest score after all the senses have been tested wins the contest.

The game of senses is easily adapted for tree study. In the case of feeling, the ridges and lenticels on the bark are made use of and for the sense of hearing the swish of the pine branch and the rustle of the oak leaves near the ear adds interest.

13. *Adventure:*

Every boy longs to be Robinhood, in the depths of his Sherwood forest. Bold Robin is a good ideal as he tried to stamp out tyranny and typified democracy. Every boy wishes to repeat the experiences of his ancestors and to dare the dangers of night. Girls have the same desires to a less degree. To give them the most adventures take them on a hike. Wander out into a storm or across the stream. The smaller the group the greater seems the adventure. Give them the enjoyment of relating their experiences.

This is the time of year for a foraging expedition and it well may be introduced by the game that brings in the edible plants. It is usually wise to carry along someone thing as a "filler," such as bread or potatoes to bake, in case the foraging is not successful. If the expedition is well planned one may usually count upon apples and possibly a few fish. The bark from the root of the sassafras may be steeped for Sassafras tea and the root of the sweet flag steeped in sugar for candy. It is well to have all these courses to carry out the idea. When "en route" gather white acorns. Tell them how the Indians ground the white acorns by pounding them on a large rock with a round stone. If they used the same place for a long time they finally wore a hollow and many of these old corn or acorn pestles and mortars have been found. One may be on exhibition at the museum. Later this was done by the colonists by millstones run by water power. We may see one of these on the trip. It is usually better to prepare the acorn flour before the trip as tannin may be bleached out by filtering water through the starchy mixture. The flour is then dried and when taken on the trip it is made into a thick batter with water and baked as small cakes in the glowing ashes.

The purpose of this chapter has been to show how the subject of nature-study may be introduced by utilizing the interests of the child. Fall flowers have been used for an example but the same interests will apply in other units of study. Eleven instinctive interests have been used. They have been arranged in the usual order of occupation in teaching but any one of these interests may be used in introducing the subject. The higher grades attack more difficult problems, collections, social activities and games. They become more expert in the use of books. Their interest may lead to long hard hours of work. They realize more and more their responsibilities to the community. In building up these activity interests they carry along the "fun of the game" so that whether they become plumbers or teachers they are in it because it is interesting and great fun.

"I'd rather lay out here among the trees
With the singin-birds and bumblebees,
A-knowin' that I can do as I please,
Than live what folks call a life of ease
    Up thar in the city.—*Riley*.

"All are needed by each one;
Nothing is fair or good alone.
I thought the sparrow's note from heaven,
Singing at dawn on the alder bough;
I brought him home, in his nest, at even;
He sings the song, but it cheers not now,
For I did not bring home the river and sky;
He sang to my ear,—they sang to my eye."
                    —*Ralph Waldo Emerson*.

Every clod feels a stir of might. . . .
and climbs to a soul in grass and flowers.
                    —*Lowell*.

Earth laughs in flowers.—*Emerson*.

QUOTATIONS

"Through every happy line I sing
I feel the tonic of the spring.
The day is like an old-time face
That gleams across some grassy place—
An old-time face —an old-time chum
Who rises from the grave to come
And lure me back along the ways
Of time's all-golden yesterdays.
Sweet day! to thus remind me of
That truant boy I used to love—
To set, once more, his finger-tips
Against the blossom of his lips,
And pipe for me the signal known
By none but him and me alone!"

—James Whitcombe Riley.

I watch the snowflakes as they fall
On bank and briar and broken wall;
Over the orchard, waste and brown,
All noiselessly they settle down,
Tipping the apple-boughs, and each
Light quivering twig of plum and peach
*   *   *
The ragged bramble, dwarfed and old,
Shrinks like a beggar in tne cold;
In surplice white the cedar stands,
And blesses him with priestly hands.

—J. T. Trowbridge.

CHAPTER XVIII

SOME GENERAL PRINCIPLES OF NATURE TEACHING

It must be evident that the first chapters of Nature Guiding are arranged in a seasonal order. We commence with Nature Guiding in the summer camp and call it Nature-Lore. The Nature Club begins in September with insects as a study and the fall flowers are studied in October as a psychological basis of organization. There is no closed season in Nature-study. With the coming of winter we turn our attention to indoor nature guiding or the teaching lesson. The school room is not the ideal place for carrying out the aims of nature-study but it is usually a necessity in our large cities. We must accept the situation and make the best of it. A nature teacher is an indoor nature guide. If her nature opportunity is less, her child opportunity is greater. Whether children as a whole make any nature contacts still depends more upon the nature lesson than any other source. The teacher must remember that, even though the temptation now becomes stronger, her duty is not to cram ideas into the minds but to guide their nature activities.

Nature-study is more than recognizing objects. Fifty kinds of birds, flowers, trees, or insects has no more meaning to a growing naturalist than fifty kinds of paint brushes, papers, or ochres would have to an artist who has had no experience in his work. Does the child love these things less or more is the test of the teacher. The teacher must stop working so hard on lists and work a little more with inspiring. How much of the time does she make nature-study a real joy? She must bring to the foreground of the child's consciousness the inspiring atmospheres of outdoor nature-study.

In transferring one's activities from the outdoors to the schoolroom it is well to take an account of stock. What is this elusive subject called Nature-study? It is where the heart is in nature. It is not trees nor the forest floor; not lacey ferns, nor rich, deep, mossy carpets. Nature-study is a spirit, not a material thing. Nature-study is an impulse, a pleasure of the senses, a sympathy with living things, a perception of natural laws. When a nature guide tries to make you "feel at home" in the open he is not thinking in terms of nomenclature or how much do you know; he is thinking of your natural desires. It is *you* that "make yourself at home" in the wild places; it is *you* that defines what nature-study is and means. A nature teacher can only help with such landscapes and nature contacts as he can command, to bring about a realization of the heart's desire. Nature teachers must have a heart. And when they sit behind a desk this heart must still be aglow with the fire of outdoor enthusiasm.

To develop this sensitiveness to nature the teacher must recognize the nature instinct, which is inborn, and the laws upon which the growth of this instinct is dependable.

215

Nature is the teacher of all animals and was the first teacher of the human race. Gravity, water, and fire; hard stones, rough bark, and the brier rose; soft fur, bright colors, and claws, are early acquaintances. These outside influences treat all comers alike. Some learn early and others require repeated experiences. The animal must adjust himself to his surroundings or pay a penalty. Such experiences go to make up his education.

The call of the wild is a natural impulse. If we do not establish these "ties that bind" early they are gone forever. If we hold back this instinct of migrating to the woods it is similar to holding back the desire to walk or play. We lose the desire. Instead of holding back we should give opportunity for these sense experiences. Whether the nature instinct develops or not is entirely dependable upon the number and variety of nature stimuli from the outside. The child's desire for pets, for growing things, for the fields, and for camp are natural impulses. These desires become regulated as he grows older. If his desires attach themselves to natural objects, we can influence the activities that will take place by providing the object. The tendency to go to the fields may attach itself to several things—to fishing, to eat wild strawberries, or possibly to see hatching birds. He yearns to do the thing that he did before. The never-to-be-forgotten trip to the sea-beach. The tendency to go to the forest gets stronger in the same way that we educate ourselves to certain foods, whether it be beans, macaroni, whale blubber, or sauer kraut. Along with the fondness for certain foods (stomach patriotism) may be developed a fondness for certain outdoor stimuli (I love thy rocks and rills). The provisions for stomach patriotism seem to be pretty well taken care of. A Bostonian is so stimulated that beans become his Saturday night diet. He demands them. That he will be as regular in going to the fields Saturday afternoon is not so certain. Our nature habits are pretty much a matter of chance.

We must think of our program of nature stimuli as a tool by which we release certain personalities. If we plant seed and cultivate the crop we raise corn or cabbages. In like manner we can guarantee the quality of manhood from these inborn seeds of interest plus culture. We must be aware that there are many external stimuli (weeds) to interfere. This is the lolly pop and chewing gum age. The movie is still a liability. The raising of corn and pigs is still better understood by those who raise corn and pigs than children by those raising children. Superstitions, Nursery rhyme, the Sunday supplement, and jazzy first grade readers play their part in denaturing the child. The modern child has about as much chance in nature as Adam and Eve had in the first days of nature-study. It is a short stay in either case. The nature instinct is not being given a fair chance. We need to give pathways of release for this high voltage of nature sentiment. The sort of sentiment that furnishes the motive power that drives one out into the fields and forest.

The teacher must not only recognize the law of nature instincts but the law of purpose. The child must have a reason for wanting nature-study. The success of early man in the wilderness meant success in nature. His food, shelter, clothing, and drink were sought in nature. These experiences were necessary. When he was hungry his mind set his whole person to work to get food. He was like a cat after a mouse. Great satisfaction accompanied the doing of the thing. It has always been so. And the stronger the purpose the more likelihood there is of success. Every great advance by the human race has been due to the harnessing of natural forces. An inner urge, purpose, or impulse, has been a necessary part of the advance.

We owe much to experimental psychology for the making of this law of purpose clear. A cat placed in a box accidentally pulled the string and opened the door which enabled her to get the fish. By trial and error she finally learned to do it. If you took hold of her paw and pulled the string it would not be a short cut. We do not learn that way, yet we have a habit of making the children do what we want them to do. We must realize that it is the child's own responses that educate him. He must react to his nature stimuli in the same way that he must eat and digest his own food. We can determine what nature stimuli will play on him but the actual change is in making the response. The function of the nature teacher is to furnish nature stimuli.

The child must want to learn to play baseball or he will never learn. He cannot learn by proxy. And going through the motions are not enough,—he must practice. In school, the desire to play baseball already exists but the chances are that he has been killed off as a naturalist. His nature instinct has atrophied. The teacher must create a desire. It is not fair to ask her to compete with the baseball coach. Camp Directors must also recognize this. The child must be reawakened. He must be given opportunity to practice nature-study. He must become anxious for nature-study. He must have the inner urge so strong that he will find time for it as naturally and as certainly as for baseball.

Just as the child must practice baseball he must practice nature-study. This is called the law of frequency. Some families look forward to going to the country every vacation. Looking country-ward becomes a habit. If we get a thrill in visiting the woods today we will be more apt to do it tomorrow. It is the old saying that "practice makes perfect." Repetition fixes it. Those who have not played in the woods when suddenly brought to them do not know what to do. They are lost. It is like the case of the New York philanthropist who took some news boys up the Hudson on an outing. When he went ashore to see how they were getting along he found them shooting craps in back of a tree. Making one contact with nature will not do. It must be frequent.

So far we have considered the fundamental conceptions necessary to teach nature-study. Every teacher must realize that nature-study is a spirit, that every child is a born naturalist, that the culture of this impulse depends upon the law of purpose and the law of frequency. These are necessary as an approach to nature-study. We must next know the values of nature-study.

Fig. 1.   Harvesting the Rights of Childhood

I. The Values of Nature-Study

1. *The Doctrine of Utility.* There is a present day tendency to emphasize the practical, and there are those who would only teach that which has an economic relation. They would teach how to destroy the tomato worm but do not care for the milkweed butterfly. They would look at squash blossoms but not the arbutus. They would know the poison sumach but not the flowering dogwood. But who shall say that the arbutus and flowering dogwood will not some day increase the farm value? And who shall say that enjoyment is not time well spent?

A naturalist friend of mine studied the foraminifera because he was interested in these microscopic shell animals. Through this interest he finally discovered that they form the alphabet of the oil strata. Whereas before, oil could not be located within a hundred feet, he can locate it within three feet. He was offered $100 a day to assist in locating oil wells in Mexico. When he understood the foraminifera he was able to set them to work for him.

The man who gets the most from the farm goes farther than the dollars and cents. He enjoys the bobwhite in his fields and protects him. He is awake to the beauty of the mountain laurel and would have it grown untorn. He understands the nuthatch and encourages him to work in his orchard. He enjoys the song of the robin as he hoes the long rows of corn. His sympathies are wide and his profits are realized in his thoughts.

Knowledge and attitude in respect to birds, insects, flowers, forests, pests, forest fires, leisure time, gardens, domestic animals, shrubbery,

Fig. 2. NATURE-STUDY TRAINS IN THE HABIT OF OBSERVATION. Can you find 30 frogs looking at you from this water garden?

conservation, and many other things in nature affect the common welfare. Ignorance in these things has caused the spread of disease and the destruction of life and property. Ignorance imprisoned Copernicus, Galileo, and Roger Bacon. Ignorance made possible the Salem Witchcraft. The same ignorance is at work today. Ignorance of nature's laws makes the sale of patent medicine possible. Toads and snakes still suffer from superstition. When evalued on an economic basis, Nature-study is worth while. The economic interest, as raising garden products for profit, increases in the upper grades.

2. *Nature-study makes for the Habit of Observation.* Nature-study demands observation. The person who has the observation habit has a lively and permanent interest in nature. He is mentally alert in the forest. He is continually making delightful contacts with plants and animals. He is better equipped to travel for it does not mean just to be able to say that he has been to the Black Forest or

that he has seen the edelweiss. He is not willing to dash into the Yosemite Valley one day and out the next. Listing places is not his aim. He is not willing to just look. He is there to investigate what nature exhibits. He has the same deep seated interest that the engineer has in engines or the old skipper in sailing craft.

And the training of the powers of observation does not mean the indorsement of the idea "never tell a child what he can see" to the extent that he need not be guided. A nature lesson may well be preceeded by six to ten written questions to direct the observation. The child should get the answer from his own investigation. The questions must be just right—difficult enough to require thought, simple enough for them to be able to answer. He acquires the mental habit of "seeing is believing." This is the laboratory method. Silent study is of great importance in mastering the acquisition of ideas and power in Nature-study.

3. *Nature-study stimulates all the Senses*. We have not begun to mine the possibilities in Nature-study. We have neglected—almost criminally so—to develop the latent powers of the child. Most of our education is through the eye. The education of the ear is in a pioneer stage. We have done a little in music appreciation but we are for the most part "Nature deaf." The sounds in nature have to be loud or unusual to provide a sensory stimulus. The rhythmic sounds of nature are unknown. The existence of the man of the wild depended on detecting sound but the civilized man does not know that they exist. The high appreciation of nature by Helen Keller shows what awaits us in touch education. As a race we are in the bright color stage. We do not see the soft colors—the harmony of colors— the dainty touches everywhere in nature. The importance of color, in school work, has been too little appreciated. The education of our nose and taste is mainly as a sentinel to our stomach. The odors of the field and the tastes of wild fruits play a small part in our patriotism. The smell of the tenement house and of paved streets means home.

Nature-study should be a pleasure of the senses. The song of birds, roar of the surf, fragrance of honeysuckle, smell of new ploughed earth, new mown hay, or the ozone from the sea, flavor of a Bartlett pear, warm sunshine, bare feet on the sand beach or in the soft mud, autumn colors, colored butterflies, the browns and grays of a winter landscape, love of pets, care for plants, wading in a brook, cloud effects, tints of forest and sea, lights and shades. All of these should enrich the life of the child.

4. *A Good Nature Lesson is a Good Language Lesson*, but the language lesson should not kill the spirit of the nature lesson. Use the nature method and not the language method else you will defeat the aim of both. I was starting off with an VIIIth Grade on a field trip. Everyone was happy with expectations. The teacher, through lust of duty or force of habit, "threw a monkey wrench in the wheel" by this parting shot: "I want each one of you to write this up for

your English lesson tomorrow." If she had let them enjoy the trip and then requested an essay the chances are that they would have gladly written the story. To be "taken advantage of" bred contempt.

When a pupil observes an object and tells others what he sees he is acquiring clearness and precision in expression. Children write better compositions when they select their own subject. When given this freedom they usually return to their nature experiences, and a natural interest means greater mental activity. If we enrich this experience we give opportunity for continuing thought and expression.

Fig. 3. "Nellie Bly" has Confidence in Her Rider

If the child observes, for a period of time, a woodpecker on a tree he gets connected observations. Continuous thought makes continous discourse possible. If he is interested his ideas are cumulative whereas poverty of thought means disconnected and unrelated statements.

Likewise if his observations have been followed by a good scattering of "whys and hows" his essay will contain reasoning. He is then using the same train of thought that is followed by the scientist. The awakening of the art of expressing cause and effect is the best investment that can be made in the written composition. We are paving the way for the development of the mind of the child in the scientific method of thinking which has distinguished the great minds of all ages. This kind of training cannot be obtained from the encyclopedia, any textbook, or any other artificial assignment.

5. *Nature's Laws are Immutable.* It is often said that the children of today do not have much respect for law. Many parents have given

Fig. 4. NATURE PRODUCES ABUNDANTLY: If every horse-chestnut blossom produced fruit it would resemble a bunch of grapes. What other examples of the law of overproduction can you find in this picture?

it up. Nature's laws have not changed. If we put a finger in the fire we get it burnt. If we catch a cold it is due to carelessness or ignorance. If we plant a garden and do not care for it we do not get a crop. If we leave filth around the kitchen sink nature provides scavengers. These scavengers may be bacteria or even cockroaches. Cockroaches cannot live without food. These laws apply to the rich or the poor. It rains on the just and the unjust. Respect for man made laws can best be acquired by a knowledge of the law and order of Nature.

6. *Nature-study is an excellent Hobby for Leisure Time.* What do the grown-ups of your community do with their leisure time? Is it loafing, gossiping, bridge whist, and the movies? What do they bring home as a result? How much do the people of your neighborhood spend for commercial recreation? How much do they spend for nature recreation? Which gives the best returns on the investment? Leisure time is the time for nature hobbies. There are always a few boys in every city that have a nature hobby in spite of a lack of encouragement. Enlightened communities should have field clubs, hiking schedules, nature games, school and home gardening, camping, boating, swimming, skating, and fishing parties. Leadership for leisure is more necessary than leadership for work. Nature activities provide recreation for all citizens and not merely for the specialized athlete. The nature guide should have, at least, an equal opportunity to the skilled coach.

7. *Nature-study cultivates some of the higher qualities of life.* The age of reason, the love of explaining things, and the spirit of invention become prominent in the upper grades. These follow the ability to observe. Professor L. H. Bailey says that "Nature-study is seeing what one looks at and drawing proper conclusions from what one sees."

Nature-study encourages imagination, not the fanciful kind of hobgoblins and fairies, but the healthy kind that leads to invention, painting, art, music, and literature. If we took all the nature-study out of these studies there would be nothing left but the framework.

Nature-study makes possible an intelligent understanding of these arts. Many literature teachers have a class read the Chambered Nautilus when they would not recognize one if they saw it. The teachers dwell upon the beautiful thought expressed yet how many times do they read the very same poem outside of the class room for its beauty? Nature-study makes for sincerity of purpose.

8. *Nature-study is an aid in School Discipline.* A student teacher was recently taken from one of my classes to substitute. A boy came to school the first day—not with the proverbial snake—but with a string tied around the hind leg of a grasshopper. It was the old story of trying out the teacher. The boy had probably given new teachers his psychological test before. He knew just about how it would work. He knew that teachers are wonderful in action. This teacher did not do the expected thing. Fortunately this embryo teacher had just been studying the grasshopper and had a fund of information. She was glad that he had brought the grasshopper. She proceeded to have a nature lesson on grasshoppers. She asked him to bring in something new for the next day.

Then again, the child is not a desk animal. He is full of animal spirits. His naughtiness is often due to his inability to sit still for hours like an adult. Nature-study furnishes opportunity for short field trips—a chance to alternate periods of quietness with activity.

II. *The Method of Nature-Study Lessons.*

1. *Nature-study is Positive.* A Boy Scout was coloring in the outline of a bird. He knew that his scoutmaster would not accept it unless it was good. He went to his art teacher for criticisms. He did not realize that the art teacher wanted him to learn to use colors. The whole scheme of scouting is built on the positive. There is not a single *not* in the Scout Law.

Who ever heard of a boy tying a tin can on his dog's tail? And why doesn't he? Because he loves his dog. Yet the method of many leaders in the humane education field is to say, "Now boys it is cruel to tie a tin can on a dog's tail. You must not do it." This usually gives a new idea to some boy and he proceeds to carry it out.

It is not good pedagogy to say, "Now boys do not steal tomatoes." Set them to work raising tomatoes. The boy who raises tomatoes is not going to steal them.

Fig. 5. "Every Dog should have a Boy"

Fig. 6 THE LOVE OF WILD FLOWERS IS IN THE HEART. Signs may help but the conservation of our wild flowers will never be successful until all our boys and girls have the opportunity of nature-study.

What the child owns he protects. And how are we to teach the children to leave our wild flowers for others to see? Not by signs prohibiting their picking them. They must appreciate the woodrose on the stalk. They must know that it takes seven years for the fawn lily (Dog-toothed violet) to grow seed. They must realize that wild flowers wilt before you get them home. They must feel that a spray is more delicate than a bunch. They must experience the secret knowledge of where this orchid or that one grows.

2. *Nature-study is good science* but all science is not good nature-study. The subject matter is the same but the method is not. Professor L. H. Bailey has said that "When the teacher's attention is focused on the subject matter he is likely teaching science; when on the child, he may be teaching nature-study." The difference has been difficult for most teachers to conceive in practice.

Any difference between nature-study and science is a matter of degree rather than being sharp cut.

In material used,—nature-study is usually concerned with outdoor wild nature, of the immediate environment, rather than pickled nature which is largely indoor, in a laboratory.

Each nature lesson is new in subject and plan whereas the laboratory period is an intricate scheme that continues over a considerable stretch of time.

Nature-study is for all people and science for the specialized few. Nature-study is for the child who is just opening his eyes to the wonders of the universe and for the adult who can walk humbly with Shakespeare and say "In Nature's infinite book of secrecy, a little can I read." We usually think of nature-study in the lower grades and science in the VII, VIII grades, high school, and college, yet nature-study can be successful in any grade. Nature-study has proven to be attractive to all, whether it be tourist in the parks or children in the grades. As difficult as it is for some of us to understand, the study of science is attractive to a small per cent of college students, and we venture to guess that one reason is that the elementary study of the earth and its inhabitants was not made attractive in the grades. Nature-study should be the concrete experience out of which science grows.

I have in mind a friend who was naturally interested in the out-of-doors. In college he elected a Botany class because he was anxious to go deeper into the secrets that he enjoyed. It proved to be a class in pathological botany. This was the only Botany course that he ever took and the effect nearly destroyed his spirit. Today he is recognized as an authority on orchids. We will probably never know the number of nature enthusiasts killed off by the scientific method which says that we must always start with the simplest, microscopic forms and proceed in the line of evolution.

A little of the Nature-study method in Tennessee might have made it easier for the community to comprehend the larger aspect of evolution. Some college professors of science are inclined to smile at the informality of nature-study yet lament at the lack of interest in science exhibited by their own children. What a tragedy! There are thousands who are apparently teaching science successfully yet cannot interest their own children.

An example of the difference in method may make it clearer. In a lesson on a frog the nature teacher usually starts with experiences with a live frog in its native environment or in an aquarium. There may be difficulty in discovering the frog, in catching him, and in holding

Fig. 8. Enjoying the Rights of Childhood

Fig. 7. By Their Chips Ye Shall Know Them. Beaver were introduced into Palisades Park a few years ago. They made themselves at home. We must bring back the wild game that we nearly exterminated.

him. When he is placed in an aquarium he may change his color. If we watch him closely we may note that he winks. We may go a step further and see that the eyelid rises from below. We may note that the throat, or the sides of the body, are moving or that the nostrils are closed when he is beneath the water. Every observation is followed by why or how? As many of these observations and reasons are dealt with as the lesson time permits. We can stop at any place and the nature lesson on the frog is complete. The next lesson may be a toad or a tadpole. It will not be a review or a continuation of frog study. It will be a continuity of interest.

The nature student has had the fun of going frogging. He has wandered by a brook side and been surprised by a sudden splash. Perhaps he watched the swimmer poke his green head up by a lily pad which was "exactly the same color." Perhaps he "waded in" and made a grab at the frog and the green mass slipped out of his grasp. Perhaps at a third trial he caught it by the hind leg and heard it croak. As he wandered away with his prize, for an interested teacher, just as he was going over the hill, he may have heard the frog pond chorus. He has had the enjoyment of seeing frogs live rather than watching them die.

And then at school he learns that the frog eats insects. He becomes aware that the boy who thinks it necessary to kill frogs on sight is serving a fool trick on a real friend. He reads with his own eyes the wonderful ways in which the frog is adjusted to its environment and his whole life has become richer through the experience.

In science he is apt to have a dead frog in formaldehyde. His experience is limited to dissection. He spends no time on the activity of the frog, a very short time on its external features, and an endless amount on its anatomy. If he does not finish the dissection today he continues it in the next period. He probably does not consider its anatomy in relation to its habits. He may review at the end of the term and acquire a few principles and relationships of amphibians to reptiles based on the plain facts of dissection, but he has missed some of the higher aspects of nature-study.

Continued nature-study may shade into science. System was the chief aim of Linnaeus. Families and orders was a need of the time. In the same way there may arise a need of arranging a stamp collection. A boy may wish to keep his Canadian stamps separate from those of the United States. As the collection grows he may have different series of United States stamps. Instead of stamps he may collect minerals, soils, leaves, insects, or twigs. The nature teacher will not worry about their classification but stands ready to lend a helping hand or to make suggestions as the occasion demands. When the collector begins to arrange his museum into families and orders he is becoming a scientist. His spirit has been fixed but not killed. His growth has been natural.

3. *Nature Lessons should be in Season and not begin with the Amoeba* and go up the scale of evolution to mammals. The opening

buds in spring, seeds forming in the fall, and evergreens in the winter. The return of the birds, the first violet, the emerging black swallow tail, frost in the lowlands, cherries are ripe, the morning star, or the eclipse are studied as they appear on the outdoor stage.

4. *Select the Common Objects of the Environment*, the wonders near home in the common things rather than the remarkable ant-eater or the freak gorilla. We have a tendency to teach about the wonderful chambered nautilus and not recognize it when we see it. Better be

Fig. 9. PLAYING WITH A BEAR instead of being brought up on the idea that bears eat naughty children.

Wordsworth's "one impulse from the vernal wood." Brighten the corner where you are.

As far as possible the selection should be by the child. What does he bring in? The teacher's interest should not interfere. Let a boy bring his rabbit. The success of that lesson may be judged by the number of offers to bring in other pets. A whole train of offerings will follow a successful lesson. This, of course, is not the only criterion in selecting the topics. The subject may depend upon the experience of the class as well as predominant interest, or a school room condition as well as season.

5. *The Object should be present.* Nature-study is a study of nature and not about it. This means a specimen in the room, an experiment, or a field excursion. If the plant or animal is small there should be one on every desk and it should be a participating object in the lesson. Often times the object might just as well be in a basement as far as any use is made of it. At least 20% of the questions should require the presence of the object, for an answer. Every pupil should have the time and opportunity to observe and reason correctly. Lazy teachers will fall down here as it means work to provide new material for each lesson. The school museum should be an important source

for material. Borrowing a course of study from district one to use in district three is out of order. The nature-study of any school is an individual responsibility of that school.

6. *Activities rather than Structure.* The stress on the study of structure is usually with a view to classification. To a child animate objects are more interesting than inanimate. He is more interested in animal activities which most resemble his own activities. He is particularly interested in young animals. His delight with the puppy and the kitten are well known. They are full of activity. The growth of a plant is rather slow and therefore the interest is not sustained in the early years in gardening. Plants should be grown in the room however, as a lesson in care.

7. *"The Truth is Stranger than Fiction."* To say in December that the tulip tree has all the leaves that will be on the tree next summer stirs one's imagination. Yet the winter buds contain these leaves neatly arranged and packed. We may remove the scales and see these baby leaves with our own eyes. How much more interesting then to say that the bud is a sleeping fairy with woolen blankets and that when old Mother Nature touches the wonderful covering with her magic wand that the fairy will jump out and shake out her emerald tresses. Sentimentality was the cause of the last serious decline in nature-study. Nature-study aids in seeing and expressing the truth in simple language.

8. *The Nature Lesson is not an assignment*, but a study period when all the room are active in the lesson. There is no textbook except the object itself. There should, however, be a reference or reading shelf. This is not to be construed to mean that questions for observation should not be put on the board in advance. It is better to have the object around for several days before the lesson. It also does not mean that a child should not carry out a special investigation to report to the class.

9. *A Review is unnecessary in Nature-study.* There should not be a lesson on the rabbit today and an examination tomorrow. It is not necessary to ask what sort of tracks does the rabbit make today if the boy observed the tracks yesterday. Words do not properly describe the odor of the strawberry. Its a matter for the nose. A review in nature-study should be a matter of repeating those nature experiences which are called for by the children, and after all that is a test of the teacher.

10. *The Nature Lesson must come regularly on the program* and not be left until the spirit of the room needs revival through nature-study. The latter method usually means that nature-study will be crowded out. One teacher told me once that she did not have time for nature-study as her little children were Italian and would leave school at an early age. She needed all the time for arithmetic, reading, and penmanship. To read about what, and to write about what, I wondered. She said that she spent two hours per day on arithmetic.

11. *The forest and field are the textbooks of nature-study.* The child has as much right to read trees as about trees. The money spent for textbooks in other subjects may well be spent for field trips in nature-study. In many European schools the field trip is a recognized part of the schedule.

12. *Nature-study may be adventure.* One time I had a Boy Scout patrol on a winter camping trip. I found a partially eaten squirrel

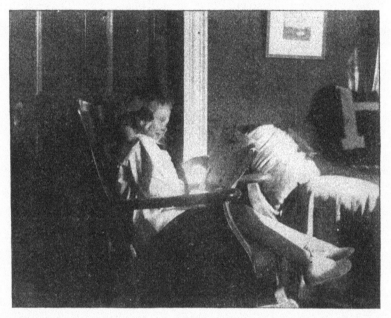

Fig. 10. A COMRADE FROM THE WOODS. This little girl is tickled in two senses of the word.

and showed it to the boys saying that it might be due to an owl, hawk, or some night prowling mammal. I asked, how many would like to stand watch tonight and try to discover the enemy? They were all eager. Later in the day I found them getting knives and axes sharpened. I told them that as they were as likely to get hurt as the enemy we would have to have a rule that only clubs be used. It is needless to say that they all had a club. They voted to stand watch by twos rather than by ones. Each couple was on duty for an hour. Their reports were interesting. They did not know that shadows could be such queer shapes nor so dark. They heard more animals and "things" dropping from the trees, than they believed were in the encyclopedia. They didn't appreciate the extent of an hour before. Every detail of the experience has been stored up in their minds for years to come and they all pronounce it as a great experience.

13. *Every child has a right to pets.* The writer fully realizes that rabbits are cumbersome when it comes time to make a visit to grandfather's and that dogs are not popular with landlords. He knows that squirrels may hide in the darning bag or may even turn on the gas. He can even say that father may be called upon at times to take care of the chickens or that mother will have to feed the white rats. But when considering pets we are thinking of the pleasures of childhood and not the inconvenience to parents. Every child should

care for, train, and grow up with a dog. Patience, good will, companionship, protection, and responsibility are taught and exemplified by both.

We should let every child raise plants. It will give him a new meaning of ownership, protection, and property rights.

14. *The Nature Story should follow the Nature Lesson.* We cannot appreciate the Jerusalem Locust Plague, the saving of the crops in Utah from the locusts, or the story of Grasshopper Green's Garden unless we know the grasshopper. We must know how he eats and his marvellous appetite, we must know that he is born from an egg laid in the ground in the fall, we must know how he flies. The child who has rabbits is better equipped to enjoy rabbit stories. The student who knows the chambered nautilus can better interpret and enjoy Holmes' poem.

15. *Nature-study saves time in the Curriculum* by sharpening the perceptive powers, by giving a better understanding for care of the health, by making geography real, by offering a richer basis for interest in drawing, and by furnishing ideas to write for language exercises. It livens the work for all studies.

When the teacher called the apple class,
  they gathered round to see
What question deep in apple lore their
  task that day might be.
"Now tell me," said the teacher to little
  Polly Brown,
"Do apple seeds grow pointing up or are
  they pointing down?"
Poor Polly didn't know; for she had
  never thought to look,
And that's the kind of questions you
  can't find in a book.
And of the whole big apple class not one
  small pupil knew
If apple seeds point up or down! But,
  then, my dear, do you?

> *—Carolyn Wells in "St. Nicholas."*

A QUEER PUSSY

ANNA E. SAMPLE

I know a Pussy soft as silk,
  Who never drinks a drop of milk.
She never makes a single sound,
  Or looks at me with eyes so round.

She never caught a single mouse,
  Or crept about our big brick house.
But, she's a pussy just the same,—
  For Pussy Willow is her name.

> *—Progressive Teacher.*

CHAPTER XIX

The Teaching Lesson in Nature-Study

For the sake of simplicity the approach to Nature-study has been described by separate methods; the Nature-lore, the Nature Club, and psychological interests. These methods have been presented separately as a matter of convenience in understanding certain principles and relations involved in organizing a nature course. As a matter of fact nature experiences may occur through any one or by a combination of all of these means. The Nature-lore method underlies the essence of Nature-study and gives the greatest nature returns on the investment. This method, however, is not always possible in the city school or in the winter weather. We have dwelt upon the general principles of indoor Nature Guiding and will now proceed to the more specific teaching lesson.

The subjects best adapted to winter Nature-study lessons are mammals, birds, reptiles, and fish. These lessons provide for a "close up" experience with the animals that later on will be met in the fields. Birds have been selected as an example of a series of indoor teaching lessons. Each lesson will be based upon certain types of birds: a live hen as a type of scratching bird, a live duck as a type of swimming bird, a stuffed flicker as a basis for knowing the climbing bird, a heron for a wading bird, and the owl or hawk as a bird of prey. We will study the gulls of our bay because they are well adapted for flying. They are also useful and need protection. At the same time our inland friends may be studying the quail which is well adapted for scratching, is useful, and needs protection. We will study the perching birds not only because they are adapted to their mode of life and are of great economic importance but also because we consider it practical to be able to admire their song and beauty. We may note that the ostrich depends upon his strong legs for running but we will conserve time to become better acquainted with the birds about our own home. If we read Coleridge's Ancient Mariner we will have occasion to learn about the Albatross, the largest seabird, measuring 12 feet from tip to tip of the wings. Then there is the curious pelican with a pouch beneath the bill for scooping and storing food. We will not have a lesson on pelicans. Should there be one at the zoo we will hope that our training in observation and reasoning will enable us to easily recognize this remarkable adaptation. We study the particular kind of heron that haunts our rivers and ponds and hear the story of the egrets rather than stories of storks which belong to the Old World. To summarize: We will select for our nature lessons those things which are common in our environment and can be brought into the school room. We will have varied the lessons upon those things which best typify the marvelous way in which the varied structure and habits of the organism are adapted to their

233

surroundings. We will also keep in mind the economic necessity for becoming acquainted with our out-door neighbors.

It makes no difference which type is selected first but it preferably should be a live bird, comparatively well known, and with evident adaptations. The pupil first needs to know the birds home and food, for where he lives and what he eats is what he is. Then comes the observation of the following topics: beak, eyes, color, feet, tail, and wings, in relation to the home and food. The lesson may start or end with any topic. The first questions pertaining to any topic require the pupil to observe, analyze, and describe the structure or habit accurately. The next questions require the pupil to explain how these facts enable the bird to live in its home and to obtain its food. In this series of lessons the idea of adaptation to environment grows upon the pupil unconsciously. It is therefore not so essential to cover a certain number of adaptations as it is that the pupil make progressive growth in the appreciation and recognition of the great principle of adaptation. It represents mental habits and attitudes rather than facts to be acquired. It leaves the child with a healthy desire to learn more about that particular object. The nature lesson is never complete but at all stages is a unit.

Closely allied to the rule that a Nature-study lesson can never be complete is the necessity of paying attention to the rate of questioning. In a Nature lesson as in piano lessons the first aim is to develop movement, accuracy, and appreciation. In piano study there is a great deal of drill in technique. In Nature-study there is a great deal of drill in the technique of observing and thinking but almost never in memorizing the subject matter. The rate of movement depends on the speed of accurate observation and reasoning. As the pupil gains in power the teacher moves to finer and more difficult observations. She does not give a question or lesson because it comes next but because the pupils have reached that stage of development. If she moves too slow or too fast it is an injury to the pupil. The pupils should practice as fine and complex movements as they can with efficiency. When the class has reached a certain degree of skill a few questions may be written on the board a day or two before the lesson so that each one may have an individual opportunity of working out the answers. These questions should not be exactly the same as those to be asked in class. Some individuals take to questioning more eagerly and more intelligently than others. Some individuals are handicapped by physical or mental hindrances. In class the questions should be simple and at a rate slow enough for every individual yet new enough to arouse the quicker ones to activity.

I. *Pupil-Questions in a Nature-Study Lesson*

The most ordinary mistake in teaching the nature lesson is made in training the pupil to wait for an invitation to think. He is taught to build with the straw carried in the teacher's basket. This is a "flapper" method of teaching. "This flapper (teacher) is likewise employed diligently to attend his master (pupil) in his walks (thoughts), and upon occasion to give him a soft flap (easy question) on his eyes, because he is always so wrapped up in cogitation (invention and discovery), that he is in manifest danger of falling down every precipice, and bouncing his head against every post (learning by experience)."* Nature-study offers the teacher an opportunity to break away from this flapper method for the nature lesson is admirably adapted to keeping the pupil active minded.

Younger pupils are inquisitive. They are bubbling over with questions. Parents and teachers usually set about to put on the quietus. They preach the doctrine that little children should be seen and not heard. By the time the pupil gets to college he is probably subdued or dead enough to be called good. Goodness is not deadness. He has became a daydreamer or star gazer and does not search for an answer unless questioned. We should reverse this order. We should furnish problems to discover as well as to solve, purposes to be thought out as well as to serve. Teach the child to use his own mind so that he will go on using it.

The first thing then is to put the ING into every Nature-study lesson. We learn by doing with emphasis on the ING. One may study swimming and know the rules perfectly yet not know how to swim. One may study English Grammar through all the grades— know all the rules—and not be able to apply them either correctly or gracefully. How long would one have to watch basketball or tennis in order to qualify on the team? We learn to swim by swimming, to talk by talking, to play a game by playing the game. In the same way we learn to think by thinking and not by looking on. We learn to enjoy nature problems by enjoying steps in nature problems. We learn to appreciate the order of nature by appreciating the orderliness of the individual. We learn to sense problems by focusing our attention upon discovering them.

Bird study teaches us the psychology of attention. Everyone has had the experience on a bird trip of having eyes and seeing not or having ears and hearing not. If one would find the bird from whence comes the burst of song he should keep his eyes focused on that spot. If one would discover the bird that has just made a certain track it will not do to let his eyes wander across the fields, looking now upon the pines and now upon the clouds. If one is trying to find a bird's nest one should keep his eyes upon the returning mother bird and not upon the tree. As soon as we begin to examine the shape of its bill, its color, or its size, our attention is shaped to these ends. The pupil sets the question or problem. If he selects the shape of the bill it is his business to keep his eye on the right spot and to let his question form itself accordingly. If one cannot learn to observe correctly indoors—to keep his eye on the mark—it is likely that he cannot learn

---

*Swift Works, Gulliver pt. iii., p. 165. (W. P. N. 1871.) Words in parenthesis have been inserted.

to observe and interpret out-of-doors. "Seek and ye shall find" is the motto for achievement in Nature-study.

The second phase of a nature lesson is the why of things. The observation plus the why is the scientific method of attack and accomplishment. Newton, Watt, and Edison held their attention on the mark that interested them. They were seeking the answers to questions. They asked many questions and discovered answers in certain facts related to their problem. After persistent work over a long period of time they solved the larger problem. It was not the mere falling of an apple that enabled Newton to discover gravitation. The moving teakettle lid was observed by Watt and he reasoned the why, but that was only one question and answer in the invention of the steam engine. The boy Edison was continually asking questions by experiment. His success today is based on the same method of interest and attention. Search out and write down the steps in the invention or discovery of any great contribution to human progress and you will have before you the method of developing a nature lesson. It at once becomes the teacher's function to arouse problems and not to retail questions.

How can this simple process of discovery in the wide world be applied in the school room? If it is bird-study and the first lesson of this kind the teacher may announce that every bird is a bundle of questions. We cannot see these questions unless we hunt for them. The bird's tools are his bill, feet, tail, and wings. What question comes to your mind about this bird? The purpose of each bird study is not only problem seeking and solving but to acquire the feeling of appreciation of the adaptation of birds to their home and food. The teacher must guide the pupils to attain these desired results. Each question that serves this purpose must become a profitable experience. The teacher does not answer the question. The first privilege to answer pupil-made questions is that of the pupil who asks the question. This may be brought about by an adroit question from the class or if need be from the teacher. If it is such a question as: "What is the name of this bird?" or "Where does it live?" or "What does it eat?" a natural need has raisen. If our education permits it we solve this need in everyday life by the library method. The teacher will have anticipated this need by having a shelf-library in the room of books containing subject matter on that particular bird. This is the time for initiating library work. The teacher is a librarian and recommends a certain book for finding the name. The pupil who asked the question has found a job and goes to work. As each pupil accumulates experiences he becomes able to recommend books. He will soon become better able to recommend a book in his line of investigation than the teacher. The use of these tools of knowledge is a valuable habit to acquire. When the habit of finding knowledge in books is fixed he should next get the habitof going to the library to find books for himself. First the book habit—second the library habit. If the question had been "Why does this bird have such long legs?" some pupil in the class might see the necessity for knowing where the bird lives. The teacher will recognize that "Where the bird lives" is a question for library work but the question "Why does this bird have such long legs?" is to be thought out. The pupil who asked this question has appointed himself a problem to solve and goes about it. The part that the student takes in the work increases progressively as the lessons advance. He is learning to employ scientific procedure in solving his problems. He is getting a winter school experience to take to the fields in spring where he will meet further problems and situations. This phase of the lesson we will call the pupil-question phase.

Up to this point the Nature-study lesson has been quite different from its nearest relative: the Herbartian development lesson. The five steps of Herbart in its current form are: 1. Preparation (based on experiences of the pupil); 2. Presentation (introducing new cases and experiments); 3. Comparison and abstraction (accounting for differences); 4. Generalization (formulation); interpretation (application to everyday life). The Herbartian lesson is used in the development of principles, laws, rules and definitions as based on facts. It is well adapted to a series of such lessons as the observation and prediction of the weather; the transmission of heat in the home; the use of electricity; light in its relation to living, or machines and work. These lessons relate directly to the laws of physical nature.* Some call the study elementary science, others general science, thinking of it as having a distinctly utilitarian value whereas pure Nature-study lessons have a cultural value. Although animal and plant lessons do not follow these five steps the teacher of the nature lesson should know the steps in order to introduce them as the need arises in dealing with each adaptation. The plant or animal lesson does not start with a review. It may start or end anywhere, perhaps with an experience, or comparison, or application, or possibly in a new field, or amidst a "hundred questions" from the pupils. The nature lesson does not necessarily have application. The teacher, however, who can teach a development lesson is more apt to succeed with the Nature-study lesson.

The teacher-question phase of the Nature-study lesson is necessary to fill in the gaps of the pupil-question phase. Teacher-questions are devices for stimuli. Teacher-questioning is an art. Some say that the ability is born and not made. The writer believes that any teacher who is willing to work and give attention to the art can acquire it and with this conviction will endeavor to contribute in some measure to the mechanics of making the art natural. This will be the subject for the next discussion.

II. *Teacher-Questions in a Nature Lesson*

Teaching Nature-study or any other subject is often much like lumbering. It is said that only 40% of a tree reaches us. The pre-

---

*Physical Nature-study is decidedly worth while. Nature-study lessons in this Chapter, however, refer to Biological Nature-study, i. e., plants and animals.

ventable decay is equal to one third of the growth. The waste left in the woods increases the fire hazard. A thick saw produces one sixth as much sawdust as useful lumber. And then there is the manufacturing waste that adds to the forest drain. If this is true with precise machinery and measurements how much more true it must be in teaching. If we are willing to work there is much that we can do to improve our teaching.

An embryo teacher usually fires questions the way that she would fire a shot gun. Instead of aiming at the bull's eye she shuts both eyes and shoots at the broad side of a barn—the shot spreading without effect. In an effective development lesson one must use a rifle and aim at the target with eyes wide open. To say (1) "Observe the wing feather" is like saying "Gaze at the wing feather." There are several bull's eyes to the wing feather,—such as: color, shape, position size, weight, texture, parts, arrangement of parts, and size of parts. The teacher must decide where to aim.

If the question is (2) "What is the color of the feather?" the pupil can focus his eye and thought on the color of the feather. He may note that it is white. But why should he know the color of this feather? It came from the wing of a white hen. The color surely does not serve to protect the hen because the white hen is always white. She does not change color with the seasons. The white hen lives in an enclosure. She does not need to hide from the fox or any other enemy. The teacher must not only decide where to aim but must aim at something that will bring a score when hit.

The wing feather has a distinct shape. Its shape is different from all other feathers. The teacher asks (3) "What is the shape of the wing feather?" If the answer is given orally the benefit may fall to one pupil. If each pupil answers the question by a quick outline drawing on paper he must make his own observation. It is not the goose step method of expressing ideas. Glance over the drawings and select a good draftsman to make an outline drawing of the wing feather on the board. While this is being done say (4) "Write one word on your paper that tells the shape of the feather." When the drawing on the board has been completed send another pupil to the board to make a list of the words used to describe the shape of the feather.

While the pupils are giving their words to the secretary at the board the teacher may walk around and look at the outlines. She will say something like this: "How could you improve your drawing, John?" "What did you leave out, Mary?" "That is fine, Henry." By the time that the pupil-assistant has a complete list of words used to describe the shape of the feather written on the board the teacher arrives back at the front of the room. The list of words may be: *long, oblong, rectangle, curved,* and *smooth.* The teacher says "What words do you object to using to describe the shape?" Someone may object to the word "long" because it refers to size and not shape. When all the misused words have been eliminated have the children vote for the word or words that best describe the shape of the feather. The teacher must so frame her questions that every individual will have a shot. If he misses fire or in the aiming the teacher gives the better marksman opportunity to assist in correcting the eye-piece, the weight of the gun, the kind of ammunition, or whatever the thing is that needs correction. The teacher is the chief guide—the pupils who shoot best, assistant guides. Where there is weakness try to strengthen it and where there is strength try to make use of it.

The teacher now needs to guide the pupils in the discovery of how a curved, oblong feather aids the bird in flying. She realizes that the pupils do not know and, therefore, will not waste time by asking them the question. As she has foreseen this difficulty she has provided material to help in the discovery. She either has a stuffed specimen with the wing spread or a live bird. Hold this feather near the wing. (5) "Which is the underside of the feather"? Open an umbrella and have the pupils jerk it down against the air. (6) "What is the result?" (7) "How does a curved wing help the bird?" (8) "How does a curved wing feather help the bird?" Have the pupils bend the feather in the direction of the curve and against the curve. In which direction does it bend easier? Of what advantage is this?

The pupil may now be able to find other bull's eyes. The teacher guides in this by saying (9) "What else do you notice about this feather that helps the bird in flying?" If the pupil does not get a reaction in the form of a query—he should as time goes on—the teacher furnishes the rebound by asking about the size. An embryo teacher might ask (10) "What is the size of the wing feather?" But she does not wish to know the size of the wing feather at all. She does not care whether it is 5 or 6 inches long, whether it is ½ or 1 inch wide. She should wish to know its size compared with the other feathers. The pupil now has another experience. He must make two aims and compare the results. He decides that the wing feather is larger than the other feathers. The child should be given the opportunity to ask (11) "Why should the wing feather be larger than the other?" It is easy to recognize that a larger feather is better for beating the air in flying. A good nature lesson consists in giving the child conditions and opportunities to observe and question. This is not new in theory but rare in application.

The teacher again tries to get a new observation from the pupils. She may pass the feather to a pupil and as she does so she may lift it up and down as though weighing it. The pupil may get this gentle hint and say that it is light in weight. The teacher should commend this observation as it may be the first observation without an oral question. It may dawn on several in the class that this is what is wanted. It may be the first time that the pupil has had the sensation of originality. It may be the foreglow of greater discoveries. Have them feel that it is a game of life in which everyone is trying to make original observations and interpretations. The value of a light feather to the bird is an easy step.

Try for another discovery. As an introduction—show a down feather. (12) Ask the pupil to "observe something distinctive about it." (Fluffy.) (13) "Where are the down feathers on this bird?" (Next to body.) (14) "Of what use to the bird are these fluffy feathers?" (Keep warm.) Hold up wing feather and say: "I want another discovery." It ought not to be necessary for the teacher to point at the base of the feather or to look at the base for the pupils to discover that the base of a wing feather has fluff. Do not let the first pupil tell what he observes. Have them indicate that they have observed something by standing or raising the hand. Sometimes have a race between the boys and girls or between the left hand side of the room and the right hand side of the room. Get the spirit of competition. When there is a lull in the response call for the observation. It is poor pedagogy to allow the answer to be given right away and it is poor pedagogy to wait for "all hands" to get the answer. There is a psychological moment. Give the most observing side a score and then say: (15) "Write on your paper what you think the next question should be." Have several given and then have the class vote on which question should come next. Then say: (16) "How many had this question: What is the advantage of the fluff being at the base of the wing feather?" Add the score again, and possibly remark that the girls are one point ahead of the boys.

The last discovery was made possible through comparison. Another method of discovery is by feeling. Have pupils pull the barbs apart. Call the barbs "things" until pupils see the need of a better name. Lead them to discover that something holds the barbs together. If possible have pupils want a microscope to find out what it is that holds the barbs together. If they do not know about a microscope, if they have never used one, the teacher must show the need of this new instrument. Even when children have acquired the power of self-activity the teacher cannot become a negligible factor for such occasions as this require guiding. This is the function of the teacher. When the child sees the hooks on the barbs through the microscope he has entered a new world of observation. It will be a simple step in logical reasoning to arrive at the conclusion that the hooks help to keep a web which gives feathers their value for flying.

Every question has been a test question. It has tested the power of observation or the ability to reason. If the lesson has ended, and it may end at any time, it should end with an appreciation of the feather in particular and of its adaptation to the environment in general. The tendency in the "next lesson" has been to give an examination in the subject matter of the preceding lesson asking such questions as: What is the shape of the wing feather? Name the parts of the wing feather. The tendency should not be to test the subject matter covered but to test the facility acquired—observation, reasoning, initiative. The following questions are to test these qualities: They are written on the board to give each individual time to think out the answers. Compare the wing feather with the feathers in the wing that is spread. Which is the forward edge (1)? Which is the back edge (1)? Label these in your drawing of the wing feather. Which edge is the quill nearer (1)? Show this in your drawing. What is the direction of the barbs in the forward web of the wing feather (1)? In the back web (1)? Show the direction of the barbs in your wing feather. If you have observed correctly you will have won five points. Write a sentence telling about the front web on the wing feather. Reason out the advantage of its differing in this way from the back web.

Then there are questions which take a still longer time for deliberation. We may think of them as research questions. These too develop the scientific habit. They may be written on the board after the nature lesson, or better, before the lesson but covered with a chart or something so that they will be brand new and may be shown at the end of the lesson when the interest is supposedly at a high pitch. (1) What uses have people made of feathers? This bit of research would be amongst friends and parents. It might lead to a small museum. The exhibition spirit is a fine thing to develop. The uses of feathers are: filling pillows, and cushions; old fashioned quill pens; brushes; dusters; trimming hats; fans and toothpicks. (2) What is the story of the egret? This will require library research and a report in the form of a story. (3) Why do hens have feathers in summer? This question will require thought and possible discussion with a science teacher. (4) How and why did the ancient Peruvian and Mexicans carry on feather embroidery? This will mean reading in a history book and possibly the assistance of a history teacher. It may lead to a study of Indian feathers, possibly to the bringing in of decorated feathers, or even to the painting or dyeing of a few feathers. (5) What are the styles in hen's clothing? (6) Why are ostrich feathers so thin and loose? (7) How do feathers serve as rain coats for birds? (8) How do feathers serve as underclothing for birds? (9) How does a feather grow? (10) How many adaptations can you find in a tail feather? (11) How many kinds of feathers can you collect from one bird? Mount them with labels on cardboard. (12) Collect the feathers of different birds. Mount them on cardboard. The pupils are asked to select the topic that interests them most and get a report ready. Research questions lead to projects.

These teaching lessons on the wing feather have been given as a means of introducing the complexity of the art of questioning. If a teacher goes through such a process of planning and reasoning, by way of preparation, for each lesson, she will undoubtedly make great progress and will become a far superior teacher than though she depended upon the inspiration of the moment. In the first lesson, question number 1 is too general. Question number 2 is useless and in question number 10 the teacher asks one question and means another. As time goes on a progressive teacher will reduce the number of poor questions. In place of a thirty-minute teacher she becomes a twenty-five, a twenty or even a ten minute teacher. Her pay should be commensurate with her efficiency. The indictment that too few teachers acquire the art of questioning has always been true and now with the reaction against the memoriter-recitation on the one hand and the tendency to all-free-work on the other come other excuses for neglecting the art.

III. Some Suggestions for Teaching Nature-Study

At the risk of being misunderstood or being thought pedantic further suggestions are offered:

1. *Know your pupils:*

The headmaster of a large city school saw an Italian working in a vineyard. The following conversation took place. Note that the trend of thought in each mind was entirely different. To get the other fellow's point of view is difficult but essential for the best teaching. "Do you get many grapes?" "I got about twelve barrel of wine last year." "Do you sell the wine?" No, I have nine children at home." "Do you think it is good for them?" "Yes, it is a great deal better than beer."

2. *Learn with Your Pupils:*

Too often a pupil learns in spite of his teacher. How often this is true in the case of the small boy and his radio? Most boys could teach their teachers if they were teachable. We are still timid about learning from our pupils. This conversation was heard in a garden class in a city school room.

Teacher: "What color are radishes?"
Child: "Radishes are white."
Teacher: "No."
Child: "I know they are."
Teacher: "You probably mean horse radish."

A similiar episode is as follows:

Teacher: (A pupil teacher who has learned "up stairs" in a normal class that locusts eat grass.) "What does the locust eat?"
Pupil: "Lettuce."
Teacher: "No it doesn't eat lettuce does it?"
Pupil: "Yes, I have seen it."
Teacher: "No, grasshoppers eat grass."

What the pupil discovered and what the teacher did not discover from this conversation is the cue to the difference between respect and distrust. The lack of a wide experience can be excused in a teacher but the lack of honesty with the pupil never.

3. *Keep the Scientific Spirit:*

In contrast to the above experience is the following episode. Pupil teacher number 1 taught in the Third Grade that bats do not carry on respiration in the winter. A month later pupil teacher number 2 taught in the same grade that bees hibernate in the winter. A Third Grader asked if bees breathe in winter. The teacher said: "Yes, all animals do." The Third Grader: "Bats don't." Teacher: "That is interesting. I didn't know that. I will look it up as I still think that all animals breathe in winter." Pupil Teacher number 2 asked the normal school science teacher who said: "The bat does not breathe but he would have to carry on respiration—the exchange of oxygen and carbonic acid gas." Pupil Teacher Number 1 heard of this and looked up her authority which was an article in the Scientific American Supplement for November 29th, 1913. She brought this to the Normal School Science teacher who said: "I am glad that you have an authority for your position. However, I am not convinced. Suppose that we investigate further." Both went to the Encyclopedia Britannica. This corroborated the Scientific American. The Normal School teacher then called up the Chemical Physiologist of a neighboring university. The university man said that he doubted the statements in these two sources. He gave references on the subject in certain physiological journals. The final outcome was that Pupil teacher number 1 reported back to the Third Grade that she was glad that they had remembered what she had taught. She also said that she had since learned that scientists do not agree on this point and that everyone would have to wait until more experiments had been performed.

4. *Use Tact:*

Teacher: "Are canaries kind to each other?"

Pupil: "Yes, we had a mother bird building a nest and the father bird helped her carry cotton and straw."

Teacher: (Who had read that males are cross) persisted with her poor forms of questions until she finally told the pupil that male canaries are cross. How much better it would have been to have accepted the facts of observation and idealism instead of substituting book knowledge without an ideal. When some teachers have a plan made they cannot change it. They must have their own way no matter what happens. In such lessons we might verily say that those who are taught the least will know the most. In nature lessons it is important to specialize in children and not in Nature-study.

5. *Ideas Rather Than Complete Sentences:*

A rather common "plague" in the class room is the insistence on a "Complete sentence." If the teacher says: "What is distinctive about the heron's legs? and the child says "Long" instead of "The distinctive thing" etc. he has implied what is given in the longer expression. The time saved in the brief answer gives time to make complete sentences in summarizing rather than parroting a question for the time being. Intelligent conversation in society does not require the machinery of "complete sentences."

6. *Weigh the Value of Facts:*

The Nature-study lesson does away with the evils of the recitation period of the "one book" age. Memory questions are practically eliminated. This does not mean that there are facts that do not need to be remembered. Every child should know that birds eat insects and that flies are dangerous. The child acquires this knowledge by frequent contact in the same way that he learns the use of soap and the dangers of a cold. We do not ask him to define soap or for a review on the use of soap. If a child learns, on a visit to his grandmother's, that feathers are used for a duster, that the turkey's wing is used for a brush, and that grandfather uses a quill to pick his teeth we do not say on the next day: "Name the three ways in which feathers are used." Neither is it necessary in school in the lesson following the teaching of feathers to say: "What is the shape of the wing feather?" or "What is the advantage of feathers being light in weight?" Nature-study is not drill—it consists of nature experiences. The worthwhile experiences, such as observing, should come often. Test questions, fact questions, and drill questions are used sparingly in a Nature lesson when referring to subject matter and judiciously when pertaining to the powers of observation.

IV. Some Forms of Questions Worth Using in Nature Lessons

1. *Observation Questions:*
On what part of the tree does the chickadee feed? The nut hatch? The woodpecker? These are observation questions. They are good and should occur often in the nature lesson. They are of most value when presented to the pupils several days before the lesson, thus giving them individual opportunity to observe in the field. It also allows time for discussion ad liberatum.

2. *"What is it" Questions:*
One of the commonest questions in bird study is: "What bird is that? This is a good question. There is a difference, however, between knowing the name of a bird and knowing how to find the name of an unknown bird. The latter experience is much more fruitful. The naming of birds may be used as a game. A "what is it" shelf is a valuable adjunct to the work shop. This type of question has been overworked in the past.

3. *Comparison Questions:*
Compare the arrangement of the toes of the flicker with the robin. This calls for comparison. Comparison questions are good as they require reasoning.

4. *Judgment Questions:*
Should be frequent. The class may have given several partial answers. Call for a final judgment. Judgment without prejudice is called for in asking for criticisms. Criticism must be constructive and not destructive. It means to speak of the good points as well as those that are not good. It is given in a spirit of helpfulness. It is not faultfinding, praise, or blame. With the word of censure ought to come the word of commendation, as, "It was difficult to hear you, James. What you have to say about your pigeons is worth hearing." Do not accept an indefinite criticism, as, "Tell that once more."

5. *Organization Questions:*
Each kind of bird has feet and a beak especially adapted for getting his food. Arrange these birds according to their feet, or beak. In

arranging the birds the pupil is getting an elementary experience in classification. Pupils should have more experiences in organization.

6. *Analysis Questions:*

Of what materials are these nests composed? This is a good winter study when the nests have been abandoned. They discover that the lining may be hairs, feathers, or plant materials. This may lead to the query as to what animal they came from. Nests may be woven or plastered. They may be camouflaged. A. R. Dugmore's Bird Homes is a good reference book. An analysis question is good because it leads to other questions. It may be thought of as a questioning question. It is the most difficult to create yet the most powerful in getting a reaction.

7. *Problem Questions:*

(a) Work up a folder to attract people to a bird sanctuary for a week's camping trip. (b) Once upon a time an ornithologist was wrecked on an island near Florida. He was alone. What could he do? (c) What birds could exist if there were no plants on the earth? (d) How can we attract birds about our home? (e) How should we organize a bob white farm? (f) Plans for a bird day program? A generation ago questions were used to test results, and came at the end of the learning process. In nature-study, as in life, problems arise that require solution and they arise as an introduction to the learning process.

The value of a problem question varies directly with the extent to which it provides for motive power by the pupil. The best problems do not arise by "cranking up" but are "self starters." The answer is desired at the time. Some way is sought to solve the problem. A boy says: "How can I get birds to the window for my crippled brother? He is greatly interested in birds and cannot get out to see them. I want to help him." This is not an answer that requires a question but a real problem to be solved. A problem question is not an application from without, like a plaster to draw from within, nor is it a wet blanket thrown about one that chills within. Real problems like charity begin at home. Real problems like all people are self-made. A real teacher provides the occasion for real problems.

8. *Adaptation Questions:*

An adaptation question brings out the reasons for the things observed. They pertain to the varied structure and habits of plants and animals as adapted to their varied circumstances and conditions of existence. They usually begin with Why or How. They give the point of view of modern conceptions of plant and animal life. Adaptation questions fix ideas and make the old fashioned drill unnecessary. Present day teachers of Nature-study recognize that the enlightened thinkers in science depend upon this training. The power to reason depends to a great extent upon this form of question. The following are given as examples of the Adaptation question. Why are strong claws useful to the eagle? How does a long beak assist the humming bird? Of what advantage is the spine on the end of each tail feather of the chimney swift? Why is silent flight advantageous to the owl? There are many incorrect forms of questions heard in the Nature lesson. Some of the more frequent errors are to be discussed.

V. Common Mistakes in Questioning

1. Plan every question:
What does the flicker's bill look like? It looks like a flicker's bill. The teacher really means to have the pupil name some tool that the flicker's bill resembles. It would be better to say: The flicker is a workman. Workmen need tools. Find one tool that this workman uses. How does he use it? It would be misleading to say that the flicker is a carpenter. This would lead the child to think of the following tools: chisel, hammer, drill. The bill is a pick rather than any of these tools. The flicker does not bore holes as is often thought.

2. Questions should have sequence:
How can you tell a gull, when flying, from a duck? How does a gull carry its food? In what direction does a gull rise from the ground when a strong wind is blowing? Why is it a sign of ignorance for people to shoot gulls? Why does a gull drop suddenly toward the water? How does a gull hold its wings when soaring? What is the food of the gull? In what direction do gulls fly? The sequence is lacking in these questions. If the teacher says: What is the food of the gull? and then follows it by: Why is it a sign of ignorance for people to shoot gulls? the second question has a direct bearing on the first. One question prepares the way for the next.

3. Questions should not commence in the affirmative form:
An egg consists of what? The English Starling was introduced into America when? The red winged blackbird is apt to live where? This form of question usually indicates that the teacher has not thought out the question before beginning it.

4. Catch questions may be well for a joke but not for serious business:
What month has 28 days? Answer: They all do. Where is your gizzard in relation to your stomach? Answer: We do not have a gizzard.

5. "Yes" and "no" answers waste time:
They require another question such as, why? "Does the web unite all the toes?" instead of "How many toes does the web unite?" "Are all the toes alike?" instead of "What is the difference between the toes?" "Are the feet of the hen like those of the duck?" instead of "Compare the feet of the hen with those of the duck." Is it the practice now to wear feathers on one's hat? This is nothing but a question in lecture form.

6. Too many questions do not give the children time to think:
A nature lesson should not consist of short steps, short questions, and short answers. The tendency to ask too many questions is usually characteristic of the teachers who are trying to have the pupils

acquire more facts or are preparing for an examination. They miss the aim of the lesson.

7. *Ask one question at a time:*

Describe the bluebird's song? Do they sing all summer? The teacher asking the second question before the pupil has opportunity to answer the first. Sometimes teachers do not like the form of the first question and get out a second or third edition. This confuses the pupil. If the first question is not quite in the best form it is better to let it stand as given.

8. *Do not ask useless questions:*

Why are radishes red? How long is the wing bar? What is the color of the left eye? How many tail feathers are there? Useless questions have no aim beyond the asking of them. No use is made of the answer. They are mere time killers. Do not say "What do chickens eat?" when everyone has chickens to feed at home.

9. *Do not go through meaningless formalities:*

Such as—calling the roll when you know that they are all there or asking if Johnny Jones is sick when you know that he is. Dull questions kill the spirit.

10. *Some facts are to be stated:*

The cat-bird is a mocking bird. The turkey is a native of America. The crop of the hen is a storehouse for grain. The heath hen is now limited to Nantucket Island. The passenger pigeon is extinct. These facts cannot be observed. On the other hand there are some facts that should be observed. This means careful preparation on the part of the teacher. Some teachers "beat about the bush" or try the "pumping process" or "pedagogical dentistry" when there is nothing within to be extracted. She should know that some facts are to be observed, some to be told, some to be reasoned out, and others to be forgotten.

11. *Questions should be definite:*

What happens to feathers in the spring? A question like this leads to guessing. They may get wet, soiled, or disarranged, they may grow, etc. What is meant is: What becomes of the old feathers in the spring? What is the shape of the flicker's tail? In this question the whole tail is mentioned when the end of each feather in the tail is meant. The question should be: What is the shape of the end of each tail feather? Other forms of indefinite questions often heard are such as: What about the feet? What can you tell about the feathers on the owl? What do you think of this topic?

12. *Questions should show a scientific background:*

Such questions as, "Where are the eyes placed" in referring to the owl or rabbit evidence an inherited concept. Eyes, ears, legs, feathers, and hair grow. They are not placed or stuck onto the body as may be the method in making animals for a Noah's Ark. This form of question is the relic of mediaevalism. The attachment of muscles probably arose from the same source. The surprising thing is the durability of a stock question when once it gets into a scholastic vernacular. "Where are the eyes?" is sufficient.

13. *A knowledge of the origin of adaptations is necessary for clear questioning:*

A teacher should first of all know that every characteristic is not an adaptation. Some structures are not as efficient as the environment demands and some are more perfect or highly developed than the environment could bring about. A characteristic might arise through artificial or natural selection; mutation; disease; arrested development; or by accident. In most cases we do not know how a certain structure originated.

Why do you think that the quill is hollow?

Why is the quill hollow?

What are the advantages of a hollow quill?

Each of these questions is distinct from the others. To fail to distinguish their aim leads to a confusion of mind. The first question should never be given when the answer to the third is the one sought. One may think that it is hollow because of the weight or transparency. It may be hollow because of natural selection. A hollow quill is lighter in weight and is stronger than a solid quill. As a matter of close observation the quill is not hollow, in the sense of being empty, any more than are heads or eyes which are said to be hollow.

Why does the cactus have spines?

Of what use are the spines to the cactus?

The first question implies that the spines are "purposeful" to protect the plant from browsing animals. If such is the case, why are there so many cacti on the plains of Texas without spines? The second question allows the possibility of spines having originated by mutation and yet is not offensive to those who feel duty-bound to the idea of "creative design." Questions are the red corpuscles in the flow of nature-study ideas. The kind of question determines whether these ideas burn brightly or whether there be left a quantity of smoke and ashes.

14. *Do not be effusive:*

Will you please tell me the use of feathers to birds? A polite person will answer the question asked. The answer should be "yes" or "no." The teacher should expect the pupil to answer. If she does not show that in voice or words, the pupil is not going to exert himself to answer the question. The please is not necessary. It should be the pupil's pleasure as well as the teacher's. To answer a question is an opportunity. A pupil should tell the class as well as the teacher.

15. *Use questions in the present tense in Nature-study:*

Nature-study is not drill or review. It is *now*. Where did we say the ears were yesterday? What shape would you say each egg was? Where were the king fisher's toes? The correct form is evident.

16. *Do not use alternatives:*

Are the chimney swift's legs long or short? The question intimates that the legs are either long or short. Even if the pupil is not a good guesser he has 50 percent chance of getting the answer correct. We should not encourage guess work.

17. *Do not use definitions:*

What is a bird? This type of question demands a definition and is suitable for a course in comparative anatomy in college. Since everyone uses such terms with accuracy in conversation this type of question in the grades is mere jargon. It sounds like the chatter of ye old school ma'am.

18. *Do not ask questions where the answer is perfectly well known:*

How many legs has the cat? is a waste of time.

19. *Do not ask questions which are too difficult to answer:*

Some teachers ask questions when it is well known that the pupil cannot answer. What is the Archaeopteryx? A good question must be on the level with the person questioned.

20. *Do not repeat answers:*

The teacher who repeats answers is acknowledging that the pupils do not talk loudly enough, clearly enough, or correctly. Teachers are apt to display their mannerisms at this point such as: "That will do," "yes," and "'er," "now," "well," "what about." If a teacher says "That will do" twenty times in a lesson and for five lessons per day, that amount of time is wasted. These expressions are sort of post mortems.

"It is due to every child that his mind be opened to the voices of nature. The world is always quick with sounds, although our ears are closed to them. Every person hears the loud songs of birds, the sweep of heavy winds and the rush of rapid rivers or the sea; but the small voices with which we live are known not to one in ten thousand. To be able to distinguish the notes of the different birds is one of the choicest resources in life, and it should be one of the first results of a good education. It is but a step from this to the other small voices,—of the insects, the frogs and toads, the mice, the domestic animals, the flow of quiet waters, and the noises of the little winds. It is a great thing when one learns how to listen. At least once, every young person should sleep far out in the open, preferably in a wood or the margin of a wood, that he may know the spirit and the voices of the night and thereafter be free and unafraid."

—*L. H. Bailey in "The Nature-Study Idea."*

"That which is first worth knowing is that which is nearest at hand. The nearest at hand, in the natural environment, is the weather. Every day of our lives, on land or sea, whether we will or no, the air and the clouds and the sky surround us. So variable is this environment, from morning till evening and from evening till morning and from season to season, that we are always conscious of it. It is to the changes in this environment that we apply the folkword "weather,"—weather, that is akin to wind. No man is efficient who is at cross-purposes with the main currents of his life; no man is content and happy who is out of sympathy with the environment in which he is born to live: so the habit of grumbling at the weather is the most senseless and futile of all expenditures of human effort. Day by day we complain and fret at the weather, and when we are done with it we have—the weather."

L. H. Bailey in "Outlook to Nature."

President Coolidge, on the occasion of the departure of the American delegation to the international gathering of Boy Scouts at Copenhagen. (July 25, 1924.)
"The first is a reverence for Nature. Boys should never lose their love of the fields and the streams, the mountains and the plains, the open places and the forests. That love will be a priceless possession as your years lengthen out. There is an instructive myth about the giant Antaeus. Whenever, in a contest, he was thrown down, he drew fresh strength from his mother, the earth, and so was thought invincible. But Hercules lifted him away from the earth and so destroyed him. There is new life in the soil for every man. There is healing in the trees for tired minds, and for our overburdened spirits there is strength in the hills, if only we will lift up our eyes. Remember that nature is your great restorer."

CHAPTER XX

NATURE-STUDY AND THE PROJECT METHOD

The scientific method with its series of questions and answers, as related to specific problems, has been described in the last Chapter. The next step in the complexity of organization of nature-study is the project method. A *project* is made up of a series of related *problems* which in turn consist of *questions*. In preparing to meet projects a child must first learn to meet the challenge of questions when they arise. If he can overcome the simple question-answer units of difficulty he is ready to meet more complex situations. If the child has been guided so that he uses his initiative in asking and answering questions he naturally comes to the necessity of carrying out larger social activities. The project should be used when the child has reached that stage of development. It is the attitude of mind of the child that determines the introduction of the project rather than his number of years or his grade. A project in its final analysis is nothing more than a purposeful continuity of questions.

The advance to the project stage should be gradual, the question always being an important feature. In a teaching lesson the majority of questions are asked of the group of children whereas in a project lesson the majority of questions are asked by the individual pupil, varying according to the individual. There is no prescribed list of questions from the standpoint of the student. Each asks the questions that he meets—that challenge his interest and ability—those with which he makes a personal point of contact—those in the field of his particular enterprise—those which appear on his horizon or in his zenith.

A project launches a practical enterprise under natural, life conditions. Its growth is much like that of a tree. The main trunk gives birth to a series of branches which develop as they may. The roots of the tree are in the soil and those of the project are in the fundamental studies. The growth of each branch is stimulated by the surrounding light and warmth. Every step is a kindred vital problem, rich in content. Every twig and detail finds its course by self-determination, free in scope. There finally develops a rich crop of genuine usable parts. The idea becomes a complete reality. It is concrete in that it can be thought of as a great structure. It ends in the stage of maturity, capable of reproducing, each its own kind.

Nature-study is particularly well adapted to the project method. Like geography, it was at first a complexity of studies. The large list of minerals, plants, animals, and physical phenomena in nearly every course of study deluged each grade until the machinery of nature instruction was thrown into confusion and despair. Nature education until recently considered the individual child and how many trees he knew rather than the child as a member of a group which demanded that the forests be protected. Emphasis is changing from individual ends to an education for forest citizenship. Until the

251

present time there has been room in the woods for hermits to live unto themselves but with the crowding of population to our timber-lines there are demands for the rights of the community. We are sending our children to the woods in groups. This means that the child must be taught to live in the woods in the right way. He must respect the rights of his group and the groups to come. The tendency in modern nature-study, to teach large natural units to groups instead of minor facts and statements, to individuals is in harmony with the project method.

Nature-study is rich in manual projects. Manual projects are fundamental. Our physical, mental, and aesthetic development must be based upon a material foundation. There is a certain attractiveness to a project that enables the child to release muscular energy. The muscles are at the same time being trained in the use of tools and in the handling of materials. The simplest projects are those in which the child presents a concrete object as a result of his experience. They have something to exhibit. Visible results are more stimulating. As the study of the natural environment demands an actual experience with materials and forces, the majority of nature projects will deal with concrete materials.

Manual projects out of the natural environments were the projects of the colonial home. Bayberries for candles, corn husks for mats, sweet grass for baskets, cattails for chair bottoms, herbs for medicine, flax for spinning, beach plums for jelly, flint for fire-building, hickory for axe handles, berries for dyes, wood ashes for lye, sand for scouring, and straw for thatching were a few of the industries. One by one these projects dropped out of the home and went to the factory. The schools went on merely emphasizing the three R's—reading, writing, and arithmetic. Thus for a long time the arts of the old homestead were lost. Now they are coming back to enrich the curriculum in the form of projects. These are the common products in our artistically furnished homes. These simple projects are the prominent features of our early American literature. The fundamental social needs and activities of present day society are represented in simple form by the arts of the old home.

Nature-study not only furnishes situations for vital projects but the project method is a means by which nature-study can realize several present needs: (1) To have purposeful aims. (2) To evolve ideas around a central unit. (3) To organize larger units. (4) To practice nature-study ideas in the open. (5) To enrich each lesson with appreciation to be continued into later life.

In developing these nature projects the teacher must distinguish between the individual projects of the child—the care of pets, a camping or fishing trip, building a shack, collecting, camera hunting,—and the larger social projects of the country such as the protection of birds, conservation of the forests, or the control of the spread of insects. The individual project aids the child to understand the full meaning of world projects. The teacher must recognize and guide each project.

The first step in the development of a project is to furnish material background. Classes usually suffer at the initial stage of the nature project because of a lack of material. Teachers who do not prepare lessons or who make teaching a daily routine will say that the children will gain in power if they find out for themselves. Progressive teachers will provide, living material, a working library of books, pictures, drawings, charts, models, and pamphlets. The children will emulate the teacher's example and add to the richness of the material. Textbooks are too scant in material to be of any use in

Troop 27, Greater Providence Council, Boy Scouts of America, decided to build a cabin

project study. Fortunately there are very few textbooks in nature-study. There are none for the child. On the other hand there are plenty of books and magazines rich in nature material. The only textbook for nature-study is nature. In no other subject is the absurdity of textbook work more evident.

We will develop as an example of a nature project the building of a cabin. First there is a real desire for a cabin. The idea of a cabin becomes basal for the organization and development of a large unit of study. The plan is dynamic in that it is to be carried to a successful conclusion. All knowledge to be used in the enterprise is organized with the building of a cabin in mind. Each problem introduces other problems. Finally the idea grows to maturity and the cabin is realized. The cabin has been the thread of thought which has tied together all the subunits into a complete story.

For descriptive materials we will have photographs, local legends and stories of cabins; plans of fireplaces, bunks, and stoves in cabins; maps showing possible sights; and models of cabins. We may visit some person in the vicinity who lives in a cabin. We may read about Lincoln, Robinson Crusoe, and other cabin builders.

This series of lessons must differ from formal lessons in that it does not furnish a list of important facts to be memorized. Nevertheless facts are important and necessary. The proportion of sand and cement, the laying of shingles, the size of the hearth, the location of the windows, the drainage from the sink, require definite consideration. These facts are all related to the cabin as a central thought. A ten minute drill on these facts is foolish. An examination on the names, dates, and lists of material is a waste of time. Definitions of shingles, cement, and hearthstones, are trivial. A textbook containing generalized knowledge in stereotyped sentences about cabin building is useless.

The Teacher Must Have the Organization of the Project in Mind.

The planning of a project is shown above. The central idea is the building of a cabin, and has a great power of growth. Each minor problem radiates from the center until the cabin is completed. Each new problem or thought enterprise brings on a new complex of questions to be met and solved. These questions deal not so much with the objects of nature as with the processes of nature. If we understand the action of frost we will set the foundation posts below the frost line. We cannot mix cement to good advantage in freezing weather. We have to know how to lay shingles in order to keep out the rain. The slopes and angles of window and door frames are important to keep out driving storms. The problems of overcoming the forces of nature were the problems of the pioneers. With the development of the cabin project comes a realization of the pioneer development of our nation. It is a manual project which results in an objective accomplishment—the cabin. It is a social project in that several individuals are cooperating in a common undertaking. At the end of the series of lessons on cabin building the children should come out into the "clearing" with a broad view. Make it a special day program. Their cabin facts and cabin experiences should give a new interpretation of the world. Their cabin knowledge should have organized itself into a well-rounded unit.

The completion of the cabin again furnishes a center of new needs. The cabin must be made attractive. It furnishes a center for our main physical needs. It is a center for social needs—work and play. It furnishes a simplified environment in which to live and form camp habits and ideals of,—fire-building, cooking, cleaning up, eating and sleeping. This is in line with a social life which continues. The child is to soon step out into the world as a camper, as a tourist, as a cabin builder for himself. His cabin life furnishes training in a real need.

Then will come the need of a local cabin museum to show the poisonous plants; to exhibit the edible plants; there will be need of tree conservation; nature dens; firebuilding places to demonstrate methods of out door cooking; storage cellars; signal towers; a fern walk; and no cabin site is complete without its clear sparkling spring. This is not "playing store" or an "imaginary trip around the world," or "seeing America First" through the movies. It is the real thing. It is a nature-study of concrete participation and experience rather than a list of facts. The fourth project or round of problems in a well regulated cabin-home may be the planning and development of the grounds into a bird sanctuary with feeding stations, nesting boxes, and shrubbery.

This naturally leads to a bird banding club which is more instructive than the hopeless task of acquiring the names, colors, and songs of a hundred listed birds. Mere facts unrelated to a central idea are worthless. It is necessary to focus our attention upon a broad organization with related strategic centers.

The large units—cabin building, cabin life, and life about the cabin—offer freedom of organization and lead to large thought-movements rather than daily lessons which choke the victim with petty details. The mind becomes occupied with big topics such as building nesting boxes or attracting birds and will call in the little facts only as needed. This is a training in the way of the world, both in thinking and in the organizing of knowledge. The teacher does not fuss with each day's lesson as a separate unit but as a big thought-movement for the whole project.

The project also differs from the formal recitation which runs on a time table—such as a twenty minute schedule. The cabin project and its surrounding problems cannot be cramped by class-room bells. The amount of time required cannot be predicted. It is a matter of free growth. It will not stop until the purpose has been achieved. The shrub cannot be left in the midst of transplanting. The time will depend upon the ability of the pupils and the nature of the project. If it become necessary to stop when the bell rings the next meeting continues on the same thought movement branching out like a tree as the occasion demands. A real nature project, however, does not respect artificial time limits or any other Chinese wall for the separation of studies.

The discovery of nature-study and the project method is like the discovery of gold in California or in the Klondike. When teachers really hear the news and sense the possibilities there will be much fine gold and rubies added to the curriculum.

Nature-lore, the Nature Club, and the Psychological Approach have been described as basic ways of acquiring interest in Nature-study. There is a definite school procedure based on these interests: The teaching lesson, consisting of the question and the problem, and the project lesson. In order to make the methods explicit we have described them analytically. These methods, however, are simply explanatory from the standpoint of pedagogy. They cannot be separated in practice. They represent the sources of paints to be mixed for the canvass. The finished production may consist of daubs or it may be harmonious in colors. The result depends upon the artist. The excessive use of one method may become tiresome. No teacher can use the project and separate it from the problems, questions, and interests of the young naturalist. The exact amount of each method to be mixed into the paint pot cannot be prescribed. The proper balance must depend upon the taste of the teacher. Failure in this has probably been the greatest stumbling block in the progress of nature-study. The vertical strata in the teaching of nature-study may be thought of as methods in nature-study and the horizontal strata as *the nature-study way*. The nature-study way is the project method. The Nature-study way is a good method for all studies. It is adapted to the modern school.

A Bayberry Candle Project

Miss Mabel T. Gardner, Grade IV, Henry Barnard School, Rhode Island College of Education, in Cooperation with the Nature-study Department.

It was reading hour in the fourth grade one day last September. Interest was keen in the new books, among which was "A Little Maid of Narragansettt Bay," by Alice Turner Curtis. Agnes, reading this book, became absorbed in the account of Penelope Balfour's plan to gather bayberries, that her mother might begin candle-making. When the period for silent reading ended, Agnes read to the class the different steps involved in the process of making candles.

Questions followed in rapid succession:
"Are bayberry candles used now?"
"Do bayberries grow near here?"
"How do they look?"
"Is it hard to pick them?"
"Can't we make some candles?"
"Does it take ever so many berries?"
"Why couldn't we make some for Christmas?"

These are a few of the questions raised. As a result of the discussion it was decided that we should make candles for Christmas gifts. Our plan was to dip a sufficient number that each pupil might have one boxed with an appropriate verse enclosed to take home as his Christmas gift. Therefore, the question of the various materials required for the carrying out of the project was our first consideration. The organization of committees was begun.

Naturally the gathering of the berries was our first serious problem; and so the location of bayberry pastures was discussed. We found that the waxy berries grow best along the shore, preferably near salt water. The distance of these pastures from the school made

Saturdays the only days available for picking. The work began. How great a proposition we had undertaken was to be learned as time went on and we gathered information on the subject. It was said that a bushel of berries is needed to make one candle. We did not doubt the statement as matters progressed. Nevertheless, everyone vied with his neighbor to contribute even a handful of tiny berries to the collection. The first berries gathered were not rich in wax. Frost tends to thicken the coating. Therefore early picking is not advisable if any large quantity of wax is desired. One pound of berries will average about four ounces of wax.

Word came to us indirectly that quantities of the fragrant waxy berries were growing adjacent to one of the State training schools. Through correspondence between the two schools the co-operation of the children of the training school was enlisted and the candle-making project was inaugurated in full swing.

Some of the correspondence follows:

Henry Barnard School,
Providence, R. I.,
October 4, 1923.

Dear Friends:

We children of the fourth grade want to make bayberry candles for Christmas. We have heard that there are many berries near your school. Would you like to gather some for us? Your friends,

FOURTH GRADE CHILDREN,
S. R.

Bayside School,
Warwick, R. I.,
October 9, 1923.

Dear Friends:

We have started gathering the berries. Already we have quite a box full. We are glad to help you and hope you will enjoy making the candles. Your friends,

FOURTH GRADE CHILDREN,
M. M.

Henry Barnard School,
Providence, R. I.,
October 18, 1923.

Dear Friends:

We are pleased with the effort you have made in helping us gather bayberries. We shall send someone after them Wednesday afternoon if it is convenient for you. Your friends,

FOURTH GRADE CHILDREN,
L. A.

Henry Barnard School,
Providence, R. I.,
October 24, 1923.

Dear Friends:

We received the bayberries that Dr. Alger brought us and we thank you for them. We should be glad to have you gather more if you want to. Your friends,

FOURTH GRADE CHILDREN,
L. A. A.

The class soon discovered that certain equipment and tools, besides berries, were required to begin the candle-making process. Large kettles for the long boiling, a dipper, a saucepan, a knife, and a long-handled spoon were procured from our kitchen supplies at the school for immediate use. A wire sieve was thought to be too coarse for good results in straining the wax, so the mesh was removed from the frame and cheesecloth substituted.   Aprons, and holders to lift the hot kettles, were soon in demand.

The size of the proposed candles was considered early, for upon the size other matters depended, such as the length of the wicking, and of the rods which were to hold the dips, and the dimensions of the boxes for the finished product.   Class groups were formed into the following committees:

Cooking.   This committee consisted of three members whose duty it was to measure the proper quantity of berries and water, light the gas stove in the domestic science room, which was at our disposal four days a week, and to see that boiling was continuous for at least three hours.   At the end of that time the gas was turned off and the mixture was allowed to stand until the next morning.   The cake of wax formed on top was skimmed off, melted, and strained into another kettle in which the dipping was to take place weeks hence. Great care and responsibility were necessary to prevent the burning of the precious wax.   (The use of a double boiler for this and very careful straining through several thicknesses of cheesecloth are advisable.)

On account of the preponderance of the palmatin element in the bayberry wax, it is best to add either mutton or beef tallow or paraffin in proportion varying from one-third to one-half.   The odor of mutton makes it less desirable than beef or other fat bodies less rich in the palmatin base.   The bayberry wax will still keep its green and give its particular aroma if not too much diluted by other fats.

The work of the cooking group was done during the regular period for free activities which occurs at the beginning of each day's session. So seriously did the pupils take their responsibility that they were always at school before their teacher and visited the cooking room to get a peep into the big kettle to find out how much wax had formed from the previous day's boiling.   Nobody in that group was ever late or even nearly so.   Approximately five bushels of berries were used, and the gatherers were kept busy until about two weeks before Christmas.

Sewing.   There was need of this group from the very beginning, that aprons and holders might be ready for immediate use as soon as boiling began.   The chairman of this committee was a little girl, who offered her services and presented proper qualifications.   She chose her fellow workers with the advice of the class.   This group of children learned to cut from a pattern and made six aprons, several holders, and a needle book for each sewer.

Wicks and Dipping Rods.   This group was engaged in measuring. One member cut the wicking twice the desired length of the finished candles, allowing sufficient surplus for the necessary twisting over the rods.   Two others measured and cut the rods which were later to hold four candles each.   The first rods were of wood but later wire ones proved more satisfactory, as they did not spread the wicking so much.

Construction. Six children composed this group. It was their duty to decide upon the dimensions for the boxes and the materials to be used. Patterns were made and offered for class approval. The box finally chosen was made of green construction paper and lined with manila tag. This committee constructed the required number of boxes. It was an excellent project in hand work.

Decorations. This committee consisted of four members, who submitted suitable designs for box cover corners and for cards to be enclosed in each box The bayberry spray was used as a motive. India ink was employed.

Pasting. The work of folding and pasting the boxes was done by three children after the decorating committee had completed its work with them.

Rhymes. The desire for simple verses to be printed on the gift cards created keen competition and furnished material for several language lessons touching slightly on the subject of metre. Some of the final results are quoted.

> "This bayb'ry candle burning bright
> Will bring you luck on Christmas night."

> "This bayb'ry candle you receive
> Will bring you luck on Christmas Eve."

> "These bayb'ry candles we give to you
> With best of wishes the whole year through."

Printing. This group was needed as soon as the verses were completed. It comprised three members, one of whom measured and cut cards.

Woodwork. Several boys desired to make candle sticks. All who so wished were encouraged to bring one made at home of whatever material available. Various sizes and shapes came to school. After much discussion a standard design was decided upon, and nine boys worked persistently with modest equipment. The results were amazing. Several pairs of large candle holders were neatly made and stained, in addition to the regulation ones for the small candles.

Molding. Two children worked at stringing up an old-time tin mould holding a dozen candles. As a result twenty-four large candles were obtained by this method. Two of the mantels in our building are at present decorated with candles and holders presented at Christmas with the greetings of the fourth grade pupils.

Boxing. Six children composed this group, two of whom wrapped the candle in wax paper and fastened it with Christmas seals. The other four boxed and tied each with red ribbon previously measured and cut to the proper length.

When the great day for the first dipping arrived excitement was high. Four pieces of wicking had been looped over each rod and twisted. The rods were hung across runners supported on two ordinary chairs turned back to back a few feet apart. Notches had been previously cut in the light strips of wood to hold the candle rods in place. Newspapers were used to protect the schoolroom floor. The big kettle of hot tallow was brought in and placed on a box to bring it to a convenient height. Certain children were designated as "dippers," all sharing in turn in the dipping of the wicking into the hot wax. As they cooled, the dips were shaped with the thumb and forefinger to make them hang straight. This little precaution is well worth while, that the finished candle may be satisfactory. The dipper having returned his rod to the rack, the next child repeated the process. This continued until all the wicks had been immersed. Other pupils continued the dipping, adding layers of wax to the growing candles. Each candle was dipped about fifteen times before it reached the desired size. After hardening for a few days in a cool place, the candles were slipped from the rods and each was trimmed with a sharp knife at the lower end.

On the whole, the results were eminently satisfactory, many of the fifty-eight candles showing distinctly the wavy appearance peculiar to the hand-dipped product. A molded candle, of course, does not have this characteristic, being straight and smooth.

A tiny pair of candles for the fireplace is in the first grade doll house was dipped and sticks made to hold them. This gift was boxed and given to the little children as a Christmas gift. Among our most treasured souvenirs of the work is a letter of thanks from them and also from the little folks in the Children's School for the gift to their mantel.

Dear Children:
We thank you for your gift of bayberry candles and candle holders. They look very pretty over the fireplace. We would like to have you come to see them in our house. Your little friends,

THE FIRST GRADE.

Dear Miss Gardner:
The little children in the Children's School thank you and the children of the fourth grade for the candles and candlesticks you made for us.

JULIA.

This project meant much work and great care, but when the candles were all boxed and tied something significant had been accomplished by the children. Their joy in the work of their own hands was a recompense worthy of the effort of all concerned.

Two posters were made. A photograph of a bayberry spray was taken for picture post cards. These were tinted with Japanese water colors and sent to many friends who were interested in the work.

After Christmas several pupils collected greeting cards on which the candle was used as a unit of design. In considering the various Christmas symbols it was found that the candle played a very important part. Keen interest was aroused by an exhibition of these cards.

In the carrying on of this project a love of nature was awakened in many of the children, who had never before roamed through the fields on such a quest. Who can estimate the initiative developed or the value derived from solving such problems as were met? Surely in the reviving of the old custom of candle-making these children have met some important problems of daily life. Moreover, they have lived and grown through delightful experience.

> "How far that little candle throws his beams!
> So shines a good deed in a naughty world."
>
> —*Merchant of Venice, Act. V, Scene 1.*

"To him who in the love of nature holds communion with her visible forms she speaks a various language."—*Wm. Cullen Bryant.*

THE CHICKADEE

A long time ago, in a clump of small trees,
Was a little bird college conferring degrees;
And so one little fellow, so learned was he,
And so pious withal, they conferred a "D.D."

The name of the birdie thus honored was "Chick."
His body was small, but his intellect quick.
I never have learned what the reason could be,
But the other birds smiled, and said, "Chick, a D.D.?"

But "Chick" didn't know they were smiling, and he
Was as happy a bird as there was in the tree;
And oft (to himself, not to others), in glee
He chuckled, and said, "I am Chick, a D.D."

What! "Chick," a D.D.? Little Chick, a D.D.?
Oh, yes, and a very good preacher is he;
And many a message of profit to me
Have I heard in the church of the Chick-a-dee-dee.

> —*Anonymous.*

TO AN ENGLISH SPARROW

Little feathered tuft of gray,
Skipping blithely through the day—
   Never resting,
   Yet protesting
In your querulous, quick way—
   Little sparrow;
You who would the woodland spurn—
Bird-prized haunts of leaf and fern—
Ever grace the crowded street,
Seeking man's companionship
With your chippy-chip chip-chip—
On your wing or tripping feet—
   Ever lightly,
   Always sprightly
Comes your note so nearly sweet,
   English sparrow.

As you peck with greedy bill
Crevices about my sill—
   Hither, yonder,
   I would ponder,
Do you never get your fill,
   Little sparrow?
Are they true, my busy bird,
All those stories I have heard—
That there's nothing good in you,
That the stains upon your chest
Match the heart within your breast;
That no chirp you chirp is true?
   Should I shoo you
   If I knew you—
Are you blackguard through and through,
   English sparrow?

Though you make no chirped defence,
Yet the absence of pretence
   Casts about you,
   As I doubt you,
Something tempting confidence,
   Little sparrow.
For you've such an honest lay—
Such a frankly flaunting way;
And though man would fain disclose
All your past, incarnadined
With the sins that you have sinned,
Yet mayhap your record shows
   No more starkly,
   Not more darkly,
Than man's own, as just God knows,
   English sparrow!

       —*Hazel Hall, In the "Boston Transcript."*

CHAPTER XXI
KNOWING NATURE FACTS

Some nature facts are acquired in the woods and fields. Others are discovered in the laboratory by careful research. Those facts which stand the acid test of experience are handed down in print,— the science heritage of the race. Pestilence, famine, unnecessary labor, and enormous suffering have been the price of discovery and often continue as the tax of ignorance. Out of this wealth of accumulation there comes a certain amount of nature knowledge essential to a democracy. To this extent nature education should be demanded by public opinion. (We must stamp out nature illiteracy.)

Among the thousands of things happening on the nature stage— among the infinite number of problems—our schools are only able to present a few types. At the end of the nature course we may well inquire as to what are the strategic facts and principles which all should be expected to have acquired. The nature teacher who possesses the belief that appreciation is the essential aim may not have stated his position concerning fact nature-study. Certain it is that the sentimentalist placed the wrong emphasis. The aim of this chapter is to serve as sort of a clearing house for nature facts that everyone should know.

There is a large group of teachers who will be tempted to think of this list of facts as the minimum essentials in nature-study. They may agree passively with the methods of nature-study which have already been described yet when this list appears they will see a chance to teach the subject by rote. To such teachers this printed list is an obstacle. If they have not tried out the methods described they are apt to start the learning of the list by the phonograph or parrot method. These facts are presented as a result of nature experiences rather than an introduction—as a working attitude rather than a last word. Nature facts should not be a smoked glass through which the student studies nature.

The listing of miscellaneous facts as minimum essentials is also undesirable from the standpoint of the best pedagogy. It is a temptation for the teacher to demand fact-cramming. Some teachers have not learned that their duty is to provide organization and growth in the learning process rather than to train children to act as storage receptacles for a collection of miscellaneous facts. Again this is not a list for drill. If nature-study is but so many required facts—200 for instance—then nature-study in all the grades is a waste of time.

As Huxley put it, "Science is trained and organized common sense." The training, the living, and progressing with the times is ever essential. The cold scientific facts in nature knowledge correspond to the multiplication table in arithmetic, to health habits in hygiene, and the 200 demons in spelling. They are the salient points which determine our every day habits and attitudes.

Suppose that we move to a new community.   If we are purchasing
a home we use such knowledge as we have in determining the desirable
qualities and construction of the house and surroundings, and these
in relation to the price.   If we know the structure of oak quite
well our information helps us in determining whether the mantle
piece and finishing is oak or imitation.   If we are ignorant we may be
disappointed.   If the grounds are ornamented with various vines
and shrubs our appreciation of the effect depends upon what we know.
Whether we allow the weeds of our garden to go to seed and scatter
into the gardens of the community depends upon our education.   How
we vote upon an appropriation for a bird reservation in this same
village depends, in part at least, upon what we know of the actual
meaning of the issue and its probable consequence.   Whether we
select a given nature story among the many offered at the public
library depends upon what we know of the character, content, and
quality of nature stories.   What we select as a place to visit for
enjoyment depends on what we know respectively of nature, art
music, etc.   If we decide on the woods, how we leave the camp fire
depends upon our knowledge of the results of a forest fire.   Such
knowledge is not acquired from a dictionary.   The method of acquiring knowledge for effective use is by experiencing it under situations which require that particular knowledge.

Furthermore—remembering nature facts is quite different from
practicing nature knowledge.   Passing an examination on the causes
of a cold infection is not of itself assurance that the individual will
take precautions when and as he should.   Merely knowing that
potted plants should be watered and that they should have drainage
to get the best results does not mean that the learner will take care
of plants in the same way his information tells him.   Merely knowing
that it is scientific to weigh the evidence before deciding does not
mean that this will always be the method of procedure.   Nature
facts must be dynamic as well as static.   They must work automati-
cally.

Nature knowledge then means a working knowledge.   It means
the power to use the facts and principles of nature-study in the
interpreting of the news items of the day.   It means that this knowledge will be applied in reading magazines and literature.   It means
that the statements of others will be tested for accuracy.   It means
civic cooperation in applying the laws of nature.   It means a wise
selection and use of nature's riches.

There is a large body of nature facts that must be taken into
account.   There may be disagreement as to the number of facts
and as to the particular facts that should be included but we can
all probably agree as to the inclusion of certain facts.   Some of
those generally agreed upon are included in this list.   They are arranged alphabetically as a matter of convenience.

Adaptation: Plants and animals are adapted to life in various
conditions—water, mud, mountains, and deserts.   The teeth of

horses are adapted to cutting off grass and chewing it. The front teeth of the rabbit are adapted to gnawing food and the canine teeth of the cat are adapted to tearing meat. In the case of the fig the seeds will not form unless the fig is pollinated by the fig insect (Blastophaga grossorum.) Perfect figs could not be raised in Cali-

A STORY OF THE BALANCE OF NATURE
Fig. 1. In 1848 the Mormon Pioneers planted the fields.

fornia until this insect was imported. This is an adaptation between an insect and a flower.

Bacteria: The smallest plants are bacteria. They live in the air, soil, and water. The germs of contagious diseases, such as diptheria, tuberculosis, and influenza escape from the mouth on tiny drops of water when sneezing or talking, and skin diseases such as measles, scarlet fever, and small pox, escape with the excretion from sores. Diseases are spread by using the same drinking cup, or towel, by placing a pencil in the mouth, handling food without washing the hands, and by putting the fingers in the mouth.

Balanced Aquarium: All living organisms give off carbon dioxide. In an aquarium this gas is dissolved in the water. Green plants take in this carbon dioxide in the daytime and with other raw materials manufacture starch. These green plants give off oxygen as a by-product which is used in respiration by animals. This partnership between plants and animals in the aquarium exists between all green plants and all animals. Green plants are the food makers for animals. Without green plants no animal could live.

Balance of Nature: Our fur coat may come from a polar bear which depends on fish for food. These fish may eat smaller fish which in turn eat certain insects that depend on a few kinds of plants. These plants require a definite depth of water and a particular kind of soil. The soil has its origin. There must be a proper supply

Fig. 2. As the crops were maturing an army of locusts descended from the mountains and destroyed all green herbs in their path. In vain did the farmers guard their crops, until the sea gulls came to their aid.

of each to maintain the other. Our shelter, food, and pleasure,—our very existence—depends in someway upon nature. We must guard the balance of nature in order that our children and our children's children shall enjoy the same privileges.

Birds: Birds are of economic necessity for our existence. They destroy insect pests and weed seeds. Some act as scavengers, as the sea gull. Song and game birds are protected by law. The passenger pigeon has become extinct. The last known of the species died in the Cincinnati Zoo on January 21, 1915. The wild turkey and the wild swan are very rare. The heath hen is now only found on Martha's vineyard. The snowy egret has been nearly exterminated by milliners. Unless we practice conservation other birds will meet the same fate. The "closed season" laws protect birds during and after breeding periods. The "bag limit" laws are to curb the game hog. A "bird reserve" is a breeding ground where the practically defenseless young are protected. The Federal Migratory bird law is a protective measure due to the efforts of the sportsmen.

The Biological Survey is studying the economic relations of birds. The Audubon Society aims to protect birds.

Bordeaux Mixture: This fungicide was named after the city of Bordeaux, France, and was invented by Millardet, in 1883, to control the grape blight. It is a mixture of copper sulphate and lime used

Fig. 3. The Mormons were thus enabled to harvest a good crop of wheat. The gulls had prevented famine and pestilence.

to prevent many of the fungus diseases of plants. It is a preventive and not a cure. In 1845–47 the potato blight took away the food on which the majority of the Irish people depended and nearly a million people starved. This is typical of other plant diseases. The discovery of the Bordeaux mixture will prevent the occurrence of another Irish famine due to the potato blight. This illustrates the value of scientific research.

Cabbage Butterfly: This pest was introduced from Europe in 1860, and rapidly spread over the country. Its natural enemy, the Ichneumon fly was introduced in 1883. The Ichneumon fly lays its eggs in the caterpillar where they hatch and feed upon the substance of the larva—killing it. This is an example of what usually results when an injurious insect is introduced. It multiplies rapidly, devastates the crops, and is only checked by introducing its natural enemies from the original home. The ladybird beetle was introduced to save the orange crop from the scale insect and the calasoma beetle to fight the gypsy moth. It is important to know these allies of man in order to maintain the balance of nature, a thing that is essential to our being.

Cells: In 1839 Schleiden and Schwann concluded that all plants and all animals are made up of cells. A cell is a unit structure of life. It contains a substance called protoplasm. They vary in size from

Fig. 4. It is said that the crops of the Mormons received, thereafter, an annual visit by the gulls. The first visit was supposed to be miraculous.

microscopic bacteria to an ostrich egg. They multiply by division. They contain a central part called the nucleus within which are bodies called chromosomes. The number of chromosomes in a given plant or animal is always constant. These chromosomes carry the inheritable qualities. It is a remarkable thing when we stop to consider that all life processes, growth, reproduction, digestion, etc., are the results of cell activities.

Chestnut Blight: This disease is killing our chestnut trees. It produces spores which are blown from tree to tree. This parasite girdles the trees. The loss of our chestnuts would not have occurred if the disease had been confined to Japan.

Cotton Boll Weevil: Introduced from Mexico in about 1894. In 1922 it was estimated that for every bale produced another bale was destroyed by the weevil. For every dollar that we pay for cotton we pay another in the form of a tax to the cotton boll weevil. It is difficult to exterminate as it is protected by the cotton bolls.

Experiment Stations are operated by the various states. If any practical question arises, such as how to destroy a certain insect, one may inquire at this station. Up-to-date information will be sent promptly, and without cost to the individual.

Fish: Fish are an important source of food. They are in danger of extermination from improper legislation. Laws should guarantee to posterity the opportunity of fishing. Fishing is a good healthy

Fig. 5. This monument was thus erected to the sea gull (Franklin Gull). It is possibly the only monument erected to a bird.

adventure. It is the inherent right of every boy and girl. The boy who goes fishing is not a corner loafer planning crime. Fish may be artificially propagated at hatcheries and transplanted to non-producing public waters. The cost should be borne by the market and the sportsman.

Flowers: Plants have flowers to produce fruit. They may consist of several parts. The green sepals cover the bud. The petals are usually bright colored and probably attract the insects. The insects come for nectar. The anther—at the top of the stamen—holds the pollen. The stigma is usually sticky. The pollen grains are caught and grow in this sitcky fluid. Pollination is the transfer of pollen from the anther to the stigma. If pollination takes place in the same flower it is called self-pollination. Pollination is necessary for the production of good fruit. That is why bees are kept in a green house used for raising cucumbers.

Forests in the United States are being destroyed four times as fast as the annual growth. A lumber famine is upon us. Forests prevent erosion and dumping into our navigable streams our best soil. They regulate stream flow thereby preventing floods and drought. We must plant trees on our mountains and steep hillsides. The stage in civilization of a country may be judged by its methods of forestry. The Bureau of Forestry, Washington, D. C., publishes bulletins on forestry which may be obtained upon request.

Fig. 6. A FIRE TOWER, Moose Hill, Sharon, Mass. In dry weather the fire warden is on the lookout for forest fires. He has a great deal in common with Nature Scouting. It will pay to visit a fire tower.

Forest Fires are preventable. They cost annually 50 million dollars worth of lumber besides the loss of lives and destruction of soil fertility. Camp fires should be built near water, some distance from logs and leaves and enclosed with stones. Camp fires must be extinguished before leaving. Burn brush heaps when the snow is on the ground. Forest fires may be extinguished by the use of soil and water, and by back firing.

Forest Recreation: Is as much a part of Forest Management as forest crops, or water conservation. We need more recreational forest areas, not only national and state parks, but in every community where they will be handled on the same broad basis. Beautiful views and certain wild landscapes must be preserved for the recreation and inspiration of future generations.

Fungi make up over one-fourth of our known plants. They have no green coloring matter and therefore have to live on living things (parasites) or on dead things (saprophytes.) Some fungi cause dead organic matter to decay so that the elements can again be used in plant or animal growth. Otherwise the elements available for growth would finally become exhausted. Without fungi there would be no decay, humus soil, and fermentation, and very few contagious diseases. Fungi require food, moisture, oxygen, and warmth for growth. Dried foods, sealing preserves, and cold storage, are based upon this knowledge.

Fur Bearing Animals such as the beaver, otter, sable, muskrat, and marten are becoming very scarce. The bison were nearly exterminated until the government and the American Bison Society protected it. The fur seal has furnished an international problem. The domestication of the silver fox, muskrat, and skunk has been successful. Game preserves also help in the conservation of our fur bearing animals.

Gypsy Moth was introduced into Medford, Massachusetts in 1869. It has destroyed hundreds of acres of forests. The female does not fly, therefore, it is distributed in the caterpillar stage. The most effective remedy is to paint the egg clusters in winter with creosote.

Grafting Fruit Trees: A Baldwin apple seed would probably not produce a Baldwin apple tree. A bud or twig from a Baldwin apple tree grafted onto any kind of apple tree root or branch will produce Baldwin apples. The same thing is true of all other fruit trees. Without grafting all the varieties that we have gained through fruit growing would revert to the original wild type with the disappearance of the present trees.

Honeybee: The yield of our home fruit trees, berry bushes and many other crops depends to a great extent upon the proper pollination of the fruit blossoms by the honeybee. Tons of nectar go to waste in the fields annually because there are not enough honey bees to gather it.

Hookworm: It has been estimated that there are two million people in the southern states infested with the hookworm parasite. These people are anemic, stunted, and shiftless. The disease is usually found in localities where the people go barefooted. The parasite enters the body through the skin of the feet. Thymol weakens the hold of the worm on the intestine wall and Epsom salts expel it. This scientific discovery has saved millions of dollars and thousands of lives.

Housefly: Dr. Howard suggests the name Typhoid Fly. The fly carries most all disease germs and leaves them in fly specks, fly spit, and wherever it walks. The housefly usually breeds in horse manure. It lays about 150 eggs at a batch. The young become adults in about two weeks. We may prevent the fly by cleaning up the filth of a community. The number of flies in a community depends upon the number of people in that community who do not know and practice fly control. The slogan for a fly campaign should be "Prevent the Fly" rather than "Swat the Fly."

Insects and Flowers: Insect-visited flowers are usually attractive and scented. Many flowers have foot holds. They offer insects food in the form of nectar and pollen. The insects carry pollen from the anther to the stigma. The pollen grain grows a tube down the style of the pistil to an ovule. The substance of the pollen grain unites with the substance of the ovule and the ovule becomes a fertilized seed.

Landscape Gardening: Beautiful homes and roads mean beautiful towns and cities. Beautiful civic centres mean a beautiful America. A beautiful America must be wrought by the whole citizenship. An important factor in patriotism is the love of our native land.

Molds and Mildews reproduced by spores. The spores must have a host as—clothing, food or books. They must also have moisture and warm air. Darkness favors their growth. They are prevented by keeping objects dry, by keeping out air and thereby the spores. A paper wrapped around fruit aids in keeping out the spores. Our preserves are made air-tight to keep out the spores. Dried fruits do not spoil because there is no moisture for spore growth.

Mosquitoes: A lack of the knowledge that mosquitoes carry the yellow fever organism prevented the French from building the Panama Canal. In 1878 there were 12,000 deaths in the United States due to yellow fever germs being carried by the mosquito. In the summer of 1900 Dr. Jesse Lazear, an U. S. Army Officer, gave up his life to prove experimentally that yellow fever is caused by the mosquito. The malarial parasite is also carried by the mosquito. Mosquitoes lay their eggs in stagnant water. The larvae or wrigglers breathe thru a long respiratory tube which they protrude at the surface. Crude petroleum suffocates the larvae. A better method is to drain the marshes. Birds, bats, dragonflies, and fishes are enemies of the mosquito. Many communities have exterminated the mosquito by a systematic campaign.

National Parks: The first national park in the world was created on March 1, 1872—The Yellowstone National Park. In 1916 President Wilson approved the National Park Service act "to conserve the scenery and the natural and historic objects and the wild life therein and to provide for the enjoyment of the same in such manner and by such means as will leave them unimpaired for the enjoyment of future generations." There have been many attempts to exploit the parks commercially. Railroads have tried to build, irrigation interests have sought to build reservoirs, and herders have tried to admit cattle and sheep for grazing. The national parks are doing educational work of immense value in nature conservation. There are nineteen national parks. They are scenic, outdoor museums. They have camp sites, trail areas, and wild areas.

Native Plants: Our most beautiful woodland plants are in danger of extermination. Mountain laurel, holly, trailing arbutus, maiden hair fern, columbine, dogwood, lady's slipper, sabbatia, cardinal flower, and fringed gentian need protection. There are societies for the protection of native plants. For literature address the New England Society for the Preservation of Wild Flowers, Horticultural Hall, Boston, Massachusetts.

Nitrogen is the most important element in soil productivity. The only way that it gets into the soil is through the decay of plants and animals, by their waste, and by bacterial action. Plants which bear their seed in pods, like beans and clover, have nodules on their roots which contain bacteria. These bacteria are able to take nitrogen from the air and make it available for growing plants. Pod bearing plants are, therefore, valuable soil enrichers.

Parasite: A parasite lives at the expense of a living plant or animal. Parasitic fungi destroy millions of plants annually. Such diseases as malaria, yellow fever, dysentery, smallpox, and rabies are caused by parasites known as protozoa. The human tape worm is parasitic. There are thousands of insects which are parasitic. The parasite usually has a degenerate nervous system and sense organs and often loses organs of locomotion and digestion. The life history of a parasite affords an interesting illustration of the needs of educating senses and exercising all parts in order to get the highest development.

Pasteurized Milk is heated at 170 degrees Fahrenheit for 20 minutes. This process kills the germs of tuberculosis, typhoid, dysentery, diphtheria, tonsillitis, and cholera, and lengthens the keeping qualities of the milk. Pasteurization is named after the French bacteriologist Louis Pasteur.

Pests Introduced from Foreign Countries increase rapidly up to the limit of their particular food supply. Some examples are: cabbage butterfly, elm-leaf beetle, codling moth, Hessian fly, corn borer, San Jose scale, gypsy and brown tail moths, potato blight, white pine blister rust, chestnut blight, garden weeds, English sparrow and starling. Each pest has cost the government several million dollars. The most economical way to combat such a pest is to introduce its natural enemies. There are several hundred more European pests waiting to be introduced. Prevention is the best policy. It is an absolute necessity for this country to prevent the introduction of such pests.

Photosynthesis: The substance which gives the green coloring to plants is called chlorophyll. By means of the energy received from the sun this leaf-green is able to manufacture starch out of soil, water, and carbon dioxide. This is one of the most important

chemical changes and without it there would be no life. It may
be compared to a manufactory in which the leaf is the factory, the
machinery is chlorophyll, the power sunlight, the raw material carbon
dioxide and soil water, the product starch, and the waste oxygen.

Poisonous Plants: Poison ivy and poison sumach cause blisters
on the skin of some people if they come in contact with it. If anyone
has touched one of these plants he should wash the exposd parts

Fig. 7.   The White Pine Blister Rust

with soap and water and then apply alcohol or a paste of soda bicar-
bonate.   Poison ivy may be distinguished from woodbine by its
white berries instead of blue and by the three parts to its leaf instead
five.   The poison sumach may be distinguished from the non-poisonous
sumachs by its white berries and by the fact that it grows in swamps.
It may be easily remembered that both of these poisonous plants
have white berries.

Some plants are not poisonous to touch but are poisonous when
eaten.   The common ones are: Jimson weed, Poke weed, water
hemlock, poison hemlock, and certain mushrooms.

Pollution: We have more inland water than any other country.
The pollution of inland waterways, harbors, and bays is increasing

very rapidly and is taking a great toll of small organisms, fish, and water fowl. It is unsightly, unsanitary, and the emanating odors are offensive. The sewerage emptying into these waterways often contains valuable nitrates which become locked up with other chemicals in the water. We are throwing away the fertility of the land, one of our greatest national assets. Pollution is not only a menace to public health but destroys recreational facilities.

Protoplasm: Huxley defined protoplasm as "the physical basis of life." It is the *only* living substance. Chemists can analyze it but have not been able to make it. Because protoplasm is alive it can take in food and digest it and throw off the waste. It breathes, grows, moves, and reproduces. One of the remarkable things about protoplasm is that it tends to mass itself into cells. The growth of plants and animals depends upon the multiplication of these cells.

Rabies: In the case of a bite by any rabid domestic animal the brain of the animal should be sent at once to a physician who will have the animal examined at the nearest laboratory for Negri bodies. If the examination is positive the patient should be given the Pasteur treatment. The rabies may develop in 14–60 days in man. When hydrophobia appears in a locality it should be stamped out by muzzling all dogs. Before Louis Pasteur discovered the cure for rabies the patient had no hopes of escaping an awful death.

Rats: The common brown rat was introduced into this country from Europe. The rat destroys thousands of dollars worth of food and property annually. They carry the bubonic plague germ, they spread the germs of intestinal disease, and frequently are hosts of trichina, a round worm which may be a parasite on man. The bubonic plague is estimated to have killed 25 million people in the 14th Century. The control of the brown rat must depend upon each citizen actually doing his part. Cement walls and floors are effective barriers.

Rotation of Crops: By planting different crops in a certain field on different years no one crop will exhaust the soil by its own extra demands. Clover replaces the nitrogen used by the corn. Wheat demands a large amount of phosphorus and grass a small amount. The rotation of crops is also a safeguard against injurious diseases and insects, and often aids in keeping down certain weeds.

San Jose Scale: This insect does more damage to our fruit trees than any other insect. Government entomologists located its original home in China and introduced a lady-bird beetle which preys upon it. The insect has lost its eyes, legs, feelers, and wings in the scale stage through disuse. It is practically a bag with a sucking beak. It kills our orchard trees by sucking out the sap destroying millions of dollars worth of property each year. It is killed by spraying with a lime sulphur solution in the winter.

Shellfish: Thousands of acres of flats along the sea shore that use to produce the soft shelled clam, quahaugs, and oysters are now barren and unproductive. This is due to the old custom of free public fish rights. In colonial days our land became privately owned and the crops were protected by law. The clam flats are still public property. When all our coastal states have passed laws to protect the individual with his shellfish garden we may expect a perpetual rich harvest of shellfish.

Snakes: There are probably more false ideas and misunderstanding about snakes than any other group of animals. As a class they are useful. The only dangerous ones to handle are the rattlesnake, the copperhead, and the water moccasin. Snakes destroy a large number of injurious insects, mice, and other mammals. The black water snake is injurious in that it kills birds. Most snakes can swallow prey larger than the diameter of their own bodies. They cannot jump from the ground, the tongue is a feeler and not a stinger, the death has nothing to do with the setting of the sun, the "hoop snake" exists only in fiction, horsehairs do not turn into snakes, rattlers do not add one rattle per year, and the teeth are not for chewing or biting but to prevent the escape of the prey.

Toads also suffer from a false reputation. They do not cause warts, they do not rain down, they do not eat their own tails, and they are never found alive in solid rock. Toads are extremely useful in that they eat insects—mostly injurious ones, cut-worms, slugs, and other garden pests.

Trees are necessary. They furnish shelter and homes for birds. Without birds our gardens would be destroyed by insects. Trees cool our streets, increase the value of real estate, and beautify our cities. They are needed individually as well as in forests. Many cities have laws which protect the trees from horses.

Variations: No two animals or plants are alike. No two noses or eyes, or leaves are alike. Some of these differences are inherited and others are due to food, climate, and other environmental causes. Those who have recognized these variations have been able to get new breeds of plants and animals. Race and draft horses; milk and beef cattle; fan-tails, pouters, tumblers, and other varieties of pigeons; the bull-dog, greyhound, and terrier; the stoneless plum, and Burbank potato are all the results of selecting these variations.

Weeds: It has been estimated that there are over 700 kinds of weeds in the United States and new ones are being introduced. Most of these weeds came from Europe and their annual tax amounts to millions of dollars. Much of this could have been prevented by intelligent coöperation. Weeds rob crops of sunlight, water, and mineral food. A community can do much to reduce the number of weeds by eliminating them along the fence rows and by seeing that they are mowed along the roadsides before they go to seed. If they have gone to seed they should be cut, dried, and burned. Clean garden seed and rotation of crops is another means of weed control.

Wild Life Areas: 1. Every community should maintain a wild life area similar to our National Parks.

2. Wild Life Areas are of great value as outdoor school rooms,— for their scenic value, to demonstrate forestry, to study geology, wild plants, and animals, to preserve natural features, as game refuges, for fishing, for preservation of native wild flowers for camping, for scouting, for nature photography, for tourists, and for picnics. All of these projects are important and interrelated.

3. We must preserve all our native wild life for future generations as well as the present. Most of our large mammals, game birds, and many native wild flowers are in danger of extermination.

4. We must preserve the balance of nature in this wild life area. To do this we must prevent pollution, fish diseases, fire hazards, and the introduction of new pests. Some pests which have been recently introduced into our wild areas are chestnut blight, white pine blister rust, gypsy and brown tail moths, and English starlings. The introduction of these foreign species from Europe has upset the balance of nature wherever they occur.

5. These areas must be governed for the good of the people, free from politics. They should be rendered accessible by scenic roads and trail systems, campers should use only dead wood for fires in designated places; there should be free nature guide service; approved sanitary arrangements are indispensable; outdoor sports should be encouraged. "Artificial" amusements such as dancing, bridge whist, and merry-go-rounds should be prohibited.

Yeast or "sugar fungus" is the most useful of house plants. In bread making the carbon dioxide produced by the yeast makes the bread porous. Yeast plants break down sugars by fermentation which results in alcohol and other products. Yeast is therefore used in making alcohol, beer, and vinegar.

> See thou bring out to field or stone
> The fancies found in books.
> *—Emerson.*

> Sweet is the lore which Nature brings.
> *—Wordsworth.*

GOOD MORNING

Mawin, Mistah Blue Jay.
How yo ah this wintah day?
My but yo is lookin' pert!
Whar yo git that foxy shirt?
Stripes o' white wid dots o' black,
Alice Blue all down yo back!
Gee, but yo's a busy bird,
Quite a hustlah, on mah word.
What yo tryin' foh ter git?
Haint yo had yer breakfa' yit?
Lovely weathah hain't it tho!
Sun a-shinin' on the snow,
Fields all decked in dazzlin' white,
It sho am a gorgeous sight!
Hain't yo glad now, Mistah Jay,
Yo and me's alive terday?
Well, old spo't, I'se got ter mose—
Wisht I had a shirt like yo'se.

—P. D. Gog, The "Chicago Tribune."

Summer or Winter, day or night,
The woods are an ever-new delight;
They give us peace, and they make us strong,
Such wonderful balms to them belong.

—Stoddard.

"Wheresoe'er I turn my eyes
Around the earth or toward the skies,
I see Thee in the starry field,
I see Thee in the harvest's yield,
In every breath, in every sound,
An echo of Thy name is found.

The blade of grass, the simple flower,
Bear witness of Thy matchless power.
My every thought, Eternal God of Heaven,
Ascends to Thee, to whom all praise be given."

—Abraham Ibn Ezra.

CHAPTER XXII

OBSERVATION TESTS IN NATURE-STUDY.

If the power of observation is one aim in nature-study to what extent is it realized? This is one of the occasions when a test is justified in nature-study. It may serve as a stimulus to the student to introduce the test as a game. The following tables are given not so much as standards by which one may measure the results of science teaching in one school as compared with another, but more especially to enable an individual to know his own standing with reference to his class. It is a scale for rating individual achievement in observing. The following exercise entitled *A. Reading Trees* was given to the pupils without further explanation.

A. Reading Trees.

Observation is an ancient game. The Indians developed alertness and keen observation, by following trail signs. Twig study is more recent but requires the same degree of alert attention. This paper is planned to test your powers of observation. Nearly everyone can read about trees but very few can read trees. This test includes the elementary language of tree talk as used by Tree Guides.

Fill in the following tables. Only one word is necessary for each blank. Each word, which correctly describes the *leaf* given you, counts as one point. The total score will show how good a guide you are compared with your classmates. Work carefully and cautiously. Quality and not quantity is the motto of a Tree Guide.

It is not expected that you will complete the paper. In fact you will probably not be able to complete any one table. Each table is a ruler which measures the "length and quality" of your ability to observe.

The pupils examined were high school graduates who had studied winter twigs for ten weeks but had had no experience with evergreens. This examination was not only a test of their powers of observation but also served as an introduction to the study of evergreens. It will be noted that the achievements range from 3 to 22 observations. Test one was given to 8 classes and test 2 to 6 classes. The lowest class median was 9.9 points and the highest class median was 14.2 points showing that classes as well as individuals vary in their powers of observation. The median for table one is 11.3 and for table two 11.6 showing that the two tables give very similar results when used to test the power to observe. A perfect score for table one would be 28 points and for table two 30 points. The results were posted, as written in table four, so that each pupil would have a chance to know his own standing.

Most students are able to see that both trees have long leaves and that the white pine has longer leaves than the spruce; that both have narrow leaves with the white pine needles being narrower of the two;

TABLE 1. LEAVES OF WHITE PINE AND NORWAY SPRUCE

TOPIC	RESEMBLANCES	DIFFERENCES White Pine	Norway Spruce
Size—Length	*Long*	*Long*	*Short*
Size—Width	*Narrow*	*Narrow*	*Wide*
Color	*Green*	*Light*	*Dark*
Shape Lengthwise	*Needle*	*Fine*	*Coarse*
In Cross Section	*Rectilinear*	*Triangular*	*4-sided*
Arrangement		*Bundle*	*Scattered*
Base	*Tapers*	*Sheath*	*No sheath*
Flexibility	*Flexible*	*Flexible*	*Stiff*
Surface (Feel)		*Teeth*	*Smooth*
Surface Markings (Hand Lens)	*White Dots*	*Inner Surface*	*Both surfaces*
Possible Score by Columns	8	10	10
Grand Total—28	Name	Time Used	

TABLE 2

TOPIC	RESEMBLANCES White and Pitch Pine Leaves	DIFFERENCES White Pine Leaves	Pitch Pine Leaves
Size—Length	*Long*	*Short*	*Long*
Size—Width	*Narrow*	*Narrow*	*Wide*
Color	*Green*	*Dark Green with White Bloom*	*Light Green*
Shape Lengthwise	*Needle*	*Fine*	*Coarse and Twisted*
In Cross Section	*Triangular*	*Triangular*	*Flattened*
Arrangement	*Bundles*	*5 in bundle*	*3 in a bundle*
Base	*Sheath*	*Short, deciduous sheath*	*Long, deciduous sheath*
Flexibility	*Flexible*	*Very Flexible*	*Stiff*
Surface (Feel)	*Teeth*	*Fine*	*Coarse*
Surface Markings (Hand Lens)	*White Dots*	*On Two Inner surfaces*	*On all sides*
Total Score by Columns	10	10	10
Grand Total—30	Name	Time Used	

TABLE 3

TOPIC	RESEMBLANCES Norway Spruce and Hemlock Leaves	DIFFERENCES Norway Spruce Leaves	Hemlock Leaves
Size—Length	*Long*	*Long*	*Short*
Size—Width	*Narrow*	*Narrow*	*Wide*
Color	*Green*	*Dark Green*	*Light Green*
Shape Lengthwise	*Needle*	*Tend to be round*	*Flattened*
In Cross Section	*Flat, elliptical*	*4-sided, flat*	*Thin*
Arrangement	*Scattered*	*Scattered*	*2-ranked*
Base		*Not stalked*	*Stalked*
Flexibility	*Flexible*	*Flexible*	*Stiff*
Surface (Feel)	*Smooth*	*Smooth*	*Slightly toothed*
Surface Markings (Hand Lens)	*White Dots*	*On both sides*	*On lower sides*
Total Score by Columns	9	10	10
Grand Total—29	Name	Time Used	

and that both have green leaves, the white pine being a lighter green. From this simple beginning the scale becomes gradually more complex. The observation of surface markings is introduced as a more difficult feature of the test. Poor pupils are usually able to note that the leaves have a bloom or white appearance. Several observe closely enough to say that there are white lines and it is an exceptional student who examines the surface minutely enough to discover that the white lines are made up of dots.

B. *Observation Test With Birds.*

It should be emphasized that these various tests can only test the ability to observe the color, or some other characteristic of that particular object. A person who is color blind might note the markings on a twig but not the display of colors on a bird. A person who is interested in birds, for example, would obtain a higher score than one not interested. Further investigation in the ability to observe is needed, following which we may devise ways of testing the "progress in ability to observe."

The pupils tested on the observation of colors of the wood duck were given a drawing to show the topography of the bird. They were asked to sit in a circle and then the bird was taken out of a box and placed in the middle of the circle. As the light came from one side of the room they were asked to rotate a quarter of the way around the circle every minute. They were given five minutes to observe the bird. The specimen was then put back in the box and the pupils were given 10 minutes to write their description. The following shows the method of scoring the results.

Observation Test on the Wood Duck. (*Male*)

(*5 minutes to observe and 10 to write description*)

1. *Length.* 17 inches. (1 point allowed if within 2 inches of being correct)

2. *Crown.* Iridescent, green, or purple. (1 point for any one of these words)

3. *Throat.* White. (1 point).

4. *Breast.* Chestnut. (1 point) Spotted with white. (1 point)

5. *Underneath.* White. (1 point).

6. *Sides.* Yellow gray, fawn, olive, brown gray (1 point for word that approximately describes the color) Black lines. (1 point) Wavy black lines. (1 point) Back of flanks—vertical white bands (1 point), vertical black bands (1 point).

7. *Upper Parts.* Iridescent, green, purple, or black. (1 point for any one of these words)

8. *Wings.* Iridescent, green, or purple. (1 point for any one word that describes this color). Front of wings—vertical white band (1 point), vertical black band (1 point), crescent shape (1 point).

9. *Other White Lines*—from base of bill over eye (1 point), from ear along crest (1 point), from ear to throat (1 point), from back of neck to throat (1 point).

10. *Bill.* Yellow at base (1 point), black at tip (1 point).

Possible score—23 points.

The wood duck is distinctly an American bird. It is peculiar in that it nests in trees and does not quack. It is our most beautiful duck and as it is in demand for food it is in danger of extermination. It should be protected at all times in all states.

The wood duck was used in this examination as the test was given in a Scout Leadership course and many of the leaders were teachers of experience. The wood duck has many variations of color and is not so familiar to most students. It is also a bird that should be protected. The following results were obtained. (See table at the bottom of this page.)

The value of this observation test as compared with the usual scout test of looking into a store window and then naming two black shoes, ten tan shoes, two cans of shoe polish, a shoe horn, etc., is evident. A much simpler and commoner bird should be chosen for children.

With younger people the following test serves better as an introduction to observation. Hold up a robin for a short time and then hide it. Ask,—what color is the breast? Some will say chestnut brown and others red. Hold it up again and ask if Robin red breast has a red breast? This will be a surprise to many of them. Hide again and ask, what is the color of the bill? There will be various answers. Ask them to repeat the poem about the Birdie with a yellow bill. Show the robin again. Now ask, how many white places are there on the robin. Put the numbers 1, 2, 3, 4, 5, 6, 7, 8, on the board and have the class vote as to how many white places they think that there are. Register the number of voters under each number. Bring out the

TABLE 4

To Show the Distribution of Points Observed by Pupils in Observation Test on Evergreens

Number of Correct Observations	1	2	3	4	5	6	7	8	9	10	11	12	13	14	15	16	17	18	19	20	21	22	Number Taking Test
*Table 1	0	0	2	1	5	5	8	13	16	25	15	18	10	16	10	8	5	7	2	1	2	2	171
*Table 2	0	0	0	2	5	3	10	8	19	13	19	13	12	8	4	5	4	4	4	3	0	1	137

*Numbers of pupils.

Number of Correct Observations	5	6	7	8	9	10	11	12	13	14	15	16	17	18	19	20
Number of Scouts Making Observations	0	1	3	3	5	4	2	4	6	8	3	1	1	1	0	0

point that the way people vote does not prove anything about the subject. It does tell something about the voters. Showing the robin for the second time will not bring complete or correct results. It has white on the throat, underneath, around the eye, and on the outer tail feathers. Continue in this way until the description of the robin is complete. It will not be necessary to draw the conclusion that we do not really see our commonest birds.

ROBIN

"The drifts along the fences are settling. The brooks are brimming full. The open fields are bare. A warm knoll here and there is tinged with green. A smell of earth is in the air. A shadow darts through the apple tree: it is the robin!

"Robin! You and I were lovers when yet my years were few. We roamed the fields and hills together. We explored the brook that ran up into the great dark woods and away over the edge of the world. We knew the old squirrel who lived in the maple tree. We heard the first frog peep. We knew the minnows that lay under the mossy log. We knew how the cowslips bloomed in the lushy swale. We heard the first soft roll of thunder in the liquid April sky.

"Robin! The fields are yonder! You are my better self. I care not for the birds of paradise; for whether here or there, I shall listen for your carol in the apple tree."—*L. H. Bailey.*

"Self-reverence, self-knowledge, self-control,—
These three alone lead life to sovereign power."
                                        —*Alfred Tennyson.*

"You cannot dream yourself into a character;
You must hammer and forge yourself one."
                                        —*Henry D. Thoreau.*

Give fools their gold, and knaves their power,
Let fortune's bubbles rise and fall;
Who sows a field, or trains a flower,
or plants a tree, is more than all.

—Whittier.

Now I see the secret of the making of the best persons,
It is to grow in the open air, and to eat and sleep with the earth.

—Walt Whitman.

APRIL

The lyric sound of laughter
Fills all the April hills,
The joy-song of the crocus,
The mirth of daffodils.

—Clinton Scollard.

Happy hearts and happy faces,
Happy play in grassy places.

—R. L. Stevenson.

Then let me joy to be
Alive with bird and tree.

—Edmund Gosse.

"Such a starved bank of moss
Till that May-morn,
Blue ran the flash across;
Violets were born."

—Browning.

CHAPTER XXIII

Mainly the Pedagogy of Seeds with Some Seeds of Pedagogy

Fashions change in nature-study as in reading or spelling, millinery or military, and what-not. There was a time when the wild apple trees in the pasture were selected for stock-plants in grafting. They were said to be thriftier. Why not? They had proven their worth through a struggle for existence with blue-berry bushes, sumac, white pine, and a host of other woody colonizers. On the farm, this last autumn, it became clear to me that every wild apple tree in that locality (Norwell, Massachusetts) is affected with the bacterial disease known as crown gall (*Pseudomonas tumefaciens*). Future grafts of apple trees, therefore, will be made on nursery stock raised from seed. The wild black cherry and the cultivated cherry which had escaped from the garden were also infected with the same disease. The choke cherry showed no symptoms of the disease. Here, then, is another indictment against the wild black cherry, already well known as the scourge of the fence-row and the harborage of the tent caterpillar. To the choke cherry, the forest weed which reproduces so rapidly by underground roots or by seeds, may come the honor of being the stock-plant for grafting cherries.

I. *Educational Measurements With Seeds*

Standardized tests for measuring arithmetical abilities, scores for reading tests, measuring scales for spelling, penmanship standards, etc., are quite fashionable in the grades. Not much has been written concerning standardized tests in nature-study. The method of studying science, however, is admirably suited to the problem of measuring abilities. The writer has found that laboratory work on the seed is particularly well adapted to this study.

In a larger sense, the phrase "laboratory work" may include silent reading, the studying of a spelling lesson, or the written operations of arithmetic. Time and effort are required for obtaining skill in laboratory work. The correct habits of study in a science laboratory are remunerative in all subjects, and efforts and results are often more clearly visible in science than in the socalled academic courses. For example: time and effort are required to make observations, to reason, etc. If directions are given orally the learner increases his power of exact hearing; if the directions are in a laboratory guide the student acquires skill in concise reading; if he follows directions he will observe closely; if whys and wherefores are asked he acquires the habit of solving problems. A nature-study scale does even more than this—the understanding of which requires a definite example.

The seed of bur-marigold (*Bidens*) or beggar-ticks readily lends itself to this purpose. Fifteen species of Bidens are described in the Seventh Edition of Gray's New Manual of Botany. Britton and

Brown write that there are "about 75 species of wide geographic distribution." The seeds of the flower cluster of these species vary in several characteristics. It is only the inner seeds that are typical. The outer ones are apt to be short and undeveloped. The length of the seed varies from 4mm. to 1.5cm. The seed may be smooth, slightly hairy, or very hairy. These hairs may point upward or downward. The surface of the seed may have a simple vein or be four-angled. There may be two, three, or four awns which vary from blunt knobs to the length of the seed. There are barbs on the awns which may project upward or downward, and they may extend onto the seed. Plate I consists of the drawings of the seeds of twelve species of bur marigold which show the wide range of characteristics.

Although the following explanation of method is given the classes. teachers should devise their own schemes and will no doubt find many improvements in material and method. A seed of one or more species of Bidens may be given to each pupil (*Bidens vulgata* was used in this experiment). Each one in the class is also equipped with a reading glass or hand lens, drawing paper, and pencil. Directions given: "You are going to give yourselves a test in nature-study. You are going to find some of the abilities that you have and some of the abilities that you may need to develop. On your desk is the seed of bur-marigold. Observe it very closely and make an enlarged drawing (2.5–3 inches long) to show the characteristics of your particular seed. Do careful individual work, and proceed as rapidly as possible without interfering with the quality of the work. When you think that you have a good drawing of that seed raise your hand and the drawing will be marked with a blue pencil. It will be marked O. K. with the number of minutes you worked, if it is successful, and simply the number of minutes if it is not satisfactory. Each new drawing will be numbered until a good one is attained. If there are any questions to be asked about the method, ask them now as no questions are to be asked after we start."

As an introduction to the characteristics of the seed the pupils are asked to draw one of the awns lightly along the tip of the fore-finger, the teacher doing the act at the same time to show the manner of performance. They are then asked to stick the awns into their coat or shirt and pull them out again. Lint will cling to the barbs. The pupil now has four reasons for suspecting that there are barbs pointing downward. The barbs may be seen, they may be felt, they go into clothing easier than they come out, and they pull the threads out as is done by a crochet hook. The pupils should also be told that they need not place their names on the paper. This gives opportunity for freedom and honesty of expression.

As the seed of the beggar-ticks consists of simple straight lines the papers may be evaluated with the line as a unit. First, the pupil must see the line. As the hairs and barbs are absent in some places it gives the imagination a chance to play. The teacher may mark this drawing 10 (meaning ten minutes) and say: "Your imagination is working" or "You see more than there is there" or "You do not see enough." The proof of seeing the lines is the representation, by drawing, of the direction in which the line projects. As there is a simple progression in the perception of these lines a scale has been evolved (Plate II, figures 1–6). This scale for testing pupils has,

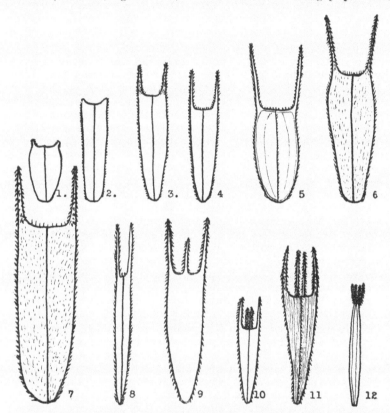

Fig. 1. BUR MARIGOLD or BEGGAR-TICKS (BIDENS).

A Key to the more Common Species (enlarged 12 times).
1. Southern Tickseed-Sunflower. (*Bidens coronata*).
2. Long-bracted Tickseed-Sunflower. (*B. involucrata*).
3. Tall Tickseed-Sunflower. (*B. trichosperma*).
4. Small Beggar-ticks. (*B. discoidea*).
5. Western Tickseed-Sunflower. (*B. aristosa*).
6. Beggar-ticks. Stick-tight. (*B. frondosa*).
7. Tall Beggar-ticks. (*B. vulgata*).
8. Swamp Beggar-ticks. (*B. bidentoides*).
9. Leafy-bracted Tickseed. (*B. comosa*).
10. Purple-stemmed Swamp Beggar-ticks. (*B. connata*).
11. Nodding Bur-Marigold. (*B. cernua*).
12. Spanish Needles. (*B. bipinnata*).

therefore, been evolved by the pupils themselves. At the end of a definite amount of time, perhaps the whole period, the papers are collected.

The interpretation of these papers is most interesting. The series of drawings (Plate II, figures 1–6) are placed on the board. This is the first time that the class has seen the scale of drawings. Different papers are held up for comparison with the series. By means of this method each pupil is enabled to compare himself with other pupils in reference to several items.

If the pupil discovers that he has only half the power of observation of some others he will coöperate in overcoming that fault. He must learn that 50 per cent of promptness will not get him very far in any line of business. In one case a girl made six drawings before she accomplished what another girl did in the same amount of time but with one drawing. Some work rapidly and make many mistakes, others slowly and accurately. Pupils should learn that as in driving an automobile, in banking, or in typewriting so it is here—speed is secondary to accuracy. We must first of all learn *how* to do it and then how to do it quickly. It does not take long for the student-teacher to arrive at the conclusion that the common assignment in the laboratory is a wasteful method. If the work is adapted to the slowest the strong ones will not need to put forth their best effort. Therefore, while the slow are observing the seed of beggar-ticks the quicker ones may be working on a second seed such as the dandelion, wild carrot, or parsnip. Those who succeeded with the beggar-ticks have gained in ability and usually get the dandelion seed correct at the first effort and at a great saving of time. The speed-accuracy test makes the whole thing worth while to the student. The power of observation gains great momentum when properly trained.

The aberrant types (Plate II, figures 7–12) indicate cases where the student needs the most help and where the teacher should give the most assistance. Figure 7 represents the point of lowest ability. The seed is enlarged but the awns remain the same size and the barbs and hairs are not seen at all. It is a question whether to include this in the aberrant list or as the lowest point in the scale. Quite a large per cent of the pupils start at this stage. It has been my experience that an astonishingly large per cent of normal school pupils are in the childhood stage of observation and expression.

Figure 8 represents a desultory manner of observing and a care-free method of drawing. The series of drawings made by this pupil show no improvement in regard to the awns. As each drawing was numbered she showed by her face that she thought the judgment of the teacher was wrong and that it was entirely the teacher's fault that each drawing was not marked O. K. The interpretations of the class must have dispelled that illusion. It might be said that class criticisms were very free and full of spice as no one knew the author of the paper but the originator. In regard to this paper such statements as,—"Has never been made to do things for herself" were made.

The teacher could soften such a remark by saying that possibly this pupil has never had a chance before, (how often that is true!) and this is her opportunity. Another girl said: "That isn't always true as the girl next to me made a satisfactory drawing the first time and she doesn't do things for herself at home." The girl referred to spoke up and said: "Well that shows what you can do when you try

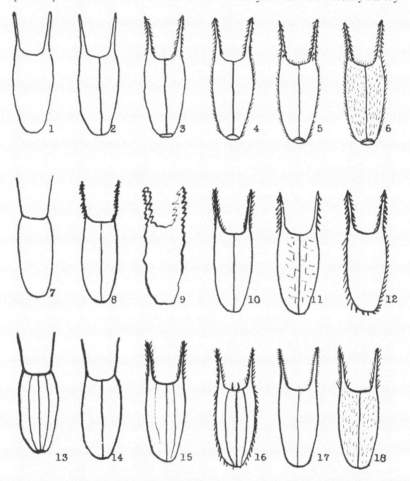

Fig. 2. MEASURING THE ABILITY OF PUPILS BY MEANS OF A BUR MARIGOLD SEED SCALE

Figures 1–6. The bur marigold seed scale arranged from pupils' drawings of the Tall Beggar-ticks (*B. vulgata*).

Figures 7–12. Aberrant pupil drawings of seed of the Tall Beggar-tick (*B. vulgata*).

Figures 13–16 and 17–18 showing progression of two different pupils.

even if your mother has 'spoilt you.'" Teacher summing up: "The pupil who made these drawings then may not have had many chances to do things for herself. At least she did not improve at this opportunity. We have seen from this discussion that it is possible even, under the handicap of always having had things done for you, to go ahead and do things. No one can grow for us mentally any more than they can eat for us or do our physical growing." The criticisms, to be sure, may often be unjust. Such a frank discussion must stimulate the whole class and may stimulate the individuality of the pupil. She may see for the first time her unawakened powers of personality, the realization of which may be a foreglow of greatness.

Figure 9 is slovenly done. One would hesitate to employ such a person for a project in engineering, and teaching is one of the most delicate problems of engineering. Figure 10 would come between 2 and 3 in the scale. The pupil saw the spines but it must have been a glimpse as she did not get the direction in which they project. The same traits and value appear in figure 11 but are represented differently. This pupil notes that there is something on the surface of the seed but does not see what it is. Figure 12 ranks between 3 and 4. The pupil observed that there are barbs on the seed but did not take in that they have a definite direction as shown by her representing them projecting downward on the left and upward on the right.

Figures 13–16 show the development of the conception of the seed in the mind of pupil A. Note that she wavered in regard to the number of ribs on the surface of the seed. Each new point came slowly, almost painfully, but there is a successful progression. Her fourth attempt (Figure 16) is still between 2 and 3. The first attempt of pupil B (Figure 17) starts where A left off. The second attempt of B shows a slight advance—sees hairs on the seed—, but not enough to move to stage 3 in the scale. Why is it that these two pupils see that there are barbs yet represent them as pointing in the opposite direction to which they really go?

The teaching of the seed should not stop with observation or representation. There is a more advanced test in which the observer not only cultivates the knack of seeing structure but reasons out the attendant thought of its relation to its environment. There is one error which the teacher must use great care to avoid. The seed of the beggar-ticks, for example, does not necessarily have barbs to enable it to catch onto clothing or the fur of animals, as some of the spines point in the wrong direction. Barbs are not purposeful structures but are the result of variation.* Since most seeds of Bidens have barbs that will hook onto passing animals, it may be said that the barbs aid in the distribution of the seed. There may be other struc-

*The idea of a special creation is seen in the following quotation from a book published 100 years ago: "The numerous organs evidently constructed for this peculiar purpose are alone sufficient to prove, beyond a possibility of doubt, that the creation is the product of superior intelligence and design."—"Outline of Botany" (1819), Dr. John Locke.

tures which are not of advantage. The pupil, therefore, must not be over zealous and imagine advantages that do not exist. With this caution the pupil has an excellent opportunity for reasoning.

The seed of the dandelion is well adapted for the introduction of the reasoning test. Everyone knows or can reason out the advantage of the down in the distribution of the seed. This fact usually forms the starting point. The pupils are asked to arrange their material in two columns,—the first being the characteristic and the second the advantage. In valuing the paper one point is given for each adaptation. (The characteristic plus one advantage.) The results are then tabulated or plotted on coordinate paper.

Number of Adaptations: 1 2 3 4 5 6 7 8 9 10 11 12 13 14 15 etc.

Number of Pupils...... 0 0 1 1 2 1 3 4 6 5 3 2 2 0 1

A pupil may be near the median ability or at one extreme. He is now able to see his relative position. Although the power of good observation is usually the basis of reasoning the good reasoning does not always follow. This test then may bring out new characteristics in diagnosing the pupil and therefore a change in the line of treatment.

A list of the more evident adaptations of the dandelion seed to its environment are hereby given. In any of this work the seed should never be isolated from the flower or fruit.

Some Adaptations of the Dandelion Seed

Characteristic	*Advantage*
1. Bracts enclose the seeds.......	A protection from the weather.
until they mature	
at night	
in rainy weather	
2. Bracts turn down............	Wind has free opportunity to sweep away the seeds.
when seeds are mature......	
in clear weather...........	Sun ripens and cures the seeds.
3. Disk changes form	
becomes concave...........	Enables bracts to enclose seeds.
becomes convex...........	Seeds more easily "caught up" by the wind.
4. Flower and fruit stem is short	
on lawns and in pastures....	Not so easily cut (Not necessary to be tall to get light).
5. Flower and fruit stem long in	
the meadows and tall grass...	Seeds more apt to be scattered.
6. Stem bends down and then	
straightens when seeds are ripe	Seeds are dispersed at right time.
7. A tuft of hairs attached to seed..	Wind can grasp easily.
8. Hairs spread out, umbrella-like.	Wind can carry easily.
9. A stem connecting fruit and	
hairs.....................	Similar to a parachute.
	Works into the ground by swaying.
10. Brown color................	Not easily seen by birds.
11. Small.....................	Not easily seen by birds.
	More easily carried by the wind.
12. Spines pointing upward from	
upper end of seed...........	Forms an anchor or grappling irons which hold the seed in place in the grass.
13. Lower end of the seed pointed...	Easier to enter the ground.

A second test in the power of observation is in the identification of species. The pupils are given Plate I, a key to the more common species of Bidens, slips of paper, and seeds of several species some of which may be duplicates. When a pupil has identified a species a hole is punched in the paper or ticket. If it is a case of mistaken identity one point is taken off the total. Pupils will thus be careful and not be too venturesome. The results are tabulated and each one can see his relative position.

Observation tests may be worked out along the following lines: Mix several kinds of seeds: beans of all sorts, peas, locust seed (Pea family); pepper, tomato, egg plant, Jimson weed (Potato family); pumpkin, melon, citron, gourd, cucumber (Cucumber family). These mixtures may be used for sorting exercises by the younger children or as a basis for family relationship with older pupils. Have the seeds separated as to kinds and the kinds grouped according to likeness. Give the names of the family groups. A mixture of nuts, stone fruits, horse-chestnut, acorns, chestnuts, beans, peas, etc., may be grouped according to their covering. Those with a hard covering have to be planted in the fall so that the winter frost will crack the covering. The thick-skinned seeds are broken by the growing force of the young plant. The thin skinned seeds often burst by swelling due to the absorption of water. Other seeds may be studied in relation to the mechanics of war. The cocoanut shell, impervious to water; bladder nut, water tight compartments; milkweed seed, corky margin for floating, tufts of hair for aviation; weeds with wings shaped like a propeller; burdock, grapling hooks. etc.

A seed scale then is a standardized test by which the pupil may determine for himself the kind of work which he is doing in seed study. He knows what he must do to qualify in his work. The use of such a standardized measurement has an important educational significance. To the pupil it means that the judgment of his work is not a matter of prejudice or merely the personal opinion of the teacher. It is not a moral obligation or evasion with the teacher but with himself. He can measure his own achievement as successfully as the teacher. He can discover for himself whether the trouble in work lies in observation, expression, imagination, nearsightedness, or other eye trouble, good morals, carelessness, speed, etc. Pupils are going to have many occasions to use or misuse these items in their practical activities outside of school work. The demand for the use of these tools in school and outside of school justifies their consideration. Nature-study, therefore, presents a fertile field for the investigation of educational measurements and standards.

To the nature-study teacher has come something definite for measuring her accomplishments. Instead of trusting to luck that the pupil sees for himself she now knows where each pupil stands and in what he needs guidance. A comparable advance has been made in agriculture. Formerly it has been the hit-or-miss method. If the farmer made a "hit" it was due to the way the ground hog appeared or the hanging of the powder horn on the moon. If it was a "miss" it was due to the devil or some other occult force. Today it is the application of scientific knowledge and methods. Much of the "hit-or-miss" method of teaching nature-study may be overcome by the use of standard scales and measurements.

II. Seed Germination

Seed terminology may be advantageously changed, although it is with some fear and tribulation that the word seed has been used in place of achene in describing beggar-ticks. However, it is almost always true that when a normal school student gets hold of a "big word" he is sure to use it in place of a simple word or phrase. Too often it has been the fate of these technical terms to trickle down from institutions of higher learning into the grades. Ideas, and not words, is what is needed. In talking about seeds one can conscientiously omit the idea of two seed coats. Pupils are always willing to imagine but rarely see the two seed coats. And if they did see them, what then? There is no purpose involved. The terms radicle, caulicle, and epicotyl are superfluous. The chalaza is not functional, at least here. The following simplifications, phrases which are self explanatory, are suggested for normal school students and grade pupils.

Testa	seed coat	Plumule	seed bud
Tegmen			
Hilum	seed scar	Micropyle	doorway
Cotyledon	seed leaf	Epicotyl, sprout or "shoot" in corn and pea. The part that shoots up.	
Hypocotyl	seed stem		

The evolution of seed terms from the Latin with their meanings is an interesting study. Such a study shows the words and phrases in vogue in former periods and how slowly our language has been Hooverized. Space does not give room for quotations from but two sources.

"An Introduction to Botany" from the works of Linnaeus, by James Lee. 1799. "The seed, according to the definition of Linnaeus, is a deciduous part of the vegetable, the rudiment of a new one, quickened for vegetation by the sprinkling of the pollen. Its distinctions are: A seed, properly so-called, which is a rudiment of a new vegetable, furnished with sap, and covered with a bladdery coat or tunic. It consists of, 1. Corculum, the first principle of the new plant within the seed. 2. Plumule, a scaly part of the corculum, which ascends, etc."

The definitions as given by Mr. Lee are:

"Semen, Seed, the Rudiment of a new Plant; Seeds are known according to their Number, Figure, Superficies, and Consistence.

Hilum, the Eye, an external Scar of the Seed, where it has been fixed to the Fruit or Receptacle.

Corculum, the Essence of a new Plant within the Seed.

Plumula, Part of the Corculum, the ascending scaly part of the Plant.

Rostellum, the descending part of the Corculum that forms the Root.

Cotyledon, the side Lobes of the Seed of a Porous Substance, and perishing."

Outlines of Botany, 1819, by Dr. John Locke, gives the following: "Corcula (*Corculum*). This is the chick or embryo of the future plant. It is the essential part of the seed, to which all the rest are wholly subservient, and without which no seed will vegetate.

The Corcule consists of two parts.

1. Radicle (*Radicula*), the descending part, which unfolds itself into roots.

2. Plume (*Plumula*), the ascending part, which unfolds itself into herbage.

3. Cotyledon (*Cotyledones*), seed-lobes. They usually constitute the principal bulk of the seed. They are attached immediately to the Corcule, which they nourish until it has taken sufficient root to support itself from the earth."

In studying germination it is believed that the best results are obtained by beginning with the seedling, where the parts are easily seen, and proceeding to the seed, where the same parts are "in miniature." It is not denied that good results may be secured by following what is called the "order of development," beginning with the seed.

Whatever the order followed, the series of seedlings should be grown by the pupil. A sand-box divided by cross strings into as many areas as there are students forms an excellent garden. The pupils are asked to take 16 seeds each, of beans, peas, squash, and corn. Four seeds of each kind are soaked over night and planted. When these appear above ground four more seeds of each kind are planted, and so on, until each pupil has a series of seedlings. Each pupil cares for his own garden. He must not neglect it nor be "too kind" and drown the seeds. The pupils are warned about the unfairness of competition if they plant more than 16 seeds and that it might be embarassing to have more than the required number appear above ground. The ability of the pupil is judged by the per cent of germination. The grade shows relative and not absolute ability as the power of germination of the seeds used may be low.

Each pupil should make an individual study of the material. Directions written on the board will help him to make the best use of his time and to guide him in such a way that he will become more independent. As the writer has already written such suggestions in considerable detail in a Guide for Laboratory and Field Studies in Botany (1911) it is not deemed advisable to take space here. Directions in any case, can only be suggestive, and each teacher must solve his special problem for himself.

The seedlings may be drawn on paper and the seed may be modelled with plastocine. A clear notion of scale may be developed at this point. Students asked to make a drawing twice the natural size

(X2) are apt to make the sketch twice as long and twice as wide as the object which results in a representation of X4. If they take an amount of plastocine equal to two, three, four, etc., beans they can model the seed and get a correct idea of the scale. Then, again, the pupil who cannot draw readily can oftentimes model the embryo to good advantage and then follow by a correct drawing.

The study of germination is a basis for illuminating the class discussion. The pupil must have some understanding of the appearance of seeds and seedlings before he can think clearly regarding the economics of seeds and their relation to their environment. In the class discussion certain probelms will present themselves, as—"Can seeds germinate in the dark?" It does not occur to most students that seeds are planted in the ground where it is dark. It may seem best for some pupil to perform an experiment i. e., ask the seed if it can germinate in the dark. In discussion, bring out how we can ask the seed the question. Another sort of question, such as,— "Why do not geraniums have seed?" will have to be interpreted by reading. A third kind of project may be the construction of a germination box, rag-doll seed tester, etc. Assign these last undertakings to those who enjoy construction work. Test the seed for farmers of the neighborhood. In arithmetic it may be figured out that seed which costs the most may be the cheapest in the end. As a written lesson send for government publications. Some of the interesting articles in this connection are:

School Home Garden Circular No. 13, Dept. of Interior, Bureau of Education Garden Projects in Seed Planting.

Seed Improvement. H. G. Bell and C. A. Waugh. Soil Improvement Committee, 916 Postal Telegraph Bldg., Chicago.

Making Out the Seed Order. F. F. Rockwell, *Garden Magazine*, Jan. 1914.

III.   Seed Dispersal

1. *Classification:*
  A. *By Wind.*
      (1) Seeds small and light,—pigweed, many cultivated plants.
      (2) Seeds with down,—milkweed, cotton, clematis, aster, goldenrod, most composites, willow, poplar, cattails, buttonwood.
      (3) Seeds winged,—ash, maple, pine, catalpa, elm, hornbeam, spruce, curled dock.
      (4) Adaptations for rolling,—round, inflated pod of radish; inflated pod of hop; flower cluster a light ball for rolling in case of hydrangea; blowing around of whole plant as in spread of the Russian tumbleweed.
      (5) Seeds shaken from swaying plants—jimson weed, poppy, lily, larkspur.
  B. *By Animals.*
      (1) Fruits adapted for clinging—burdock, cocklebur, agrimony, bur-marigold, bedstraw, wild carrot, motherwort.

(2) Fruits edible—poke berries, poison ivy, sumac, red cedar, bayberry, barberry, nuts, apples, peaches.

C. *By Mechanical Contrivances.*

(1) Twisting of pods—pea, bean.

(2) Explosive mechanism—witch-hazel, violet, jewel weed.

D. *By Water.*

(1) Buoyant covering—sedge, butternut, cocoanut.

2. *Suggestions for Teaching a few Types:*

Have the class collect several kinds of seeds. Place in vials or boxes. Arrange according to method of dispersal, in primary grades after studying types, in grammar grades before studying types. Do not simply make a collection.

Have children keep booklets. Write the story of making wheat into flour; how the cherry trees came to grow along the fence row; the journeys of a sedge seed down a stream; how the burdock stole a ride; etc. Seeds may be kept in the booklet by pasting them upon the paper or by keeping them in little envelopes pasted into the book. Make drawings of types. It is better to teach a type of each means of distribution thoroughly than to teach many kinds superficially.

(1) *Seeds small and light.* Hold some grass seeds in the palm of the hand and expose to the wind.

(2) *Seeds with down.* Illustration, Milkweed.

Keep ripe pods in room for several days. Place under a glass jar so that they will not get about the room.

Describe the stem of the pod. What is the advantage of each pod having a spring-like stem? What is the result of the drying of the pod? Does the pod open toward the outside of the plant or toward the inside? Do the seeds show when the pod first cracks open? What advantage? How are the seeds arranged? Why? Are they scattered at the same time? Why? Which seeds would leave first? What reasons do you have for your answer? Do the seeds or the hairs leave the pod first? Why should it do that? Is there any milk in the stem after the seeds form? Why? Describe the texture of the margin of the seeds. Place a seed in water. Result? When would the corky margin be advantageous? What is the use of the silk? Compare with cotton.

(3) *Seeds with Wings.* Illustration, Pine seeds.

Show the class the cones of several trees. Have pupils give the distinguishing characteristics. Stand in a chair and shake out a few seeds. Class describe how they fall. What enables those seeds to keep in the air for such a long time? What is the advantage of this? What trees, besides conifers, have winged seeds? Drop a maple seed and have the class note which end lands first. Why is this fortunate for the seed? Why do not trees produce burr-fruits such as the burdock?

In primary grades the swaying of the cones might be compared to Rock-a-bye Baby in the Tree-top.

(4) *Fruits adapted for clinging.* Illustration, burdocks.

Throw against a piece of cloth. Advantages of catching. How would fur bearers help scatter these seeds? Class observe closely to see how burrs catch hold and cling? Send some to board to draw one spine. Fifty per cent will omit the hook. Tell them that they have missed something important. It is more important to see and draw one spine correctly than to draw the burr with all the spines incorrectly. Burdocks grow beside paths. Does the plant grow there so as to be where people will brush by it or is it there because people walk that way and scatter it?

In lower grade, let pupils make a basket. See Poem pp. 388. Nature Study and the Child by C. B. Scott.

(5) *Fruits Edible.* Illustration, Apple.

What is the advantage of immature apples being green and sour? Cut a cross section of the apple. Advantage of the seeds in the center? What is the use of the skin? Seeds smooth and slippery? Bitter?

(6) *Mechanical Contrivance.* Illustration, Jewelweed.

The pod splits into ribbons which twist upward. The seeds are flung to some distance from the parent plant. This is the origin of the name Touch-me-not and also the scientific name Impatiens. Hang on the school room wall a branch of witch-hazel which has ripened but unopened nuts. Observe distances to which these seeds are thrown. Try to fit seeds back in cases.

(7) *Pepperbox arrangement.* Illustration, Jimson Weed.

Hold the ripe pod vertically and jar the stalk by snapping the finger against it. Also sway as it would in the wind. How are the seeds of the Jimson Weed scattered?

3. *Suggestions for Field Work:*

Find a parent tree with its many seedlings. How are its seeds distributed? How far from the parent did they go? Advantages of this? Is there any evidence that seeds were produced every year? How account for this observation? Which seedlings are flourishing? Why?

Find birds eating berries. Where did you find? What were they eating? Did they swallow the whole berry? Which birds were distributing seeds?

Find animals storing seeds. Give the name of the animal and the seed. Where did he store the seed? How did he store the seed?

Visit a weedy area. Many or few seeds? How are the seeds protected before they ripen? Note damage caused by the scattering of weed seed.

How are grains and grasses spread?

4. *Foreign Seed and Plant Introduction:*

Send to the Office of Foreign Seed and Plant Introduction, Bureau of Plant Industry, U. S. Department of Agriculture for their various publications and inventories of seeds and plants imported from abroad. Records have been kept for about 21 years and some very interesting data can be obtained. As early as 1743 the British Parliament granted $600,000 to promote Indigo Culture and other crops in the American Colonies. In 1898, James Wilson, the Secretary of Agriculture, got an appropriation of $20,000 for introduction work. Just a hint of the work is tabulated here.

Date	Introduced by	What	From	To
1898	Dr. S. A. Knapp	Rice	Japan	Louisiana
?	Thomas Jefferson	Rice	France	Carolinas
1900	Walter T. Swingle	Date Palm	Algeria	South west
	M. A. Carlton	Rust resistant wheat	Russia	Western states
	N. E. Hansen	Drought resistant alfalfa	Russia	Northwest
1905–08	Frank N. Meyer	Seedless persimmons	Tamopan China	Southern states

5. *California Seed Farms:*

The horse latitudes; a dry climate; a heavy climate; a home for the wealthy who can afford to go there; an educated class of people who read and study. It is said that more good books are read in southern California than in any other part of the country. So much for the climate and people.

How is this region adapted to raising seed? A rainless season; irrigation; constant sunshine for ripening and curing seeds (no need of buildings for curing); level land for use of improved machinery; Chinese labor which is reliable and at a low cost.

A few statistics: About 95 per cent of world's lettuce seed raised here; 500 pounds of lettuce seed are grown on an acre; there are 1,025 varieties of lettuce seed grown there; over 3,000 acres are used for raising lettuce seed; and there is an annual yield of nearly 2,000,-000 pounds of lettuce seed. What has been said of lettuce could be written about most of our seeds that we use in large quantities.

The resume given above is enough to show the possibilities for a correlated "Control-response" lesson in geography.

6. *References:*

Beal, W. J. Seed Dispersal. Ginn & Co., 1899. A small book, well illustrated.

Bryant. Planting of the Apple Tree.

Dana, Mrs. William Starr, 1896. Plants and their Children. American Book Company. Written for Children. Interesting Reading.

Darwin. Origin of Species, Chapter XI, Dispersal. Dissemination of plants by ocean currents and by Migrating Birds. Scientific.

Daulton, Agnes McClelland, 1903. Wings and Stings. Rand, McNally & Co. Gadabouts, pp. 90–98. Not as interesting as the plain facts.

Fultz, Francis M., 1909. Public School Pub. Co., Bloomington, Ill. Written for children. Gives no facts but what can be easily discovered by the children. Similar stories could be produced in the grades. Not complete enough for teachers.

Gibson, William Hamilton, 1898. Sharp Eyes. A Ramblers Calendar. Harper & Bros. "A cordial recommendation and invitation to walk the woods and fields with me and reap the perpetual harvest of a quiet eye."

Lubbock, Sir John, 1896. Flowers, Fruits and Leaves. Macmillan. The subject treated at length.

Poulsson, Emilie. In the Child's World. How West Wind Helped Dandelion. A story.

Little Chestnut Boys. A story.

Sleeping Apple. A story.

Apple Seed John. A poem.

Scott, C. B., 1900. Nature study and the child, pp. 370–395. Good suggestions for teachers.

Smith, Eleanor. Dandelion Fashions.

Milkweed Babies. Easy poems.

Thoreau, H. D. Succession of Forest Trees. Interesting.

Weed, Clarence. Seed Travellers. Good to read when observing.

Whittier. The Fruit Gift.

IV. Mounting Seeds for Class Use

1 *What seeds to collect:*

Weed seeds to show how plants travel.

Crop seeds to show common ones used in school and home garden.

2 *How collect:*

Use old envelopes in the field, writing name of seed on outside.

3 *Preparation for Mounting:*

Clean of impurities.

Kill injurious insects—carbon bisulphid or Formaldehyde fumes.

4 *Mounting:*

A. Screw-top vials, 2 or 3 dram size.

B. Glass boxes.

Obtain a pane of clear window-glass of size desired and two pieces of stout cardboard of the same dimensions as the pane of glass. In one of these cardboards cut 1 inch holes which are one-half inch apart. Between these cardboards place a thin layer of cotton batting. In the first row of pockets place seeds that are carried by the wind. In another row, place burrs, berries, etc., that are carried by animals, and in another row, seeds that are spread by mechanical contrivance. Cut appropriate sized numbers from a calendar and paste a number beneath each pocket. A little paste may be used to hold some of the seeds in place. The lower parts of the bulky seeds may be cut away so that they will not project above the pockets. Place the pane of glass over the pockets to hold the seeds in place. The various parts are then held in place by fastening them in a picture frame or by passe-partout binding. On the back of the mount should be numbers and names to correspond with the numbering below the seeds.

In the case of an identification chart the seeds should be arranged according to form, color or size.

5 *Use of the Mounts:*

Seeds mounted this way may be studied by use of a simple microscope.

The pupil should not merely make a collection. The collection is only a means to an end.

See outline for sudy of seed dispersal.

The pupil may wish to refer to the crop seeds when planting his garden.

6 *References:*

Farmers' Bulletin 586.

Nature-Study Review, November 1914. Page 292.

V.   *Seed Folklore, Omens and Charms.*

The many knockings on wood by people who say that they have not had influenza, and the camphor charms carried in their pockets would lead one to infer that all interest in signs and omens has not passed away. There is a fertile field for studying seed-lore. A few are given to show the possibilities for research and interpretation:

Carry a chestnut in the pocket to prevent rheumatism.

A child blows the seeds on a dandelion head three times. If he blows them all away his mother wants him. Or he can tell the time by the number of seeds left.

Plant corn when the beech leaves are as large as a mouse's ear.

Thick corn husks means a severe winter.

Pumpkin-seed tea used as diuretic.

"The year's at the spring
The day's at the morn;
Morning's at seven;
The hill-side's dew-pearled;
The lark's on the wing;
The snail's on the thorn;
God's in his heaven—.
All's right with the world."

—Browning.

THE HEN AND CHICKENS VISIT SCHOOL. It is not necessary to have the hen wearing overalls and a beaver hat, or to have her baking bread for an Ugly Duckling, to be interesting.

CHAPTER XXIV

FIRST GRADE READERS

A SURVEY AND CRITICISM

Nature-study teachers will concede that first grade readers are not supposed to be nature readers or for nature-study. However, as the large majority of stories in these books are about plants and animals the books have a decided relation to nature-study. Librarians, kindergartners, and literature teachers will not agree with this article. Their experiences certainly cannot be ignored. What position do science teachers take in regard to first grade readers?

The following statistics were gathered from thirty-three of the best and most used first grade readers. Over 50 per cent of the stories in these readers are about animals. Only 33 per cent of these animal stories are true to nature. The other 66 per cent are either fables or personifications where the animals act and talk as human beings. For every five animal stories there is one plant story and one geography story, such as:—The wind, the rain, or the Eskimo. Out of fourteen stories related to some special day nine are devoted to Christmas, three to Thanksgiving, one to St. Valentine's, and one to Hallowe'en. Only two books have any biography and these honor Dupré, Millet, Abraham Lincoln, Confucius, Washington, King

Midas, and Columbus. The other subjects consist of a little fiction, such as:—Dolly's Ride, or Helping Father, and stories about inanimate objects, such as:—A top, a drum, or a sled. Only four books out of the thirty-three give any attention to our flag or to patriotism.

A study of the accompanying table will be of interest to all naturalists. For instance: The cat is written about more than any other animal. Out of thirty-eight cat stories fourteen are of the fairy type (includes myths, legends, mother goose, and fables), fourteen are about cats who talk as human beings, and ten are true to cats and their doings. It is only fair to state that the majority of these cat stories which cannot be classified as personifications or fables are of the following style: "See mamma. See kitty. Mamma, see kitty. See kitty, mamma. My kitty. See my kitty. See my kitty, mamma, etc." As soon as possible these books launch out from this stage into cat-lore,—

TABLE SHOWING NATURE CONTENT OF THIRTY-THREE OF THE BEST AND
MOST USED FIRST-GRADE READERS

ANIMALS				PLANTS					GEOGRAPHY			
Name	Total number	True to nature	Personification	Fairy Tale, etc.	Name	Total number	True to nature	Personification	Fairy Tale, etc.	Name	Total number	True to nature	Impersonation
Cat.........	38	10	14	14	Apple.......	10	7	3	0	Snow........	9	8	1
Dog.........	20	9	9	2	Maple tree..	5	3	2	0	Wind........	8	3	5
Cow.........	17	9	3	5	Daisy.......	4	3	1	0	Rain........	7	4	3
Mouse.......	13	2	4	7	Dandelion...	4	2	2	0	Frost.......	5	1	4
Hens........	17	9	8	0	Seeds.......	4	3	1	0	Dutch.......	4	4	0
Rabbit......	13	4	8	1	Seedlings....	4	3	1	0	Eskimo......	4	4	0
Fox.........	11	1	7	3	Leaves......	4	3	1	0	Sun.........	4	1	3
Bee.........	10	2	4	4	Cherries.....	3	3	0	0	Farm........	3	3	0
Horse.......	10	8	2	0	Orange......	3	3	0	0	Moon........	3	1	2
Pig.........	10	3	2	5	Pussy Willow	3	3	0	0	Cloud........	1	1	0
Squirrel.....	10	3	7	0	Milkweed...	2	2	0	0	Japan Girl....	1	1	0
Chickens.....	9	4	5	0	Oak........	2	1	1	0	Spring.......	1	1	0
Robin.......	6	4	2	0	Tulip.......	1	1	0	0	Wool........	1	1	0

SUMMARY	Total	True	Personification	Fairy Type	Books*
Animals..................	339	120	126	93°	33
Plants....................	79	57	22	0	17
Geography...............	63	41	22	0	18
Special days.............	14	14	0	0	10
Biography...............	6	6	0	0	2
Our flag.................	4	4	0	0	4

*Indicates the number of books out of the 33 examined in which these kind of stories appear. For example: The 22 personified stories about plants occurred in 17 books.

"Ding, Dong, Bell;" the kitten-mitten episode; the Pussy Cat and the Queen; etc. There is an intermediate stage of true stories which is omitted. What has been said about the cat is typical of other animals. One must conclude that the "True to Nature" stories are of rare occurrence in first grade readers.

The child's written composition is too often a reflection of his reader. It should be vice versa. The method of the teacher is also influenced by the method of the readers. This influence permeates the primary grades and the following essays written by primary grade children represent the work of three different teachers and three distinct methods.

1. *Goldenrod.* "My name is goldenrod. I live by the side of the road. I have purple aster nearby. I am dressed in yellow. It is like the sun. My friend has a purple and yellow dress. I have a lot of gold-dust. I come in the autumn when the trees are turning brown and yellow. My leaves are green and narrow." This teacher saw nothing interesting in goldenrods and had had no scientific training. The child did not learn anything about goldenrods. The plant was humanized. Nature is interesting enough in itself and needs no fluffiness.

2. *Nasturtium.* "My flower has a green stem. My flower's name is nasturtium. It has petals and is yellow and brown. It smells nice. My flower grows in my garden. Bees go in it. They make honey for us. There is some honey in the end." This paper is a decided advance over type one. The plant has a relation to the life of the child. Each flower fact however, is isolated and has nothing to do with the life of the plant. Bees do not gather honey. They gather nectar to make honey.

3. *Dandelion.* "The dandelion grows in the meadow and on the lawn. The root is long so that it can get the food in the ground. The leaves are pointed. It is called the dandelion because the leaves are like a lion's tooth. There are many dandelions because when the lawn mower goes over them it cuts the leaves but it does not hurt the roots. The animals do not eat them because they are too bitter." Although there is much that could be improved in a second draft the results attained in this composition are very creditable. This pupil saw a relation between the form of the plant and its life. The facts have a meaning. The adaptation of a plant or animal to its environment should play an important part in every story of its life. It is this form of story that is lacking in the first grade readers.

What literary development has the child at the end of the first year? He has had a year of exciting reading. He probably prefers this style, but, does that make it right? Most children prefer candy but we do not make it 50 per cent of their diet. "Is ding dong bell" the means or is it the object of education? Education is a process of living. The child lives with a cat that catches mice and laps milk with its rough tongue. These every day subjects of child life are natural, fundamental subjects. It is sort of a "See America first" idea. Why go by the fable-route to tell about interesting things at home?

If the child does not get true stories now, when will he? The majority of children never will get true animal tales. The only ones I know of, about cats, are the leaflets written for the Humane Education Society and many of these are the "Do not tie a tin can on the cat's tail" kind. (It would be better pedagogy to get an interest in cats.) The writers of real nature literature may be counted on one's fingers. They are Bradford Torrey, Frank Chapman, Ernest Ingersoll, Dallas Lore Sharp, Gilbert White, Henry D. Thoreau, and John Burroughs. And, after all, were not our great writers worth while naturalists? Emerson, Longfellow, Whittier, and Tennyson were not nature fakirs. What has a child in his early literature to prepare him for these writers? Nothing. His early training prepares him to make such books as Long's and others successful. These writers tell about animals, not as they are but as their readers like to think they are. I well remember my bitter disappointment in the learning powers of two pet rabbits after having read about rabbit intelligence as "observed" by these two men. The average first grade reader is the initiator of the appetite which enjoys nature faking, the colored supplement of the Sunday newspaper, and shall we add, ante-nuptial love stories,—situations remote from real life.

The existance of gullibility as to fairy stories was illustrated in class recently. I asked how deep in the ocean are growing plants found? The answer was rather fabulous. Questioning brought out the information that the pupil knew it was so because she had read about it in Jules Verne's "20,000 Leagues Under the Sea." The pupil did not know the meaning of a league but on looking it up returned with the same amount of credulity even though 20,000 leagues means 60,000 miles and the diameter of the earth is about 8000 miles. Any one entertaining any doubt as to the universal faith in such matters should question farmers in regard to the Old Farmers' Almanac and their belief in weather folk-lore.

The reasons given for the fairy-land route are more traditional than real. The supporters might as well say,—"The more remote an idea is to the life of a pupil the greater is the value of the idea." One purpose of the fairy story is to teach a moral truth. I would give more for one moral taught by the realities of the Boy Scout method than for one hundred morals dramatized by the gaseous vertebrates of fairydom. But, someone says, the children like the poetical and rhymes. Cannot the truth be made as poetical as the untruth?

The fairy story trains the imagination. If a little more than 50 years ago someone had told General Grant that some of his men would live to see the day when battles would be fought in the air and beneath the surface of the ocean he would have smiled incredulously. His imagination could not have conceived the things of the great war of today. These wonderful inventions of science are due to the imagination based on realities. Study the life of any of these great inventors and you will find them based on realities. When the imagination wanders about unrealities the real environment does not change. When imagination is based on reality it accomplishes things. And so it is with the child. His imagination should not be started on the unrealities of life but should be allowed to start from the higher plane of truth. This is what the child does when he uses the stick for a horse and the chair for a boat. Fairy stories are the inventions of adult imaginations forced upon the child and yet we inconsistently feel pleased when we can say that "he no longer believes in fairies."

Then the recapitulation argument. The literature of primitive man was folk-lore and the childhood of the individual should parallel the childhood of the race. The mother of the howling savage stilled her papoose with a folk-lore story. Does that excuse the modern (?) mother who awes her little one with "The boogy man will get you" or "I will shut you up in a *dark* closet?" That child is to live in the future. He may have to perform some duty in the dark but lacks the courage because fiction has preceded fact. The Br'er Rabbit stories of Joel Chandler Harris are valuable records of the folk-lore of the African Americans. They were told by negro mammies to silence their picaninnies or to rouse them into delight. The triumph of these imaginations over fact in the mind of the negro has made him superstitious and awry. That method of child discipline is not necessary or desirable. Neither is it a prerequisite for raising rabbits. Folk-lore stories are illustrative of old methods and are rich in material for the mature student. I believe that they have no place in the primary grades and above all things should never precede the true story.

As the criticism of these readers may be thought to be severe, verbatim samples of their nonsense have been selected.

"*A little bee cried, 'Buzz, Buzz,' all the way home.*" The author was so impressed with the little pig and wee wee that the same style was extended to bees and other animals.

"*Can I pat the fat cat? Did the dog see the frog on a log?*" Here the writer is obsessed with the phonic idea at the expense of good sense. I am not an authority on the matter but have been taught that *can* means to be able or to be competent to do something.

" '*Why, why, why!' said Goosey Loosey.*" This reminds me of a friend who took great pleasure in using baby talk with his baby girl. He would say: "De boomps want wa wa!" Translation: Does baby want water? Cute? Yes. But, why waste time in learning two vocabularies? Why keep the little one in that early confused stage of acquiring words? The aim of the parent or teacher should be to help the child over complex places and not to prolong the period of difficulty.

> "If all the world were apple pie,
> And all the sea were ink,
> And all the trees were bread and cheese,
> What should we have for drink."

Children read and listen to this foolishness while Hooverism is a vital part of our existence.

"*Once a cat ate a kid.*" Quite a fete-cham-petre for a cat.

" '*Little Mouse, will you have some milk?*' says Mrs. Cat." How modest as compared with the cat mentioned in the preceding quotation. Cats are not usually so considerate with mice.

"*Mrs. Tabby Gray told stories to her kittens.*" She could also tell interesting stories to the person who can observe closely.

"*Help! Help!*" he cried; "*I am buried alive.*" This is said by a squash seed. A squash seed must be buried alive in order to grow into a healthy plant.

> "Bees don't care about the sno
> I can tell you why that's so
> Once I caught a little bee
> Who was much too warm for me."

Rather of a pointless point.

> "And the milkweed laughed and laughed,
> It laughed so long that the pod burst."

Some interesting truths about milkweed pods. Contrast, with milkweeds that laugh, for educational value. Each pod has a spring-like stem. When the wind blows the spring shakes the seeds out and they are blown away. As long as the pod is green and growing it cannot burst even though it laughed harder than ever. When the pod stops growing it begins to dry. It then shrinks and splits at the seam. The pods open toward the outside of the plant instead of toward the stem. The seeds, therefore, are not apt to catch onto any part of the plant and they are ready to be taken in any direction. When the pod first opens its exposes the seed. As these seeds ripen the silk dries and becomes exposed to the wind. As the seeds are arranged on top of each other, like shingles, they are not distributed at one time. Some of the seeds thus have a greater chance of being distributed at a favorable time.

> "The world is so full of a number of things,
> I'm sure we should all be as happy as kings."

This has become ironical. School readers in the future should be 100 per cent democratic.

"*Quack, quack, quack!*" said the little duck.

This was on the day that it hatched. Little ducks do not *quack.*

"*But the Half-chick only laughed,*" and said,—"*I'm off to see the King.*" The sum and epitome of absurdity.

The pictures of first grade readers are fashioned after the style of the stories. Many include such pictures as: pig crying; kitten baking pie; hen wearing a thinking cap; wolf blowing a house down; a fairy pulling the seed coat off a squash seedling; kittens knitting mittens; duck with an umbrella; fly weeping; pansy with a human face and wearing a dress. Another kind of picture, the one that is supposed to tell the truth and does not, is really the most mischievous. Fortunately there are not many of them. In one reader there is a story of the mother blue bird setting on her nest of eggs. On the opposite page is an excellent colored plate of the blue bird. The blueness of the picture shows it to be the male bird. The child naturally supposes that it is the mother bird. In another book the story of the ox finding the cross dog in the manger is illustrated by a good drawing of a cow and a dog. Some country children know that a cow is not an ox. Many first grade readers are beautifully illustrated and this is rather remarkable when we consider the very low price at which they are sold.

Suppose a child escaped the vicissitudes of a first grade reader, and acquired the vocabulary of an (un)ordinary young citizen. If suddenly ushered into the inner shrine of a first reader he would meet a foreign language which for our convenience may be classified as follows:

Strange animals: Magpie, roebuck, falcon, civet, vampire, unau, newt, ibex, dickey-birds, terrapin, yak, zebu, cockhorse, etc.

Unusual words: Slink, huffed, dillar, diddle, dickory, malt, whey, ply, tuffet, curds, elf, pease, etc.

Baby talk: Lammie, mousie, henny, goosie, etc.

Deceased (?) words: Lambkin, prig, ye, dame, ewe, etc.

Marvellous creations: A-riddle-ma-ro, chinchopper, cherri-o, jiggityjog, clumpety, higglety, pigglety, etc.

As reading is the main occupation of a first grader the verdict of a rigid course of study would be quick and merciless. The child with the vocabulary of ordinary folks must remain in the first grade.

Conclusions regarding first grade readers from the point of view of a naturalist:

1. True to nature stories are decidedly lacking in first grade readers.

2. Stories about the ordinary everyday doings of animals are missing in first grade readers.

3. First grade readers belittle the mind of the child by giving him foolish stories a world of hobgoblins, void of thought.

4. True nature stories are interesting enough and do not need to be artificialized.

5. The fairy stories of the ordinary child primer are for the study of the mature student.

6. The true story should precede the myth.

7. The writers of first grade readers have the preconceived idea that the unreal is a necessary preparation for literature. This inherited custom hinders the writers from proceeding in a natural way and so they distort the mind of the child for months with the literary method of the savage. Should we keep the child on all fours to be sure that he passes through that stage? Education should help the child pass through these stages and is not a device to keep them in them.

THE COMING AMERICAN

Bring me men to match my mountains;
   Bring me men to match my plains,—
Men with empires in their purpose,
   And new eras in their brains.
Bring me men to match my prairies,
   Men to match my inland seas,
Men whose thought shall pave a highway
   Up to ampler destinies;
Pioneers to clear Thought's marshlands,
And to cleanse old Error's fen;
Bring me men to match my mountains—
      Bring me men.

Bring me men to match my forests,
   Strong to fight the storm and blast,
Branching toward the skyey future,
   Rooted in the fertile past.
Bring me men to match my valleys,
   Tolerant of sun and snow,
Men within whose fruitful purpose
   Time's consummate blooms shall grow.
Men to tame the tigerish instincts
   Of the lair and cave and den,
Cleanse the dragon slime of Nature—
      Bring me men.

Bring me men to match my rivers,
   Continent cleavers, flowing free,
Drawn by the eternal madness
   To be mingled with the sea;
Men of oceanic impulse,
   Men whose moral currents sweep
Toward the wide-infolding ocean
   Of an undiscovered deep;
Men who feel the strong pulsation
   Of the Central Sea, and then
Time their currents to its earth throb—
      Bring me men.

—Sam Walter Foss

CHAPTER XXV

SUGGESTIONS FOR ARBOR DAY PROGRAM

TREES IN WINTER

Project 1. See Something

Every school citizen is a member of the Committee for the Arbor Day program. The first duty of a Committeeman is to *see something*. He will, therefore, be given a horse-chestnut twig and a last summer's leaf stem which is easily gathered beneath a horse-chestnut tree. If he is a working member of the committee he will sharpen his pencil and eyes and fill in the blanks left for this purpose.

Look carefully at the large end of the leaf stem and find where it fitted onto the twig. This place on the twig is called a *leaf scar*.

When did this leaf fall off from the twig?................

The "nails" in the leaf scars are the ends of *veins*. How many veins went out into the leaf?..................

What did these veins carry to the leaf?..........and......

How does the leaf scar differ in color from the rest of the bark?....

Which has been exposed to the weather longer?,.........

What do you find just above the leaf scars?............

Where on the twig is the largest bud?..................

This large bud holds both leaves and flowers. When will the leaves come out?..............................

What must the *bud scales* do before the leaves can come out?......

Find the ring-marks which were left when the bud scales fell off last spring.

How often do these *scale scars* form?............

How old is your twig?......................

How many leaves were on your twig last year?...............

Why would it always be an even number of leaves?............

Find small dots not in the leaf scars. These are *breathing pores*.

The "varnish" on the bud scales prevents the small leaves from drying out.

A temporary chairman now calls a meeting of the committee. The first matter of business is to call for the written report (Arbor Day Booklet with Blanks filled in) of the members. These reports are placed on the table. The chairman then reads the first question for discussion. "When did this leaf fall off from the twig?" The opinions of various members are given. If they agree the answer is accepted and the chairman goes on to the next question. If they do not agree each one is given a chance to defend his opinion, and then when the assembly is ready for the question a vote is taken. An appeal may be made to the chairman. When the questions have been answered satisfactorily a committee should be appointed to rate the written reports. At the next meeting the committee reports the names of those who did the most efficient work. A permanent chairman and committees may be elected from this group.

Committees for stories, music, pantomimes, debate, election of state tree, games, decoration of room and community welfare are in order.

Project 2. Reading Signs

Every tree keeps a diary. If we can read this diary we will know its life history. Following is the autobiography of the horse-chestnut twig shown in the drawing. Fill in the blanks and you will agree that it is more fun to read trees than to read about trees.

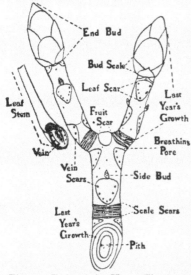

Fig. 1. Twig of the Horse Chestnut

I was born in the spring of 192..... from an end bud. The weather being warm I shed my bud scales and put forth....... leaves. I also had a great many pinkish-white............ My flowers were fragrant to attract the bees. By autumn one of my blossoms had developed a green prickly ball or................. I did not have food supply enough to afford to grow any more. After a few frosty nights my husk broke down the seams and showed a smooth, shiny................ I produced this............so as to grow other horse-chestnut trees.

As I had fruit on the end of my twig this year I could not grow an end bud. However, I had two buds on the side which I carried along for just such an emergency. When the........came up my...... the next spring I shed the............on my side buds and put forth two branches. My branch on the left had more........and therefore made better growth than the opposite branch. I had...... leaves this summer. As I did not have to manufacture starch for horse-chestnuts I was able to make two large............buds.

My leaves for next summer are securely wrapped with bud scales. These scales are arranged like the..........on a house. Then, to be further sure that the tender leaves do not dry out I secreted some varnish. I am going to stay this way all winter.

Project 3. Do Something in Song

Oliver Wendell Holmes saw the Chambered Nautilus. From its diary he was able to read something about its life. He then wrote a poem. Let every member of the committee write a poem about the horse-chestnut twig. Then vote which poem is the best. That is

what the boys and girls did in the third and fourth grades of the
Henry Barnard School. And then they sang the poem. They
changed the words and song until they had it just the way they wanted
it. They did so well we are going to print their original songs and

SPRING SONG

Words and music by Third Grade Pupils.   Henry Barnard School.
    Miss Lina F. Bates, Critic Teacher.   Miss Alice Maguire, Student Teacher.
        Miss Grace Gormley, Student Teacher.

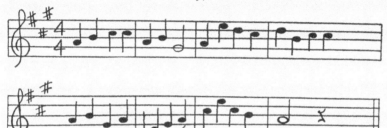

Spring is here with op'ning buds,
  Soon they will be blossoms gay;
The breezes rock them to and fro,
  All the live long day.

Birds are coming from the south,
  Singing us a cherry song;
Bees are buzzing in the flowers;
  Happy all day long.

Little bud, wake from your sleep,
  Tell to me your dream so sweet;
Put on your dress so bright and green,
  Birds and bees to greet.

THE OPENING BUDS

Words and music by pupils of Fourth Grade.   Henry Barnard School.
Miss Mabel T.Gardner, Critic Teacher.   Miss Jessie Molansky, Student Teacher.

O buds, you've slept the winter thru,
'Tis time for you to wake;
To shed your tiny blankets warm
And pretty blossoms make.

'Tis time to show your colors gay,
For now the bluebirds sing;
The earth is full of happiness
Because you tell of spring.

And when your blossoms gay shall fade,
And fruits begin to swing,
We'll still be glad for we are sure
That you'll come back next spring.

Fig. 2. The Buds.

Fig. 3. The Scales drop off.

Fig. 4. Butterfly comes.

Fig. 5. Buds asleep.

Fig. 6. Sunshine awakens Buds.

Fig. 7. Leaves and Buds come.

music. If you compose a good one send it to the Commissioner of Education, who is the editor of the Arbor Day booklet, and perhaps he may publish yours next year. We also hope that many people will enjoy the song that you write and sing for Arbor Day.

Fig. 8.   White Pine Blister Pantomime.

Project 4.   Do Something in Pantomime

The first and second grades of the Henry Barnard School had a lesson on the horse-chestnut twig and then told in pantomime how it grew.

In Grade I the following pantomime was worked out by Miss Theresa Barone, critic teacher, and Miss Alice F. Kelley, student teacher.

Grade I, Scene 1. Two Horse-chestnut Buds. The children on the outside represent bud scales. They are dressed in brown. Figure 2.

Grade I, Scene 2. The brown bud scales have fallen off and the green leaves are dancing about the pinkish flower buds. Figure 3.

Grade I, Scene 3. The blossoms come forth and receive a visit from a butterfly. Figure 4.

In Grade II the pantomime was worked out by Miss Emma G. Pierce, critic teacher, and Miss Virginia Nass, student teacher. Note how differently the same idea may be expressed.

Grade II, Scene 1. The brown buds are fast asleep. One little bud is just beginning to awaken. Dark grey clouds are dancing about. They bring moisture to the buds. Figure 5.

Grade II, Scene 2. The yellow sunshine brings warmth to the buds. The buds begin to shed their brown scales and the green leaves show underneath. Figure 6.

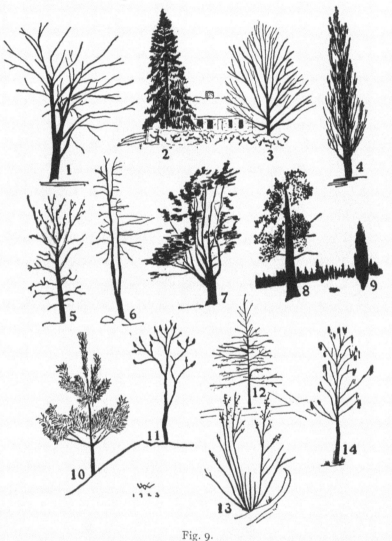

Fig. 9.

Grade II, Scene 3. The buds have shed their brown scales, which may be seen on the ground. The green leaves are now fully exposed to the sunlight. Some of the boys have bright colored caps which represent the flowers that came out with the leaves. Figure 7.

Project 5. Play a Game

This game may be played indoors or out-of-doors. When indoors, someone steps to the blackboard and with the side of a crayon draws a tree, such as the Norway Spruce or Lombardy Poplar. In the Lombardy Poplar all the limbs reach up to the sky. In the Spruce the lower limbs droop toward the ground. The audience then guesses the name of the tree. (See Figure 9.)

Trees may be recognized from a distance, in much the same way that you can recognize other friends. Each tree friend has its characteristics. How many tree friends can you recognize in the drawings? The fourteen trees shown are: (1) American Elm; (2) Norway Spruce; (3) Sugar Maple; (4) Lombardy Poplar; (5) Sycamore; (6) Sour Gum or Tupelo; (7) White Pine; (8) Red Cedar (Veteran); (9) Red Cedar (Young); (10) White Pine (Young); (11) Sumac; (12) Pin Oak; (13) Alder; (14) Catalpa.

See who can recognize the greatest number of trees from any good view point, out-of-doors.

These trees may be cut out of black paper to make tree silhouettes. White silhouettes may be used for a blackboard border. The group of red cedars seen in the background (8 and 9) might also be used for a border. Their thick dark foliage makes spires against the sky line.

Project 6. Our State Tree

In 1895 the Maple Tree was elected as the Tree of State by the school children of Rhode Island. The whole number of votes cast was 16,776. The results were as follows:

Maple Trees	5,750	Hickory	262
Elm	5,260	Buttonwood	210
Oak	3,707	Ash	196
Chestnut	632	Cedar	191
Pine	369	Birch	189

Now it so happens that there are eight Maple Trees growing in Rhode Island and we would like to have our school citizens of to-day decide which Maple Tree they wish to fill this important office.

There are then eight candidates for the office of the Tree of State. In order to vote intelligently each school citizen must know the candidates by name and sight, and something about their characteristics. This makes desirable a caucus or primary convention to be held before the decisive vote is taken. It is the duty of every good school citizen to take part in the primary election. The purpose of the caucus is to learn the relative merits of the candidates and to nominate the two most promising representatives of Mapledom.

In order to help our school citizens become acquainted with the various tree candidates in this campaign we have printed twig profiles. We suggest that each voter color in these outlines according to the color shown in the various maple twigs that he can find.

Find out if these candidates have always been citizens of this state. We know two that were introduced from Europe. We know two others that are common in the North but occur only occasionally in Rhode Island, and then in the northern part. Two others are more common in the South and are less frequent in this state. And the remaining two are native to Rhode Island and grow abundantly throughout the state.

You will also wish to know how useful these citizens are. One of these maples is a favorite food for deer and moose and is known as

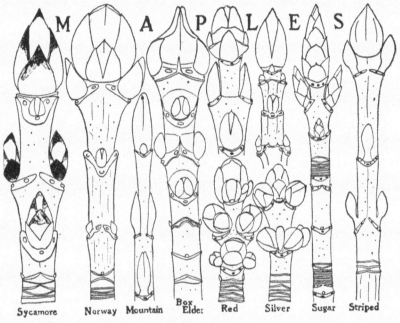

| Sycamore | Norway | Mountain | Box Elder | Red | Silver | Sugar | Striped |

Fig. 10.

Moosewood in northern New England. Sometimes the wood of a tree has another name when spoken of by the cabinet maker. Curly Maple and Bird's Eye Maple are examples of this. What maple is used for maple floors? Maple furniture? What food is obtained from one of the maples? Which are best for shade trees? Which maple would be a good state tree for Vermont? Why? Which maple is the most useful to us in Rhode Island?

The maples in literature must also be thought of in this connection. The following poem was written for the Arbor Day Program of 1895, and dedicated to the public schools of Rhode Island by George Shepard Burleigh of Little Compton. There are seven different lines in this poem that tell me which maple Mr. Burleigh had in mind when he wrote the poem. How many of these lines can you find? See how many quotations you can find about the Maples? Which Maple do the poets like to write about best?

THE MAPLE

Exalt who will the oak and pine
  To flutter in their banner's folds,
The goodly Maple shall be mine,
  The glory of our rocky wolds,
That fires the spring with reddening buds,
And blazes in the autumn woods.

Its stainless blood draws sweetness in
  From shining sun and dusky marl;
Its thickening foliage shrouds the din
  Of whistling blackbirds, and the snarl
Of catbirds, holding back the notes
Of every songster, in their throats.

Observe the two-leaved germ beneath
  Its horny shelter locked secure;
Unwind it gently from its sheath,
  And lo, a tree in miniature!
So in the boy the future man
Is wrapped, in Nature's perfect plan.

The sturdy trunk has gnarls and coils
  That gladden the aesthetic eye,
When polished by the carver's toils
  To serve the boudoir's marquetry,—
For all is good, from bud to core,
To win our praise and crown our store.

Then lift our banners to the breeze;
  Our symbol Maple proudly sing;
The brightest of autumnal trees,
  The first to prophesy the spring;
The State's fit emblem where began
Full freedom for the soul of man!

After the good points of each tree candidate have been presented the school audience should be ready to cast a ballot for the Maples that they wish to nominate. Instruction in preparing a ballot should be given at this time. The two Maples receiving the largest vote are nominated for final candidates to be elected by a plurality vote on Arbor Day, 1924. That is, your school or room will vote upon the two nominations on Arbor Day, 1923, and will send the result to the Commissioner of Education. The final vote will be taken on Arbor Day, 1924. It might enliven interest to have a debate before the balloting, on the merits of each of the Maple tree candidates.

Project 7. Community Welfare

The white pine blister rust was chosen as a community welfare problem by the Blue Bird Troop of Girl Scouts, Henry Barnard School, Miss Gertrude Evans, Captain. The picture illustrates how the life history of the rust was presented in pantomime. The tall girl on the left represents a white pine that has become infected with the disease. The next three girls are healthy white pines. Their five fingers indicate the five needles found in a cluster in the white pine.

The girl with the cape is the wind which takes the spores (pieces of paper) from the orange sac on the right and carries them to the pine. The wind then carries spores from the pine to the gooseberry where sacs are again formed, such as shown enlarged on the right.

A group of Girl Scouts come along and discover that the White Pine is infected with a disease. They look it up in their guide book and find that the disease is the White Pine blister rust. They read further and learn that it is prevented by digging up all the currant and gooseberry bushes within 900 feet of the pine. They placed the bushes on the camp fire which they used to cook their dinner. The healthy trees lived but the infected pine was finally killed. The scouts went away happy that they had learned something and that they had done their "good turn" for the community.

Leaflets containing colored pictures of the white pine blister rust and its hosts together with descriptive material may be obtained from the United States Department of Agriculture.

Project 8. National Project

An Arbor Day program should not only contain community and state projects but some national project. The one that seems most appropriate just now is "Our National Parks."

In 1922 the fiftieth anniversary of the founding of the first national park in the world was held in the Yellowstone National Park. The United States has added seventeen national parks since the creation of the Yellowstone National Park in 1872. The story of our national parks is most interesting.

In this connection it would be well to tell about the great naturalist, Enos Mills, who did so much for our national parks of the West. The early death of Enos Mills was a great loss to the country.

TWIGS IN WINTER

In project 2 we wrote the autobiography of a horse-chestnut twig. We have also made some drawings which will enable you to become acquainted with other twigs in winter. If you will look at the drawing of the twig of the horse-chestnut Figure 1, you will see the names of the various markings on the twig. Every tree has similar signs. In the same way that you recognize your friends by the size and shape of their nose and eyes, or their complexion and freckles, so you identify various trees by their characteristics. The leaf scars vary in size and shape. The vein scars differ in number and arrangement. The breathing pores may be round or long. The pith often differs in color and shape. If you can see these simple differences in color, size, shape number and arrangement of parts, you will be able to name our common twigs by comparing them with these drawings.

In some schools we have intelligence tests. Twig study is also an intelligence test. It may easily be made into a game. Take two twigs and see who can write the greatest number of differences in ten minutes. Twig study is one of the best means of testing the power of observation.

Twig study has always been profitable. The Indians knew that the ash makes a good bow. They knew the sumac, because they made Indian lemonade from its red berries. The frontiersman knew the common locusts, because they made good fence posts. The landscape gardener knows the flowering dogwood and magnolia, because

TWIGS IN WINTER

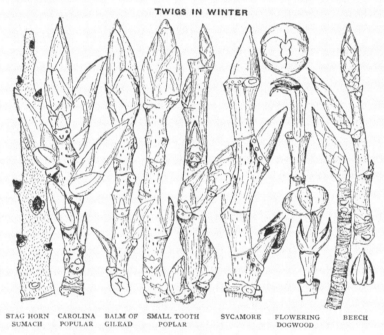

| STAG HORN SUMACH | CAROLINA POPULAR | BALM OF GILEAD | SMALL TOOTH POPLAR | SYCAMORE | FLOWERING DOGWOOD | BEECH |

Fig. 11.

he uses them to decorate the landscape. He may select the hawthorne or ailanthus, because of its showy fruit, or the beech and yellow birch, because of its beautiful bark. The poet knows the twigs as evidenced by his productive appreciation. To the boy scout and the girl scout they are more than mere sticks. A real scout (not a parlor scout) knows that the bark from a dead gray birch, the fibres from the bark of the dead chestnuts, and the matting from the red cedar make good tinder. The experienced woodsman gathers dead twigs from the evergreen trees for kindling. He avoids the white berried sumac because of its poisonous effect. He may steep the twigs of the witch-hazel for its healing extract. Each tree is a friend which contributes to his welfare when he is in the forest.

Some boys and girls may enjoy making a twig collection and mounting them on a board. The labels should be placed near them, and if the various uses of the tree are written alongside of the twig it will be doubly interesting. How many kinds of trees grow within one mile of your school building?

OUR STATE TREE

The suggestion was made in the Arbor Day pamphlet for 1923 that, as the school children of Rhode Island in 1895 selected the maple as our state tree, the school children of 1924 should identify the state tree among the eight varieties of maples to be found in Rhode Island,—Sugar, Silver, Striped, Mountain, Norway, Sycamore and Box Elder. Request was made that on Arbor Day, 1923, nominations should be made for an election to be held on Arbor Day, 1924. Nominations have been made by Walley School, Bristol; Meshanticut Park School, Cranston; Thomas A. Doyle School, Providence, and Plainville School, Alton. The preferences, as indicated on the total vote for each tree place the nominations in the following order: Red, Norway, Sugar, White, Sycamore, and Box Elder. Under the conditions prescribed for nominations in 1923, the Sugar Maple and Red Maple are candidates for election. Schools should send the results of the ballot on these trees to the Commissioner of Education. The Red Maple was elected the Rhode Island State Tree on May 9, 1924.

THE RED MAPLE VS. THE SUGAR MAPLE

A debate by the pupils of the seventh grade of the Henry Barnard School on Arbor Day, 1923.

Myles Sydney, chairman: We, the children of the seventh grade of the Henry Barnard School, have been studying the different kinds of trees and have gone on field trips observing them. We have made a special study of the maple. The school children of Rhode Island have already voted that the maple shall be the state tree. However, there are eight different varieties of maples, and we have selected the two which we think best suited for that purpose. The girls will uphold the sugar maple and the boys the red maple. Each debater will be allowed four minutes and each side will have two minutes for rebuttal. The judges will consider the strength of the argument, the use of English, and the delivery.

Florence Urquhart: The girls of the seventh grade believe that the sugar maple should be the state tree of Rhode Island, first, because it is stately, majestic, beautiful, and symmetrical. By the last I mean that all the branches tend to make the tree a perfect egg-shape. Second, this tree can be easily distinguished by its leaves. There are five letters in sugar and there are five lobes in the leaf. Therefore, since this is so, we can easily distinguish the sugar maple in summer. The buds of the sugar maple are easily identified. They are long and narrow, whereas the buds of the other maples are short and broad. Third, the tree has many uses. Its sap can be made into sugar. Are there any children in Rhode Island that do not like maple sugar? I do not think so.

Now let me repeat the points I have mentioned. First, the tree is stately, majestic, beautiful, and symmetrical. Second, the tree can be easily identified. Third, its sap can be made into maple sugar. Because these things are true, the girls of the seventh grade assert that the sugar maple should be the state tree of Rhode Island.

Jack Dolan: The boys of the seventh grade believe that the red maple should be the state tree of Rhode Island because it can grow easily in Rhode Island. Rhode Island is a state of many rivers, lakes, and swamps, and the red maple can grow easily in moist soil. The buds of the red maple are small compared with the buds of the other maples. Our state is small compared with the other states. Therefore the red maple typifies our state. You can distinguish the red maple in summer because the leaves have three distinct lobes and the leaves of other maples have five distinct lobes. You can easily distinguish the red maple in winter because the buds are red. The wood of the red maple can be made into furniture and household articles.

Now I will enumerate the points I have given. First, the red maple can grow easily in Rhode Island. Second, its bud typifies the size of our state. Third, it can easily be distinguished. Fourth, it is useful. The boys of the seventh grade believe that the red maple should be the state tree of Rhode Island.

Meriel Vinal: We, the girls of the seventh grade, think that the sugar maple should be the state tree of Rhode Island because it is valuable in many ways. The wood of the sugar maple is hard, strong, heavy, fine-grained, and takes a high polish well. Think of its value then for furniture! From the sugar maple is produced bird's eye maple. Have you ever walked by a furniture store in Rhode Island without seeing furniture made of bird's eye maple? No, you have not. The sugar maple is used for hardwood floors because there are no soft places in it to be worn away and make splinters which may get into your feet and possibly cause serious illness. Do you wish this to happen to you? No, you do not. Then let us make our floors out of sugar maple and have no splinters. Another point in its favor is that it is an excellent fuel, and, after it is burned, potash can be obtained from the ashes.

Let us review all my points. The sugar maple supplies valuable furniture, excellent hardwood floors, a multitude of wooden household articles, and fuel and potash. The sugar maple is entirely devoted to the service of man. Therefore the girls of the seventh grade think that the sugar maple should be chosen as the state tree of Rhode Island.

Andrew Rougvie: The boys of the seventh grade think that the red maple should be the state tree of Rhode Island, because it typifies our state in another way besides the one my colleague mentioned. The red maple is the first to bloom in the spring and the first to color in the fall. It is said that Rhode Island is either first or last in any movement of national importance. Thus the red maple illustrates a striking characteristic of our state. The red maple benefits our state's chief industry, manufacturing, because its wood is used for furniture, toys, and all kinds of wooden articles. Valuable substances are extracted from the bark and made into ink and dyes. My colleague stated that the red maple can grow easily in moist soil. It can also grow in city streets and provides shade during the hot summer months. Is there a street in Providence, where there are any trees at all, on which you cannot see the red maple growing?

Therefore, because the red maple is like our state in being first and last in many things, because it benefits our state's chief industry, because substances are extracted from its bark, and because it is a valuable shade tree, the boys of the seventh grade assert that the red maple should be the state tree of Rhode Island.

Rebuttal

Andrew Rougvie: The red maple is the most beautiful of all maples. Its wonderful crimson is unequalled. It is also one of the most beautiful trees in our country. One of my opponents has stated that the leaf of the sugar maple has five lobes and that there are five letters in sugar. So there are, but there are also five lobes in the leaf of the white maple and five letters in white. Does it distinguish the sugar maple because there are five letters in the word and five lobes in the leaf? It certainly does not.

Meriel Vinal: The sugar maple is used as a shade tree in Rhode Island more than the red maple because it is more beautiful than the latter. The silhouette of this tree has a regular outline, but that of the red maple is irregular. The sugar maple is more valuable than the red maple because more than five hundred different articles are manufactured from its wood. Therefore the girls of the seventh grade contend that the sugar maple should be chosen as the state tree of Rhode Island.

IF ALL THE SKIES

If all the skies were sunshine,
  Our faces would be fain
To feel once more upon them
  The cooling plash of rain.

If all the world were music,
  Our hearts would often long
For one sweet strain of silence,
  To break the endless song.

If life were always merry,
  Our souls would seek relief,
And rest from weary laughter
  In the quiet arms of grief.

—*Henry Van Dyke in "Songs Out of Doors."*

CHAPTER XXVI

Some Mechanical Aids in Nature-Study

That there should be artificial devices for teaching nature-study ideas seems rather contradictory. The writer fully realizes that mechanical aids are powerless to educate but believes them to be convenient tools for the skilled workman. A large supply of nature-study machinery, therefore is not to be confused with progress.

Originality is not claimed for the following schemes. Some are inherited with various changes from past instructors while others are the sum and epitome of several years of experience. Such well known contrivances as the germination box with sloping glass front to exhibit root characters are not included in this article. Stories and anecdotes to illuminate certain points, as in the case of the notes scientist who made the difference between the terms specie and species clear by saying that specie is something with which biologists have little to do, are left for a separate sketch.

GENERAL AIDS

1. *Aquaria.*

The salt water aquarium should be used to a greater extent. Sea-salt may be purchased or obtained by evaporation of sea water. Sea lettuce (Ulva) may be floated on the surface by means of corks. Do not over-populate the jar.

A metal plug with a pin-hole bore, placed in the opening of the faucet, makes a fine stream which is excellent for aeration.

2. *Blackboard.*

Linoleum, framed and reënforced at the back, makes an excelient blackboard.

Wires stretched along the top of the blackboard and also five or six inches above the ledge are very serviceable. The latter is convenient for holding pictures on the sill. The pictures can be placed there or removed very quickly without injury by patent fasteners.

Frames the size of the blackboard sections, covered with cloth, can rest upon the sill and be attached at the top by means of screw eyes. These are useful for pinning on clippings, announcements, etc. They are also handy for exhibitions. Fall fruits look very attractive when pinned upon them.

3. *Calendars.*

The spring calendar is made in the following form:

WASHINGTON COUNTY									PROVI- DENCE, etc.	
BIRDS				TREES		FLOWERS		Other Signs	BIRDS	
Name	Date	Nest	Young	Flowers	Leaves	Name	Date	Date	Sign	Name	Date
				Name	Date	Name	Date					

323

Since our girls come from all parts of the state we are able to have a competition by counties and by classes. The observations are written upon a blank form, as:

Name of collector......................................Class.........
Object..Date.........
County..................City or Town............Locality.........
Remarks...

These slips are dropped into a box made after the fashion of a letter box. Each day the observations are entered upon the calendar. The first blue bird reported in Providence County is credited to that county and to the class of which that student is a member. The class records are kept distinct by using a different colored ink for each class. In graded schools the competition might be by districts, between classes, or between the boys and girls. The whole thing can be taken charge of by the pupils.

The numbers upon the monthly calendars may often be used in nature-study, as in labeling specimens for identification. I always save each sheet as I remove it.

4. *Examinations.*

Examinations which invite cramming have no place in nature-study. The true test of knowledge of transplanting, making cuttings, grafting apple trees, etc., is in the success of *doing* the thing. Unlike most examinations the nature-study test means *observation* at the beginning rather than a written paper at the end of the course, as— Note the direction in which the "leaders" (Top shoots) bend on the pine trees of the vicinity. How account for this? Rather than,— What *did* John Smith *write* about the trees in his first chapter?

There are three forms of examinations which I commonly use in Nature-study: (1). *The five minute quiz.* Questions are asked to test whether pupils have worked out the assignment. (2). *The Identification test.* The identification of flowers, shells, etc. The point of view in this test is different from other subjects in that it is to show what one knows rather than to find out what they (cannot?) stuff in for the occasion. This is shown by writing of two lines of figures on the board, the top line for instance representing the number of shells and the lower line representing the number in the class recognizing 10 shells, 11 shells, etc. Each pupil is thus able to see how he stands with the average, the lowest, and the highest of the class. To facilitate the mechanics of these two tests I have slips of printed paper with twenty lined spaces which are numbered as:

1...
2...
3...
etc.

The questions require one or two-word answers. By holding the papers beside a correct copy the mistakes are soon noticed. (3). *Power to do.* In this type of examination the questions may be answered by the use of notes and text. Suppose the class has studied the seeds and seedlings of the bean, squash and corn. To test the power gained in this study the pupils may be given peas and pea seedlings, to work out independently without suggestions. It is not knowledge but the use of knowledge that is power.

5. *The making and uses of a hectograph.*

(1). Obtain a rectangular tin about an inch deep and an inch larger, each way, than the paper to be used.

(2). Measure the pan with water to see how much material is needed. (3 oz. powdered gelatine, and 1½ lb. glycerine to ½ pt. of water is sufficient for a hectograph 9 x 12 inches in size.)

(3). The proportions are 6 oz. of ground white glue to one pt. of glycerine. (Gelatine may be substituted for glue. It does better work but does not last as long.)

a. Add a little hot water to the glue and dissolve in a double boiler to prevent the glue from burning.

b. When thoroughly dissolved add the glycerine.

c. A few drops of carbolic acid will prevent the pad from moulding.

d. If too soft, melt and add more glue.

e. If too hard, melt and add more glycerine.

f. If there are bubbles on the surface, prick with a pin.

(4). If surface becomes rough set in moderately warm place, as on the radiator. Pan must be level.

(5). Make writing or drawings with hectograph ink, and a heavy pen on hard paper. Make a broad line.

(6). Allow ink to dry. Do not use a blotter.

(7). Dip a sponge in cold water. Press nearly dry. Moisten surface of the pad.

(8). Place written sheet of paper, face downward on pad. Smooth paper with hand so that all parts stick to the pad.

(9). Leave paper on pad for about three minutes and then gently remove.

(10). Moisten the pad again.

(11). Place a sheet of hard paper on inked surface and smooth it as in original copy.

(12). Remove the paper carefully and repeat the process as many times as desired.

(13). If the pad is covered with strips of paper to the edge of the writing the paper used in copying can be more easily removed.

(14). When through copying, wipe the surface with a sponge dipped in slightly-warmed water.

The uses of the hectograph pad are too well known to dwell upon. Test slips, blanks for calendar observations, outline drawings of birds, shells, etc., are useful for identification sheets, for coloring, etc.

6. *Maps.*

I find that maps of regions over which we go on certain field trips are most serviceable. On bird trips, for example, the class marks out the swamps, fields, scrub land, etc. Others mark the location in which we observe the red-winged blackbird, the meadow-lark, the towhee, etc. After a few trips they are able to draw certain conclusions and make predictions as to the places where they would expect to find certain species. The best map for this purpose is the enlarged contour map. It can be made from the government topographical map. It can be enlarged by use of the pantograph or by dividing the original into small squares with pencil lines and then copying the map onto a chart with larger squares. The latter method is more accurate.

7. *Material.*

(1). Shoe boxes labeled on the end, arranged alphabetically, and stacked as in shoe stores, make a systematic arrangement for storing dry material such as: seeds, varieties of corn, leaf stems of the horse-chestnut, bayberries, etc. This material lasts year after year and is one of the best nature libraries. Anything that does not pay rent by being useful at least once a year should be delegated to the waste basket.

(2). Mounted birds, preserved specimens, etc., are numbered and a corresponding number is placed on the shelf so that whenever they are taken for class work or as loan material they may be returned to their particular niche.

8. *Pictures.*

Picture study does not take the place of field work. The highest inspiration comes from seeing the actual place or object. Pictures necessarily take the foreground when speaking of the Big Trees of California and the scenes of the Yosemite Valley. Such important lessons as the Enemies of the Forest, and Uses of the Forest, are made effective by pictures. The Forest Service has a traveling photograph exhibit to be loaned to schools. Pictures of beautiful yards and country homes, taken from such magazines as *Country Life in America,* form an important part in the lesson on landscape gardening. The pupils are given a list of things that show good taste in ornamenting a front yard. They are then asked to pick out the desirable effects and arrangements that are illustrated in the pictures. These pictures are classified and placed in drawers or filed vertically in cabinets.

9. *Plasticine.*

This material plays an important part in nature-study. Not many years ago I felt that a drawing was the only sure test that a pupil had seen what he had been studying. In many cases the plasticine has supplanted the drawing. I find that the pupils get a clearer notion of the bean embryo when they make an enlarged model. Plasticine also has the advantage of being in different colors and it can be used several times. The colors are of especial value in showing the annual rings in a tree and how they would look in quartered sections and in other cuts.

Permanent models for the school room may be made out of papier maché. Soak small bits of newspaper in water for two or three days. Stir this mixture until it becomes a thick mass. Varnish or shellac the model after it has dried. The salt and flour model may be made by mixing two parts of common salt to one of flour and then add water slowly until it resembles wet sand. Work this mixture on a smooth board. Varnish or shellac after it has thoroughly dried. Cement flower pots and bird houses furnish interesting lessons in construction.

10. *Quotations.*

I have these printed in large type upon cardboard. They are displayed at various times during the year. These cards should be changed frequently. Such pithy statements as follow are used for the Normal School classes:

"Our children have as much right to be taught to read a roadside as a book."—Slogan of the Sacramento Playground Commission.

"Nature-study is the alphabet of agriculture."—From a *Nature-Study Review* circular.

"Nature-study is the fundamental basis for the conservation of our natural resources."—From a *Nature-Study Review* circular.

Quotations from the poets are also displayed according to the season, as:

"The wee willow pussies are climbing the trees."

"By the flowing river the alder catkins swing."—*Celia Thaxter.*

"The redwing flutes his o-ka-lee."—*Emerson.*

11. *Sand Box Gardening.*

I consider this the most valuable nature-study equipment that I have. Its use is unlimited.

(1). *Individual Projects.* The box may be divided by cross strings for individual gardening. Each garden is then about 6 x 8 inches in size. Slips of the geranium, etc., may be rooted and transplanted to a pot for Christmas; a series of seedlings may be raised for study; tree seedlings may be raised for Arbor Day; privet cuttings can be started for a hedge; a class exchange of valuable grape cuttings, etc., to be started for early spring; the raising of orange seedlings and the cotton seeds are interesting and suggestive.

(2). *Community Projects.* In this case the strings are removed and groups of students work out certain social and community ideas. All sorts of resources are brought out in this work. Trees may be represented by small bushy twigs which have been covered with a sticky substance and dipped into a box of finely cut green paper. Evergreens may be represented by small twigs of cedar. Green meadows and lawns may be represented by sawdust colored with a green dye. In representing an ideal lawn real grass seed gives a good effect. Tree seeds may be planted to raise miniature shade trees. Color the sawdust with a brown dye to show plowed ground. Use white beach sand for roads. A mirror makes a clear lake with reflecting surface. Glass over a blue surface gives the effect of a large sheet of water. Animals may be represented by plasticine models. Use cardboard for buildings. Such equipment may be used to illustrate many things, as: The yards of a tidy family and of an untidy family; a railroad station with well-kept grounds; a practical school ground that is beautiful; a miniature rockery, etc.

(3). *Forestry.* For teaching the value of a forest the Forestry Bureau has suggested some such scheme as this: Make a fairly good sized sand-hill. Cover one side with moss to represent the leaf mould beneath the forest. Use twigs of cedar, etc., for the forest. Pour water from a watering-pot onto the hill and note that erosion takes place on the side that has been deforestated. The rapid run off not only washes away the rich soil but would cause floods. These conditions are unfavorable to navigation and to manufacturing. Some of the readers may be interested in two other methods of raising plants, perhaps more of a novel than a practical idea. The wandering Jew may be grown in a test tube of water tied to the curtain string. Cotton batting stuffed loosely in the mouth of the tube will prevent too rapid evaporation of the water. Conch-shell flower pots give an interesting variety to the window-shelf. The use of the egg shell for seedlings is described in Comstock's *Handbook of Nature-Study.*

12. *Written Papers.*

Nature-study is one subject which may serve to extend the period of the child's writing and drawing because he enjoys it. One method of killing this natural pleasure is to spread red ink on poorly constructed sentences and misspelled words, and to point out defects in the drawings. I mention this for fear of the misuse of the following scheme. In the concentrated curriculum of the normal school, especially with city bred girls who have had no nature work, and many of whom never saw a "pollywog" or "mushrat" I have hesitated in altogether carrying out what seems ideal. To facilitate the correction of laboratory work I have a series of rubber stamps which read: Make lines definite; label all parts; follow directions, etc. I give them a copy of the rubber stamp readings in the beginning and tell them that my experience has shown the following mistakes so common that I have had rubber stamps made to save time in correcting papers. A few periods result in the almost total disappearance of these common mistakes. Having once attained a certain standard the pupil is then ready to show originality with a just pride.

SPECIFIC AIDS

1. *Birds.*

a. The mounted specimen is perhaps one of the first class-room aids that one thinks of in bird-study. There are many accessory objects that may be used. For instance, besides a series of mounted woodpeckers we have: a portion of a dead limb split to show where the grub entered the tree and the hole made by the woodpecker when obtaining his breakfast; an old portion of a trunk to show the abandoned home of the downy woodpecker; a section of the trunk of a white birch to show the drillings of the sapsucker; a piece of wood in which the California woodpecker stored some acorns. By exchanging with a school in Maine we expect to obtain specimens to show the work of the pileated woodpecker, etc. These are the tools for a live lesson on woodpeckers.

b. *Audubon Charts*. Realizing the risk of breaking two fundamental laws in teaching nature-study, one to study nature afield and the other to avoid drill, I believe that a great deal can be obtained from quick, brief drills with these charts. Besides pointing to the birds for the class to name in unison the teacher may take a book and cover up individual birds and have members of the class describe them. Try the common ones, as—the robin. What is the color of his breast? Where is there white on him, etc.? This will bring out surprising results for both the teacher and the pupil. Have a pupil face the back of the room and describe the goldfinch.

c. Outlines of the birds made on the hectograph may be colored with crayons. All the sparrows might be drawn on one sheet. Outlines of the birds may also be made by tracing around cardboard patterns.

d. Field work. Besides the contour map the pupil finds it interesting to have a list of the common birds of the vicinity and check off what he saw or heard on the trip. These records are placed on file for comparison with other records.

2. *Fish*.

The way a fish rises and sinks in the water is well illustrated by the Cartesian Diver which is explained in any Physic's text-book.

3. *Plants for the Class Room*.

The "Fern Dish," so-called, may be any glass dish with a glass top. On the bottom are placed various mosses, ferns, partridge berry, possibly a pitcher plant, and other bog members. The bog is then saturated with water. The moisture evaporates, collects on the glass cover and falls back into the swamp. New plants spring up from the seeds that were in the moss. The "fern dish" presents an interesting center and requires little care.

The following list was worked out by a girl in the Senior nature-study class:—Miss Susie Cooper of Newport, R. I.:

I. *Plants suitable for sunny places*

1. Geraniums : old plants or cuttings.
2. Petunia: stalks of old plant cut nearly to ground, plant potted, put in cool place to sprout, then in window.
3. Pineapple: grown by breaking top from pineapple, putting the top in water or sand until rooted.
4. Jerusalem Cherry (Solanum Melvim) decorative.

5. Plants from seed.
Marigold, "Dwarf French." Bloom after 5 or 6 weeks.
Sweet Pea, "Earliest of All." Bloom after 10 weeks.
Sweet Alyssum. Bloom after 6 or 8 weeks.
Candytuft.
Lobelia: Plant seeds in spring, excellent for borders of gardens of window-boxes. Transplant strongest of seedlings.
6. Beet, Carrot, Parsley: Ornamental as well as useful.

II. Plants requiring a medium light (north, east or west windows).

1. Begonia, rubra: bloom for several years; new plants or cuttings.
2. Abutilon: bloom when quite small.
3. Pandanus: decorative foliage.
4. Coleus: decorative foliage.
5. Sedum, stonecrop.
6. English and German ivys, wandering jew. Suitable for edges of window boxes, or training branches over wire.
7. Cyclamen; bloom best if not kept in very warm place.
8. Cactus.
9. Ferns: dwarf ones preferable.
10. Dracaena: ornamental.
11. Pepromia: ornamental.
12. Primula: Obconica and Chinese. Sow seeds in spring, blossom by December.

III. Wild flowers suitable for sun or shade.

1. Jack-in-the-pulpit. Bulbs or corns taken up in fall, potted, placed in window, bloom by Christmas.
2. Mullein: common weed in America. Ornamental plant in England, called American velvet plant.
3. Columbine, violet, taken up in fall, bloom during winter. If wild plants are frozen once, bloom better.
4. Evergreen plants: beautiful window box. Juniper, cedars, pine, arbor vitae, cypress. Use young plants.

IV. Plants growing in a weak light (corners of rooms, halls, etc.)

1. Rubber plants. When frozen, cut back until white sap appears. Keep end bandaged until healed.
2. Aspidistra, "Iron plant."
3. Palms.

V. Bulbs.

Bloom from Christmas until Easter.
Plant in early fall, put in cool dark place (corner in cellar or trench in ground), cover with soil; remain there 6 to 10 weeks, then bring to light.

Suitable varieties: Daffodils, tulips, Roman hyacinths, narcissus, crocuses, jonquils.

For growing in pebbles and water. Chinese lilies, paper white narcissus.

VI. Soil.

From woods if possible, or a mixture of common soil, loam and leaf mold.

4. *Seeds.*

Mounting seeds for class use is well described in Farmers' Bulletin 580.

I usually save old paper bags to take in my pocket when out in the fields. One is thus always equipped to gather seeds, etc., which he may wish for class work.

5. *Trees.*

a. *Cross section of a large tree.* This is important to show the time that it has taken our large trees to grow. Important historical dates of the vicinity may be pinned onto the section to show the size of the tree at corresponding times. This makes an interesting correlated lesson.

b. *Tree surgery and dentistry.* A very practical collection of specimens may be obtained for this work by sawing off dead stubs, unpainted wounds which have started to decay, wounds which have partially healed, and those that have healed entirely.

c. *Blocks of wood.* Pine and oak blocks cut and polished to show the annual rings, silver grain, etc., give a good basis for understanding wood structure.

d. *Evergreens.* Obtain a pane of clear window glass of the size desired and a piece of stout cardboard of the same dimensions. On the cardboard place a thin layer of cotton batting. Arrange the evergreens on the cotton in a definite order and place calendar numbers beside them. Place the pane of glass over them and hold the various parts in place by fastening them in a picture frame or by passe-partout binding. This will always be ready for use.

e. A blank for tree observations may be obtained from the Bureau of Plant Industry by sending for Form 219. Obtain Tree Observation sheets from Comstock Publishing Co. loose leaf material.

f. The following is a key for the trees in winter. Similar ones may be worked out for birds, flowers, insects etc. Pupils are not expected to learn the differences for examination but this form is a convenient way for teachers to identify specimens brought in. The natural way to learn any new vocabulary is through use and should the teacher develop an interest in tress in her students she will eventually find that she is acquiring a large fund of information for herself. In a list of this kind it is taken for granted that the teacher knows a maple tree etc., but is not sure of the species.

Winter Tree List

Locality .... Class ....
Date.................... Total ....

Yew Family
  *Ginkgo: not evergreen; spurs: leaf scars far apart.

Pine Family
  White Pine: 5 leaves in a bundle; cone scales thin.
  Pitch Pine: 3 leaves in a bundle; scales thick, spine.
  *Red Pine: 2 leaves in a bundle, slender, flexible.
  Austrian Pine: 2 leaves in a bundle, stiff, thick.
  *Scotch Pine: 2 leaves in a bundle; cones point backwards.
  American Larch: deciduous; spurs; cone ¾", few scales.
  *European Larch: many leaves in cluster; cone 1", many scales.
  Black Spruce: cone nearly spherical, persistent.
  *Norway Spruce: cone more than 3"; leaves all around stem.
  Hemlock: leaves 2 ranked, whitened beneath, flattened.
  White Cedar: leaves awk-like; globular cone; swamps.
  *Arbor Vitae: 2 ranked spray; small tree; hedges.
  Red Cedar: bluish berry; dry hills; red heart-wood.
  Juniper: berry axillary; leaves prickly pointed, 3 in whorl.

Willow Family
  Yellow Willow: yellow twig.
  *Weeping Willow: drooping branches; ornamental.
  *White Poplar; retains whitish green bark; woolly twigs.
  American Aspen: buds, reddish slightly sticky, appressed.
  Large-toothed Aspen: buds downy, small.
  *Balm of Gilead: fragrant buds, resinous, large.
  *Carolina Poplar: light yellow twigs; appresssed buds.
  *Lombardy Poplar: sharply up-curved branches.

Walnut Family
  Butternut: terminal bud longer than broad; brown pith.
  Black Walnut: nut spherical, 4 celled; pale buff pith.
  Shag-bark Hickory: white, thin shelled nut; shaggy bark.
  Mockernut: brown, thick shelled nut; smooth ridges.
  Pignut: nut nearly spherical, thick shell; slightly shaggy.
  Bitternut: thin shell; superposed buds, yellow, dotted.

Birch Family
  Hop Hornbeam: fruit like hop; flaky bark.
  American Hornbeam: twisted appearance to bark.
  Black Birch: dark bark, not peeling, aromatic.
  Yellow Birch: yellow bark, peeling, aromatic.
  American Canoe Birch: white bark, papery layers.
  Gray Birch: dingy, gray bark, not papery.
  Speckled Alder: cone-like catkins, lateral, erect.
  Smooth Alder: cone-like catkins, terminal, erect.

*Commonly cultivated.

Beech Family
　　American Beech: bluish gray bark; long buds; dead leaves.
　　Chestnut: burs; 2 ranked buds, oblique to leaf scar.
　　White Oak: young branches not peeling; large buds.
　　Swamp White Oak: young branches scaly.
　　Chestnut Oak: bark not flaky, round ridged; cup thin.
　　Red Oak: rusty hairs apex bud; large flat cup; flat flutings.
　　Pin Oak: smooth, continuous trunk; saucer cup; low grounds.
　　Scarlet Oak: bark gray, red inside; cup conical base.
　　Black Oak: bark gray, yellow inside; buds pale, woolly.
　　Scrub Oak: shrub; sterile ground; smooth bark; many acorns.

Nettle Family
　　Slippery Elm: inner bark mucilaginous, rough gray twigs.
　　American Elm: drooping branches; whitish layers of bark.
　　Hackberry: drupe; chambered pith; galls; bark ridged, warts.
　　*White Mulberry: many small branches, cut twig milky.
　　*Red Mulberry: darker twigs; dark margined bud-scales.

Magnolia Family
　　Tulip Tree: fruiting cones; winged seeds; twigs curve up.

Laurel Family
　　Sassafras; twigs aromatic, green, mucilaginous.

Witch Hazel Family
　　Witch Hazel: fruit 2 chambered capsule; flowers in fall.
　　*Sweet Gum: fruit cluster spherical, sphiny; corky ridges.

Plane Tree Family
　　Buttonwood: upper trunk white areas; no terminal bud.

Rose Family
　　Shad Bush: long narrow buds; berry-like fruit; diseased.
　　English Hawthorne: unbranched thorns; drupe-like pome.
　　Wild Black Cherry: develops scaly bark; lenticels lengthened.
　　Choke Cherry: smooth bark; lenticels do not elongate.
　　Wild Red Cherry: many small terminal buds; recent clearings.

Pulse Family
　　Honey Locust: branched thorns; smooth buds.
　　Common Locust: paired thorns; downy buds.
　　*Kentucky Coffee Tree: narrow ridged bark; stout twigs.
　　*Redbud: shrub flower buds on old growth; ornamental.

Quassia Family
　　*Ailanthus: seed in centre of wing; solitary buds.

Cashew Family
　　Staghorn Sumach: velvet hairs; red berried.
　　Smooth Sumach: smooth twigs; red berried.
　　Poison Sumach: terminal bud; white berries; swamps.

---

*Commonly cultivated.

Maple Family
 Rock Maple: twigs fork; keys persist; bud sharp, many scales.
 White Maple: flaky bark; collateral buds; twigs curve up.
 Red Maple: red twigs; collateral buds; usually swamps.
 *Norway Maple: ridged bark; red, hairy buds; leaf-scars meet.
 *Sycamore Maple: flaky bark; green buds; leaf-scars not meet.
 *Box Elder: green twigs; downy buds; leaf-scars V-shaped.

Soapberry Family
 *Horse Chestnut: large twigs, leaf-scars and sticky buds.

Linden Family
 *Linden: buds bright, 2–3 scales show; twigs flat, zigzag.

Dogwood Family
 Flowering Dogwood: opposite leaf-scars; large flower buds.
 Tupelo: horizontal branches; pith partitioned; low lands.

Olive Family
 White Ash: leaf-scars concave; ridged bark.
 Red Ash: downy twigs; upper margin leaf-scar not concave.
 Black Ash: scaly bark; black buds.

Bignonia Family
 *Catalpa: three leaf-scars at node; bundle scars meet.

This list in loose leaf form, can be obtained from the Comstock Publishing Co.

---

*Commonly cultivated.

Camp Chequessat - Pirates - 1915.

CHAPTER XXVII

SUGGESTIONS FOR PICTURE STUDY IN NATURE-STUDY

1. *What pictures should be selected?* Animals and plants; beautiful yards and country homes; enemies of the forest; uses of the forest; care of the forest; famous naturalists; scenery. Industries, as fishing, irrigation, lumbering, and raising of products, as cotton, silk, wool, flax, hemp, coffee, cocoa, tea, wheat, corn, rice, rubber, tobacco. Pictures must be lifelike, accurate and attractive.

2. *How obtain pictures?* Brought in by pupils; advertisements (railroad pamphlets); magazines as, Country Life in America, Geographical Magazine, Harper's, Century and Scribner's; text-books; postcards; actual photographs; and stereoscopic views. The Forest Service has a traveling photograph exhibit to be loaned to schools.

3. *How keep pictures?* Mount on square, grey cardboard with photomount. Do not merely stick corners as pictures will become torn. Title should be placed on the back of the cardboard so as to be easily read when showing the picture, or so as to be out of the way when identifying pictures. Several may be mounted on a card if closely related. Pictures should be classified and placed in drawers or filed vertically in cabinets. Each picture should be numbered and card-catalogued according to subject.

4. *Value of Picture Study.* First pictoral text-book, "Orbis Pictus" made by Comenius. Value of pictures recently recognized by business men who use them in commerce and trade; by lecturers and by publishers as seen by comparing old text-books and new ones. In many texts the pictures are now accompanied by questions rather than statements. Photography is a study in many schools. "Hunting" with a camera is a favorite out-of-door study.

335

Picture study does not take the place of field work. The highest inspiration comes from seeing the actual place or object. In studying the Big trees of California, Yosemite Valley, foreign scenes, etc., pictures necessarily take the place of field trips. Pictures make the text work clear. They aid in recognizing and interpreting forms in the field. They prepare breadth and quality which inspires the student to travel. They enable him to disentangle the wealth of detail which he sees and arrange in an orderly fashion. They give a correct idea quickly, hold attention and lessen discipline. The best picture is the stereoscopic view which gives three dimensions instead of two. The hood of the stereoscope aids in concentrated study.

5. *How Use Pictures?* In the primary grades the pictures may be used as the source of knowledge. The pictures used should have a story telling quality and represent their message in a clear, broad way. The teacher tells the story, as lumbering, by the use of pictures. The pupils use the material furnished as a basis for an oral language lesson. There are certain new words that the pupil will need to use, such as skid, log-jam etc., which should be written on the board. The pictures make the work interesting and a pleasure as well as educative.

In the grammar grades, field work, the object, or the text-book should be used as the scource of knowledge. The pictures are used to supplement, to review and to make clear the conclusions which have already been made. The pictures should be placed in order on the blackboard sill, on a cloth-covered frame or hung from a line. The pupil can inspect the pictures without soiling or damaging in other ways. In this position they are convenient for explanation to the class by the pupil or by the teacher. Never pass pictures around a class. It divides the attention. There is a danger of having too many pictures at one time.

A few good pictures may be selected to cover all the facts which are to be interpreted and thus train the pupil in the power of concentration. Each picture represents a truth which the pupil should discover. He must learn that seeing a picture is not studying a picture. The first step is for the pupil to interpret the picture from his own standpoint using the knowledge which he has already acquired. He should interpret the conditions illustrated, the past conditions which led to the present form and the probable changes in the future, as in the picture of a tree trunk which has a decayed hole. This hole probably indicated the past location of a branch. The branch was broken off and the remaining stub decayed into the heartwood of the tree. The decayed part should be removed and the cavity filled with cement. In the second step the subject has to be developed by careful questioning of the teacher, and, thirdly, important points which have not been deduced should be added by the class and the teacher. Whenever possible emphasize the influence of the particular scene upon plants, animals, and human affairs.

For seat work or individual study, each picture should be accompanied by directions and a carefully prepared set of questions. To thrash the ideas out without direction of thought may be an ideal way but it is not practical.

The arrangement of the pictures in a logical order is a valuable study. The pupil should learn how to hold the picture so that the whole class can see. It is valuable training to show a picture and talk at the same time. The school exhibit of these products is of utmost value and they should be used in connection with the pictures.

The use of pictures in an identification test of such things as leaves, buds, birds, fishes, etc., has long been recognized. This method forms a quick, convenient way of reviewing.

"When country roads begin to thaw
  In mottled spots of damp and dust,
And fences by the margin draw
  Along the frosty crust
  Their graphic silhouettes, I say,
  The Spring is coming round this way."
  —*James Whitcombe Riley.*

"In the urgent solitudes
  Lies the spur to larger moods;
In the friendship of the trees
  Dwell all sweet serenities."
  —*Ethelwyn Wetherald.*

My soul is full of longing
For the secret of the sea.
  —*H. W. Longfellow.*

God makes sech nights, all white and still
  Fur 'z you can look or listen,
Moonshine an' snow on field an' hill,
  All silence an' all glisten.
  —*James Russell Lowell.*

"You ask me for the sweetest sound mine ears have ever heard?
A sweeter than the ripples' plash, or trilling of a bird,
Than tapping of the rain-drops upon the roof at night,
Than the sighing of the pine trees on yonder mountain height?
And I tell you, these are tender, yet never quite so sweet
As the murmur and the cadence of the wind across the wheat."

—*Margaret E. Sangster.*

"The grass so little has to do—
  A sphere of simple green,
With only butterflies to brood,
  And bees to entertain.
*   *   *
"And then to dwell in sovereign barns,
  And dream the days away,—
The grass so little has to do,
  I wish I were the hay!"

—*Emily Dickinson.*

"I think the first virtue is to restrain the tongue; he approaches nearest to the gods who knows how to be silent even though he is in the right."—*Cato.*

CHAPTER XXVIII

NATURE-STUDY BY GRADES

The following course of study aims to overcome obstacles of selection, organization, and presentation and to prevent duplication of subject matter in the grades. It has been reduced to a minimum of ten units for each grade, with topics arranged in order for the months of the school year, beginning with September. If it is necessary to reduce this still further, the teacher may omit the portions printed with an asterisk in the margin.

The aim of the course is to stimulate interest and respect for common things. The questioning spirit is turned toward practical problems.

The method to be used may be summarized as follows: To give time for observation, a few questions, such as are given in the outline, may be written on the board a few days before the lesson is due. The class work should consist of reports and discussion leading to further investigation.

Grade 1

1. Gardening. Arrange wild flowers in bottles according to color. How many colors? Play games of seeing, smelling, feeling. Tell stories. Hold up a flower, and have a pupil give its name and one interesting fact about it. (Children must be warned against trespassing in their search for flowers.)

Transplant a geranium and a Jack-in-the-Pulpit to the room before frost. Slip a begonia. Plant Dwarf French marigold seeds in pots or boxes. Grow Chinese lily in pebbles and water. A rubber plant is excellent if it can be obtained and kept. The children appreciate having something to care for.

2. Autumn Coloring of Leaves. Have the children gather the leaves and group them according to colors. How many colors? Tell why the leaves change color. (Not due to frost, but to the return of sap to the trunk.) Children may trace around leaves on paper, and fill in the outline with color. Cut these out and use them as a blackboard border. Make leaf booklets.

Make landscape drawings to show changes in color of trees or hills or lawn.

3. Collect Seeds. How seeds travel. (Sail, steal rides, roll, carried by birds, etc.) Have the seeds named and used in games of touch, sight, smell, and sound. Make designs with them.

4. Caterpillars. This lesson may extend over several months. Have the caterpillars or cocoons collected in September. Feed the caterpillars until they spin cocoons. Keep the cocoons in a cool moist place. What do the caterpillars eat? (Leaves.) How? How much? How do they spin the cocoon? (Spin silken cocoon or make a mummy case.) Where? (Protected spots.)

5. Squirrel. What does it eat? (Nuts, buds.) How? (Gnaws. Does not crack the nuts.) How does it carry its food? How does it climb? How does it go down a tree? (Head first.) How does it get from branch to branch? How does it hide? How does it get ready for winter? What is its winter home?

References: John Burrough's *Squirrels and Other Fur-bearers;* Long's *Secret of the Woods;* Schwartz's *Wilderness Babies.*

6. The Cat. Observations at home and school. Feel of its tongue. What does it eat? How does it catch mice? (Claws.) When does it use its claws? What is the shape of the pupil in daylight? In the dark? Where are its sharp teeth? (At the side.) Their use? The use of its whiskers? (To feel its way in the dark.) When does it wash? How? Have children tell experience with pets, and bring picture and stories from home. The Humane Society.

7. The Cow. Tell the story of its origin and domestication. Get a little cream and make butter in school. Children's observations: hoof (split, two toes); use of tail; method of eating.

References: *Farmers' Bulletins 55 and 63;* Burkett, Stevens, and Hill's *Agriculture for Beginners.*

8. The Spider. Find a cobweb. Where was it? Where does the spider hide? How does the spider know when a fly gets caught in the web? (It shakes the web.) How does he fasten the insect? (Sometimes spins a thread around it.)

9. Gardening. March through June. Divide a sand-box by cross strings, so that each child will have an individual garden. You may obtain seeds from Members of Congress, or from the Department of Agriculture. Competitive raising of radish and beans. Which comes up first? Which backs out of the ground? How do the first radish leaves differ from the later ones? How do the first bean leaves differ from the later ones? How do the bean plants differ from the radish plants? Plant some seeds on moist blotters, to show the growth of underground parts.

Plant a carrot and a sweet potato, and watch developments. Transplant a geranium to the school grounds.

10. Signs of Spring. March through June. Force as many as convenient of the following buds in water: horse-chestnut, pussy-willow, red maple, apple, magnolia, lilac. Collect and recognize: dandelion, buttercup, Jack-in-the-pulpit, violet, daisy. Observe garden flowers: tulip, jonquil, crocus, hyacinth, narcissus. Observe the emergence of insects from cocoons.

If possible, observe birds that return. Observe tadpoles in a glass jar.

Grade 2

1. Grasshopper. Have children collect grasshoppers and care for them in a cage. Feed fresh clover, moistened, so that they may have water to drink. Its relative, the cricket, may be kept in another cage. Feed the cricket apple cores. What does the grasshopper eat? (Grass and clover.) How? (Holds it with its front legs. Move its head from top to bottom, cutting off bits of leaf.) How many legs has it? (Six.) Different ways of moving? (Jumps, walks, climbs, flies.) How does it climb? (By means of its claws.) How does it feel about? How many wings has it? (Four.) Advantage of the different colors? (Birds cannot see it so well.)

Reference: Schwartz's *Grasshopper Green*.

2. *Fall Flowers and Animals.* September through November. Repeat associations of first grade. Extend problems. Make a fuller study of the dandelion. What does the flower do at night? To what does the flower change? What carries away the seeds? Does the plume end or the seed end land first? Why is it hard for birds to find dandelion seeds? (Color of the soil.) Why does the plant have such thick roots? (It stores food.) How are the leaves arranged? What is the advantage in this arrangement?

3. *Flowers Changing to Fruit.* Collect plants that show transition from flower to fruit. (Butter-and-eggs, shepherd's-purse, pumpkin, bean.) Use these in sense games. What fruits and vegetables are in the markets? An old seed catalogue may be used for pictures of vegetables. Some of these may be colored by the children. Cut them out and arrange according to some plan, as, those that grow on trees, in gardens, above ground, below ground, etc. Which grow near home? Where are they kept for the winter? Where are the seeds for next year?

4. *Gardening.* This subject will need to run through the fall months. Collect and prepare squash seeds for use in the spring. Take in the house plants. Collect pictures of autumn landscapes. The children should care for the school plants, giving them water regularly washing the dust from the leaves, loosening the soil around them so that they can breathe, picking off dead leaves, and occasionally giving them food. (Fertilizer.) Keep those that need sunlight near the windows.

September. Rake up the leaves. Plant crocus bulbs out of doors. The paper narcissus may be planted in pots and kept in a cold room until February. Transplant petunias to pot for the windows. Plant seeds of sweet alyssum for a hanging basket. Grow abutilon in the window, and palms in shady corners.

5. *Evergreens.* Study at least one pine. Foliage is evergreen, but the leaves shed gradually. Why are the leaves called needles? On what part of the branches are the leaves? (Near ends.) Why? What becomes of the old leaves? Observe a cone. Where are the seeds? (Two under each scale.) How are the seeds scattered? (Wings for the wind.) Draw a bundle of leaves, a cone, a scale, and a seed. Draw a whole tree.

Have a window-box of evergreens. This is possible even if the fire goes out over the week-end.

6. *The Hen.* Observations or study of pictures. Difference in clothes between the hen and the chickens. Why cannot chickens fly? How does the hen eat? Does it chew its food? How does it drink? Where do the chickens sleep? How do they talk to other chickens? How do they scratch? Why? Where are their ears?

Reference: Miller's *True Bird Stories*.

7. The Dog. What does he eat? How? Describe his teeth. How does he drink? How well does he hear? How well does he smell? If you hide a bone, how does he find it? Compare with the cat as to claws, and as to its method of hunting and catching its prey. Uses of dogs. (Pets, watch-dogs, hunting-dogs, dogs for pulling sled and carts, etc.) The children may bring photographs of their dogs, and tell stories of their experiences with dogs. Collect pictures of different kinds of dogs. Play games for recognition of the kinds.

8. The Woodpecker. The woodpecker is a workman with wood. Have the pupils discover what he does and how he does it. Describe the tail, toes, and bill. How is each fitted for its work? Collect pieces of wood to show where the woodpeckers have dug for grubs, and where they have made a home. Show pictures of different kinds of woodpeckers.

References: Eckstorm's *The Woodpeckers;* Burrough's *Winter Neighbors.*

9. Increased Acquaintance with Birds and Flowers. Early in the spring place in bottles of water twigs of elm, cherry, forsythia, hickory, blueberry, and some of those used in grade one. Keep a simple calendar of signs of spring. Hang up pictures of spring landscapes.

Dip up and transplant skunk cabbage, bluet, and clover, cutting with pieces of sod around them. Cut flowers of anemone, lady's-slipper, arbutus, and columbine.

10. Gardening. April. Germinate squash, pea, lettuce, and morning-glory seeds in pots or window-boxes, or in individual gardens marked off by strings in the sand-box. Watch the spring bulbs out of doors. How do you recognize them? Make a blackboard border of flowers, with colored chalk.

Plant the top of a pineapple. Plant some cotton seeds.

Grade 3

1. Tomato Worm or Cabbage Butterfly. Life history and habits studied in a breeding cage. On what is the caterpillar feeding? What is its color? Advantage? What position does it take when disturbed? How does it eat? What flowers does the moth visit? Where is the nectar in these flowers? How long is the moth's tongue? Where does it carry the tongue?

2. Migration of Birds. Scientists do not know why birds migrate. They migrate when food is most abundant, and before cold weather. Some of the migrants occasionally remain all winter. Children should observe that birds gather in flocks in the fall to go south. Winter feeding is worth while, even if it is for the English sparrow.

Watch for the wild geese going south. Obtain Audubon charts.

3. Gardening. Plant Dutch hyacinths in pots or boxes, and others in beds in the school grounds. Plant sweet peas ("Earliest of all") in pots. Make slips of coleus as a border for the window-box. Grow Jerusalem cherry and pandanus. Transplant mullein from the fields. This is sold as American velvet plant in England. If convenient, grow aspidistra (iron plant) in a shady corner.

4. Earthworm. Observe in class room. How does it crawl? Which is the head end? (Nearest the ring.) Can it crawl in either direction? Can it feel? (Touch it with a pencil.) Describe its home. What are its enemies? The earthworm disfigures the lawn, but is useful in grinding the soil. Can you find any evidences of this?

5. Goldfish. Have children care for and observe goldfish for a week before the lesson. How does the shape of the fish help it in swimming? How is the tail-fin used? The paired fins? What is the covering of the fish? Watch specks of dirt in the water near the mouth. What do they tell you? (Water enters the mouth.) Watch specks near the openings of the side of the head. (Water comes out.) Explain that in this way the fish breathes the air in the water.

6. Rabbit. Teach as type of gnawing animal. Children are acquainted with the squirrel. If possible, have a rabbit brought to school. Feed carrots or potatoes. Note position of ears when resting. When startled? Describe movements of nose. What is peculiar about the upper lip? How does this help in gnawing? Compare the teeth with those of the squirrel and dog. Does the rabbit sleep with its eyes open? What protects the bottom of its feet? How does it clean its head, face, and paws? Use Easter booklet with rabbit and chicken.

References: Thompson Seton's *Raggylug*, in *Wild Animals I Have Known;* Sharp's *Watchers in the Woods;* Joel Chandler Harris's *Uncle Remus Stories.*

7. The Crow. Does it have any color besides black when seen in the sunshine? What is its food? Where does it build its nest? Of what material? Does it walk or hop? Did you ever see one that seemed to be a sentinel? How did it act?

References: Thompson Seton's *Silver Spot*, in *Wild Animals I Have Known;* Celia Thaxter's *Scare Crow;* Long's *Ways of Wood Folk;* Needham's *Outdoor Studies; Bulletin 6, Division of Ornithology, U. S. Department of Agriculture.*

8. The Toad. Start a tadpole aquarium in April. In the bottom of a two-quart glass jar, arrange carefully some mud, leaves, stones, and plants, from the bottom of a pond. Fill with water. Keep out of the sunshine. When it has settled, put in half a dozen tadpoles. Every week add a slimy stone from the pond. The slime is food for the tadpole. Toads' eggs are in strings, and frogs' eggs are in masses. Where are the eggs found? When? (April and June.) What is the size and shape of the eggs? Make drawings to show the changes. How is the head end first recognized? (Larger.) Which legs appear first? What becomes of the tail? (Absorbed.)

In a jar containing some gravel and wet moss, place a toad for study. Cover the top with a netting. Feed it flies, other insects, and worms. What advantage has it in its color? Does it feel cold or warm? In how many directions can it see? Find the ear. Count the fingers. Why are the hind legs so large? What does it eat? Did you ever see it drink? (No, it takes moisture through the skin.)

Notice the movement of its throat. (Breathing.) What sound does it make? (Trills.) How? (Throat.)

References: *Farmers' Bulletin 196;* Dickerson's *The Frog Book;* Sharp's *A Watcher in the Woods;* Long's *K'dunk, the Fat One, A Little Brother to the Bear.*

9. *Trees.* Keep a bud-calendar. Make a leaf collection.

Make a booklet of leaf drawings, or of leaf prints. References: Matthew's *How to Know the Trees by Their Leaves;* Collins and Preston's *Key to Trees.*

10. *Gardening.* Become familiar with the shrubs: lilac, syringa, forsythia. and bridal-wreath. Germinate corn.

Germinate tree seeds which were gathered in the autumn. Tree seedlings are to be used on Arbor Day. Use a box with glass sides to test depth to plant corn.

Grade 4

1. Ants. Have an observation-nest in the schoolroom. (See Comstock, p. 423.) Feed some syrup. How do ants eat? Where are the feelers? How does the ant keep them clean? How does it carry objects? How does it care for its young?

Reference: Schwartz's *Grasshopper Green's Garden,* p. 147.

2. Beetles. If possible, secure some potato-beetles early in the fall, and some of the larvae. Sprout pieces of potato in pots, and let the larvae feed upon the leaves. Feed the adults on raw potato. Do they bite or suck the food? What enables them to climb the plant? Notice the disagreeable odor when handled. Is there any advantage in this to the beetle? When you pick it up, can you discover any trick it has for deceiving its enemies? Collect other beetles. Are they injurious or useful? Why? How do they differ from the potato-beetle? How may the injurious one be destroyed?

3. Gardening. Fall months. Encourage home planting of bulbs. Observe the fall work with flowers and shrubs in parks or home grounds. Bring in tender plants from out-of-doors. Beets, carrots, and parsley may be transplanted to window-boxes for ornament. Plant apple seeds for seedlings to be used next year for grafting.

Plant seeds of candytuft. Make slips of inch plant. If convenient, grow various cacti.

4. Annuals, Biennials, and Perennials. Have the class collect plants and classify them as follows:

Annuals: those that have died below ground, such as jimson weed, ragweed, shepherd's purse.

Biennials: those that store food in the root the first season, produce seeds the second season, and then die, such as wild carrot, mullein, thistle, burdock.

Perennials: those that live year after year, such as milkweed, dandelion, poison ivy. Those that die down to the ground each year are called herbs.

Classify plants that are growing in the room. References: *Farmers' Bulletins 195 and 660.*

5. Carniverous Animals. (Flesh-eaters.) Collect pictures of members of this group. Learn to recognize from the picture the

name, distinguishing characteristics, use, and domestication. Powerful jaws, stout canine teeth, strong claws. Cats (such as panther, tiger, leopard); sharp, retractile claws; climbers; pounce on prey. Dogs (wolves, foxes); blunt fixed claws; strong sense of smell; chase prey. Musk animals (skunk, mink, otter). Bear. Raccoon. Seal.

6. Poultry. Review work of second grade.

See *Farmers' Bulletins 41, 51, 236, 287, 357, 528, 562, 656, 594.* Use for a home project. A careful account should be kept to see if the hens pay their board. The children should study carefully the bulletins and other sources of information for the best methods of feeding and caring for their hens. All will be interested in pictures of different varieties or breeds of fowls.

7. Potato. Plant in a box. Keep a diary, and make drawings to show different stages of growth. Where do the sprouts start from? Where do the roots start from? Where do the new potatoes form? Remove the skin from a potato and observe changes in appearance and weight of the potato after a few days. (Nearly four-fifths of a potato is water.) See *Farmers' Bulletins 35, 91, 295, 324, 386, 407, and 410.*

8. Fly. Use as a slogan "Prevent the Fly," rather than "Swat the Fly." See *Farmers' Bulletins 200, 340, 459, and 679.* Make fly-traps in manual training. Study the life history of the fly, and the dangers from allowing flies to enter our houses. Abundant material may be found for study.

9. The Parts of a Flower. The object of a flower is to produce seeds. Use the apple blossom as a type. From what do the leaves and blossoms appear? Which appear first? (Blossoms.) The green outside portion of the blossom bud is the calyx, and the parts are sepals. The petals are pink outside and white inside. How many sepals? How many petals? (Five.) The innermost organs are pistils. How many? The white threads tipped with yellow knobs are stamens. How many? The yellow knobs are anthers, and contain a yellow pollen. Look through the microscope at some of the pollen. Inside the base of the pistil are eggs, or ovules. Cut through the base and find the ovules. The apple will not grow unless its seeds are fertile, and a seed will not be fertile unless the substance of the pollen grain unites with the ovules. Bees and other insects help to bring the pollen to the end of the pistil, because the pollen dust gets on their hairy bodies, and is then scattered over the pistils when they enter the flowers. To attract the insects, the flower secretes nectar at the base of the stamens. Compare other flowers with this type. In the clover blossoms you may easily find some of the nectar.

10. How an Apple Grows. Observe through May and June. How many blossoms come from one bud? What is the advantage in this? What part of the blossom falls off? What parts form the apple? Bring growing apples to school, and cut them open at different stages to study the development. Why are some apples one-sided? (Some of the seeds are not fertile.) What are the uses of the skin of an apple? How many seeds has an apple? Taste a seed. What is the advantage to the apple tree of having bitter seeds?

Grade 5

1. Insects and Pollination. Have the class collect blossoms of different kinds, like butter-and-eggs, violet, nasturtium, salvia, pea, and squash. Have the children find what insects visit these blossoms. Find the nectar. How do the insects get the pollen and leave it on the pistil?

Reference: Dana's *How to Know the Wild Flowers*.

It will be possible for some schools to secure observation hives. Dried specimens of the honey bee may easily be secured for close observation. Try to distinguish queen, worker, and drones. Find the feelers. Describe the feet. Find the pollen basket on the hind leg of the worker. How does the drone differ from the worker? (The drone is larger. No pollen baskets or sting. Cannot reach nectar). Describe the young bees. (White grubs.) What is the shape of the cells in the comb? Why do the bees store honey? How do they build the comb? Tell stories about the bee.

2. Wood Structure. Make a collection of blocks or pieces of wood showing the rings of growth, and, if possible, obtain a specimen showing a cross-section from center to circumference of a piece of oak. This may usually be gotten from a piece of firewood. Have it smoothed to show the grain. The porous portion of the annual rings is the spring growth; the compact portion is the summer wood. The light lines are called the silver grain. Study the rings and the silver grain as you find them in the blocks, in desks, and other furniture. What is quartered oak?

3. Pruning. Early spring is a good time for pruning trees. Fruit trees that are not pruned grow to bushy heads and produce small fruit. Try to find pictures showing well-pruned trees. A limb should be pruned smoothly where it joins the trunk or the parent branch. The wound should be painted to prevent decay. The children will be able to find specimens of good and poor pruning, and of decay following from lack of paint or from leaving surfaces rough or broken. They may also find specimens showing how the bark has grown over a wound that had been properly treated.

4. Cuttings. Divide a sand-box by strings so that each child will have a garden. If this is not available, it should be possible to have a window-box. Make cuttings of currants, willow, poplar, privet, grape, and of some house plants, such as carnation, geranium, begonia. Make the cut just below the node or joint. One-third of the cutting should project above the soil. In the case of house plants, remove a part of the leaves to prevent wilting. The cuttings may be transplanted to the homes late in the spring.

5. Grafting. The seed of a Baldwin apple does not produce a Baldwin tree. Varieties of fruit trees are maintained by grafting or budding. Furnish each pupil with drawings and directions. With sharp knives brought from their homes have them practice on sprouts from an old tree. As they become proficient, allow them to use the seedlings raised in Grade 4, and set them out at home. See *Farmers' Bulletins 157, 218, 408.*

6. Transplanting. Hectograph a sheet, or write on the board, directions for transplanting. The child learns the method by transplanting something himself, according to directions. Give demonstrations if necessary.

7. Bulbs. Many florists will give away bulbs after Easter. From a vertical section of a bulb, find where the food is stored, where the blossom comes from, and where the new bulbs are formed.

Grow a Chinese lily in a dish of pebbles, with water just reaching the bottom of the bulb. Each child may pot a bulb, keeping it in a cool, dark cellar until the roots form. In early spring, bring it into a warm room.

8. Birds. How is a bird built for flight? (Shape, breastbone in form of a keel, and many bones hollow for lightness.) How does a bird steer? (The wings as the oars of a boat and the tail serves as rudder.) Collect pictures or borrow specimens to show the most important forms of beaks and feet. How can you tell a swimming bird? What is peculiar about the feet of wading birds? Why? About the legs? Why? Why do the hawks and owls have such strong claws? What else can you observe about their feet that helps them in capturing their prey? Why does the chimney-swift have weak feet? How do the toes of the woodpecker differ from most birds? Why is this? Describe the beaks of the birds of prey. Why do they need this sort of a beak? What kind of beaks do wading birds have? (Long, for probing.) Birds that eat seeds have short, thick beaks for crushing their food. Find some birds that you think eat seeds. Find some birds that you think have bills fitted for digging into the wood of trees. Examine a wing-feather and find ways in which it is adapted for its use.

Lead the class on field trips or invite someone to lead them. Obtain the Fuertes Outline Drawings for Color Work from the Loose Leaf Material of Comstock Publishing Co. Obtain the Audubon Charts. Drill on recognition of birds. Visit some bird collection.

9. Protection of Native Plants. Obtain free leaflets from the Society for the Protection of Native Plants, 234 Berkeley Street, Boston. Obtain the Outlines of Flower Plates which the Comstock Publishing Company supplies to the Wild Flower Preservation Society. Many species are threatened with extermination. Urge moderation in picking, and careful cutting instead of breaking off the blossoms. Teach the greater enjoyment of living flowers than of withered, discarded flowers, selfishly gathered. Some flowers, as daisies, buttercups, and asters may be gathered with freedom. "Hast thou named all the birds without a gun? Loved the wood-rose and left it on the stalk?"—*Emerson.*

10. Tomato. Intensive study. Hectograph, or have the pupils copy the steps in cultivation. Give interesting historical facts about the tomato. Provide enough plants to furnish fruit for the canning experiments in the fall. Have the children prepared to watch for the enemies of the plant. See *Farmers' Bulletins 230, 185, 521.*

Grade 6

1. Mosquito. Keep wrigglers in a jar of water. Notice their resting position at the surface of the water for the purpose of breathing.
Pour a very slight amount of kerosene, perhaps half a teaspoonful, on
the water, and note the result. The surface will be so covered that
the larvae can no longer breathe. Oil sprayed upon the surface of
stagnant pools has proved a most efficient method for destroying the
young of the mosquito. See *Farmers' Bulletin 155.* Tell the story
Mischievous Madam Mosquito from Schwartz's *Grasshopper Green's
Garden.* What relation did the mosquito have to the building of the
Panama Canal? (Until scientists found that the mosquito carried the
germ of malaria and yellow fever, and learned how to destroy the
mosquito, the canal could not be built.) Find pictures of the malaria
carrying mosquito.

2. Orchard Insects. Study the brown tail, gypsy-moth, tent-
caterpillar, and coddling moth. When new insects are introduced
that are injurious, their natural enemies are often found in their home
lands. Obtain pamphlets and specimens from the State Board of
Agriculture. See *Farmers' Bulletin 453.* All citizens should be informed concerning these orchard pests.

Have the entire class make an intensive study of one of the insects, or divide
the class into groups, assigning to each group an insect for study. The caterpillar of the brown tail is poisonous. Breed the insects in a cage. Study the life
history, and determine at which stage it is most easily exterminated. Study
wormy apples. Where did the worm enter the apple? (At the blossom end.)
Is the worm still in the apple? (A doorway is not always a sign that it has left.)
You may find a case where decay has started at the worm-hole. Read about the
introduction of the brown tail and the gypsy-moths.

3. Weeds. Have the class collect different weeds from gardens.
Arrange these in bottles, naming each. Why are weeds harmful?
(They take food, sunlight, and moisture from the plants we want.)
Are these weeds annual, biennial, or perennial? Some weeds are
useful. Find how. Write on the board a list of ways in which weeds
make their way into gardens. Exhibit the weeds in turn, and have the
pupils tell the name of each, its particular method of making its way,
and the manner in which it may most easily be destroyed. See
Farmers' Bulletins 17, 28, 86, 188, 195, 660.

4. Yeasts and Molds. Expose pieces of the same bread to dry air,
both warm and cold, as in the schoolroom and on a window-ledge, and
to moist air, both warm and cold. For this it may be moistened and
placed under a glass dish. Notice what conditions are favorable to
the growth of molds. It is possible to expose the same bread in
different houses and to secure different varieties of mold. The black
dots on the ends of the threads are spore cases, and give off spores in
the form of very fine powder. Try to examine spores under a microscope. What things become moldy? When are molds our friends?
When are they our enemies? Why do we place paraffine over jelly?
Why should we wrap fruit in paper? Why should wooden houses rest
on stone, brick, or cement? Expose a weak solution of sugar to the

air for several days. Note its odor. The scum at the top is made up of yeast plants. Show these under the microscope. What are the uses of yeast? (Fermenting and bread-making.)

5. Shell-fish. Secure oyster shells for study, or if possible secure from the Shell-fish Commission a supply of oysters. The shells will open quickly if dipped in hot water. Cut away the flat shell. Note the lines of growth. The deepest grooves mark the ends of a year's growth. Find how old the oysters are. Study the differences in the two halves or valves of the shell, the muscle that holds the parts together, and the membrane that lines the shell. It is this lining that forms or secretes the shell. Find the oldest and the newest parts of the shell. The thin, lacelike organs next to the lining are the gills, and are used for breathing. Read about the shell-fish industry in Rhode Island. Read about useful shells. Collect different kinds of shells and try to find their names.

6. Value of Birds. Topics for individual investigation and report to the class: birds and insects, birds and weed-seeds, birds as scavengers, birds of prey, the passenger pigeon, Audubon, the snowy egret, the McLane bill, the English sparrow, the English starling, the Audubon Society, domesticating the bob-white, the value of birds for their music, the wild turkey. Have someone give the class a bird talk. See *Farmers' Bulletins 54, 497, 506, 630; Biological Survey Papers, 61, 107.*

7. How Plants Use Water. With care to avoid breaking the membrane, pick the shell from a part of the large end of an egg. Then with a pin or toothpick break a small hole in the small end of the egg. Stand the egg, with the small end up, in a glass containing water, having the water cover about half the egg. In about an hour see what has happened. The water enters the denser liquid of the egg through the membrane in the same way that the dilute solutions of the soil enter the root hairs of the plant.

Try an experiment to show that plants take moisture from the soil and give it out to the air. Place a cardboard over the top of a flower-pot containing a small plant, having the stem of the plant extending through the cardboard. Invert a glass over the plant. Stand it in bright sunlight. Observe that moisture gathers on the inside of the glass. Or take two flower-pots of the same size, one with a growing plant and the other with the same amount of soil and moisture, but without a plant. Cover the surface of the soil in each case with paraffine. Weigh carefully and in a few days weigh again. Cut the stem of a milkweed, or cut a maple twig in early spring, and observe the result.

8. Plant Breeding. Review lessons that pertain to the reproduction of plants. Read in magazines and books about Burbank, "Chinese" Wilson, and the Bureau of Plant industry. Have pupils investigate the domestication of some plant in which they are especially interested.

9. Strawberry. Each pupil should, if possible, have a complete plant. Does the new plant form before or after the runner takes root? How many petals has the blossom? What insects visit the blossoms? What parts of the blossom fall off? What part forms the hull of the fruit? What part forms the berry? Where are the seeds formed? In what two ways are strawberry plants reproduced?

10. Corn. Grow seedlings in a box or flower-pot. How are the leaves fastened to the stalk? What prevents the water from running down between the clasping sheath and the stalk? How is the plant adapted to resist the wind? Study the roots. Make a cross-section of the stalk. Which is stronger, a solid rod or a hollow rod of the same weight? What kind of rod is the corn-stalk? (Practically hollow.) As a preparation for summer study raise such questions as these: How do husks differ from leaves? What part of the flower is the silk? The kernels? The pollen is a fine powder. Try to find some. Is it carried by the wind or by insects? It may be possible to obtain the exhibit and literature of the Corn Products Refining Company, of New York. Write to the Extension Department of the State College. *See Farmers' Bulletins 229, 292, 303, 313, 325, 400, 409, 414, 415, 537, 554, 617.* Send to the Bureau of Plant Industry for circulars 104 and 644.

"President Eliot once said that he had often reflected on the problem of why one person is a successful teacher, while another of equal knowledge, talent and character fails. As the result of much observation he had concluded that what makes a teacher successful is the power to impart joy. The end of a teacher's work should be to inspire in the pupil joy in learning, joy in the possession of truth. This is not a lower aim than some other conceivable one, but the highest of all, for joy is the highest end of the universe, the final purpose of God himself."

—F. C. Porter.

"Your power of seeing mountains cannot be developed either by your vanity or your curiosity, or your love of muscular exercise. It depends on the cultivation of sight itself, and of the soul that uses it."—*Ruskin.*

"To my thinking the real reason for the unsatisfactory condition of nature-study in American schools is general in that it is practically impossible in many places to find teachers who are competent to direct the study in an intelligent manner."

*—Charles W. Eliot, Nature-Study
Review, 3, page 52, Jan. 1907.*

PART III

SUGGESTIONS FOR TEACHING SOME NATURE LESSONS

"And where'er they found the top
  Of a wheat-stalk droop and lop,
They chucked it underneath the chin
  And praised the lavish crop."

"And the great green pear they shook
  Till the sallow hue forsook
Its features, and the gleam of gold
  Laughed out in every look."

"Through the woven ambuscade
  That the twining vines had made,
They found the grapes, in clusters,
  Drinking up the shine and shade."

From "The South Wind and the Sun."
—By James Whitcomb Riley.

CHAPTER XXIX

The division of the elementary schools into primary and grammar grades is merely a matter of convenience. There is no break in the lines of study, the only difference being in the mental and in the bodily conditions of the pupils. The following lessons have been based upon differences which exist in degree rather than kind. The methods employed developed from the following comparison of geography in the primary and in the grammar grades.

Primary Grades

> First three years called nature-study.
> No text-book used. (Songs, poems, stories.)
> No memorizing of words or definitions.
> Aim to cultivate self-reliance; work, therefore, voluntary.
> Observational.
> Informal expression of thought.
> Real object, picture, and language.

Grammar Grade

> More distinctly geographical.
> Secondary use; for review and to increase information.
> Definitions used sparingly to make general concepts clear.
> Aim to cultivate power of observation, thought, and correct speech; less voluntary.
> Explanatory (why and how?).
> Formal expression, principles, and classification.
> Model, diagram, and experiment added.

Lesson I—The Wind

PLAN FOR TEACHING IN THE PRIMARY GRADES

1. *Idea of object of thought.* (Equals introduction.)
Open the windows in a warm room. What is coming in the window. (Air.)

2. *Ideas of qualities of the object of thought.*
Brought out by conversation but discovered and expressed by the pupil. (The following qualities are to become a familiar part of the pupils' vocabulary; strong wind, breeze, calm, cold, cool, warm, hot; wet, moist, damp, dry; unseen, heard; felt; moves; presses; direction—smoke, dust; leaves, weather-vanes.)
How did you know that the wind came in the window? (Felt, heard, effects.)
Could you see the wind come in the window? (Unseen.)
How does the wind feel? (Warm or cold; strong or weak; moist or dry.)
What kind of a wind makes Nature feel glad in spring? (Warm.)
From what direction does a warm wind come? (South.)

What does a warm wind do to the muddy streets? (Dries them.)

Why does the farmer like to have plenty of wind in the spring? (To dry the gardens so that he can plow.)

What month is the windy month? (March.)

What kind of a wind makes you wrap up in winter? (North.) Why? (Cold.)

Is the north wind gentle or strong? (Strong.)

What kind of a wind is lazy? (South.)

Why do the people like the seashore in the summer? (Cool.)

What makes it cool at the seashore in summer? (Wind.)

From what direction does the sea breeze come? (East.)

How does the wind sound in the wires? Sails? Oaks? Pines? Storms? (Howl, whistle, shriek, groan, sigh, wail, whisper, moan, rustle.)

Make the sound that the wind makes.

What does the wind do to the snow? Ocean? Fallen leaves? Tall grass? White fluffy dandelions? Apple blossoms? Apples? Wet clothes? Sail-boats? Windmills? Weather-vanes?

How does the wind help some plants in the fall? (Shakes off leaves and fruit, scatters seed.)

What toys do you like best on a windy day? (Kites, boats, wind-mills.)

Do firemen like the wind? Why? (No, because it spreads the fire.)

Do sailors like the wind? Why? (No, when it causes wrecks. Yes, when it helps him sail.)

What does the wind do that is mischievous? (Spreads fires, causes wrecks, blows down trees and fences, etc.)

Why is the wind useful? (Cools us in summer, dries streets and gardens, turns wind mills, helps ships to sail, etc.)

3. *Pictorial expression*

Collect pictures to illustrate, the direction, temperature, moisture, strength, and uses of the wind. Have children give stories orally.

4. *Information by songs, poems, and stories*

Andrews, Jane—"The Talk of the Trees that Stand in the Village Street."

Bryant—"The Evening Wind."

Coolidge, Susan—"The North Wind."

Field, Eugene—"The High Wind."

Longfellow—"Maiden and Weathercock." "The Windmill."

Mason—"Whichever Way the Wind Doth Blow."

Pratt, Mara L.—"A Legend of the South Wind."

Proctor—"The Wind."

Rossetti, Christina—"Who has seen the Wind?"

Stevenson, R. L.—"Windy Nights." "The Wind."

Tennyson—"Sweet and Low."

Thomas, Edith—"The Weather Vane."

Whittier—"The Wind of March."

"The Weathercock's Complaint"—Longmans' Pictorial Geographical Reader. Book I.

"Naughty North Wind"—Through the Year. Book II. Sliver, Burdett.

"The Wind's Frolic"—Earth and Sky. A First Reader. Ginn & Co.

5. *Review—Oral and Written*

PLAN FOR TEACHING IN THE GRAMMAR GRADES

The experimental method is good in presenting the effect of forces. In studying the wind each experiment is a queston asked of the air and the air is allowed to speak for itself. In this way the pupils get direct information. The experiments are performed before the pupils that they may be conclusive. The pupil should be able to give the method, observations and inferences taught by the experiment, otherwise the purpose of the work is lost. The correct expression of these three steps is a test of the success of the experiment. Oftentimes the over-zealous teacher spoils the aims of the experiment by telling the class what they should see.

1. *Experiments*

Experiment 1—Method. Have two test tubes; one partly full of water, the other containing a small amount of mercury. How does the weight of mercury compare with that of water? (Heavier.) Pour the mercury into the test tube containing water.

Observation. Mercury goes to the bottom.

Inference. What does the heavier mercury do to the lighter water in order to get on the bottom of the tube? (Pushes it up.)

What will heavier liquids and gases do to lighter liquids and gases when they are mixed? (The heavier liquid or gas pushes up the lighter liquid or gas.)

Exp. 2—Method. Fit a test tube with a rubber stopper through which passes a small-sized glass tube. Hold the test tube in closed hand and place end of tube under water.

Observation. Bubbles of air pass out of the test tube.

Inference. Heat causes air to expand.

Exp. 3—Method. Weigh a bottle of warm air and a bottle of cold air.

Observation. Cold air is heavier than warm air.

Inference. Warm or expanded air is lighter than cold air.

Exp. 4—Method. Cut two holes in a paper box. Stand a lighted candle in one hole. Place a glass chimney over each hole. Hold a smoking paper over the cold chimney (the chimney without the candle in it) and over the warm chimney.

Observation. Smoke goes down the cold chimney and up the warm chimney.

Inference. The heavy cold air is going down and the lighter warm air is going up. Compare the action of the cold air and the warm air with the action of the mercury and the water. (The cold air pushes up the warm air.)   Avoid the expression,"Warm air rises and cold air rushes in to take its place.")

Sum up what you have learned about the air.   (Heat causes air to expand and become lighter than cold air.   The heavier cold air pushes up the lighter warm air.)

2. *Application*

a.   Make a diagram to represent the vertical section of a room with a stove in the centre.   A window, at each end of the room, is open at the top and at the bottom.   Indicate, by arrows, the direction of the currents of air in the room.

(Pupils draw arrows.)   How should we open our windows in order to ventilate the room?   (At top and bottom so as to let the fresh air in at the bottom and the impure air out at the top.)   In what direction do you think the air is circulating in this room? (Test theory by means of a lighted candle or by watching dust particles which are floating in the room.)

b.   *Forest and land breeze.*   Diagram to represents a forest in the center of a field.   In what direction would the wind blow on a hot day in July?   (From the forest to the field because the forest is cooler than the field.)   How do you know that this is true from experience.   (Felt the cool breeze as went toward the woods.   Cattle seek the shelter of the trees on a hot day.   Noticed the odor of pines before reaching the pine grove.)   In what direction would you expect the wind to blow in this area in the winter?   (From field to the forest as the woods are warmer in winter.)   Sum up the effects of a forest upon the temperature of a locality.   (Forests make the summers cooler and the winters warmer.)   This might be expressed in a briefer way by saying that forests make the temperature more equable.

c.   *Sea and land breezes.*

Remember how the sand felt to your bare feet on a hot summer day and how the water felt in contrast.   In what direction would the wind blow under those conditions?   (From the ocean to the land, because the ocean was cooler than the land.)   Do we name winds from whence they come or whither they go?   (Whence they come.)

What name could we give to the breeze at the seashore on a hot summer noonday?   (Sea breeze.)

In what other way do you remember the direction of the seashore wind at noonday?   (Smell the salt air.   Ocean breezes are cool and refreshing.   The waves came toward the shore.   The sand blew onto the east piazza.)   When would you expect a land breeze?   (At night when the land is cooler than the water.)   In Peru, the fishermen go out in the morning by aid of the land breeze and come back in the afternoon by aid of the sea breeze.

d.   *Monsoons.*   Which freezes first, the ground or the water in the harbor?   (Ground.)   Which shows a lower temperature in winter, the land or the water? (Land.) What would be the consequent direction of the wind in winter? (Land to ocean.) When the wind blows from the land to the sea in winter and from the sea to the land in the summer they are called monsoons. The sailors of India carry out merchandise in the winter and return with the turn of the monsoon. Why may flooding a cranberry bog save the crop? (Water cools less readily than the land.) Why would you expect the interior of a continent to be colder in winter than the sea-coast? (There is no warm water to raise the temperature.) Which has the more extreme temperature a place which has continental climate or one which has oceanic climate? (The continental climate, as there is no water to warm it in winter or to cool it in summer.) Why does England have such a mild climate? (Has an oceanic climate.) Why do the coldest winters on the earth exist in northeastern Asia? (Extreme distance from ocean favors continental climate.)

e. Terrestial Winds. Where is the hot zone on the globe? (It is a belt located about the equator.) Would the wind blow toward or from the equatorial belt? (Toward, as it is the region of light expanded air.) These are called trade winds, not from the fact that they are favorable, for trade, but that they maintain a given path. (Unfortunately the remaining facts of atmospheric circulation must be given to the class without much explanation, as they are far beyond the understanding of elementary pupils.)

"The sullen day grew darker, and anon
   Dim flashes of pent anger lit the sky;
With rumbling wheels of wrath came rolling on
   The storm's artillery.

"The cloud put on its blackest frown,
   And then, as with a vengeful cry of pain,
The lightning snatched it, ripped and flung it down
   In ravelled shreds of rain."

—Descriptions by James Whitcomb Riley
From "The Shower".

"Then laugh on happy Rain, laugh louder yet!—
Laugh out in torrent-bursts of watery mirth;
   Unlock thy lips of purple cloud, and let
Thy liquid merriment baptize the earth,
   And wash the sad face of the world, and set
   The universe to music dripping-wet!"

—From "The Rain."

CHAPTER XXX

GEOGRAPHY LESSONS IN THE PRIMARY GRADES

Lesson II—The Rain

PLAN FOR TEACHING IN THE PRIMARY GRADES

1. *Idea of object of thought.*
Teach on a rainy day.

2. *Idea of qualities of object of thought.*
The following qualities are to become a familiar part of the pupil's vocabulary:
Effects of sun and wind. Fine and heavy. Size and shape of drops. Storm, shower, sprinkle.
Beat against face, window-pane, ground. Warm, cold.
Fair, cloudy, stormy.
Watch the raindrops fall. From where do they come? (Sky.)
What is the form of the rain-drop? (Round like a ball.)
How do you know when it is going to rain? (Cloudy.)
How do you sometimes know that it is raining even though you do not see it? (Hear it beating against the window-pane.)
Why do we sometimes hold our head to one side when we are out in the rain? (So the rain will not beat in our face.)
How would you know that the rain comes down with some force by looking at the surface of a pond? (See the effect of the rain on the surface.)
What do you think the rain does to the soil when it strikes the bare ground? (Washes it away.)
What kind of a rain makes you "wrap up"? (Cold.)
What kind of a rain makes nature feel glad in spring? (Warm.)
Why is it good to have plenty of rain in the summer? (Makes the plants grow.)
Does the farmer like to have it rain? Why? (Makes his crops grow, fills his wells and springs, and settles the dust.)
What birds like to have it rain? (Ducks, geese, sparrows, robins, all.)
What favorite food do birds hunt for after a rain? (Worms.)
Why do earthworms come out after a rain? (Ground is too wet.)
Name other animals, besides birds, that like to have it rain. (Frogs, turtles.)
What does the rain do to the snow? (Melts it and washes it away.)
What does the rain do to the grass? Streets? Brooks?
Why do we like to have it rain on a hot day? (Makes it cooler.)
What toys do you like best after a rain? (Boats, waterwheels.)
What can the rain do that is mischievous? (Wash away rich soil, cause floods, make gullies in the street.)
Could we get along without the rain? (Sum up the uses.)

3. *Pictorial expression*

Make a collection of pictures to illustrate the typical conditions in different parts of the earth of plants and animals, and their relations to the amount of rainfall. Show relation to human life—dwellings, clothing, food, and occupations. Keep a weather calendar.

4. *Information by songs, poems, and stories*
"Before the Rain"—*T. B. Aldrich.*
"After the Rain"—*T. B. Aldrich.*
"The Little Cloud of Liberty"—*J. H. Bryant.*
"To a Cloud"—*Wm. C. Bryant.*
"Through the Year," Book I.—*Clyde and Wallace.*
  "The Water-Drop's Journey."
  "The Little Lazy Cloud."
  "Three Drops of Water and What Became of Them."
  "The Endless Story."
"The Sunset City"—*H. S. Cornwall.*
"Signs of Rain"—*Jenner.*
"The Rain"—*Caroline Mason.*
"The Cloud"—*Shelley.*
"Two Little Clouds"—*R. L. Stevenson.*
"Earth and Sky,"— A First Reader—*Stickney.*
"The Water Drops."
"A Tempest."

5. *Review—Oral and Written.*

A GEOGRAPHY LESSON

Rain

PLAN FOR TEACHING IN THE GRAMMAR GRADES

I Evaporation

a Experiments.

Exp. 1. *Method.* Place same amount of water in four separate dishes of equal size. Place under following conditions: warm, cold, wind, and calm. Place the same amount of water in a shallow dish and in a deep dish, and leave under same conditions.

Observations. What changes do you perceive? (*Less amount of water in each dish.*) What became of the water? (Went into the air.) What was the size of the particles of water leaving the dish? (Invisible.) Call this form of water *vapor* and the change *evaporation.* From which dishes did the most water evaporate? (Warm, wind, and shallow.)

Inference. What conditions favor evaporation? (Heat, wind, and large exposure of surface.)

Exp. 2 *Method.* Heat same amount of water, as was placed in the dishes, over an alcohol lamp.

Observation. Water evaporates more quickly.

Inference. An increase of heat increases the rate of evaporation.

b. Questions for Thought.

1 What becomes of the water after a shower? (Runs off, soaks in, evaporates.)

2 Name other instances where you think water must have evaporated. (Wet blackboards, drying clothes, ink, sidewalks, ponds, lakes, rivers, and the ocean.)

3 What is meant by the expression "Water dries up"? (Evaporation.)

4 Why do brooks dry up in summer? (Evaporation greater than supply.)

5 Why do not all bodies of water dry up? (Supplied with rain.)

6 On what kind of a day will most water evaporate from the ocean? (Warm, sunny, clear and windy.)

7 What do farmers do when they want their hay to dry? Why? (Turn it over so as to expose new surfaces. Shake it up so as to expose more surface.)

8 What is meant by the expression "The sun is drawing water"? (The heat of the sun is causing the water to evaporate.)

9 What becomes of the dew? Frost? (Evaporates.)

10 Place fresh leaves under a tumbler. What collects on the inside of the tumbler? (Moisture.) What is passing off from the leaves of plants? (Vapor.)

11 What becomes of perspiration? (Evaporates.)

12 Does evaporation of perspiration make us feel cooler or warmer? (Cooler. Evaporation is a cooling process.)

13 Why is it cooler after a summer shower? (Moisture is evaporating.)

14 Explain cause of feelings on a "muggy day." (Perspiration is not evaporated and we have an oppressed feeling.)

15 What are the uses of evaporation? (Dry streets, sidewalks, ink, clothes, hay, etc., helps leaves to give off vapor and us to perspire; cools the air.)

II. Condensation

a. Experiments.

Exp. 1 *Method.* Leave a tumbler of ice in a warm room.

Observation. What collects on the outside of the glass? (Moisture.) Where does the moisture come from? (Air.)

Inference. How does the temperature of the glass compare with the surrounding air? (Cooler.)

How did the temperature of the air change when it came in contact with the tumbler? (Became cooler.) What do you think caused the moisture to form on the tumbler? (Cooling of the air.) Call this form of water *dew* and the change *condensation. Dew* forms on objects at a temperature above freezing and *frost* forms on objects at a temperature below the freezing point.

Exp. 2 *Method.* Heat some water in a test tube over an alcohol lamp.

Observations. What do you observe above the test tube? (Steam.) How does steam differ from vapor? (Visible.)

Inference. What must be between the steam and the water in the tube? (Water vapor.) How do you know? (Because heat causes water to pass off as a vapor.) How does the temperature of the air in the room differ from the temperature of the air in the tube? (Cooler.) What happened when the vapor struck the cooler air? (Condensed.) What kind of air can carry the most water vapor? (Warm.) When water vapor is condensed near the earth it is called *fog*, and when it is high above the earth it is called a *cloud*.)

b. *Questions for Thought*

    1 What collects on the window on wash day? (Moisture.)

    2 What causes moisture to collect on the window? (The hot water evaporates and the cold window-pane causes the vapor to condense.)

    3 Name other instances where you have seen vapor condensing?

    4 When does an ice pitcher "sweat"? (When it is cooler than the surrounding air.)

    5 Why is there more dew on vegetation than on rocks? (Plants give off water vapor.)

    6 What are the uses of dew to vegetation? (Gives moisture.)

    7 When can you see your breath? (On a cold day.)

    8 Why does it happen only on cold days? (Cold causes vapor to condense.)

    9 Why does fog form over low places? (More moisture and colder.)

    10 At what time of day do you usually see fog? (Morning.)

    11 When does it usually begin to clear away? (When the sun warms the air.)

    12 Of what is a cloud made? (Condensed water vapor.)

    13 Is air ever perfectly dry? (No.)

III. *Precipitation*

Recapitulate history of rain formation by experiment on distillation. (See text-book in Physics for method.)

Questions for Thought

    1 Since dust stays in the air how must its weight compare with that of air? (Lighter.)

    2 Why will water vapor stay in the air? (Lighter.)

    3 What causes the water vapor to form into drops? (Cooling.)

    4 When will the drops of water fall as rain? (When they become heavier than the air.)

    5 Why do clouds settle before a rain? (Moisture is condensing and becoming heavier.)

    6 Why does it not always rain when it is cloudy? (Moisture does not condense enough.)

    7 We know that air coming out of a bicycle tire is expanding and that it is becoming cool. Air which is pushed up is expanding

and becoming cool. What is the result when very moist air is cooled? (Rain.)

8 Would prevailing westerlies deposit moisture on the leeward or on the windward side of mountains? Why? (Windward, as they have to ascend to cross mountains. When they ascend they expand, become chilled and deposit rain on western coasts or slopes.)

9 What is relation of trade winds to rainfall? (Trade winds are growing warmer and absorb moisture. They yield rain on windward slopes of mountains.)

10 Why does it rain in the belt of calms? (Lighter warm air is pushed up, expands, becomes cool, and vapor condenses into rain.)

IV Review by writing the life history of a rain drop. Teacher represent the same by a diagram.

The Desert Shall Rejoice, and Blossom as the Rose.
—*Isaiah.*

What Nature Asks, That Nature Also Grants.
—*James Russell Lowell.*

"Rain! Rain!
Oh, sweet Spring rain!
The world has been calling for thee in vain
Till now, and at last thou art with us again.
Oh, how shall we welcome the gentle showers,
The baby-drink of the first-born flowers,
That falls out of heaven as falleth the dew,
And touches the world to beauty anew?
Oh, rain! rain! dost thou feel and see
How the hungering world has been waiting for thee?"
—*James Brown Selkirk (Rain).*

The woods were made for the hunters of dreams,
  The brooks for the fishers of song;
To the hunters who hunt for the gunless game
  The streams and the woods belong.
They are thoughts that moan from the soul of a pine,
  And thoughts in flower bell curled;
And the thoughts that are blown with the scent of the fern
  Are as new and as old as the world.

—Sam Walter Foss.

CHAPTER XXXI

Field Work for Fourth Grades

I. *The Hill*

The aims of this work may be stated as follows:

1. To recognize natural objects and processes in the home environment.

2. To know the causes and variations of the facts and activities.

3. To discover the significance of the immediate outside world in its relations to man.

4. To appreciate the usefulness, beauty, and truth of these natural features.

5. To give a foundation for the understanding of nature about other homes.

Lesson I

The field work may often be possible from the window of the school. Many teachers believe in field trips, but never take one. This may be due to a lack of training, to a non-appreciation, to want of time, or to the supposed inability to handle a class in the field. The suggestions in this outline are written to meet the first objection. Appreciation develops with practice. The best time is to take the last period of the morning and a part of the noon hour or the last period of the afternoon and additional time. Discipline will take care of itself after one or two trips. If possible, it is better to divide the class into two sections, taking about twenty pupils at a time.

In any textbook so far printed the material is a mere guide for the study of these objects and processes in all homes. Since each school has its own environment, which is different from that of all other schools, the treatment of the subject can never be the same. The trip which is written about in this paper, therefore, is not so general that you would see the points in the study of all hills, but suggests how the teacher may make a selection of the fact to be studied and the method of studying it.

It will be noted that the descriptive vocabulary of a hill has been left in italics. These terms should not be emphasized. They are acquired by use. Contrive to have them needed and used frequently. The study of the form, height, slope, base, structure, etc., of a hill is not important, but form, height, etc., in relation to the daily life of the people is important. The children should make observations under the guidance of the teacher; as location of streets, absence of stores, etc. This ought to suggest *why* each time. The method here is from effect to cause or observation to inference. The important thing to keep in mind is the definition of geography, a study of the earth in relation to man.

Trip to Prospect Hill by way of Star Street and Jenckes Street. Teach the *cause* of hills on way by noting where the rain has carved out miniature hills in the gutters. Explain that the Woonsquatucket and Moshassuck Rivers carved out Capitol Hill in the same way. Show how the Capitol is located on a favorable site. Note where railroads pass in relation to the hills. Explain. Note location of grain store, warehouses, factories, and stores. What street follows the *base* of the hill? Why does it do that? Note relation of electric car lines to the hill. Observe that streets go up the hill and around the hill. Which are the main streets? Why? It is said that Benefit Street used to be an old cow path back of the barns which belonged to the houses situated on North Main Street. The posts and fences on Star Street are worn where people take hold to prevent falling when it is icy. Why does the class walk so slowly on Star Street? The part of Jenckes Street on the western *slope* of the hill has a rough, stony surface, because it is too *steep* for the use of the steam roller. Study characteristic way in which people relax their muscles as they walk down the hill. A stranger can be recognized by the way he braces back when going down the hill. Study the gutters. What has become of the fine soil that we usually see in gutters? How large are smallest pebbles which have been left? What does this indicate as to the swiftness of the "run off?" The steepness of the slope? What changes have to be made in order that the houses can cling to the sides of the hills. Bring out the idea of *terraces*. How are the houses able to have entrances at two different stories?

The class is weary by the time it reaches top of the hill. Follow Congdon Street to the south and rest in the second vacant lot on the right, which gives a fine *view* of the Narragansett Basin. Why called Prospect Hill? Prospect Street runs along the *summit* of the hill. The *height* of the hill is 206 feet above the level of the Providence River. Enjoy the *beauty* of the landscape. The distant hills have an even *crest*. The place where they seem to meet the sky is called the *horizon*. Note the distant forests; the spires and roofs outlined against the sky; the majestic appearance of the State House. Discover by aid of the compass that the steep slope is the western slope; and that the hill *extends* north and south. Some of the class have been on the eastern slope of the hill. Have them tell that the eastern slope is *gentle*. Observe that there are no warehouses or business blocks on the hill. The western slope is too steep for that kind of traffic. Observe the apple trees on the slope of the hill. They are the remnants of the orchards of the early settlers. This orchard extended around the hill above Benefit Street. The hill was too steep for the gardens. The pastures and gardens were on the gentle eastern slope. The trees on the side of the hill have the advantage of good sunlight. Explain. Why are there so many vacant lots on this side of the hill? It is interesting to know that there were fifty-two "home lots," each about five acres in size, which ran back to the old "highway" now known as Hope Street. Each settler also had six acres for planting on the east side of the hill. Listen for distant noises in the city below. Compare with quietness of the hilltop. Observe smoke and dust in city. Compare with freshness of the air on the hilltop. Sum up reasons why Prospect Hill is a residential district. Test the temperature at the *foot* and at the top of the hill. It is usually cooler on top. An advantage in summer as one gets a good breeze, but a disadvantage in winter, as houses are hard to heat. Lead pupils to see that marshy places are not good places for dwelling houses. North of the railroad station is a considerable area not settled. This was a large marsh land the central part of which was known as the Cove. It has now been filled in. The hills are well drained and therefore settled.

Locate the Normal School, Moshassuck River, Woonasquatucket River and principal buildings. Locate direction of East Providence, Saylesville and Pawtucket. Follow the course of freight trains by their smoke. Note trains which come from the tunnel.

Prospect Hill forms the backbone of a peninsula which is bounded on the east by the Seekonk River and on the west by the Moshassuck River. This peninsula was called *Moshassuc* by the Indians. The hill at the southern end of this peninsula has been carted away and it is now known as Fox Point. This is opposite to the filling in of the Cove and illustrates two ways in which man overcomes geographical obstacles

Lesson 2

This lesson may be a correlated lesson in construction work. The interest of the child has been aroused in a neighboring hill. He now makes models, writes letters, and interprets pictures because he has an end in view. The teacher does not have to search for busy work, just for keeping the hands busy. The following ways are suggestions for making a model of the hill. The first method is for the teacher and the others for the children.

Plaster Paris Model A model can be made from the contour map of the home locality. Make board patterns according to the form of the contour lines. The thickness of the board represents the contour interval. Nail the boards together in proper relative position and mold clay, plaster, or whatever material is desired, over the boards. Paint the surface and represent the drainage systems, roads, etc., with various colors. Plaster Paris will not stick to wood unless mixed with gum or glue.

Sand Models Use damp beach sand on a modeling board. (The modeling board should be about three feet square with a two or three inch rim.) Sand has many advantages over clay—it is always clean and ready, and is more easily worked. Water may be poured on to show erosion.

Papier Maché Soak small bits of newspaper in water for two or three days. Stir mixture until it becomes a thick mass. Varnish or shellac the model after it has dried.

Salt and Flour Model Mix two parts of common salt to one of flour and add water slowly until it resembles wet sand. Work on a smooth board. Varnish or shellac after it has thoroughly dried.

Plasticine Model This material has the advantage of being in different colors and can be used several times. It is not affected by pouring water into it to show rainfall, drainage or tides.

Lesson 3

Teach the use of the model.

Since we cannot visit all the countries that we are to study we have to use representations. The model and map, then, become a basis for all future geography lessons and therefore the method of using this geographical tool should be made an equipment of the child at an early period. The fundamental ideas of maps and models can best be understood by comparing what is represented upon the model and map with the familiar features of the home locality, while on the other hand a model or map gives a better idea of the home environment.

At first, the model is kept in a horizontal position with the northern end toward the north. The transition from out-of-door work to the model consists of testing the pupils' understanding of direction, distance, and scale. Point north. Point south. What river has the direction north and south? (Moshassuck.) Find the Moshassuck River on the model. (Easily recognized from blue color, position and name.) Draw pointer along the Moshassuck River from north to south. What part of the model is north? South? What direction is Prospect Hill from the Moshassuck River? Point to it. The Capitol? Point to the east side of the model. The west side. Draw pointer along a street that runs east and west. Drill in location of places—Normal School, streets, etc. Point to Prospect Hill. What direction is Prospect Hill from us? How far is it from us? Point to Prospect Hill on the model. How far is Prospect Hill from Normal Hill on the model? Prospect Hill is slightly over two hundred feet high. How high is it on the model? Drill until children understand that the model represents to scale a bird's eye view of Providence and vicinity.

Lesson 4

Review of the trip to the hill by use of the model. The teacher should add points when the opportunity suggests itself. Correlate with history as far as possible.

Lesson 5

The study of other hills.

Other hills should be studied in comparison. Neutaconkanut Hill affords a magnificent view. It is 299 feet high. Why does the path zigzag up the hill? This hill has not been built upon by houses. Why? It has been set aside as a park.

Distant hills might be studied from pictures, as Pueblo Indians on table-like hills.

A *mountain* is not an overgrown hill.   It differs from the hill in other ways than size.   It is better to study a picture of a mountain than to create an imaginary mountain from a hill.

"Tell you what I like the best—
  'Long about knee-deep in June,
   'Bout the time strawberries melt
  On the vine,—some afternoon
Like to jes' git out and rest,
  And not work at nothin' else!"
"Pee-wees' singin', to express
  My opinion, 's second class,
Yit you'll hear 'em more or less,
  Sapsucks gittin' down to biz
Weedin' out the lonesomeness,
  Mr. Bluejay, full o' sass,
   In them base-ball clothes o' his,
Sportin' round the orchard jes'
Like he owned the premises!
  Sun out in the fields kin sizz,
But flat on yer back, I guess,
  In the shade's where glory is!
That's jes' what I'd like to do
Stiddy fer a year er two!"

"Knee-deep in June."
—*By James Whitcomb Riley.*

Fig. 1. Lesson 9. Free Expression Work. Representations were exhibited from the free expression work of the third grade. The papers were not chosen because of their perfection, for they might be improved in many ways. They showed, first of all, that a child illustrates his observations in different ways. These various modes of expression come naturally and not as a task. Why does a child lose his initiative in higher grades? Because he does not draw from self-expression. He is required to draw from a specific object and is then criticised because he does not see the object in the same way that the teacher does. Even though the child cannot be an artist he should not be robbed of the enjoyment of drawing. I indicated the interest of the child in the canoe trip of Roger Williams—across the Seekonk and up a "Great Salt River" (the Moshassuck), where he landed on the west side of Prospect Hill and found a clear, sparkling spring. 2 pointed toward an artistic temperament. A little sketch to show trees beside the river—casting their reflection upon the clear water. This was made with colored crayons. 3 This child associated the river with the hills. It is a hachure map. The short lines indicate a steep slope and the long, separated lines a gentle slope. The Woonasquatucket and Moshassuck Rivers are represented as flowing along the valleys into Narragansett Bay.

370

CHAPTER XXXII

II. The River

I believe that the following story is told by Professor Thorndike. A teacher who had just given a lesson about the Amazon River took the class for a walk along the banks of the Hudson. When she asked, "What is this?" the class responded in unison, "The Amazon." Dr. Herbert E. Walter, of Brown University, relates an authentic case that took place in Chicago. A high school girl, who walked daily across a bridge that went over the river, maintained that she had never seen a river. The Chicago River was so full of boats that she never associated it with what she had learned from map study. Although these cases are rather extreme, they are typical of much that is being taught in the grades to-day.

That the knowledge of local objects is necessary for the understanding of distant regions is not a new idea. Carl Ritter said that, "Wherever our home is, there lie all the materials which we need for the study of the entire globe." It was a century and a half ago that Rousseau said in "Emile:" "Let him learn his first geography in the town he inhabits, stimulating his imagination by expressing wonder as to natural phenomena."

Even a century and a half before Rousseau, an equivalent idea was given by Comenius: "We know the elements of geography when we learn the nature of the mountains, the plains, the rivers, citadels, or State, according to the section of the place in which we are reared."

In the study of the river the same aims are kept in mind as in the study of the hill.

Lesson I

Field trip along the Woonasquatucket River. Class begin observations at the corner of Promenade and Park Streets. Recognize characteristic appearance in different parts of the course. Terms come in naturally as needed. The mouth of the river is near the railroad station. This is the *lower course* of the river. Class look about and note what relation the lower course of the river has to man. Each observation should be followed by the question, Why? Note that streets follow on either side by *artificial banks* (levees or dikes); *bridges;* grease from sewerage; drain-pipes entering; the exhaust pipe pouring out steam from the Brown & Sharpe shops. Recall that mills were formerly run by water wheels. Determine by color of the water what kind of sediment the river is carrying. Where does the river get this sediment? This is the process of valley making. Lower a bottle from the bridge to get a sample of water. Carry back to school, allow sediment to settle. Use in class work on river. Throw in sticks to observe swift *current* in center and slower current on the side. Note fine mud deposit along sides where water is less

rapid. The river is swifter after a heavy rainfall. Why? Would it
deposit along the banks then? Find where the banks have been
worn away. Why is the stone wall *caving in* on the opposite side?
Find other proofs of a *higher stage* of the river, as markings on wall,
wearing of rocks and plant life. Just above the wooden embankment are *rapids*. What is in the bed of the river that causes the
rapids? Observe the flat-bottomed boat just below the *foot bridge*.
Why flat-bottomed? This is characteristic of *river boats*. Above the
foot bridge there are *gullies* or miniature *canyons*. Cause?

This forms a basis for understanding the action of the larger river.
The sediment of these *washouts* has been carried to the river. This is
not true, however, on the *terrace* beyond the honey locust tree.
Why? Distinguish between the *building up process* at the lower course
of a stream and the *tearing down process* at the upper course. Call
such a branch stream a *tributary*. This is a small tributary. We are
going to explore a larger tributary. This larger tributary used to
wind its way across this region. (Point to area filled in by ashes
and refuse.) This was a swamp. Why are there few houses here?
Why are they filling it in? Some day this will become a very beautiful spot. What will make it beautiful? The tributary now flows
underground through this part of its course. Look along the opposite border of this old swamp region and see if you can discover where
this side stream used to enter the swamp. Reasons for thinking so?
We will walk up Rathbone Street to Valley Street. Why called
Valley Street? Recess until get to head of Rathbone Street.

Compare the tributary stream with the trunk stream as to rapidity
of current, size of material, depth, clearness, width of the stream, and
width of valley. Note the three slopes of a valley. Compare the
beauty of the stream here with that of the Woonasquatucket. Follow
the winding course of the stream. Note that it is wearing away on
the outer curves and building up on the inner. Why? The rocks in
the bed are a great deal larger. Class observe what relation the
middle course of a river has to man. A four-foot water-fall is discovered near the approach to Chalkstone Avenue. If possible, bring
a miniature water wheel, constructed in the industrial laboratory, and
illustrate the use of water wheels. The class can realize the forces of
water used for large wheels by watching the upper falls from the
bridge on Chalkstone Avenue. Observe other uses of the *creek*
along this part of the course as, the dam, pond, farm, and icehouse.
Bring out the reasons after each observation. Look back into Davis
Park and call attention of class to the *lowland plain*, which is damp
and without houses, and the *upland plain* with houses. Follow the
brook above the ponds. This is the upper course of the river. Compare current, sediment, depth, clearness, bed rocks, width of stream,
sounds of stream, and width of valley with parts already studied.
Note cascades, absence of bordering plain, wearing away of soil
from roots of trees. Why? Teach about the relations of such a
stream to man—not thickly settled as along the lower course, im-

portance as a scenic center, and the presence of forests. Follow each observation by question, *Why?*

If it is found desirable to make two field trips for the purpose of studying the river, the first trip could end at Chalkstone Avenue and the second trip could cover the upper course of the brook and then proceed along the meadows to the source. This brings the class, however, onto an upland plain where the river again takes on the characteristics of old age, being sluggish and carrying fine sediment. Along this part of the stream is a parkway. It is mostly filled-in land, being formerly a marsh area. This is the reason why no houses were built there and the region has been saved for the city to use as a parkway and recreation ground. The rapid change of scenery from the smooth water of the Woonasquatucket River, with its uncouth sewerage and neighboring factories, to the turbulent stream with its picturesque waterfalls and forests, affords an opportunity for a field trip which few schools can parallel. The understanding of this little water-way, with all its problems, furnishes the fundamental knowledge for understanding all water-ways.

Lesson 2

A résumé of the field trip by use of the model of Providence. The development of the lesson will necessarily be under the careful guidance of the teacher. Later the children will be able to trace the path of the excursion on the model, giving the observations and inferences.

Lesson 3

The other drainage lines, such as the Moshassuck, the Seekonk, and the Providence Rivers, can be brought in. Distinguish the *main stream* and *tributaries*. Trace the path which rainwater would take from the various hills. Draw pointer along lines where the water will run off in two or more directions. Call a *divide*. Divides are not always high places. Introduce this idea by use of model.

Fix these ideas by a variety of ways, as, a blackboard sketch showing the similarity of a river system to a tree; compare a divide to the top of an umbrella or to the roof of a house; reasons for a crowned surface on a good road such as a macadamized street.

Make clear to class that divides may be a low land. Use Davis' Harvard Model and Providence Model to drill on main stream, tributaries, divides, river systems, and river basins.

Lesson 4

The class is now ready to study and interpret the features of distant rivers from pictures. Show views. Have pupils tell what part of course is illustrated, how recognize, swiftness, width of stream and valley, depth, size of sediment, and relation to plants, animals other than man, and man. Bring in new terms as cascade, cataract, rill, rivulet, channel, canals, water-ways, navigation, flood plains, deltas, drought, flood, slack water, shoals, irrigation. Emphasize the beauty of the rivers.

Lesson 5

Pupils take an imaginary journey down a stream by describing a series of pictures, beginning with the mountains where the river rises and tracing the course of the river to the mouth.

Lesson 6

Questions for thought. Write questions on the board where the class can study them. Each question is based upon the previous experiences of the class. The following are merely suggestive.

(1) From where does the water in the Woonasquatucket come?

(2) Where does it go?

(3) In what direction does the river flow? Why?

(4) Why is the tributary beyond Davis Park so rapid and the main stream so slow?

(5) What becomes of the rain that falls on Normal School Hill?

(6) Why is the Woonasquatucket River larger than the brook in Davis Park?

(7) In what way is it larger?

Lesson 7. Lantern Talk.

Lesson 8

If the class has had map work this lesson could come now. Otherwise, this should be postponed until after the study of the map.

Trace the courses of the Moshassuck and of the Woonasquatucket Rivers. Find where they begin. Find height of these rivers at various places. In what direction do they flow? Tell class how the sun draws water "up into the air." This moisture forms clouds which give out rain. Suppose it rains on these hills (pointing to region near the source of the river). Pupil indicate with pointer where some of the rainfall would go. (Discover that the goal of all rivers is the ocean.) Have pupils trace a river which originates from a lake or pond. How do the rivers change in size as they flow toward the bay? How is this shown on the map? Find tributaries of the river. What do they show about the slope of the land on each side of the main stream? Indicate the river systems. Trace the divide of the Moshassuc River System. Bring out uses of the river which are shown on the map. Where are cities? By use of the scale find the length of the rivers.

Teacher tell the following story and have class follow on maps with pointers. In the year 1636 Roger Williams built his cabin on the east bank of the Seekonk River at East Providence. The settlers at Plymouth asked him to move away from their land. Roger Williams and his four followers crossed the Seekonk in a canoe. Paddling around Fox Hill he entered the Great Salt River toward the north. What do we now call this river? On the right he saw a steep hill covered with trees which we now call? The little craft followed the base of this hill along what river? Until they saw a hill on the left, which is? Here the small party landed on the east bank of the stream and drank from a clear cool spring near where North Main Street is now located. They then climbed the steep wooded hill to get a view of the surrounding country. They could see the Woonsaquatucket River crossing a marsh below where it joined the Moshassuc to form the Providence River. Beyond was Narragansett Bay. These men were pleased with the fine view and decided to make their home at the foot of the hill near the spring. Why there? This was the beginning of the city of Providence.

Lesson 9

Should the school possess a stereopticon lantern the children would now be able to appreciate a river lecture. The talk might include scenes in parts of the neighborhood outside of the field trip and of distant rivers. Postcards could be used in the reflectoscope. This latter means might make it possible for some of the children to tell the rest of the class about the rivers in some place that they have visited.

III. The Valley

The pupil has already gained many ideas about the valley in the study of hills and rivers. The work has been planned so that a field trip on valleys is not necessary, although it could be taken with much profit. The children have already noticed the small *channels* and *washouts* caused by the rain, on previous trips. The water dripping from the eaves has made miniature valleys by washing out the finer soil. The last excursion was *up the valley* of the Woonasquatucket. By referring to this trip bring in terms such as: width, length, broad slope, gentle slope, steep slope. By use of pictures bring in terms gulley, ravine, gorge, canyon, dale, dell, ditch, mountain pass, gap. Bring out the human relations—kind of soil, cattle in the valley, farms, roads, railroads, and business houses. Note in particular, from school window, how the main line of railroad enters Providence from Boston by the Moshassuck Valley, which is between the Capitol and Prospect Hills. Show on the model that there is no natural passage from the city to East Providence and hence the necessity of the tunnel. Brook Street is so named because there used to be a stream of water flowing from a swamp, near the Hope Reservoir, along that line. Note this valley on the model and bring out the idea of how it was a convenient line for the building of a trolley line. The N. Y., N. H. & H. railroad sweeps around to the northeast in leaving the city so as to avoid the city proper. The railroad, however, does not leave the main valley, but goes along next to the distant hills which can be seen in the west. This broad lowland is called the Narragansett Basin. There are three such basins in New England, the Boston Basin and the Connecticut Basin being the other two. These three lowlands contain the most of the homes of the New England people. Explain.

"They's something kindo' harty-like about the atmusfere
When the heat of summer's over and the coolin' fall is here—
Of course we miss the flowers, and the blossoms on the trees,
And the mumble of the hummin'-birds and buzzin' of the bees;
But the air's so appetizin'; and the landscape thru the haze
Of a crisp and sunny morning of the airly autumn days
Is a pictur' that no painter has the colorin' to mock—
When the frost is on the punkin' and the fodder's in the shock."
"When the Frost is on the Punkin."
—*James Whitcomb Riley*

"I have jest about decided
  It 'ud keep a town-boy hoppin'
    Fer to work all winter, choppin'
Fer a' old fireplace, like I did!
Lawz! them old times wuz contrairy!—
    Blame' backbone o' winter, 'peared-like
    Wouldn't break—! and I wuz skeered-like
Clean on into *Feb'uary!*
    Nothin' ever made me madder
Than fer Pap to stomp in, layin'
On a' extry forsetick, sayin',
    "Groun'-hog's out and seed his shadder!"
"Old Winters on the Farm."
—*James Whitcomb Riley.*

CHAPTER XXXIII

Map Study

The study of maps formerly consisted of finding names, measuring distances, and mere copying. Experience has shown, however, that the student gains little real knowledge from such use of maps, while on the other hand he often acquires incorrect ideas. The cultural value of such a method is at a minimum since it merely trains the eye in finding names. Distances measured (in straight lines) on all maps are incorrect; the colors copied are without harmony; and no intellectual powers are trained unless it be that kind of perseverance which overcomes persistent drudgery.

The real aim of maps is to make regions as vivid and real as though they were examined in detail with the eye. Maps should enable the pupil to study the development of a region from the cause and effect method and should make him better able to give events their true surroundings. When one learns to understand a map to that extent he has not only trained the memory, but has disciplined the powers of reasoning, of visualizing, and of generalizing.

The art of understanding and applying map knowledge should be gained early in life. It is agreed that the geography of the home region should be taught first. Through the study and observation of our home surroundings we are able to study and interpret the geography of distant places. The map gives a better understanding of the home locality and becomes the basis of all future geographic work.

The best map, and fortunately the least expensive, is the government topographical map. This map more nearly fulfils the modern conception of geography as the study of the earth in relation to life than any other map. Have a pupil who is studying a colored physical map describe the physical features of a region where the two colors meet. Have him describe the same region from a contour map. Ask students to describe the picture that they have of the Alleghany Plateau. They invariably say that the region is a level elevated plain, because that is the definition of a plateau. Have the class look at the contour map and they will discover that it may be one of the most hilly regions in existence. It is better to see a thing than the definition of a thing. Mention the Ohio River. Does it call to mind a winding black line or a picture of the river? Try the same experiment with a city. Do you see a black dot or a picture of the city when Cincinnati is mentioned?

It is better to visualize the thing than the symbol of the thing. The contour map portrays the physical features as they are, the rivers with their valleys, flood-plains, bridges, ferries, etc., the city with its streets, houses, etc. The greater value of the contour map has not been realized by most teachers in the graded schools.

The Use of the Contour Map

To show in detail the use of a contour map let us develop a picture of the Allegheny-Cumberland plateau. This may be done by taking the map for any section of the plateau. We will use the map of the Huntington, West Virginia, Quadrangle. The irregularity of the contour lines shows that it is a hilly region. The figures tell that the Ohio River at this point is five hundred feet above sea-level and that the highest hills are about one thousand feet in elevation. This enables us to calculate that these hills are about five hundred feet high. The contour lines are near together, indicating that the hillsides are rather steep. Most of the hills are four hundred feet high and should a person stand on top of one he could look along the hilltops for a considerable distance. The numerous blue lines indicate a good drainage system and one would expect to see swift streams bounding down the steep hills. The velocity of the streams would enable them to carry away a great deal of sediment. If this material were slowly brought back, instead of being carried away, the river valleys would become filled and then the hilltops would ultimately become part of a great plain with a hill projecting, now and then, about a hundred feet above the general level. This must be the way that it looked at some remote period of the past.

We might speculate as to the future of this region as well as interpret the past. In the distance appears the Ohio River winding slowly within its wide flood-plain. The tributaries have cut deep V-shaped notches into the once level plain and the hills are being rapidly washed down to the master stream. If this process keeps up the hills will become smaller and in time the whole area will be worn down nearly to base level. It is through observing the forces in action and their results that we are able to interpret the past and to predict the future.

The contour map pictures to us, then, a hilly region separated by steep-sided valleys. The hills are of about the same height. Down the hills are rushing many streams carrying heavy loads to a larger stream which meanders in a broad flood-plain below. Such are the basic facts that influence the industries and mold the character of the inhabitants.

The black parallel lines indicate roads and we note that there are two kinds, the hill roads and the valley roads. In either case they are crooked like the contours, which indicates a rather excessive relief. We think that they follow these lines in order to keep on level ground, as it is difficult to pull heavy loads up hill. Many of the roads are parallel because the rivers have cut parallel valleys. If we were told that these wagon paths are made up of clay soil we would know that during winter and rainy periods they must be very muddy and difficult of passage.

The scarcity of black dots along the roadside indicates that the region is thinly populated. This would probably mean far-separated district schools. The impassable roads of winter would make the transportation of the pupils to a central school impractical. It has been said that good roads make better schools. On the other hand we might say that poor schools lead to poor roads. Although the roads may be an index to the development of a country the real reason lies in the soil and physical features. The Allegheny region does not have the advantage of the hard glacial soil and the more nearly level land of New England.

The absence of lakes, as shown on the map, tells us that this part of the country was not invaded by the glacier. A hilly region with poor roads invariably means a sparse population.

The black lines with short cross lines represent railroads. They follow the main rivers where they get a nearly level grade and find the larger towns and cities. The junction of the Guyandotte River with the Ohio River predestined the junction of a railroad. This focus of highways together with a wide flood-plain made a favorable site for a city. Huntington, therefore, has free communication along the rivers with flood-plains, but the bordering hills present altogether a different field of life. This sharp contrast in the distribution of population is equally distinct in the allotment of industries.

Having studied the topography and lines of transportation we are now ready to interpret the dependent industries. Since the hills are so steep, they are not adapted for agriculture and are therefore probably covered with forests. The rivers are strong enough to furnish power for the sawing of logs and in their lower courses furnish a means of transportation. Wherever the forests are cleared for agriculture the rich soil would doubtless be washed away in a season or two and that area would have to be abandoned for another. The most favorable places for tilling the soil would be along the flood-plains, where the soil is renewed each year by the spring freshets. Since only the large rivers have flood-plains farming would not be a very extensive industry.

Fruit growing might be an important industry. On the sunny slopes vineyards could be planted so as to get the sun, after the same manner as on the terraces of France. Apple orchards might also thrive and where there was danger of early frost in fall and late frost in spring they could be planted on the northern slopes.

If there are economic minerals within the hills, and the names of some of the villages indicate that there are, the plateau is well adapted for mining them. In mining it is easier to tunnel into the side of a hill than to tunnel down. The manner in which the rivers have carved the hills therefore makes the minerals available. The lumber, which is necessary in mining, is also near at hand. These conditions together with the water routes and accompanying railroads make mining a thrifty occupation.

All these things show their influence upon the character of the people. In such a rough country we are apt to find log-cabins nestled along the creeks, each settler eking out his existence from the bottom land. If you take an imaginary journey up a creek you will probably find a cabin at the head and should your trip continue along other branches there you would doubtless find similar houses, each sheltering its mountaineer family. People isolated in this fashion are apt to develop different habits. When they meet they cannot agree and hence the family feuds which correspond to the clans of Scotland. These people have their ups and downs in life similar to the physical character of the country. Physically you would expect them to have sinews as strong as the hills and their morals to be of the highest. A Swiss guide once said that "A man can have no evil thoughts here," and we would expect the same spirit in the West Virginia hills.

On the other hand, the rough country forms a good hiding place for outlaws and in such a place is the home of the Kentucky Night Riders and the Moonshiners.

Having studied and interpreted the map through the aid of the teacher, the student is now ready to read such stories as, "The Trail of the Lonesome Pine," and "The Little Shepherd of Kingdom Come." The characteristics and life of the people are well portrayed in these stories.

The class are now ready to study the political map in the textbook. This map is complementary to the contour map in that each shows features not displayed by the other. In getting this broader view of the plateau region the pupil approaches it with a clear picture in his mind and does not form erroneous ideas as to the physical features, rivers, etc., which were spoken of in an earlier paragraph.

The study of the Appalachian Plateau region through the interpretation of the contour map serves to illustrate that such a map is a far better tool than the one of the ordinary textbook. Why they are not used more in the grammar school is difficult to answer, but in a larger measure it must be due to the teacher's neglect of an opportunity to improve geographical teaching. It is hoped that some day every teacher will obtain the government contour map of the home vicinity and type maps of other regions.

St. Louis—A River City

The method of teaching cities may be well illustrated by taking St. Louis for an example of a river city. Suppose the class is using Tarr and McMurry's New Geography, Second Book. Class open the books to show Fig. 41. St. Louis is at the mouth of the Missouri River. Find its location on the map. What does the name *St. Louis* indicate as to its early settlement? What nation first settled the Mississippi Valley? Do these two facts point to the same conclusion? Name rivers which would influence the growth of the city? From how many directions do they come? What region is drained by the Missouri? What products would you expect to be shipped on the Missouri to St. Louis? What products might be brought down the Mississippi? What important raw material could easily be sent from the Gulf Coastal Plain? Sum up the list of raw materials that are probably sent by the rivers to St. Louis. What factories would you expect to find there? Examine the railroad maps in the pamphlet case of Figure 157. How many lines enter the city? Are they mainly east and west or north and south lines? Why? Influence on its size. Class turn to page 112 and read "The largest city on the lakes is St. Louis, the fourth in size among our cities (Fig. 157). It has a very favorable position in the center of the Mississippi Valley, on the Mississippi River, near the mouths of its two largest tributaries. The railway bridges across the Mississippi at this point have also had great influence on the growth of the city. It is an important shipping point both by water and by rail.

"Like Chicago, St. Louis is one of our leading markets for grain and live stock; but, being so far south, it handles Southern products also, especially cotton and tobacco. Besides this, it is a noted mule and horse market, and a great manufacturing center. It manufactures immense quantities of tobacco, beer, flour, boots, shoes, clothing, and iron and steel goods."

What the child reads in the text is certainly more interesting and means more than it would have without the map work. The class may now close their books and visualize the different scenes which they would expect to find in St. Louis. Show class pictures of the city, Fig. 156.

The map in the upper right hand corner of Fig. 157 shows the plan of the city. The government contour map shows the marked peculiarities of the city to better advantage. What direction does the city extend? Why? Where are the bridges across the Mississippi? Why? Which part is more thickly settled? Why? In which section is the street arrangement better? How explained? Locate the large parks. Reasons for their situation?

Extend knowledge by reading encyclopedias, reports of Boards of Trade, magazine articles, etc.

Teaching Cities from the Map

The first step in teaching a city should be by use of maps as in the teaching of all other geographical units. Describe the location of the city by use of political and outline maps. The name of the city may suggest its early settlement by the French, Dutch, Spanish or English. It might suggest the former presence of Indians or the discovery of a valuable ore which caused the city to spring up like a mushroom. The relief map shows whether it is in a mountainous region, a plain, or plateau; whether the coast is regular with steep cliffs or low near the shore running out into shallow water or irregular with safe harbors. The rivers of the city are the arteries which give it a new commercial life, feeding it with the raw material of the hinterland. Railroads may approach the city through a water gap or tunnel or along a river valley. Having studied these advantages follow out the trade channels to the adjoining country. What occupations would you expect there? What consequent products would you expect to be sent to the city? Fertile valleys are apt to produce grains and fruits, mountains yield minerals and lumber, regions too dry for agriculture are suitable for ranching, while the wheat of the plains, the corn of the prairies and the cotton of the gulf coastal plain pour into the natural trade centers. The principal products sent in from the hinterland determine what the occupations of the city will be. If coal or water power is near at hand the raw material will probably be manufactured. The supply of cotton means cotton factories, grain leads to flour mills, cattle require packing houses and the hides furnish leather for shoes. The city may be a port where railroads or rivers terminate. From what regions do these channels lead? Are the railroads transcontinental? From the map study the commercial routes by which steamers travel from the city. What foreign countries probably receive these exports? What products are probably imported?

The Government contour map of the city adds to the interest of the study. It shows the height of the hills and their elevation above sea level. It shows how the physical features determine the plan of the city. The main streets may follow a river or shore line and other streets run perpendicular to the main thoroughfare. Swamps and rocky situations have probably been avoided and may not be available for a metropolitan park system. The plans and development of all cities are forced upon them by their geographical setting.

The climate of a city is an important factor and should be interpreted from the map as in the study of countries.

The use of the text-book follows the interpretation of the maps. The text is now read to verify or modify conclusions. (The guidance of the pupil should have unconsciously prevented any gross errors.) The reading of various books will also give added information.

The study of a city is made more lifelike by the use of pictures. From the industries the pupil may visualize. The manufacturing of flour would mean grain elevators, the importing of cattle requires stockyards, etc. After creating an imaginative scene show the class pictures of the city. The impression resulting will certainly be more valuable than the remembering of the city as a dot on the map.

CHAPTER XXXIV

Ten Lessons on Our Food Supply.

The writer does not claim originality for any appreciable part of this article. Most of the mathematical facts were gathered by the class in general science,—high school graduates just entering the Normal School. The figures have not been verified and undoubtedly contain mistakes. It is simply written as being suggestive as to the method of teaching this extremely vital subject. All around us we hear discussions of the high cost of living. In our windows we are hanging cards to show that we are members of the United States Food Administration. What can the teachers of our public schools do in this great drive for the conservation of food? The following is a summary of the lessons taught.

Lesson 1. Organization of the Course.

At the first meeting of the class the pupils were given a mimeograph sheet telling the terms used in general science and their meaning. Most of these terms and definitions were taken, with some modification, from the General Science Bulletin of the Massachusetts Committee. A few new terms were added. When the class understood the terms they were asked to read the notes on the selection of a general unit and prepare to vote by ballot for the one that they considered most worth while. The vote was almost unanimous for *Our Food Supply*. The following is a copy of the mimeograph sheet:

A. DEFINITIONS OF TERMS.

Education is the process of living.

General Science is learning those things in the natural environment which best fits one to meet those problems in life.

A *project* is an organized undertaking to solve some problem. Projects are of two kinds according to their needs:

Individual projects are based upon the definite need or desire of a pupil. (Desk light, electric bell, leakage of gas.)

Community projects are based upon the definite need or desire of a group of individuals, (Sewerage disposal, municipal baths, prevention of infantile paralysis.)

Projects are of two kinds according to the methods of meeting these needs:

Constructive projects are those in which the student or group of students is making or assembling the parts of some mechanical device. (Making a wireless outfit or loaf of bread.)

Interpretation Projects are those in which the individual or group of individuals observes or reads to interpret some question or problem. (How the wireless works or the action of yeast.)

An experiment is an exercise performed by a pupil to obtain the answer to a problem met in a project.

A *demonstration* is an exercise performed by a pupil or teacher before a class to make clear some fact or principle met in a project.

A *general unit* is a main topic to be developed by projects, experiments, demonstrations, and discussion. (Food, water-supply, fuel, lighting.)

A *topic* is usually limited to one subject which is related to a general unit. (General unit, fuel; topics, gas, alcohol, coal, petroleum, safety matches.)

The *scientific method* is to make observations from which one may draw a conclusion.

B. SELECTION OF A GENERAL UNIT.

In selecting a general unit one should keep in mind that it must be (1) worth while, (2) interesting, (3) possible to make observations and do extensive reading, (4) a problem of this locality. The class may select the general unit which they think most interesting and most worth while at this time. Three to six weeks will be spent upon that unit and then another general unit will be selected.

Cleansing and dyeing	Street lighting
Our food supply	Home lighting
Household chemicals	Heat in the home
Baking powders and sodas	Our water supply
Metals used in the homes	Sanitation
Household electrical devices	Ventilation
Uses of electricity in the city	Photography
Building our houses	

Lesson 2. Community Projects in Food Conservation.

The needs of a community may become the needs of a nation or of the world and conversely the needs of the world become the needs of every community. Not only to make the world safe for democracy but to make democracy safe for the world or for the community the individuals who make up the group should have intelligence as to their responsibilities. As food is the deciding factor of this war the students of the country should become acquainted with the food situation. Fully understanding this the members of the class were asked to make an outline of Lesson 1, in the pamphlet entitled "TEN LESSONS ON FOOD CONSERVATION" published by the United States Food Administration. The following outline was worked out by class discussion and then they were expected to finish it outside of class.

1. *Aims of Course.*
   (1). Acquaintance with world situation.
   (2). Definite and immediate things to do.
   (3). To carry out suggestions.

2. *Causes of Universal Shortage of Food.*
   (1). *Unkindness of nature.*
      a. Late springs.   d. Poor conditions of rainfall.
      b. Droughts.   e. Unexpected frosts.
      c. Hurricanes.   f. Periods of intense heat.
   (2). Reduced productivity of soil in Europe.
      a. Bad management.   1. Withdrawal of men from
         farms. 2. Overworked women.
      b. Unskilled work.   3. Unskilled old men. 4. List-
         less prisoners.
      c. Lack of fertilizers—sunk by submarines.

3. *Conditions in Germany.*
   (1). Fats.
      a. No food is fried.
      b. Soap a luxury.
      c. Candles have disappeared.
   (2). Why Germany has power to endure.
      a. Four-fifths self supporting before the war.
      b. A nation given to overeating—reduction a benefit.
      c. Cultivating Belgium, Northern France, Roumania.
      d. Intricate food organization.

4. *Position of Allies.*
   (1). Dependent, even in peace, on importations.
   (2). Cannot get supplies from Central Europe.
   (3). Russia—disorganized railroads.
   (4). India and Australia.
      a. Shortage of tonnage.
      b. Long distance.
   (5). South America—general crop depression.
   (6). United States
      a. Greatest food-producing country.
      b. Large acreage in crops.

Lesson 3. Local Community Projects.

The study of the plan of the United States Food Administration led to the question: What is being done in Providence to meet the food situation? The class was able to give quite a list of local activities, as—

Canning Demonstrations by the Housewives League at the Arnold Biological Laboratory, Brown University.
Food Exhibit, Roger Williams Park Museum.
Home Gardens. Prizes offered by the Chamber of Commerce
Faculty Garden of Brown University. .
Cooking Demonstrations at R. L. Rose Company.
Freight Embargoes.
Etc.

The projects were listed on the board and the class told to sign their name opposite the one which they wished to investigate. No two were allowed to take the same project.

Lesson 4. *The Fundamentals of an Adequate Diet.*

A knowledge of the fundamentals of an adequate diet also becomes a community project at this time. Members of the class were thus asked to make an outline of Lesson IX in the pamphlet mentioned above. An examination was given upon these two summaries.

Lesson 5. *The Cost of Breakfast.*

This meal was chosen as it is simpler and has a smaller range of variation. This becomes an individual project. In order to standardize results for comparison the following table was presented as a basis.

Food	Average Weight	(Sept. 28, 1917) Price	Calories	Protein in Grams
Oatmeal	¼ oz.	$0.00094 (6c lb.)	28.1	1.08
Corn Flakes	½ oz.	.00625 (10c 8 oz.)	51.0	1.25
Banana	3¼ oz.	.02500 (30c doz.)	52.8	0.83
Milk	1 glass for cereal for coffee	.03000 .01500 .00300 (12c qt.)	182.5 76.0 15.2	8.25 3.43 0.68
Coffee	9 grams per cup	.02500 (40c lb.)	00.0	0.00
Sugar	heaping teaspoonful	.001716 (10c lb.)	30.0	0.00
Slice White Bread	1 oz.	.00833 (15c 1 lb. 2 oz.)	75.0	1.86
1 pat butter	¼ oz.	.00750 (48c lb.)	50.0	
1 medium potato	3 oz.	.01000 (43c pk.)	55.3	0.79
1 egg	2 oz.	.06000 (72c doz.)	93.0	8.75
Bacon, 1 slice	28 grams	.02292 (42c lb.)	75.9	2.55

The pupils were asked to tabulate what they had for breakfast for several days, including the cost, the calories (energy value), and the protein in grammes (tissue builders). From the preceeding data they figured the results for their average breakfast.

Lesson 6. *A Comparison of the Cost of Breakfast for the Different Members of the Class.*

The investigations of the class in Lesson 5 were now tabulated on the blackboard. Care was taken not to associate names with

the cost of the meal, etc., so as to obtain free discussion. A few
examples are given:

Student	Cost	Calories	Protein
A	30c	975	30 grams
B	19.9c	411	20 "
C	12.5c	450	9.81 "
D	12.0c	787	18.36 "
G	4.8c	167	18.35 "

General conclusions were derived from the table, such as:—
   (1). The price paid for food must not be measured solely in
      dollars and cents.
   (2). Thought and study is needed in planning the dietary.
   (3). We need to find what foods will supply the most energy,
      and the various materials for repairing and building the
      body at the least cost.
Interpretation projects arose, such as:—
   (1). What is the daily food requirement?
   (2). What food habits can we change?
   (3). What is the cheapest source for our food essentials?

    Lesson 7. Some of the Reports on Individual Projects.
These developed out of the class discussion in Lesson 6.
   (1). *Standard Amounts of Different Nutritive Constituents Re-*
      quired Daily (Hutchinson, Food and Dietetics).
Protein 125 grams .... 512.5 calories
Carbohydrate 500 grams .... 2050.0 calories
Fat 50 grams .... 456 0 calories

     Total   3027.5 calories.
Discussion of rations for children, normal school girls, athletes.
   (2). *Table for Estimating the Comparative Cost and Food Value*
      of Fish. Student visited the Public Market to obtain the
      prices and studied textbooks for other data. A few of the
      significant facts reported are given:

Food as Purchased	Refuse Per ct.	Protein Per ct.	Calorific Value	Market Price	Real Cost
Cod, whole, dressed	29.9	11.1	220	10c lb.	13c lb.
Cod, salt	24.9	16.9	325	13c lb.	15.5c lb.
Herring	?	11.5	825	8c apiece (¾ to ½ lb.)	
Mackerel	44.7	10.2	370	18c apiece (about 1 lb.)	27c lb.
Halibut	17.7	15.3	475	28c lb.	33c lb.
Salmon, canned	00.0	21.8	915	19c lb.	19c lb.
Salmon, fresh	40.	21.8	915	25c lb.	35c lb.

Some conclusions:
   (1). Whole fish, not dressed, are high priced due to the refuse.
   (2). The market price is not the real cost.
   (3). Herring is a cheap source of energy and protein.

A few interpretation projects that arose.
(1).   What use can be made of fish refuse for food?
(2).   May fish be used as a substitute for meat?

(3).   *Table for estimating the Comparative Cost and Food Value of Meat.*

Food as Purchased	Refuse Per ct.	Protein Per ct.	Calorific Value	Market Price	Real Cost
Chicken	41.6	58.1	295	35–40c lb.	56c lb.
Fowl	25.9	62.1	775	32–36c lb.	45c lb.
Sirloin	12.8	74.8	985	48–50c lb.	56c lb.
Frankfurts	00.0	88.9	1170	28c lb.	28c lb.
Corned Beef	21.4	65.3	1085	26–28c lb.	31–33c lb.

Conclusions:
(1).   Should consider the proportion of edible material when purchasing meat.
(2).   Some cheaper kinds are just as nutritious and often less wasteful.
(3).   A given amount of money will purchase about seven times the energy in corn beef that it will in chicken.

Interpretations:
(1).   Why is the food value of chicken so low and the cost so high compared with fowl?
(2).   How may cheaper meats be made palatable?

(4).   *Comparison of Prices in the Local Produce Market and the Retail Markets.*   Two students visited the farmers' wholesale market on Promenade Street and the retail markets.   Results were tabulated on the board, as follows:

Local Produce Vegetable	Market Price	Retail Market Price	Large quantities at the same rate
Apples	$1.50–2.75 bu.	13c qt.	$4.16 bu.
Shelled Beans	$1.75–2.25 bu.	25c ½ pk.	$2.40 bu.
Pears	$1.50 bu.	22c qt.	$7.04 bu.
Tomatoes	$0.75–1.75 bu.	10c qt.	$3.20 bu.

Conclusions; such as,—In the case of pears the same amount of money will purchase about five times as much in the wholesale market as in the retail.

(5).   *The Cost of Cereal.*   Facts tabulated as follows:

Name of Cereal	Weight per package	Price
Corn Flakes	8 ounces	10c
Force	10 "	10c
Shredded Wheat	12 "	11c
Grape Nuts	14 "	13c
Oatmeal	1 lb. in bulk	6c
Cornmeal	1 lb. in bulk	6c

Conclusions:
(1).   Ready to-eat cereal foods are more expensive.

(2). Crackers and milk give more nourishment for the same money.

(3). Cereals are one of the cheapest energy builders.

Lesson 8. New Foods and Methods.

A list of supplementary foods and methods were placed on the board, as—soy beans, cow peas, cottage cheese, home-ground wheat in the coffee grinder for a cereal, skim milk, black mussel, dogfish, puffballs, wrapping green tomatoes in paper for winter use, putting eggs in water glass, drying fruits, preserving greens with salt, cold pack method, cotton seed flour, potato flour, rye bread, etc. Some of these, as canning were demonstrated. Others might be classed as construction projects, as the making of cottage cheese which is said to have as much protein in one pound as is found in a pound and a half of fowl. The use of soy beans might be an experiment. They are richer and cheaper than other varieties but are not so palatable. Each pupil was asked to select one as an individual project and report to the class.

Lesson 9. A War Breakfast.

The individuals were now supposed to know the war situation in regard to our food supply. They had figured out the daily cost of breakfast. They had gained some knowledge as to the purchasing of foods, and as to the planning of meals that are fundamentally right. They were now asked to plan a war breakfast and compare the data with what they had tabulated for the fifth lesson.

The following was one of the best records. It was not only planned but carried out.

	Price	Calories	Protein
Ordinary Breakfast.................	20c	461.2	21.0 grams
War Breakfast.....................	10c	1046.8	5.2 grams

Another wrote: "I used bananas on my shredded wheat instead of sugar and used Challenge Condensed Milk in my coffee for the same reason. I had corn muffins instead of wheat bread. I have not used fresh bread in any meal. In this one I have made use of stale bread in a bread pudding."

Lesson 10. Organizing the School into a Working Unit.

This consists of morning talks of about five minutes before the school in the assembly hall. The first talk was by the psychology teacher about "Food Ruts." These talks were on such subjects as, *The Black Mussel.* The black mussel is seldom eaten except by our foreign population. It is more nourishing, cheaper, and more easily digested than the oyster. Volunteers were asked to pledge themselves to support a mussel menu and the shellfish was served in the school lunch room the following day. It is hoped that each object lesson is later carried out in the homes.

"The first farmer was the first man,   And all historic nobility rests on possession and use of land."

—*Ralph Waldo Emerson.*

"Plow deep while Sheperds sleep."
—*Benjamin Franklin.*

Spacious and fair is the world; yet Oh! how I thank the kind heavens
That I a garden possess, small though it be, yet mine own.
One which enticeth me homeward; why should a gardener wander?
Honor and pleasure he finds, when to his garden he looks.

—*Goethe.*

"Now is the high tide of the year....
We sit in the warm shade and feel right well
How the sap creeps up and the blossoms swell;
We may shut our eyes, but we cannot help knowing
That skies are clear and grass is growing;
The breeze comes whispering in our ear
That dandelions are blossoming near,
That maize has sprouted, that streams are flowing."

—*Lowell, (A Day in June.)*

"Whoever makes two blades of grass grow on a spot of ground where only one grew before deserves well of mankind."—*Jonathan Swift.*

CHAPTER XXXV

NATURE GARDENING

The number of school and home gardens has fallen off greatly since the World War. The present problem of gardening is to make the study interesting as well as practical. It is easy enough to find the directions for planting and hoeing beans but that is no sign that we "know beans" nor does it guarantee anything interesting about beans.

The pupils of the lower grades have been prepared for this work by observation and thought questions. If the nature lessons have been rich in content and method there will be no decided gap between this more advanced method and the less formal oral lesson. The pupils are now given opportunity to sit down and do intensive work and to express what they see by sketches. They perform simple experiments which are nothing more than questions asked of the plant. The plant gives the answer. They need this experience of individual, quiet research, still under guidance, as well as the class or mass recitation.

The following exercises are planned for older pupils. The questions are an aid to study. The answers to the questions need not be written. Whether the pupil sees the answer clearly or not will appear in the drawings. The sketches should be made by each individual and passed in at the end of the period. They should be corrected and returned by the next session. The returned work is not redrawn but it is understood that a mistake should not be repeated. The notebook is a record of progress in observing, thinking, and expression by drawing.

The aim of the questions for thought is to cultivate the scientific habit of thinking. We accept most of our knowledge as pure information. The plan is to question the student in such a way as to lead him to think out the answer with the laboratory work as a basis of thought, rather than to repeat what has been memorized from a book. He has as much right to read beans as to read about beans. Each student should be required to do this thinking that he may attain self-assurance and be an easy thinker. If he acquires the habit of approaching the plants of the garden with the scientific attitude he will find it a perennial source of pleasure.

The questions for thought may well be used as an intelligence test or one may make a game out of it. Give everyone a sheet of paper and pencil. Give them three minutes to make a list of differences between the "seed leaves" and the "second pair" of leaves in the bean seedling. Exchange papers and have the students check up the number of differences. Write the numbers one to twelve on the board and ask how many papers have one difference, two differences, etc. Return papers. Each one can then see where he stands in relation to the average.

I. *The Seed and Its Germination*

The seed is a logical beginning for the study of plants. It is also more convenient to study seeds before the leaves and the flowers appear out-of-doors. The seed, above all other parts of the plant, is of interest in the spring when it is germinating or in the fall when it is forming. Furthermore, the seed is easily experimented with and is readily obtained.

One of the rules of teaching is to begin with the known and proceed to the related unknown. The pupil should draw the adult seedling where the parts are readily seen and trace parts down through stages of growth to the corresponding parts in the seeds. This makes it necessary to have a series of seedlings of about the following ages:—three weeks, two weeks, one week, three days, and soaked twenty-four hours. The seeds should be soaked twenty-four hours before planting. These ages are approximate as the kind of seeds and variations of moisture and temperature have to be considered. Large seeds are easy to obtain and easier to study. The sprout protrudes in about three days in the pea and five days in the bean and corn.

A variety of seeds and fruits should be gathered in the fall, such as beggar-ticks, cockle-burs, milk-weed, thistle, hop, and jimson weed. The key fruits of the Maple should be collected in the spring. Pine cones and their seeds should be collected in early summer and in fall. Plant these seeds three weeks before they are needed.   Only a small per cent will germinate.

Seedlings planted in sand, sawdust, or sphagnum moss are neatest to handle and the specimens can be removed without injury. The best plan is to grow the seedlings in a window-box, dividing the garden into small plots for individual pupils. If the specimens are grown in a greenhouse instead of a window garden they should be grown in wooden boxes. Each student is then given a box for growing specimens.

The bean is better to begin with since the parts are more evident than in most seeds and it is most easily handled of those studied.

A. KINDS

1. Bean.

a. The Seedling. Note that the plant has three parts or organs; *roots*, *stem* and *leaves*. The roots serve to anchor the plant, to dissolve food material, and to take in water. The stem conducts the water and the raw elements to the leaves, which manufacture food products for the plant. The halves of the bean, or first pair of leaves, are the "seed leaves". The part of the stem below the "seed leaves" is called the "seed stem." Make careful outline drawings of the series of seedlings proceeding from the adult to the seed. Indicate all parts by printed labels. Draw a "ground line" in each case to show the relation of the surface of the soil to the plant. The general label should show what each stage illustrates; as, adult seedling, young seedling, or baby seedling of the bean.

b. The Seed. Examine the inside of the pod for the attachment of the seeds. What is the function of the "seed stalk?" Remove one of the seeds. The scar left on the seed is called the "seed scar." Look near the "seed scar" for a minute opening, the "seed doorway". Press the seed and observe the water ooze out of the "seed doorway." Sketch the bean (X2)* to show these parts. Draw side view (X1). Carefully remove the covering of the seed, beginning on the side opposite the "seed scar." What is the use of the seed coat? Call all within the covering the *embryo*. Find the "seed bud" from which the second pair of leaves develop. This bud is called the *plumule*. How many leaves in the plumule? How are they folded? Identify the seed leaf and the seed stem. Remove one seed leaf and draw (X4) showing seed leaf, plumule, and seed stem.

C. *Questions for Thought.*

1. Account for differences between a dry and soaked bean.
2. How much of the plant is formed within the seed?
3. What force causes the seed coat to burst?
4. What are the parts of the embryo? Which part of the embryo breaks through the seed coat first?
5. At what point does the embryo appear through the seed coat?
6. What part of the bean makes its way through the soil first?
7. Why is the seed stem arched when it breaks through the soil?
8. What would be the disadvantages of the "seed leaves" being pushed out?
9. Does the seed stem or the plumule develop more rapidly at first? Why? Root or plumule? Why?
10. How does the bean seed differ from the bean embryo?
11. What are the new parts in the seedling not originally in the embryo?
12. What part of the bean gives it its food value for man? For the young plant?
13. What are the changes in the seed leaves? Why?
14. What are the reasons for calling the halves of the bean leaves?
15. How do the "seed leaves" differ from the second pair of leaves?
15. How do the third pair of leaves differ from the second?
17. What becomes of the "seed leaves" in the older plants?
18. Describe the development from an embryo to a self-supporting plant.

2. Squash

a. The Seedling. Follow the same procedure of study as with the bean. Make outline drawings showing the successive stages in the

---

*The number following the directions for drawing indicates the scale to which the object is to be drawn.

life history of the seedling. Use the ground line; label. Look for the *peg* at the base of the hypocotyl. What is its use? Compare with the use of the bootjack. Draw edge view (X5) to show the use of the peg.

    b. The Seed. Make an enlarged drawing of the seed.

    c. Questions for Thought.

        1. Why do the "seed leaves" become green?

        2. What other visible changes occur in the "seed leaves?"

        3. How does the leaf which developed from the plumule differ from the seed leaves?

        4. Why is the plumule of the undeveloped squash seed very small as compared with the plumule of the bean?

        5. What part of the seedling neither grows up nor down?

3. The Pea (Pisum).

    a. Life History. Make a series of drawings to show the life history. Draw view (X1) of the seed from the seed scar end. Does the seed stem have any relation to the seed doorway? What is the advantage in this? Why does the seed bud grow more rapidly than in the bean and squash. Sketch side view (X4) with one seed leaf removed.

    b. Comparison with the Bean. Compare the following parts with the bean: (1) Outside covering of the seed. (2) Seed leaf. (3) Seed Stem. (4) Seed Bud.

4. The Corn.

    a. The Seedling. Brace roots grow from the stem of the corn. Why needed? In what direction does the *primary* root grow? What are the advantages of this? How does the direction of the *secondary* roots differ from the primary? Advantages? Find the joint (the joint is where the leaf originates and not where it appears to originate). Make outline drawings to illustrate the life history of the corn seedling.

    b. The Grain. Find where the grain was attached to the cob. Notice a light area with a central ridge on one broad face near this end. This is the embryo. Compare with a seedling and note whether the plumule is near the broad or the cob end of the grain. Identify the seed stem. Note the single cotyledon beneath the small cylindrical plumule and seed stem. Draw (X5) the grain of corn showing these parts.

The material which fills the rest of the kernel is starch. Place a drop of iodine on the starch. Note color which results. This is a test for starch since it is the only substance receiving this color from iodine. What is the use of the starch? What difference is there in amount of starch as you proceed from younger to older plants? How explain? What is the sweetest part of the corn grain? Explain. Cut a lengthwise section of the grain through the embryo. Draw the cut face (X10).

B. EXPERIMENTS. QUESTIONS ASKED OF SEEDS.

1. Amount of Water in Seeds Under Ordinary Conditions. Weigh several seeds and then dry them in an oven until they no longer lose in weight.

2. Amount of Water Absorbed by Dry Seeds. Weigh several seeds and then place them in water. Reweigh after a complete absorption of water has taken place.

3. Expansive Force in Germination. Fill a bottle with dry beans or peas. Fill spaces with water. Cork tightly.

4. Moisture and Germination. Expose seeds to different amounts of moisture. Keep other conditions the same. Tabulate results.

5. Temperature and Germination. Expose several seeds to different but nearly constant temperatures. Keep other conditions, as moisture and exposure to the air, the same. Tabulate results.

6. Light and Germination. Keep moistened seeds under dark and warm conditions. Record all growth.

7. Germination Test. An introduction to *Economic Botany*. Define. Take a box divided into compartments about one and one-half inches square and two inches deep. Put about an inch of moist sawdust in each division of the box. Number the compartments and number ears of corn to correspond with the numbers on the box. Remove six kernels from different parts of each ear, planting the kernels from ear one in division one, etc. Record results. Which ears would you select for seed corn? Why is seed testing plant selection? Sum up the value of a germinating test of seeds.

8. Test for Purity of Seeds. Illus., Grass Seed, Clover Seed, or Alfalfa Seed. Test several samples of one of the kinds mentioned above. Weigh the sample and then separate into: (1) Pure seed; (2) Weed seed; (3) Refuse. Record the results as follows:

Name of sample						
Grade of Sample	Low Grade	High Grade	Low Grade	High Grade	Low Grade	High Grade
Per cent of pure seed						
Per cent of weed seed						
Per cent of refuse						
Cost per bushel of sample						
Cost per bushel of pure seed						
Per cent of pure seed that germinated						
Actual cost per bushel of pure seed that germinated						

Why is it that the cheapest seed is sometimes the most expensive?

9. Depth to Plant Seeds. Fill a glass jar with soil, planting beans next to the glass at different depths. Keep the jar covered, when not making observations, so as to keep out the light. Plant peas, wheat, and corn in the same way. Compare results. Record in the following table the number of days it takes each to appear above ground.

Seed	Depth in inches						
	I	2	3	4	5	6	7
Beans							
Peas							
Wheat							
Corn							

10. Test for Proteids. Place a drop of nitric acid on the finger. Note color. This is a general test for proteids. Test sections of a grain of corn. What parts contain the most proteid material? Test a bean for proteid. Compare the test with the test of a grain of corn.

11. Test for Oil. Heat a few beans on a piece of paper.

12. Use of Seed Leaves. Remove the seed leaves from a few bean plants. Compare their growth with the remaining seedlings.

13. "Seed Stem Arch." Sprout some squash or bean seedlings and suspend by threads with roots in water. Compare the form of the "seed stem" with seedlings grown in earth and sawdust.

14. Direction of Growth. Invert several seedlings between a piece of moist blotting paper and a piece of glass. Show results by sketches.

15. Exhaling. Place a glass of lime water in a jar of sprouting seeds. Cover the jar tightly. Blow through a glass tube into another glass of lime water.

16. Inhaling. Place a few embryos in a glass of water. Displace the water by hydrogen gas, Seal and place in a favorable situation for growth. Make a control experiment.

C. SEED DISPERSAL

1. Wind. a. Wings (Key-fruit.) Illus., Ash, Elm, Box-elder, or Maple. Sketch to show parts.

b. Tufts of Hair. Illus., Dandelion, Milkweed, Thistle. Sketch.

c. Bladders. Illus., Hop. Sketch.

2. Water. Illus., Water-lily.

3. Animals. a. Beggar's Tick (Bidens). Make an enlarged sketch to show the shape of the terminal bristles and barbs.

b. Cockle-bur; Burdock. Make a sketch to show shape, size, and means of holding to animals.

D. CLASSIFICATION

1. Objects. a. To increase the power of searching out facts and arranging them according to some well defined plan, thus building higher and broader conceptions from concrete material.

b. To give an idea of relationship and some understanding of the principles of classification.

c. To prepare the way for appreciation of true classification of plants and animals.

2. *Tables*. To be assigned to sections of the class.

a. Divide the bean, squash, pea, and corn into two classes according to where the nourishment is stored.

b. Make a list of seeds protected against

Heat	How	Cold	How	Drought	How	Moist-ure	How	Animals	How

c. Make a list of seeds adapted to distribution by

Water	How	Wind	How	Gravitation	How	Man	How	Birds	How

d. Make a list of seeds used as

Food	Why	Medicine	Why	Poison	Why

e. Make a list of the troublesome weeds of the neighborhood.

Name	Means of Dispersal	Remedy

E. FIELD WORK

"In those vernal seasons of the year when the air is calm and pleasant, it were an injury and sullenness against nature not to go out and see her riches and partake in her rejoicing with heaven and earth."—Milton.

1. *Experiment*. Take up a little mud near the edge of a pond. Keep it under favorable conditions for plant growth. Pull up each plant as its grows. Count the number and kinds.

2. Make a seed collection to exhibit (1) Seed dispersion; (2) Seed protection; (3) Variety of edible seeds; (4) to distinguish between poor and good seed for the farm.

3. Watch for seeds germinating out-of-doors, as the buckeye, maple, and oak. Mount young seedlings to show their life history.

4. Account for the location of certain trees by reasoning how the seeds were probably planted there. Locate their parents when possible.

F. GENERAL QUESTIONS FOR THOUGHT

1. Compare the parts of a grain of corn with the parts of a hen's egg.

2. A fruit is a ripened seed vessel, with its contents and whatever parts are consolidated with it. The grain of corn is a fruit. Why has it no need for a seed coat?

3. What are the essential parts of a seed?

4. What is the fruit of the bean plant? Why does the pod twist when mature?

5. What is the seed?

6. What does the tough outer coat of the corn correspond to in the pea?

7. Give some differences between the fruit of the bean and of the corn?

8. Give resemblances in arrangement of the fruit of the corn and of the banana.

9. What is bran?

10. Does the cotyledon of the corn appear above the ground?

11. What part of the corn embryo grows up through the soil?

12. In what plants do the seed stems appear above ground first?

13. Does the corn or the bean grow more rapidly at first? Why?

14. When does a young bean plant cease to be an embryo? A seedling?

15. What do all seeds contain?

16. "The seed is a unit of plant structure." Explain.

17. What does a seed need in order to germinate?

18. What causes corn to pop?

19. Will a pound of pop corn weigh the same after popping? Explain.

20. Why does a farmer sometimes soak his seeds?

21. What is the first obvious stage of germination?

22. From where does the seedling get its nourishment?

23. Where does the root originate?

24. What would happen if the root end pointed upward in the ground?

25. What does the root do if an obstacle intervenes?

26. How does the plant increase its absorbing surface below ground? Above ground?

27. From what do seeds always develop?

28. Does the peanut develop from a flower? Explain answer.

29. Is the potato a seed? Explain.

30. Is a dry seed dead or alive? Why?

31. What are the advantages of the resting period of the dry seed?

32. Why is it better to have the seed grown in the region where it is to be planted?

33. What is the use of the seed to the plant?

34. Why does nature pack these embryos in such small bundles?

35. What advantages does a plant have in developing large seeds? Disadvantages?

36. Should a farmer allow weeds to grow up after gathering his crop? Why?

37. Is it better farming to prevent weeds or to kill weeds?

38. Name the worst weeds of the neighborhood. How are they adapted to survive?

39. What has been the influence of man on the character of weeds?

G. LITERATURE

Science does not destroy the poetry of life, but becomes a sound basis for the development of the aesthetic nature. This was the keynote of the life of Thoreau and of Ruskin. In the study of the parts of plants you should acquire that type of mind which enables you to interpret literature related to that particular part. This power should grow in richness, interest and appreciation. Throughout the course references are made to literature from which you are expected to gather a few quotations related to each part of the plant.

References: Ganntt's "Year of Miracle;" Thoreau's "The Succession of Forest Trees," from "Excursions in Field and Forest."

II. Buds

A. PREPARATORY WORK

Draw the geranium plant to show the following parts: *Blade*, *petiole* (leaf stem) and *stipules* (reduced leaves at base of the petiole), *axillary bud*, (bud in apex of the angle formed by the petiole and the main stem), stem of plant, leaf scar, and stipule scars.

B. KINDS

A series of horse-chestnut twigs should be developed in water several weeks before this exercise.

1. *Horse-chestnut Twig*, or Buckeye. What is the arrangement of the scales? Draw outline of the twig and show the terminal bud-scales. What becomes of the bud-scales? When? Find scars left after the removal of the scales of the terminal bud. Find similar scars left from last year's bud. What is the difference in the appearance of the bark above and below such a ring? Why? How often do these scars form? How many years old is the whole twig? What was the growth-length last summer? The summer before? What may have caused the difference? Compare the large horseshoe-shaped scars with the base of a leaf stalk. What inference can you make from this comparison? How many dots in each? What do the number of dots indicate? What is the cause of the dots in these scars? Were the leaves arranged oppositely or alternately? Were the leaves of the same size? How did you get this inference? If they were not of the same size account for the difference? Note the difference in color between last year's leaf scar and last year's bark.

How account for the difference? What was the position of the side bud with respect to last year's leaf? Call such a bud *axillary*. Did every leaf have a bud in the axil? Which would grow under ordinary conditions, the terminal or axillary bud? Why? When would the axillary bud be useful? Find small pores along the bark. These are breathing pores. The *flower bud scar* is in the angle produced by the forking of two twigs. Finish the sketch of the twig in detail and label the following parts: terminal buds, bud-scales, bud-scale scars, year growths, leaf scars, fibre scars, axillary buds. Write the autobiography of your twig.

2. *Tulip, Beech, Magnolia, Elm.* What is the form of the bud? How does the arrangement of the bud-scales differ from that of the buckeye? Sketch the twig and bud in detail. Find the scars left by the removal of the scales. Find stipule scars. Indicate the annual growths in the drawing. Compare this twig with that of the buckeye.

3. *Bud in Groove of Petiole.* Illus., Green Brier. Draw to show axillary bud in the groove of the petiole. What is the method of leaf casting? Why is this an advantage to the bud?

4. *Sub-Petiole Bud.* Illus., Sycamore (Plane Tree or Buttonwood). These buds are formed under the protecting leaf-stalk which in the sycamore forms a cup, very much like a candle extinguisher. Draw, showing the petiole just removed from the bud. Find stipule scars. What is the age of the oldest bud on the branch?

5. *Accessory Bud.* Illus., Red Maple, Butternut, Cherry, Box Elder, Honey Locust. Draw to show characteristics. Would the axillary or the accessory bud be nearer the leaf scar? How may accessory buds be arranged in respect to the axillary bud? How distinguish the flower bud from the leaf bud in the Butternut?

C. CLASSIFICATION

1. Name kind of buds according to

Position	Protection	Growth	Contents

2. Name buds used for

Food	Medicine	Poison

D. FIELD WORK

It is better to take a field trip on buds and stems at one time. Look for adventitious buds on the elm, the willow, and the poplar where the trunks or roots have been injured. Remove the bud-scales from several varieties of winter buds. Can the young parts withstand the exposure? Watch the growth of some bud not already studied, making successive drawings to show its stages of development. Collect twigs of various trees. Group according to resemblances and differences. Learn to recognize trees by these features. Compare the buds of the Horse-chestnut and the Buckeye, the Red Maple and the Sugar Maple, the Black Walnut and the Butternut. Note the forms of stems as in the mints and the sedges.

E. GENERAL QUESTIONS FOR THOUGHT

1. Why do trees shed their leaves?
2. Would a Maple tree shed its leaves if kept in a warm room?
3. Were ancestral trees deciduous or evergreen?
4. When are new leaves first formed?
5. How are they protected?
6. Why are leaves always borne at the end of branches?
7. Do all shoots of the plant form winter buds?
8. Why are trees early flowering?
9. What parts of the bud are only of temporary use?
10. Does an unfolding bud produce any new parts?
11. Why is it that seeds store up reserve material and buds do not?
12. What are the indications that bud-scales grow?
13. Name trees in which the terminal bud dies.
14. Why are older bud-scale scars obliterated?
15. What is the first bud of the plant called?
16. Does that bud develop continually in annuals?
17. Why do annuals have unprotected buds?
18. Why is disbudding a rational mode of directing the growth of plants?
19. Explain the use of buds in grafting.
20. Are grafted buds parasites?
21. What would be the effect on fruit growing if grafting became impossible?
22. Why does pruning somtimes cause a tree to become fruitful?
23. Why should we prune just above the bud?
24. What makes it possible for the horse-chestnut to blossom at the same time that it has leaves and the red maple before it has leaves?
25. What is the usual difference in age between a twig and its branch? Why?
26. Were the terminal buds of this spring in existence a year ago? If so, where?
27. Do terminal buds always come from terminal buds?
28. From what kind of a bud may an axillary bud develop?

F. LITERATURE

"Modern Painters," part VI, Chap. I.
Lowell's "Bigelow Papers," No. VI.
Lubbock: Buds and Stipules. D. Appleton and Co., $1.25.

III. Stems

1. *Rootstock.* Illus., Peppermint, May Apple, Solomon's Seal, Bloodroot, most any Sedge. Find leaves (reduced to scales), and axillary buds. Where are the roots distributed? Draw to illustrate parts. Why are these stems so thick for their length? Why is cutting of the rootstock by a hoe favorable rather than destructive to the plant? What is a "sucker"? How does a rootstock differ from ordinary stems? Define rootstock.

2. *Tuber.* Illus., Potato, Jerusalem Artichoke. Find a projection in each "eye" called the axillary bud. What is the tiny scale below the axillary bud homologous to? (Seen best in new potatoes.) Give reasons for inference. What is the stem? Are the buds arranged oppositely or alternately? Find where the potato was attached to the parent stem. Compare the purpose of storing food in a tuber with that of fleshy roots. Why does it not need a woody tissue? A thick bark? Examine a young potato plant. Do the roots or shoots start first? How know? Where do new tubers form? Draw a diagram of a potato plant showing last year's potato and his year's potatoes formed from it. Write a paragraph giving the complete life history of the potato plant. What is the potato ball? What are the differences between a potato seed and a "seed potato"? Why do potatoes not come true from seed? What is the advantage of the method of propagating by tuber (=asexual) over the seed method (=sexual)? Why should larger pieces produce a larger yield? When would the propagation of potatoes by seed be advisable? Why is it that the only satisfactory way to improve potatoes by selection is by "hill selection" rather than by selecting large potatoes? When do potatoes change to a green color? Why is the potato usually white? What enables potato sprouts to grow in the dark when it is said that plants cannot live without sunlight? Give the advantages and disadvantages of sprouting potatoes before planting them. Should potatoes be sprouted in a dark or in a light room before planting. Why? Take two potatoes and find the weight of each. Remove the epidermis of one. Leave both potatoes in a dry place. In a few days weigh again. The parings should be weighed with the pared potato. What is the use of the epidermis in the potato? Should potatoes be cut the day of planting or before. Explain.

3. *Corm.* Illus., Crocus, Gladiolus, Indian Turnip, Cyclamen. Draw without removing the leaves. Remove a leaf. Describe the leaf scar. Describe the character of the leaf. Remove the remaining leaves. Is there any axillary bud present? Terminal bud? Is the axillary growth a root or a branch structure? Why? Draw showing all these parts. Label by brackets what represents this year's growth, last year's and year before last. Why do corms need replanting every few years? Does the plant reproduce sexually or asexually? In how many ways may new corms be found? Explain. In what part of the corm is food stored? Define a corm.

4. *Bulb.* Illus., Hyacinth, Tulip, Onion, Garlic. Draw external view. Draw view of a vertical median section showing stem, leaves, nodes, internodes, axillary buds, terminal bud, and roots. Where is food stored in the bulb? Define a bulb. Why are the roots smaller in proportion to other parts than those of ordinary plants? Compare a bulb with a bud. Compare a bulb with a corm in size of stem, number of leaves, and place of food storage. Compare length of life of bulbs and fleshy roots.

5. *Spines.* Illus., Honey Locust. Draw and label. What reasons for thinking this thorn a modified stem? From what kind of a bud did it develop?

6. *Tendrils.* Illus., Grape. Draw. How recognize as a modified stem? Why does the grape vine not stand erect? Would you expect it to twine? Why?

7. *Elder Twig.*

(a). *One Year Old.* Cut a cross section of a stem one year old. Moisten the cut end and examine with a simple lens. Notice three distinct regions; *bark, wood, pith.* The bark is made up of three layers. Note the outside gray covering or the *epidermis.* (Replaced in a few years by *cork* in most dicotyledons. It is quite conspicuous in the oak). The greenish layer beneath the epidermis forms the *bast.* The *cambium* (LL. cambium, exchange) is a thin layer between the bast and the wood. The circle of wood represents an *annual ring.* What geometrical solid would a year's wood-growth most nearly represent? The pith is a packing tissue. Observe the fine white threads of pith extending to the cambium layer; called *medullary rays.* Draw cross and longitudinal sections of the twig (X10).

(b). *Two or More Year's Old.* Notice the pores in the woody portion. Where are they largest? (See the grape stem). What is their function? (Observe a stem which has stood in red ink a few hours.) Which parts represent fall wood? Spring wood? Notice the number of rings in the different twigs. How old is your twig? Draw a transverse section (X10). Cut off a twig of buckeye above and below a bud scale scar. Count the number of rings in each section. Inference?

8. *Oak Block.* The concentric rings represent yearly growths. Show by a diagram what part of the tree your block might come from. Which would you expect to have the most compact structure, spring or fall wood? Find the following parts in the transverse face: spring wood, fall wood, medullary rays, annual rings, and ducts. Draw and label. To find the corresponding parts in the other sections imagine yourself as a small insect traveling along certain paths. For example: travel along a medullary ray going slowly over to the radial face (=Face view of medullary ray). Find other medullary rays in the same face. Travel along another ray which leads to a tangential face (= End view). What geometrical figure is represented by a complete medullary ray? Medullary rays are called "*silver grain*" in vertical sections. Draw radial and tangential

faces. Was the growth of your block uniform? What is quartered oak?

9. *Field Work.* Note the following facts in regard to the shedding of bark: At what age of the tree does it begin? Evidences of new growth? Trees most pronounced in? Peculiarities, as in sweet gum? Cut the stem of the grape vine and notice which part "bleeds." Infer which part is most active. Cut the stem of the milkweed. Collect sticks of wood and split them longitudinally into equal parts. Have them planed at ends and on the split surface. Waste bits may be collected from a lumber yard or carpenter shop. Make a collection of twining stems. Study the healing of wounds in trees. Notice on the edge of forests whether the large limbs are next to the open fields or on the forest side. Note the size of trunks of trees in thick forests. Study stems as homes for animals. Show that the beauty of a plant does not belong to the flower alone. Note character of stems with their graceful curvatures and manner of attachment. Learn to know trees in winter from a distance by studying their form as they appear against the sky. Collect twigs and branches of various shrubs and trees before leafing stage. Place in separate jars so as to have each species a thing distinct in itself. The opening of the buds is most interesting.

10. *General Questions for Thought.*

1. Give the differences between stems and roots.
2. Does the root spring from stem or the stem from the root?
3. Is it necessary for a plant to have a stem? Explain.
4. What kind of stems do pasture plants generally have? Why?
5. What would be the disadvantage of the sunflower in the pasture?
6. Why is the dandelion so successful on the lawn?
7. How is the cactus adapted to an arid climate?
8. Why are stems modified?
9. How distinguish stems from leaves?
10. What parts of a tree have a mechanical function?
11. Why do water plants not need mechanical tissue?
12. Why do the stems of climbing plants lack in the strengthening elements?
13. How classify stems according to structure?
14. How do hard woods differ from conifers in structure?
15. Which would make better lumber for building vessels, oak or cedar? Which is better for shingles?
16. Why do we use brush for peas and poles for beans?
17. What are the oldest parts of a tree?
18. What are the living parts of a tree?
19. How much of the tree is living that was alive ten years ago? What parts of the tree are eternal?
20. Where does the stem increase in length? In diameter?
21. Why are the trunks of palms nearly equal in diameter and conifers tapering?

22. What height will a nail be ten years after it has been driven into the trunk of a tree? Into last year's twig? Into this year's growth?

23. Why is it easy to remove the bark in spring in making whistles?

24. What parts of the tree carry the sap upward? The food downward. The food to the interior?

25. Why is a tree able to develop its leaves after girdling?

26. Why would "ringing a tree" hasten the production of fruit? Is such a practice to be recommended?

27. Why do the parts above the girdle increase in thickness?

28. Why will girdling a tree kill it?

29. What would be the effect of girdling Indian corn?

30. Where may food be stored in the stem?

31. What is meant by "bark"?

32. What causes the bark to crack?

33. Why is the bark not as thick as the wood?

34. What are the advantages of scraping the trunks of trees?

35. What is sapwood? Heartwood? What are the differences?

36. What part of the tree makes the best timber?

37. In what part of the tree would you seek proper juices for medicine?

38. Which is heavier, a green or a dry cornstalk? Why?

39. What is the difference between timber and lumber?

40. How does a knot become imbedded in the wood?

41. Do knots injure lumber to any extent? How prevent the formation of knots?

42. Which grow more rapidly, annuals or perennials?

43. What are the advantages of rapid growth of plants? Disadvantages?

44. Why do annuals not produce bark?

45. What is the cause of annual rings?

46. Will annual rings be more distinct in the temperate climate or in the tropics?

47. Why are there no annual rings in the oldest geological trees?

48. Is the number of annual rings always a correct indicator of the age of a tree?

49. On which side of the tree are the thicker rings formed, the leafy or the less leafy side? North or south? Hillside or valleyside?

50. Where would you expect the annual rings to be thicker, near the heart or near the bark in a large tree?

51. Where should we count the rings of a tree, from a section near the tip or near the base?

52. Which makes better timber, a tree growing in the open or in the forest?

53. Why is Norway timber so excellent in quality?

54. What layers should come together in grafting?

55. What is a scion? A stock plant?

56. Why can a tree be vigorous and healthy with the center of the trunk gone?

57. Need a fruit tree be discarded when it has a hollow trunk? If not, how treat it?

58. What causes the lower limbs of pine trees to die?

59. What is pruning? Does nature prune?

60. What are the objects of pruning?

61. Should pruning be done in wet weather?

62. What is the best time to prune?

63. Discuss the old saying, "Prune when your knife is sharp."

64. What is the best way to remove large branches?

65. Upon what does the healing of wounds depend?

66. Discuss the protection of wounds by artificial means.

67. How does "heading in" induce fruitfulness?

68. Why head commercial fruit orchards low?

69. Why are plants with undreground stems among the earliest to bloom?

70. Many plants with underground stems are woodland species. Why must they bloom early?

71. Why do farmers bank up celery to bleach it?

I should regard the most valuable of all arts to be the deriving of a comfortable existence from the smallest area of soil."

—*Abraham Lincoln.*

That wonderful gift which some gardeners seem to have for growing anything is no magic; it comes from the love of plants. . . .

And that other gift for making a garden beautiful is no magic either; it comes of loving the garden as well as the plants.　　　　—*Tropical Agriculturist.*

CHAPTER XXXVI

An Outline for a Study of the Potato

(The so-called Irish potato, *Solanum tuberosum*, and not the Sweet Potato which belongs to the Morning-glory family.)

I. *Historical Facts*

For account of origin of the potato, read De Candolle's "Origin of Cultivated Plants," pp. 45–53.

Probably native of Chile, South America. Cultivated by Indians at time of discovery of America. A similar vegetable is eaten by natives of South Africa. Belongs to nightshade family, therefore once thought to be poisonous. First introduced into Europe 1580–1585 by Spaniards. Sir Walter Raleigh introduced "Virginian potatoes" into County Cork, Ireland, about 1584. "Potatoes are of less note than horse-radish,—beets."—Bradley, 1719. "Plant your potatoes in your worst ground."—John Evelyn. Parmentier (1737–1813) a noted chemist, popularized it in France. Potato soup is known as "potage Parmentier" in his honor. Probably introduced into the United States toward end of 16th century (Virginia and North Carolina).

Frederick the Great succeeded in introducing it into Prussia.

Louis XVI and Queen Elizabeth wore potato blossoms to help popularize it. Grown in European flower gardens in 17th century. Called Irish potato because used so generally in Ireland. Potato blight caused famine in Ireland in 1846.

II. *Nature Lesson*

A. *Subject Matter for Teacher.* An ordinary stem, as a geranium, has buds and leaves. A potato is also a stem. The projections in each "eye" are buds and near the center of the "eye-brow" may be found a tiny scale which is an undeveloped leaf. This rudimentary leaf is seen best in new potatoes. One could imagine a branch of a geranium converted into a potato by reducing the leaves and storing a great deal of starch in the stem. If this modified geranium was then placed underground it would give up its green color due to the absence of sunlight. The stem of the potato is large on account of its vast amount of stored food and the leaves have degenerated because the stem is underground.

In early spring, the "seed potatoes" are cut into pieces, each piece having an eye. These are planted in furrows and covered with soil. The bud soon sends out a sprout and from the base of this stem appear small roots. The sprout grows above ground and sends out ash-shaped leaves. The soil is usually drawn around the young plants with a hoe and this forms the "hill". This practice is not good as it drains moisture away from the potato. The leaves take carbon dioxide from the air; water and mineral sub-

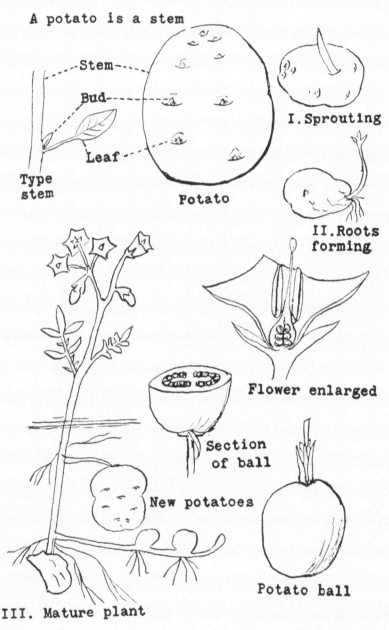

A potato is a stem

Stem

Bud

Leaf

Type
stem

Potato

I. Sprouting

II. Roots
forming

Flower enlarged

Section
of ball

New potatoes

III. Mature plant

Potato ball

Fig. 1.

stances from the soil by means of the roots; and, in the presence of sunlight, warmth and leaf-green, manufacture a liquid nourishment. In the meantime, underground branches have started from the base of the sprout which look very much like the roots but are slightly larger. These underground branches receive the sweetish food, which is manufactured by the leaves, and begin to enlarge at their ends into potatoes. The food is stored in the potato in the form of starch.

The blossoms do not usually produce fruit. This is because it has not depended upon seed but has been reproduced by man in the way just described. In colonial days, the "seed-balls" or "potato-balls" were quite abundant. After the blossoms fall off, the plant above the ground withers and dies. The potatoes are now said to be "ripe" and are dug with a hoe or potato digger. On large, level farms the potatoes are dug by machines. When the potatoes are harvested they are stored in bins in the cellar or in a dark room where it is warm enough not to freeze the buds, and cold and dark enough to keep the sprouts from starting. In the spring, sprouts start from the eye of some of the potatoes. The sprouts grow very rapidly, taking the nourishment out of the potato until it becomes wrinkled and unfit to eat.

B. *Method of Procedure.* Aim: Appreciation of common things and how plants reproduce. Preparation for intelligent and appreciative interest in potato growing in the spring.

Have pupils plant potatoes in a box containing sandy soil. Keep warm and moist. Watch how they grow. Take one up at different intervals to show changes. Draw different stages. Describe the changes. Keep a diary. Review parts of a typical stem using geranium as example. Have someone draw geranium on board to show stem, leaf, and buds.

Lead pupils to find same parts in a potato. Make drawing of potato beside sketch of geranium. Connect same parts with dotted lines. Lead class to infer that the potato is a stem. Have pupils reason out: Why leaf is reduced? Show class a potato which has been exposed to sunlight and one that has been in the dark. How account for difference in color? Conclusion: The potato is an enlarged underground stem.

Apply a weak solution of iodine to the cut surface of a potato. Observe results. Iodine colors starch blue. Infer that starch has been stored in the potato or underground stem.

About a month before this lesson remove the skin from a potato and weigh. Leave in a warm place. Pupils now observe changes in size and weight. Lead class to infer that its loss in size and weight is due to evaporation of water. The composition of potato is 78.3 per cent water, 18 per cent starch, 2.2 per cent protein, 1.0 per cent ash, and 0.1 per cent fat. Show by diagram about how much of a potato is water and how much starch.

Show class a potato which has sprouts upon it. How does it differ from a potato without sprouts? (Wrinkled). Why? New growth eating the starch which was stored in it. Infer that starch is stored in the potato as nourishment for new plants. Note that sprouts are long white stems with small white leaves. Why white? Why long stems? Why small leaves? Leave specimen exposed to light and observe changes. Why are potatoes with sprouts not desirable for the table?

Pupil dig up a piece of potato which has been planted. From what does the sprout start? From what do the roots come? Do the shoots or roots start first? In what direction do the roots go? Why? Why should the earth be kept loose about the growing plants?

Teach class why potato has large leaves above ground. Show flowers. Show "seed-balls". These should be collected the year before and preserved in formalin in a small bottle. Explain law of disuse. When the potatoes are ripe the parts above ground wither and die. Why? (They have served their function.)

If possible show a potato plant with the new potatoes. Have class review by giving life history of this plant. A diagram may be substituted if the plant is not available.

III. *Potato Gardening*

A. *Organizing Club.* Send to the U. S. Department of Agriculture and to the State Agricultural College for all literature related to the Potato Club Work.

Farmers' Bulletins 35, 91, 295, 324, 386, 407 and 410 and Bureau of Plant Industry Circular No. 113 are especially valuable.

Let pupils take home cards showing requirements for potato club work. Obtain permission of parents for children to join the club and so far as possible for land. The "landless" children should be given opportunities in the school garden, in window-box gardening, and in experimental work.

B. *Steps in Gardening.* Select an open unshaded area. Rich sandy loam is best. Drive a stake in four corners of proposed garden and connect with string. Spade up garden for depth of eight inches. Rake until surface is fine and level. Fertilize with street sweepings or buy fertilizer from a seed store. Mix well with soil. Select a well known variety that grows best in the neighborhood. Throw out any diseased potatoes. Mix one-half pint of formalin to 15 gallons of water. Put potatoes in solution for two hours and then dry them. This prevents scab disease. Cut potatoes lengthwise into quarters, each piece having two or three eyes. Plant four inches deep and about a foot apart in rows. The rows should be a foot apart unless it is possible to cultivate with a horse and harrow. Hoe once a week to keep top soil fine and to prevent weeds. Keep this up until the plants shade the whole ground. Spray with Paris green as soon as the potato-beetle is seen. Add

Bordeaux Mixture when plants are eight inches high to prevent disease. Mix one-fourth pound Paris green, one-half pound lime to 25 gallons of water. Dig potatoes as soon as tops are dead. Keep potatoes from each plant separate. Do not allow them to be exposed to the light. The best hills have a large number of potatoes which are uniform in size. Select seed potatoes from these hills for the next season. Separate the remaining potatoes into marketable potatoes and culls. The culls include all small and injured potatoes. Sell or store in a dark place.

C. *Suggestive Records for Booklet.*

Name.........................Street and number....................
Date..........................Grade...........................
Size of garden...................feet by.................feet.
Description of soil:
Calendar: Prepared soil.................; planted................; sprouts appeared above ground.................; first flower................; matured (tops died)..............; harvested................

ACCOUNTS

EXPENSE			DATE	INCOME	
Cost of seed.........	....	....	......	Amount sold or used	Market price per bus.
Cost of fertilizer......	....	....	......		
Preparation of garden, hours at 10 cents....	....	....	......	....	....
Planting potatoes, hrs. at 10 cents.........	....	....	......	....	....
Cultivation........at 10 cents...........	....	....	......	....	....
Date Hours					
Spraying mixtures....	....	....	......	....	....
Spraying at 10 cents...	....	....	......	....	....
Date Kinds of Spray Hours					
Harvesting, hrs. at 10c	....	....	......	Amount on hand	....
Rent of land.........	....	....	......	....	....
Total cost........			....	Value of crop......	

D. *Other Garden Facts.* In 1889 there were several hundred varieties. New varieties are obtained from seed. Some varieties never produce fruit; others do not even blossom.

Two crops are grown in one year on the same land in the South.

1900–1910. The average acre yield in United States was less than 93 bushels.

1900–1910. Average acre yield in Germany and Great Britain 200 bushels.

E. H. Grubb, "Potato King" of Colorado, produces 600 bushels per acre. Earl of Roseberry, Potato wizard of Scotland, 2000 bushels per acre. Germany yields one-fourth of the world's crop. We consume 3½ bushels per person a year in the United States.

We import about one-fourth of potatoes used in United States. There is no garden crop for which spraying is so necessary as for the potato crop.

The best potatoes are about 12 ounces, somewhat cylindrical shallow-eyed, and white in color.

E. *Pictures.* Collect pictures as well as specimens to show varieties, diseases, different characteristics, machinery for cultivation, and methods of caring for the crop. These may be arranged in the booklets. Obtain Millet's "The Angelus", "The Potato Gatherers", and "Going to Work".

IV. Uses of the Potato

A. *In the Home.* The potato stands next to wheat as an important food crop in the western countries.

Potato Starch. The making of starch is both valuable and interesting. Grate potatoes into a dish of water. The starch granules settle on the bottom of the dish. Keep rinsing with water until the starch is white and clean. Send for Recipes for the use of the Potatoes and Home-made Potato Starch, Form O-7, to the States Relations Service, U. S. Department of Agriculture.

Boiled Potatoes. Most economical to boil in jackets. If pared, place in cold water to prevent coloring. If of uniform size they are done at the same time. Place in hot water instead of cold—will not be soggy. Salt added to water makes the potatoes more tasty. Drain off water.

Baked Potatoes. More wholesome than boiled potatoes. Place in a hot oven until soft. Break skin to let out the steam.

B. *Commercial Uses.* Denatured alcohol; sizing for paper; thickening colors in calico printing; glucose.

V. The Potato-Beetle

Aim: To know the life history and means of protection of the potato-beetle in order to exterminate it. Teach when needed.

Place the larvae of the potato-beetle on a potted potato plant which may be kept in a breeding cage.

In what way are the larvae injurious to the plant? How might we attack the insect in this stage? Handle the larvae and note the odor left on the hands. Would birds enjoy eating the larvae? Why? (A distasteful secretion protects them from birds). Of what advantage are the bright colors of the larvae to itself? (A warning to birds.) The larvae change into the pupal stage beneath the ground. How would fall plowing help destroy this pest? In what way would rotation of crops help this problem?

Study the adult. Feed the insect on raw potato. Note whether the adult has sucking or biting mouth parts. (Biting mouth parts). Does the adult or the young larva have the greatest appetite? Compare the size of the legs with those of the ground beetle? (Smaller.) What does this indicate? (Depends more on wings for locomotion). In what way does this suggest the need of co-operation amongst farmers of a neighborhood? (If each one does not kill these pests they will soon infect clean areas). Describe the feet. (Claws and pads). Use? (Crawling and clinging). How many pairs of wings does this insect have? (Two). Note which wings are used in flying? (Inner). In what way are they adapted for flying? (Light and gauzy). Describe the outer wings. (Hard and shell-like). What do you infer is their use? (Protection to delicate under-wings). Pick up a beetle and note how it "plays possum." When one brushes against a potato plant the adult insects drop to the ground and "play possum". What is the advantage of this habit? Hold the insect rather firmly. What is another method of protection? (Acrid secretion).

The orange colored eggs are laid on the underside of the leaves. Advantage?

The potato-beetle was a native of Colorado where it lived on a wild plant of the nightshade family.

VI. Conservation of Moisture

This lesson should precede the cultivation of the potato crop. Experiment 1. Devise a rack to hold four lamp chimneys in a vertical position. Tie a piece of cloth over the bottom of each and fill within three inches of the top with gravel, sand, clay, and loam, respectively. Pour the same amount of water into each chimney and note the time it takes each soil to absorb the water. Which allows the water to pass through most quickly? Which soil should we add to a garden that loses its moisture too rapidly? What soil should we mix with garden soil that does not absorb water readily? The force which causes the moisture to go downward is called gravity. In heavy soils plant potatoes about three inches deep and in light loose soil four to five inches deep. Why?

Experiment 2. Pack the soils in the lamp chimneys rather firmly and add a loose dry layer of the same kind of soil on top. Let the lower ends of the chimneys in a pan of water. Note the rate at which the water passes upward in each soil. Compare the rate at which the moisture passes upward in the packed soil and in the loose surface soil. This upward movement of the water is due to capillary attraction. Which would you expect to be more important for the plant, gravitational water or capillary water? This experiment shows how tillage will conserve capillary water. Explain. The surface layer is called a mulch and this practice is known as dry farming in the west. If it were not for saving of the moisture they could not farm in dry regions.

Experiment 3. Place soil in a tray. Pack the surface firmly on one side and leave a loose surface on the other. Incline the tray slightly and pour water on gently from a watering pot. In which case does the water soak in and in which case does it run off the soil? Lead class to infer that a second reason for keeping a mulch is to absorb the rainfall.

The sun bakes the soil after a rain forming a hard surface. What effect would this have on evaporation of the water? (Hastens). What would be the relation of such a surface to the next rainfall? (Water would run off and not soak in). Give two reasons for hoeing the garden after a rain. Tillage to conserve soil moisture is really more important than tillage to kill weeds.

Experiment 4. Place the same weight of wet sand in two trays of the same size. Arrange soil in one so that it is made up of parallel ridges and keep the contents of the other tray flat. These two trays illustrate two methods of cultivating potatoes, one is called the ridged system or "hilling-up" potatoes and the other is known as the flat method. Weigh occasionally to see which is losing the moisture more rapidly. (The flat method exposes less surface soil and therefore conserves the moisture.) Should we "hill-up" potatoes or use the "flat method" in a dry region? In a region which has a great deal of rainfall?

"Out of its little hill faithfully rise the potatoes' dark green leaves,
Out of its little hill rises the yellow maize stalk."
—*Walt Whitman.*

. . . .If vain our toil,
We ought to blame the culture, not the soil.
—*Alexander Pope.*

CHAPTER XXXVII

Potato Geography[1]

A knowledge of the origin of our cultivated plants, experience in the use of maps to show their paths of distribution, an adequate conception of how they accumulate their pests, and a comprehension of plant geography in general makes possible a rational policy of plant control. When a plant is subjected to the influences of domestication it becomes physiologically weakened. When it is introduced into a new region modifications are brought about by the change in the climatic or other physical features of the environment. Any disturbance in the structure or habits of a plant may lead to the inroads of disease-producing organisms. A brief historical survey of potato geography may serve to portray important facts bearing upon the relations of plant geography and progressive agriculture.

Our common potato (Solanum tuberosum) grows wild in western America from Chile as far north as Mexico. There has been considerable confusion as to its origin due to related species being found in Peru and Mexico but the ordinary white potato was unknown in Mexico during the period of western exploration.[2] De Candolle[3] gives a detailed account of its origin. It is probable that the potato crossed the Atlantic three times before it reached England in about 1586. It is mentioned as having been taken by the Spanish explorers from Quito, Ecquador in about 1580. They must have carried it up the western coast and then across the Isthmus by mule train. From Spain the tuber was carried to the Georgia seaboard prior to 1585 and thence to Ireland by Sir Walter's Raleigh's expedition (Raleigh did not go to Virginia himself). It is interesting to note that the potato did not reach China until 1875. Our Department of Agriculture is now experimenting with 248 races of potato which have been imported from South America.

It is told that when the potato was first introduced into France it would not produce tubers.[4] As in the case of the clover which would not thrive in Australia until the bumblebee was introduced to aid in cross-pollination, some concluded that the potato is a gall produced by a fungus and only when the potato-producing organism gets into the soil is the gall or potato produced. Tuber formation, however, has been induced in sterile soil by a concentrated solution of sucrose and there is, therefore, a possibility that the fungi may raise the concentration of the soil media. In this way the organism is an indirect cause of tuberization.

The ancestral potato blossomed, produced seed, and died. Con-

---

[1]Seminar paper, Botany Dept., Brown University, May 24, 1919.
[2]Von Humboldt, Nouvelle Espagne, edit. 2, vol. ii p. 451; Essai sur la Geographie des Plantes, p. 29.
[3]De Candolle, Origin of Cultivated Plants, pp. 45–53.
[4]Robinson, 1917, The Botany of Crops.

sequently there were a few small potatoes. Under domestication the vine quickly lost the power of growing seed-balls, many plants not even blossoming. The domestic potato plant, then, may be said to be one that blossoms, produces potatoes, and dies. Artificial selection has resulted in the production of potatoes instead of seed. The period just following the blossoming is rather precarious as there is an inherited tendency to die at that time. If the plant can be carried over this critical period it will grow for some time. If the plant enters a decline it is easily affected by heat, drought, insects and disease. A Bordeaux spray is of great benefit at this time. The thin veil of copper shields the leaves from the glare of the sun, reduces transpiration, checks disease, and stimulates growth.[5] The domestication of the potato has brought about a complete series of changes.

The cultivated potato is rather delicately adjusted to its surroundings. Water supply and temperature are the most important environmental factors. Europe is able to secure a higher yield than the United States due to having a cooler climate and a greater rainfall. The damp climate of Maine is one of the reasons for its high average yield per acre (206 bushels in 1914) as compared with the average yield of the whole country (96 bushels). The potato itself is about 75% water and when we buy a bushel of potatoes we get about three pecks of water to one peck of starch. Its demands for soil moisture therefore are very great.

The soil is another important factor and is closely related to the water supply. Sand absorbs too much heat and the water supply is usually some distance below the surface. Clay, on the other hand, holds too much moisture and causes the tubers to decay. Then again, the root system of the potato is comparatively weak and cannot penetrate a heavy soil like clay. A rich mellow loam is best suited to potato culture.

Although the potato is very sensitive to its surroundings it is one of our most widely distributed crops. The largest centers of production on the Atlantic coast are Aroostock County, Maine, and Norfolk, Virginia. The San Joaquin and Sacramento Valleys are the best known in the Pacific district. The planting of early potatoes starts about December 21 in Florida and migrates northward with the season. The progression of planting by months gives an idea of the geographic movement of the crops: January 1, northern Florida; February 1, Georgia; March 1, Virginia; April 1, Long Island; May 1, Northern Maine. As the crop matures toward the north the distance of shipment to large centers of population lessens and the price declines. The distribution of seed potatoes is in the opposite direction, most of them coming from Maine, New York and Minnesota, showing that the higher temperature of the south injures their vitality.

A summary regarding the potato in other countries yields some

---

[5]Stuart, Vermont Agric. Experiment Station, Bulletin 179.

interesting facts.[6] The crop is grown as far north as Dawson City in the Klondike region. Austria, before the war, grew 30-70% more than the United States. Belgium production was 60 times as great per square mile as the United States. Germany was the greatest potato producing country in the world with an average yield of 200 bushels per acre for 10 years, more than double that of the United States. Professor S. N. Patten states that this enabled Germany to overthrow France in the Franco-Prussian War. Sweden with 5½ million people grows as many as Italy with 35 million people. Switzerland suffers at times from an over-production of potatoes

The political boundaries of countries offer no barrier to the insect pests and swarms of fungus spores which are seeking their hosts. The spores of the potato blight may be swept by the wind from village to village or they may live on seed potatoes and thus be distributed from country to country. We must realize that the problem is of geographic magnitude and that its solution means world-wide intelligence.

The importance of this knowledge was first thrust upon the world during the great Irish Famine in 1845-1847. Ireland with its increased population had come to depend upon the potato crop more than any other country. The harvest of 1845 promised to be a rich one when lo there came out of the west a blight which changed the luxuriant crop to a blackened waste; 600,000 people died from famine. What was the source of this great scourge?

During the first 250 years of potato culture the blight was unknown. In 1887 Jensen[7] concluded from experimentation that the blight fungus cannot exist where the mean temperature exceeds 25 degrees centigrade. From these two facts Jensen reasoned that so long as potatoes were carried by slow-going sailing vessels the excessive heat of the tropics had disinfected the tubers. In 1830-1840 the sailing vessel gave way to the steamship. This quicker means of transportation enabled the organism to exist while crossing the tropics. This conclusion was substantiated by Reed[8] in 1912 who claims that the disease is practically unknown in Virginia below an altitude of 2000 feet and that it appears earlier at high altitudes. The disease scarcely ever occurs south of Latitude 40 degrees.

The quick spread of the disease is shown by the following itinery: France, 1840, Von Martius; Norway, 1841, Westrem; Belgium, 1842; Denmark, 1842, M. Fjeldstrup; Boston, 1842, mentioned in a letter by B. Watson to Jensen; New York City, 1843. Here is a disease then originating with its host plant. The organism could have been held within bounds by as simple a check as low temperature. Had the solution of the problem been provided for there would have been a great saving of lives, suffering, and expense.

[6]J. Russell Smith, Industrial and Commercial Geography.
[7]Jensen, 1887, Mein. Soc. Nat. Agr. France, 131, 31-156.
[8]Reed, 1912, Phytopathology, 2: 250-252.

The powdery scab (Spongospora subterranea) is also native to the Andes. It was introduced from Europe to Canada and thence to the United States being discovered in Maine by Morse and Melhus[9] in 1913. A humid climate and damp soil favor its distribution. This disease was first studied and described by Thaxter of Harvard. Would it not have been a good investment to have met the enemy at home?

The story of powdery scab is repeated by silver scurf (Spondylocladium atrovirens). This was known in Europe in 1871 being first studied and described by Hertz of Austria. It was next reported by Frank, 1897–1898, in Germany; Johnson, 1903, in Ireland; Smith and Rea, 1904, in England; and recently in Connecticut by Clinton. This disease is practically limited to the northern boundary of the United States, although it has appeared in one small area in northern Florida. Infected potatoes have been planted in the coastal states but the disease did not appear in the field.[10] Soil was then transported to these northern areas from 10 states and the disease was produced in 8 cases, showing that a favorable climate is necessary for the existence of the disease.

A more recent pest, traveling the same path, is the potato wart disease (Synchitrium endobioticum). It was known in Hungary in 1896 being first reported by Schibbersky, and reached New Foundland in 1909 coming via Germany and England on seed potatoes. It crossed the line into Aroostook County and was found in Pennsylvania in 1918. This is one of our most serious potato diseases as no spray is effective and it persists five or six years in the soil even though no potatoes are grown in that area. It is hoped to keep the disease in this state by a quarantine act. It would have been a great deal simpler to have prevented it crossing the ocean.

Black-leg (Bacillus phytophthorus)[11] was introduced from Europe to Connecticut in 1904. The disease is spread by tubers and soon radiated over the country being observed in New Hampshire in 1906, in Maine in 1907, and in Virginia in 1909. It has now reached the potato growing centers of Colorado and Oregon.

The common scab (Actinomyces chromogenus) may serve as an example of a fungus disease which awaits the coming of a host. The scab organism existed in the soil before potatoes were introduced. The disease is common to turnips and beets but it also has found the potato to be a congenial host.

A survey of the origin and distribution of the fungus diseases of the potato lead one to infer that for the most part they travel along certain geographical routes. They come uninvited, propagate rapidly, and cause tremendous destruction. A knowledge of their climatic limitations would have served to keep out most of them. Instead we have maintained the open door policy and have begun

---

[9]Morse and Melhus, 1913. Sci. 38: 61–62: 133.
[10]Weekly News Letter, Department of Agriculture, Jan. 26, 1916.
[11]Maine Agric. Exp. Station, Bulletin 174, Dec. 1909.

the fight too late. It is too late to make amends for the past but our responsibility for the future has become manifold.

The story of the potato beetle serves as a good summary of the way new pests behave[12]. Until 1856 the insect fed upon an allied plant (Solanum rostratum) in Colorado. Chittenden thinks that it must have originated as a species in Colorado but Tower[13] assumes that it must be of tropical origin as is true of its principal food plant Solanum rostratum. Whatever its origin it is interesting to note that when it once got a taste of Solanum tuberosum the beetle abandoned the wild food plant which its ancestors had fed upon and feasted upon our economic species. Originating as a potato pest in 1856 the beetle marched eastward at a terrific pace, covering 1500 miles in 16 years. By 1874 it appeared on the Atlantic coast and in 1877 reached England and Germany. This was one of the first instances of a plant pest of this country moving eastward. Before this the weeds, diseases, and injurious insects of the world had been spreading westward with the progress of civilization. The isomigs (Lines of equal migration) show that the march of these hordes lagged on the north and south due to the extreme cold and heat, and today they only appear on the northern and southern boundaries when the temperature has been unusually moderate. The quick invasion of the potato beetle was aided by the prevailing westerlies.

A knowledge of the potato beetle and his climate limitations should be of value in keeping him from crossing the Rocky Mountains. Irrigation has extended the geographic range of the potato but the beetle has not found his way across the cold mountain barrier. A little thought and carefulness now will save these new potato areas from the heavy tax demanded by the potato beetle.

These undesirable citizens cannot be deported. The mistake would be in thinking that there are no more. The Bureau of Entomology of the United States of Agriculture has listed 3000 injurious insects which are likely to be introduced. The Federal Horticultural Board is backing up the Federal Embargo Act which prohibits the importation of plants without the sanction of the United States Department of Agriculture. This will guard against the dangers of new plant diseases as well as insect pests. Another safeguard is in the recently organized American Plant Pest Committee which is made up of state foresters, entomologists, agriculturists, and pathologists. It is their duty to give publicity to the campagin and to act quickly whenever any new pest is discovered. The time seems fitting to take up at once an efficient means of curbing the activities of these enemy aliens.

---

[12]Chittenden, Circular 87, U. S. Department of Agriculture.
[13]Tower. An Investigation of Evolution in Chrysomelid Beetles.

CYBELE

Spirit of th' raw and gravèd earth
Whenceforth all things have breed and birth,
From palaces and cities great
From pomp and pageantry and state
    Back I come with empty hands
    Back unto your naked lands.

                        —*L. H. Bailey.*

CHAPTER XXXVIII

The Tomato

These facts are simply told to the class as interesting knowledge in connection with their growing tomatoes. They are not to be learned.

I. Historical Facts.

First found in Peru. Belongs to same family as potato and egg plant. First used as food in 1830. Food value was discovered by accident. First grown as curiosity and was thought to be poisonous.

Copy these directions by use of hectograph. Each pupil should be given a copy as a guide, not to learn.

II. Steps in Cultivation.

Fill a flat box with good, rich soil. Plant seeds second week in March, one-half inch deep. As it originated in the tropics it needs a long growing season. Place on a sunny window sill. Young plants appear in about one week. Raise window on warm days. Water early in forenoon. When there are about four leaves transplant 2" apart. When 4–6 inches high and crowded transplant s urdy ones to paper-pots, or strawberry boxes or tin rims tied with strings. Water well and replace on sill. Turn pots every day so plants will not lean toward light. If grow spindling, pinch off tops. Transplant to garden about June 1 or when danger of frost is over. If in paper pot tear off the bottom to let roots grow. Dig a hole about six inches deep. Put in little stable manure and cover with 1 inch of dirt. Cover plant to second pair of leaves. Press soil firmly about pot. Leave sides of pot to keep away the cutworms. Make holes 3' apart in rows; rows 4' apart. Have rows run north and south to get as much sunlight as possible. Support vines with trellis or stake so berries will not rot, be eaten by slugs or wire worms, or splashed with mud. Water copiously early in forenoon. Shade with newspapers if sunny.

Keep soil mellow and free from weeds. In about a month (July 1), cut off all but three strongest branches. Tie branches to supports with raffia or strips of cloth. Blossom about middle of July. Insects fertilize blossoms.

Cut off all tops and side branches (=suckers), so nourishment will go to the fruit. Cuttings may be made the last of August.

III. Harvesting and Marketing.

Maryland is the largest tomato producing state. The tomato is a perennial in Texas and an annual in Rhode Island. If the fruit is to be shipped some distance gather when partially colored and wrap in paper. Ripen fruit on plant for home market. Cut fruit from plant as they ripen leaving a portion of the stem on the fruit. The best fruit for market or exhibition is smooth, uniformly ripened, free from disease, uniform in size and attractive in color.

IV. Canning.

Send for Farmers' Bulletin 521 on Canning Tomatoes at Home and in Club Work. This forms a good basis for a lesson in canning. No other fruit or vegetable is used so extensively. The census for 1909 shows that 12,800,000 cases of 12 cans each were packed in the United States. It is a basis for soups, sauces for fish, salad and ketchup.

V. Suggestions for Accessory Lessons.

These lessons are given if there is an occassion which demands them, otherwise they are omitted. For example,—if the vines are attacked by the tomato worm that is the time to study it or if there is danger of frosts teach how to cover the plants with paper and the principles involved.

1. The Tomato Worm (Also known as Humming-bird moth, Hawk-moth and Five Spotted Sphinx).

The first lesson should be from the point of view of interest in the life of the insect. The second phase is to protect the plant. Knowledge of the life history of the insect and its adaptations are fundamental to its extermination as a pest. When the tomato worm is first noticed have several specimens brought to school. Rear in a cage by feeding tomato leaves. Give fresh food every day. The larvae will burrow into moist earth and transform. After the class has had opportunity to watch the development of the insect have a lesson upon its life habits. The four stages should be at hand for the lesson. The class is led to make observations, by questions and then to think out the probable reason, or advantage, of each characteristic. The following list of adaptations is for the benefit of the teacher.

LARVAE

Green color and markings...............imitates a leaf.
Tail-like projection near hind end ....terrifying enemies.
Rears threateningly ....intimidates enemies.
Nine pair breathing pores...............to purify blood.
Small claws on true legs................holding on to twigs.
Jaws....................................eating vegetable food.
Chrysalis. The larvae forms a chrysalis, (*not cocoon*) in the ground.
Slender handle of chrysalis.............case for tongue.

MOTH

Tongue four inches long—sipping nectar from deep tubed flowers.
Tongue coiled like a watch-spring—economical way to carry.
Small, undeveloped legs—merely used for support while probing for nectar.
Large, strong wings—for flight and poising when extracting nectar.

EGG

Have pupils hunt for eggs and bring to class. In the construction class or elementary handwork they could make a box with glass top, fill with cotton-batten and mount to show the four stages of the insect.

A small ichneumon fly is parasitic upon the tomato worm. The ichneumon deposits eggs beneath the skin of its victim, within which the larvae feed for a few days. The larvae then emerge and spin their cocoons. These cocoons are often seen upon the tomato worm in late summer. Since the ichneumon fly renders a great service to gardeners, by keeping in check the tomato worm, it should be allowed to breed upon the pest.

2. *Transplanting.*

This lesson precedes the transplanting of the young plants. Give demonstrations in the window box indoors, and later in the school gardens. Place an inverted glass over a thrifty plant and look for drops of water inside of the glass. The leaves give off moisture through breathing pores or stoma. If the plant gives off more water than it takes through the roots it wilts. The process of taking in moisture through the roots and giving it off through the leaves is called transpiration.

Steps in Transplanting. To be used as a guide and not to be learned except by use.

Transplant when weather is cool, cloudy and damp, preferably in late afternoon.

The plant should be young and vigorous.

Transplant shrubs and trees when leaves are off. Break roots little as possible and keep them moist and wrapped.

Cut off top to balance loss of roots.

Make hole larger than extent of roots and so plant will be two inches deeper than it originally grew.

Place good soil at bottom of hole in which roots are to grow and subsoil at surface.

Make soil firm and pour water about roots.

Leave a mulch on the surface.

Shade plant for a few days.

3. *Hotbeds and Cold Frames.*

Plant houses kept warm by fermenting manure, and sun's heat.

Early crops, as lettuce, radishes, etc., raised this way.

Long season crops, such as tomatoes, peppers, etc.

Construction. Begin middle of March; pit 2′ deep and 1′ wider than frame; fill pit with manure when it is steaming; thoroughly tramp manure down; 6″ layer good soil over pit of manure; board in,— front (south), 6″ high and back (north), 12″. Hotbed sashes are 3′ x 6′ in size. Bank frame on outside with manure; cover glass with mats cold nights; give plants air in favorable weather. Cold frame is a hotbed without pit of manure.

VI. *References*

Bureau of Plant Industry Document 883.
Farmers' Bulletin, 220
Farmers' Bulletin, 185, Construction of hotbeds for tomatoes.
Nature-Study Review—Jan. 1916 pp. 27. Tomatoes for the City Gardener. Canning Tomatoes at Home and in Club Work. Farmers' Bul. 521.

"The little cares that fretted me,
    I lost them yesterday
Among the fields above the sea
    Among the winds at play
Among the lowing of the herds
    The rustling of the trees
Among the singing of the birds,
    The humming of the bees.

"The foolish fears of what may happen
    I cast them all away
Among the clover-scented grass,
    Among the new mown hay,
Among the husking of the corn
    Where drowsy poppies nod
Where ill thoughts die and good are born
    Out in the fields with God.
        —Elizabeth Barrett Browning.

CHAPTER XXXIX

Study of the Tulip and Other Bulbs

I. Nature Lesson

Have a bulb for each pupil. Many florists will give them away after Easter. This will help the florist by increasing the demand for cut flowers and bulbs. Have one plant in blossom. Obtain bulb catalogues from seed stores.

Have pupils cut a vertical section of the bulb. Nearly every plant has leaves, stems, roots and buds. The bulb is an underground plant. Find the leaves. The roots. Remember that leaves and roots are attached to stems. Find the stem of the bulb. Find the terminal bud. Are there any side buds? In which part of the bulb is food stored? Why is the stem so small?

Observe plant in blossom. From what part of the bulb did the blossom come. The sepals are green at first but later become the same color as the petals. Advantage? The new sprout is made up of leaves which later open out. What is the shape of the sprout when it comes through the ground? Advantage? What is the difference in color of the bulb leaves and the leaves above ground? Why? If tulips are allowed to form seeds they cannot form such good bulbs. Why? After blossoming the plant forms new buds or bulblets at the side of the bulb. These may be separated to form new bulbs.

II. Drawing

Make a sketch of the vertical section of a bulb. Crayon sketching of tulips in blossom is interesting for primary grades. The coloring of pictures in seed catalogues is also enjoyable.

III. Garden Study

Why Study? Bulb study is especially practical for city children. Bulbs are inexpensive and easy to grow. They appeal to the aesthetic nature and grow when flowers are a luxury. They illustrate one way in which plants prepare for winter.

Few Interesting Facts.

Most bulbs are imported—Roman hyacinths from France, Easter lilies from Bermuda, Narcissi from England and Dutch Bulbs, hyacinths and tulips from Holland. In 1912 we paid Holland nearly ½ million dollars for bulbs. Bulbs of special merit bring from $500 to $2000. Washington State is the bulb region of the United States. The department of Agriculture has a bulb garden there for the purpose of conducting experiments. The climate and soil correspond to those in the Netherlands. The largest areas devoted to bulb culture are in Virginia, Rhode Island, Washington and California.

Bulbs in Water.

1 hose adapted to this method are the Chinese Sacred Lily, Paper-White narcissus and Roman Hyacinths.

Place 1 inch pebbles in bottom glass dish. Stand bulb in center and prop up with pebbles. Pour in water until reaches bottom of bulb. Change water every two days. Keep in dim light 2–3 weeks. Give abundant sunlight. This is the best method for Kingergarten and primary grades where the pupils can watch its development.

Bulbs in Pots.

Narcissus and hyacinths are best adapted to the school room.

Plan to receive order in September. Purchase best quality (Best bulbs are sold first.) Pot soon as possible after obtaining. Soak pot in water. Put in ½ inch drainage material—pebbles or ashes. Put soil (1 part sand: 1 loam: 1 manure) on drainage material. Hold bulb in place. Place spoonful sand under bulb for drainage. Fill soil around bulb to within ½ inch of top of pot. Firm soil about bulb with hand. Bulb is just barely covered with soil. Water.

Place in dark, cool (40°–50°) cellar. Keep moist. Keep there until roots well formed (2–3 months). "Forcing"—Bring to warmer place in dim light. When leaves turn green place in sunlight. Give more moisture during flowering period. Let leaves die naturally.

Pots in Trenches.

Dig a trench 15 inches deep. Place 3 inches ashes on the bottom. Place pots on ashes. Pack leaves around and over pots. Fill with soil. Allow ground to freeze. Cover wth 6 inch litter of leaves to keep temperature even. Use boards or brush to hold leaves. Remove litter in spring. Take out pots as desired.

Bulbs in the Garden

These are the hardy bulbs: hyacinths, narcissus, tulips, crocus, snowdrops, scilla and glory-of-the-snow. They are planted in middle of October. Canna and dahlias are planted outdoors in the spring.

Bed of mixed bulbs lacks harmony and design. One variety gives harmony—wanting in variety. Best to plant informally in borders.

Pattern beds must consider time blooming and height. Make a heavy dibble length desired.

Plant three times deeper than bulb is high; place 6 inch litter or mulch over to keep temperature even; hold leaves in place with boards or brush; remove leaf covering in spring; break off seed pods as petals fall to give strength to bulb; when cutting flowers cut as few leaves as possible; allow leaves to dry up before digging up bulbs; keep each kind of bulb by itself; spread out loosely in a dry shaded place; remove and destroy diseased bulbs; place in small peach baskets lined with paper; put away in dry shady place for summer; plan arrangement for next summer; plant in October; ferns combine well with bulbs as the ferns are later in starting.

CHAPTER XL

COMMUNITY PROJECTS IN A SAND-BOX

There are three important elements which enter into a community project in a sand-box. (1) The desire of a group of individuals to represent some important fact of life. (2) The gathering of the needed information from observation, conversation, and books. (3) A worth while representation of this fact so that not only the designers but others will obtain a clearer understanding of the question at hand.

Fig. 1. Sand-box Project.—Beautiful Home Grounds.

The advantage of a sand-box over a sand-table lies in the possibility of having a real and often times a living representation. One may grow real trees instead of using paste-board images, have grass instead of green sawdust, water instead of mirrors, cement walks instead of lines on paper.

Some of the larger values of this kind of project work are quite evident. There is what may be called a sub-conscious practical knowledge. The majority of city children do not realize that grass comes from seed. They unconsciously realize this fact through growing grass from seed. The mixing of sand and cement becomes a real problem. The foreman of this committee was chosen in a third grade because he had some notions from having watched an uncle who is a builder of cement side-walks. A spirit of coöperation is developed which is often needed in older communities. The representation must not be a failure because some one does not do his part in the organized whole. Each individual representation must be up to the standard of the general scheme or it will not harmonize or "fit into things." Such an undertaking cultivates the imagination. Each

427

pupil or group of pupils finds his part of the whole truth or principle and seeks some way to express it. The lesser or immediate benefits are as varied and numerous as the projects themselves.

A few examples are described to illustrate how such a project may be carried out.

I. Beautiful Home Grounds

The preparation consisted of studying some of the general principles of landscape-gardening as outlined in Chapter XLIII. The pupils were given the pictures of beautiful homes and by studying the principles found what gave the pleasing effects to the different estates.

The organization consisted of various committees who reported their progress to the class for suggestions and criticisms. The committees and their duties were as follows:

1. *Lawn.*

The lawn was planted with grass seed. Care had to be used to distribute the seed evenly. The watering had to be done gently in order not to wash the seed into hollows. There was an interesting competition to discover the first blade of grass. The lawn was trimmed with scissors.

2. *House.*

The location of the house determined the walks, trees, and shrubbery. It presented an excellent lesson in colors. The green roof and trimmings harmonized with the grass. The white body of the house gave a very neat appearance. The construction of the card-board house furnished an excellent project in drawing. The size and number of curtains gave live problems in arithmetic.

3. *Garage.*

This building was made on the same style and color as the house. It was placed economically and conveniently for the owners.

4. *Walks.*

The committee voted to use cement. The mixing of the cement, the form for the cement, the plan and scale of the walks furnished practical problems in fractions.

5. *Tennis Court.*

The location in regard to the sun, the fence net, and rackets tested the ingenuity of the designers.

6. *Hedge.*

The hedge is determined by the location of buildings, walks and the tennis court. The committee made real cuttings from the privet hedge. The buds produced leaves and the cuttings took root in the sand. They were later transplanted to their homes.

7. *Vegetable Garden.*

This play garden caused a great deal of research for natural representations. Green buds were used to represent cabbages, pine needles for onion tops, twigs of barberries for peppers, etc.

8. *Flower Gardens.*
   This was another problem in colors and representations.
9. *Trees and Shrubs.*
   The most satisfactory trees were made by using evergreen cuttings. The deciduous trees were made by planting maple seeds and acorns.

II. Providence, 1650: A Correlation in Nature-Study and History

In nature-study there was research as to the representation of the pine forest that covered Capitol Hill, the material of thatched roofs,

Fig. 2.   Sand-box Project—Providence in 1650.

the characteristics of swamps with hummocks, the drainage to the old spring, the workings of a grist mill, the possible logs for a cabin, the material at hand from which the settlers made the fireplace and chimney.

1. *Prospect and College Hill.*
   This steep-sided hill is seen on the far end of the box. It is covered with a thick growth of trees. Each settler had a strip of land running from his homestead to the highway (Hope Street) on top of the hill. Amongst the growth on the vacant lots of this section may be seen wild apple trees. These undoubtedly have descended from the original orchards that used to be on this steep hill-side.

2. *Log Cabins.*
   The log cabins are made of plasticine. In the plasticine are placed fine sticks which represent logs. Back of the houses is a chimney, the exterior consisting of fine grains of sand to represent rocks. The thatched roofs are made of fine grass arranged like shingles.

3. *North Main Street.*
   The trail running in front of the log houses is shown in the picture as a light line. That trail is the present North Main Street.
4. *The Moshassuc River.*
   Flows down the valley parallel to North Main Street. It turned the wheel of Smith's old grist mill and passed beneath one of the first bridges of the town into the cove. The Louquassuck Trail crossed by this mill.

Fig. 3.   Destructive Lumbering.

5. *The Cove.*
   The dark central portion of the picture. A cement basin made water proof by paraffin formed the base. This was covered with clean sand to form a natural setting for the water.
6. *Smith's Hill.*
   Covered with pine trees. Its sandy bluffs formed the northern boundary of the Cove. On the summit above the mill is the site of the present State House, and a little farther to the westward the Rhode Island College of Education.
7. *Weybosset Street.*
   The dark line winding on the right of the fore-ground is the Pequot Path which led to the land of the Narragansetts on the south. Weybosset Island and the hummocks amongst the marshes formed stepping stones to the village on the east side. This detour in the trail forms the present curve in Weybosset Street.
8. *College Street.*
   A dark line showing the continuation of Weybosset Street up College Hill. It was called the Wampanoag Trail and led to the country of Massasoit and the Plymouth Bay Colony.

9. *The Woonasquatucket River.*

Hidden by the forests. An early settler had set himself away from the rest of the village and built a cabin on the banks of the river. Just west of the pine woods is a little ravine which marks the location of a small that runs through Davis Park to the Woonasquatucket.

III. A Study in Forestry: Suggested Projects for Arbor Day

1. *Destructive Lumbering* (Fig. 3).

The owner of this forest has had no regard for the future and has been very wasteful with his timber. Reckless lumbering is seen in the

Fig. 4. Constructive Forestry.

high stumps and large logs left to decay on the ground. The brush furnished food for a fire and the fire-killed timber is now falling and getting ready for the next fire. The young trees are returning but will be killed by the next fire. The wood which was exposed by the fire is now being attacked by fungi and the insects are finding good opportunities for boring into the trunks. The leaf-mould no longer holds the rainfall and the soil is washing onto the railroad track making it dangerous for transportation.

2. *Constructive Forestry.* (Fig. 4).

The golden rule of forestry is to thin often. The stumps show where the large useful trees have been cut for lumber and fire-wood. The younger trees now have room to grow. The brush has been burned as "an ounce of prevention" for forest fires. The forests are not cleared from the steep hillsides and therefore there are no floods or wasting of the rich soil. The owner of this wood-lot will always have a supply of wood.

"Who does his duty is a question
    Too complex to be solved by me,
But he, I venture the suggestion,
    Does part of his that plants a tree."

                                   —*Lowell.*

"We read and studied out of doors, preferring the sunlit woods to the house.
All my early lessons have in them the breath of the woods—the fine, resinous
odour of pine needles blended with the perfume of wild grapes. Seated in the
gracious shade of a wild tulip tree, I learned to think that everything has a
lesson, and a suggestion * * * * Indeed, everything that could hum, or buzz, or
sing, or bloom, had a part in my education—noisy-throated frogs; katydids
and crickets held in my hand until, forgetting their embarrassment, they trilled
their reedy note; little downy chickens and wild flowers; the dogwood blossoms;
meadow violets and budding fruit trees. I felt the bursting cotton-bolls and
fingered their soft fiber and fuzzy seeds; I felt the low soughing of the wind
through the cornstalks; the silky rustling of the long leaves; and the indignant
snort of my pony, as we caught him in the pasture and put the bit in his mouth—
Ah me! how well I remember the spicy, clovery smell of his breath!"

                                   —*Helen Keller.*

CHAPTER XLI

TREE SURGERY AND DENTISTRY

RULES FOR THE TREE DOCTOR

1. *To Remove Large Branches.*

Cut off the branches that are dead, or are dying or are broken.

Saw one-third through the limb from the underside and several inches from the support. This prevents splitting of outer wood and bark.

Saw on the upper side near the first cut until the limb falls.

Saw off the stump so that the cut will continue with the surface of support.

2. *To Prevent the Infection of Wounds.*

Paint to keep out the decay-producing organisms.

If cracks appear, fill and paint.

3. *To Fill Cavities.*

Dig out all the decayed portions.

Fill the cavity with cement so that the surface will be continuous with the inner bark.

Collect specimens such as shown in the picture. By observing each specimen try to interpret its past, present, and future. Every branch has a form dependent on two kinds of forces, one growth, the other decay. Whenever a limb is injured the decay-producing organisms of the air attack its surface. In time the largest branch must be reduced to dust, if the decay is not hindered by the healing of the wound. Between the beginning of decay and the completed cavity in the trunk there is a whole sequence of forms. The decay may begin as the result of old age, as a broken limb, as a stub from poor pruning, or as an unpainted surface following a good cut. Each form has certain signs by which a good scout can read its history. When the meaning of these forms is once perceived by the student he not only has learned something practical about the care of trees but he has experienced one of the most valuable methods of thinking. The child has as much right to read trees as he has to read about trees.

INTERPRETATIONS

The following interpretations are given as a basis for future work:

Figure 1.

OBSERVATION	INFERENCE
Branch has no bark and grey wood.	It must be dead.
The outer end is ragged.	Broken by wind or snow.
Weathered appearance.	The beginning of decay.

Conclusion: As the decay will continue into the trunk, this branch should be sawed off and the wound painted to prevent infection.

Figure 2.

OBSERVATION	INFERENCE
Branch nearly gone, through decay.	May have been broken or cut.
Hole entering trunk.	No care or preventive measures.
Nut shells, remains of a nest, insects.	Decay has been aided by animals.

Conclusion: As the decay has spread into the trunk, the base of the limb should be sawed off and the decayed portion cut out. To prevent further decay, the hole should now be filled with cement. This is similar to treating a tooth.

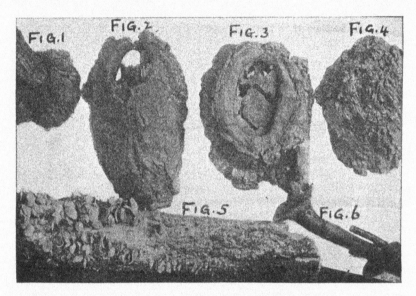

Figure 3.

OBSERVATION	INFERENCE
A smooth surface.	A limb sawed off. The surface is not smooth enough for an axe cut.
A rim of new growth over wound.	Wound has started to heal.
Hole in cut surface.	Wound was not painted.

Conclusion: This portion should be sawed off and if the decay has not entered the trunk beyond the new cut, paint the fresh surface so that it will not decay before complete healing.

This story might be reviewed by giving the steps in its life history. A branch; sawed off; partially healed; decayed; sawed off as a specimen.

Figure 4.

The steps in the life history of this specimen are as follows: Limb; broken or cut off, leaving a stub; decayed; sawed (as seen on other side); not painted (as shown by decay on inside); decayed; healed; sawed as a specimen.

Figures 5, 6.

The growth on these limbs are called fungi. They live on the substances of the tree and their presence is always a sign of ill health. Limbs having them should be removed. They produce a dust called spores. These spores are blown to exposed places on trees, where they grow new fungi.

"To live an increasingly rich and worthy life is the aim of all endeavor in both life and education."—*Bonser.*

"No blooming of roses endureth forever,
The glories of sunset not alway remain;
Yet liveth their grace in the spirit, tho' never
The senses perceive the same beauty again."
—*S. M. Newman.*

"Since we love, what need to think?
Happiness stands on a brink
Whence too easy 'tis to fall
Whither's no return at all;
Have a care, half hearted lover,
Thought would only push her over!"
—*Lowell, Love and Thought.*

"Under the greenwood tree
  Who loves to lie with me,
  And tune his merry note
  Unto the sweet bird's throat
Come hither, come hither, come hither."
—*Shakespeare* (*As You Like It*).

  "I chatter over stony ways
        In little sharps and trebles
    I bubble into eddying bays
        I babble on the pebbles."
—*Tennyson* (*The Brook*).

"And this our life, exempt from public haunt
  Finds tongues in trees, books in the running brooks,
  Sermons in stones, and good in everything."
—*Shakespeare* (*As You Like It*).

Only a little shrivelled seed,
It might be flower, or grass, or weed;
Only a box of earth on the edge
Of a narrow, dusty window-ledge;
Only a few scant summer showers;
Only a few clear shining hours;
That was all. Yet God could make
Out of these, for a sick child's sake,
A blossom wonder, as fair and sweet
As ever broke at an angel's feet.
—*Van Dyke* (*The Builders*).

CHAPTER XLII

The Uses of the Forest

(An application of knowledge of rainfall in an upper primary grade.)

1. *Aim.*

The aim of this lesson is to present the idea of the value of trees as a group and not as individuals.

2. *Introduction.* (Experience of the pupil.)

How many have seen a muddy brook?
Where did you see it? (Street, gardens.)
When did you see it? (After a rain.)
What made the brook muddy? (Rain washed in the soil.)
Are muddy brooks swift or slow? (Swift.)
Do they contain much or little water? (Much.)
Does a muddy brook flow all the time? (No, it is dry part of the time.)
Tell the story of a muddy brook. (A muddy brook is seen in the gardens and gutters after a rain. The rain washes the soil into the brook. The brook is then swift and contains much water, but it soon becomes dry.)

3. *Experiment.*

Have a tray with soil arranged as a hill. Cover half of the hill with moss. Insert small evergreen twigs to represent trees. Use a watering can to represent rain.
What is represented when I pour water from the watering pot on the hill? (Rain.)
I have placed a covering over this hill. Notice how fast the water runs off this hill with a covering on it. Notice how fast the water runs off the hill without a covering on it.
On which hill did it run off the faster? (The hill without the covering.)
This hill is covered with moss and twigs. What is the covering of a real hill? (Grass and forests. Usually forests.)
Think of the hill with a covering and how the water "ran off" from it when it rained. Think of the hill without a covering and how the water ran off from it.
From which hill would you get the large, swift, muddy brook? (The hill without the forest.)

4. *Application.*

Is it a good thing to have the rich soil washed away? Why? (One use of the forests is to prevent such washing away.
Why do muddy brooks contain much water? (No covering to hold back the water.)
What would be the result of several overflowing brooks emptying into a river? (A flood.)

437

How may floods be prevented?   (By having forests at the source of rivers.)

Which river would have the most constant water supply, one whose brooks are covered by forests or one whose brooks come from the bare hills? (Those protected by forests.)

Are forests favorable or unfavorable to navigation in a river? (Favorable, as they regulate the water supply and make navigation safer and possible for a longer period of time.)

Are forests favorable or unfavorable to the supply of water for manufacturing? (Favorable, as the water power for a factory is determined by the amount that can be depended upon for the year.)

Sum up the uses of the forests that we have mentioned.

(1) Prevents washing away of rich soil, (2) Prevents floods, (3) Regulates the water supply for navigation, and (4) For manufacturing.

Name some other uses of the forest. (The pupils will be able to think of other uses.)

Home for native plants and wild animals. Produce lumber and other products such as turpentine, maple syrup, rubber, fruit, etc. Source of beauty and pleasure. Checks velocity of winds. Increase safety of farming and fruit growing. Cools temperature in summer and warms in winter.

"They shall not hurt nor destroy in all my holy mountain: for the earth shall be full of the knowledge of the Lord, as the waters cover the sea.—*Isaiah, xi.*

Nature-study is learning those things in nature that are best worth knowing, to the end of doing those things that make life most worth the living.
—*C. F. Hodge (Nature Study and Life).*

"A little of thy steadfastness
Rounded with leafy gracefulness
Old oak, give me,
That the world's blasts may round me blow
And I yield gently to and fro
While my stout-hearted trunk below
And firm roots unshaken be."
—*Lowell (The Beggar).*

CHAPTER XLIII

DECORATION WITH PLANTS

"I do not own an inch of land—
    But all I see is mine—
The orchard and the mowing-fields,
    The lawns and gardens fine.
The winds my tax collectors are,
    They bring me tithes divine."
        —*Lucy Larcom* (*A Strip of Blue*).

Fig. 1. A WOOD PATH THROUGH A VALLEY. Beauty without Adornment. The Grotto, Butler Asylum Grounds, Providence. Such examples of the expression of Nature's forces are becoming rare. Stand very quietly under the hemlocks on the right, some day after school, and you may see the black-crowned night heron in the topmost branches.

I.   The Town or City Beautiful

Figures 1-4 in this chapter tell the story of the development of the Gladys Potter Memorial Gardens in Providence, Rhode Island.

Fig. 2.   THE PASSING OF A VALLEY. Irving Avenue, near the East Side Fire Station. An example of the effects of the spreading out of a large city. Picture taken in the late fall, 1911.

Fig. 3. RESCUING ONE OF OUR CHOICE BITS OF LANDSCAPES. SAME VIEW AS ABOVE. Photograph taken in March, 1919. Can you identify in this picture the trees shown in Figure 2.

It is typical of the way that many of our city playgrounds come about. The same story might be represented by words, as—path, road, street, boulevard; forest, field, vacant lots, park; Indians, settlers, country folks; city people; hunting, farming, trading, manufacturing; wigwam, cabin, homestead, apartments. You can think of other words that name the chapters in the history of Gladys Potter Memorial Gardens.

The following expressions apply either to Figure one or to Figure two. Underline with a pencil those that you think apply to Figure one. Perfume of hemlocks. Rubbish. Unpleasant odors. Stillness. Home of wild plants and warblers. Clatter of the English sparrow and starlings. Fern embroidered. Weeds. Noise. Dust. Rat

Photo by Dr. Marian Weston

Fig. 4. Same area as Figure 3, December 3, 1925. What has been added to this scene?

hatchery. Lofty trees. Foliated arches. Ashes and tin cans. Carpet, leaves and mosses. Clang of the trolley. Poplars, grey birches, wild black cherries. Squirrels watching from above. A muddy brook after a heavy rain. A clear, cool stream. Quiet seclusion. Making room for buildings. Oaks, elms and beech. Primitive works of nature. Unclean. Buried in smoke. Tap of the downy woodpecker. Which scene do you prefer?

An interesting project would be to study a park or playground area in your own community. If possible obtain a series of pictures to show the changes. Talk about it with some of the older people of the vicinity. Try to trace the whole length of the former valley. Show where it was on a map of the city. Whose farm did the little stream drain? What animals drank from its clear pools? What children played along its banks? Write your results in the form of a story.

In Tokio the city directory tells where to go in order to see the most beautiful snow scenes or where the chrysanthemums are at their best. Compose a directory of the beautiful spots in your town or city.

II. The Home Grounds.

The laying-out and improvement of grounds is called landscape-gardening. Landscape-gardening is something like landscape-painting. Some general conclusions regarding the principles and practice of the art are given to enable you to read and interpret the home grounds pictured here and then the grounds of other homes. For example: Read the rules for arranging the shrubbery and then call

Fig. 5. A Drive on the Senator Aldrich Estate, Warwick Neck, Rhode Island.

attention to the effects which the landscape-painter has used to bring out the pleasing driveway on the Senator Aldrich estate.

1. *General.*

"Nature unadorned is adorned most." Nature works in curves. The arrangement of plants to resemble nature is called the English style. A formal planting is the Italian or geometrical style. The formal arrangement is out of place on a small lawn. Send for Farmers' Bulletins 248 and 185.

2. *The Lawn.*

"The greensward is the canvas upon which all architecture and landscape effects are produced." The lawn should be placed where one can get a distant outside view,—as of the bay or of a river. Small lawns should be convex; large areas undulative. Keep out all weeds. There should be a strip of grass between walks and shrubs, or gardens.

3. *Walks and Drives.*

Walks and drives add no beauty to the grounds. Do not have unnecessary walks. Do not make any wider than needed. Unless the distance is short avoid straight lines. Sink the walks a few inches below the surface of the lawn so as to give the effect of a continuous lawn from a distance. The walk should be higher in the center for drainage. Gravel or cement make good walks. Paths should conform to the contour of the land.

Fig. 6.   Shrubs Covering a Curved Walk.

4. *Trees*.

a.   *Individual Trees*. Used for their individual beauty and to furnish shade. They lend distant grandeur to the landscape. Trees standing alone should have characteristic beauty, as: oak for strength; elm for arching form; purple beech, blue spruce, or golden arbor vitae for color. Fruit trees serve a double purpose. Overarching trees are effective for the enframement of a building, as the elm. The following trees are very satisfactory on the front lawn: ginkgo, Norway maple, horse-chestnut, Japanese maple, purple or copper beech, and pin oak.

b.   *Trees in Groups*. Furnish a background for ornamental material and cover unsightly objects. Groups should not give too solid an appearance. Avoid groups of same kind, color or size. The taller varieties should be in the centre or form a background. Evergreens form a good screen or shelter. The sombre effect may be taken off by deciduous trees, as the golden willow. Trees look well on both sides of the entrance from the street.

5. *Shrubs*.

Shrubs should blend the trees and the lawn. The line where the lawn and shrubbery meet should be made of curves. Evergreens should not border paths as they are apt to be injured when frost is on them in the winter. The skyline of the shrubbery should be irregular, with alternate groups of tall and shorter shrubs. Plant shrubs in

Fig. 7. THE USE OF CEDAR to emphasize an entrance and to relieve a smooth, cement surface.

masses and not singly. Plant thorny shrubs where paths meet and where cuts are likely to be made (barberry, locust, prickly ash). Taller shrubs should not shut out vistas from the windows. Low growing shrubs may conceal the foundations of the buildings. Evergreens rise as a mass from the ground. The foliage of the groups of shrubs should be carried to the ground. Use shrubs to cover the abrupt endings or curves in the walks. Shrubs may be planted on the concave side of all curves. Shrubs with berries furnish food for birds, and color for effect. Wild shrubs such as elder, sumac, dogwoods, and viburnum are good.

6. *Hedges*.

Hedges make a good background for herbaceous plants. Low hedges are more ornamental. A rounded or triangular top gives more foliage on the side. The top of the hedge should have the same curve as the ground line. A wall thickly covered with a vine, as Boston ivy, gives a similar effect.

7. *Vines.*

Vines carried over walls and pillars carry the green of the lawn upward. They make the house a part of the landscape. If too close to the building they make it damp. They form a cover for pergolas,

Fig. 8. ENTRANCE TO THE SWAN POINT CEMETERY. The landscape gardening of the Swan Point Cemetery ranks among the best of the country. Pleasing effects are produced both in winter and summer.

verandas, and arbor trellises. The Boston ivy is an excellent climbing plant for brick or stone buildings. The Virginia creeper is good for backyard fences.

8. *Hardy Herbaceous Plants.*

The hedge makes a good background for hardy herbaceous plants. Plant perennials in the bays and recesses of the shrubbery. Transplant every three or four years to get a vigorous growth. Plant bulbs in masses along the borders of the shrubbery to furnish the edge. A mixed arrangement on small places gives the most pleasure. Mix perennials in groups that bloom at different seasons. Do not have inharmonious colors bloom at the same time. Intricate designs seen on public grounds should not be attempted on small places.

9. *Annuals.*

The best place for annuals is in the flower garden. The tall-growing, large-leaved plants, as castor bean and sunflower make good screens for fences. Verbena, pansy, sweet alyssum, and sweet William make beautiful front line or border plants. Arrange according to height, color, and time of flowering.

Fig. 9.   Perennials with a Shrubby Background.

10. *The Wild Flower Garden.*

Wood Anemone: transplant rootstock in early spring to border of shrubbery. Bloodroot: transplant rootstock in early spring to moist, sheltered soil. Bluets: transplant clumps to open, moist ground. Wild columbine: transplant root masses to dry, sunny, rock-banks. Jack-in-the-pulpit: transplant early in the spring to rich, shady soil. Goldenrod: There are over 50 species. Transplant root masses. Lady's slipper: Transplant very early to rich, shady soil. Trillium: Plant rootstocks in shaded borders. Violets: place nearly as possible in environment such as they come from. Ferns: usually in moist places away from sun, as—near northern side of the house.

There are many original ways of ornamenting the home grounds. One of the most interesting is shown in the picture taken on the estate of Mr. John G. McIntosh, Pawtucket Avenue, East Providence.

Fig. 10. PLANT ORNAMENTATION. On the grounds of Mr. John McIntosh, East Providence.

It consists of the trunk of a large sycamore which has been placed over a pump (June, 1918). The handle of the pump projects through the trunk. A sign is attached to the trunk which says: "Water, free to all." The sycamore was cut down in 1916 and the annual rings tell that its age is 244 years. It therefore started its career in 1672

Fig. 11. A VISTA ACROSS THE SEEKONK. Note that the scene is enframed with shrubs and trees. The bounding lines go toward the view point. The inspector feels as though he were invited down the hill for a nearer view. As one nears the bank of the river he is pleasantly surprised with a mass of azaleas and rhododendrons. Plan to visit this spot when these shrubs are in full bloom. Look for other vistas.

and was probably planted by the colonists, as many ornamental sycamores were introduced from England. What an interesting story that seed might tell if could be here and speak. Its birthplace; how it crossed the water; how it traveled to the Narragansett country. Then the tree itself. It may well have seen Roger Williams and the Indians. Being a hundred years old at the outbreak of the Revolution it could have furnished shade for the colonial troops and it saw the stirring days of 1861 when the boys went to the front. May the old sycamore exist for a long time in the new service which it is rendering.

The ash flag-pole on top of the trunk has experienced no less interesting events. It was the pole in the old coach now owned by Senator Le Baron B. Colt. The coach was originally owned by James de Wolf (Senator, March, 1821–October, 1825). Senator de Wolf rode from Bristol to Washington in this coach with an array of coachmen and lackeys in brilliant livery. It used to take about a week for the trip, and many famous men have ridden in the old coach which, although 100 years old, is in perfect condition.

The top of the table shown in the picture was made from a cross section of the same tree. Mr. McIntosh is sitting in the foreground

III. The School Grounds

School buildings used to be built on ledges, or on ground which was considered too poor for farming. It might almost have been said, in those days, that some out-of-the-way spot where nothing else could grow was good enough for raising children. Today we feel that the children should have plenty of room to run and play and get fresh air. Even the soil must be good for gardening and for beautifying the grounds. The principles of decorating school grounds are the same as for the home grounds. It is desirable to have hardy trees and shrubs on the school grounds and also a great variety for the purpose of study. We should like to publish the best paper on the decoration of your school grounds. Take a picture of the grounds before and after planting. Send for Farmers' Bulletin 134 before planting.

IV. The School Room.

Plant decoration of the school room is an art. These two references will help you to understand the principles of arranging flowers. The School Arts Magazine, June, 1914, p. 754; and The Garden Magazine, November, 1918, p. 106.

1. Aim,—beauty in:

a. Color,—peony, pansy. May be massed.

b. Form,—calla lily. Do not mass flowers chosen for form. The Japanese never crowd their flowers. A "bunch" of flowers is an American expression.

c. Color and form,—rose buds, chrysanthemums.

2. Receptacles.

a. They should be dull colored and simple; less attractive than that which they hold.

b. The shape should be such as to hold the flowers in their natural positions.

c. They must have some element in common with the plant, as color or form. The tall, straight vase is good for the tall, single flower. Use a shallow bowl for short stemmed flowers.

3. Flowers.

a. Use only one kind with its own foliage.

b. The arrangement must present the freedom of wild nature yet maintain balance. Some flowers must be long and others short.

c. Native wild plants are disappearing. Pick moderately. Do not pull up, but cut flowers with a knife.

d. The soft dull colors of winter sprays harmonize and form picturesque bouquets. Try pussy willow sprays in January; berried branches in November; evergreens in December and twigs of forsythia, magnolia, cherry. peach. and apple in February.

Fig. 12. A Vista across Wellfleet Bay, Cape Cod.

A HYMN FOR ARBOR DAY

(To be sung by Schools to "America.")

God save this tree we plant!
And to all nature grant
    Sunshine and rain.
Let not its branches fade,
Save it from axe and spade,
Save it for joyful shade,
    Guarding the plain.

When it is ripe to fall,
Neighbored by trees as tall,
    Shape it for good.
Shape it to bench and stool,
Shape it for home and school,
Shape it to square and rule,
    God bless the wood.

Lord of the earth and sea,
Prosper our planted tree,
    Save with thy might.
Save us from indolence,
Waste and improvidence,
And in thy excellence,
    Lead us aright.

—Henry Hanby Hay.

IF THERE WERE NO TREES

What would the birds say,
The squirrels and chipmunks and all
The little wild folk
If there were no trees straight and tall?
No place to cling to,
Or wing to, or sing to,
No place—spring or fall—
To build houses small,
Elm, apple or oak?

What would they all say—
The children so eager and sweet—
If there were no trees
To make them a leafy retreat,

No place to stray in,
Or stray in, or play in,
Away from the sleet and the heat?
No fruit, and no nuts, nor the fleet
Stir of leaves on the breeze?

And what would the earth do
While bright season rolled
If God all the charms
Of the trees should withhold?
Without them to dress her,
Or bless or caress her,
To sing her their songs, dear and old,
Around and about her to fold
Their strong, tender arms?

"Jock, when ye hae naething else to do, ye may be sticking in a tree; it will be growing, Jock, when ye're sleeping."
—*Sir Walter Scott.*

Great, wide, beautiful, wonderful world,
With the wonderful water around you curled,
And the wonderful grass upon your breast—
World, you are beautifully drest."
—*Wm. B. Rands.*

How snug seemed everything, and neat and trim:
. . . .
With little paint-keg, vases and teapots
Of wee-blossoms and forgetmenots:
And in the windows, either side the door,
Were ranged as many little boxes more
Of like old-fashioned larkspur, pinks and moss
And fern and phlox; while up and down across
Them rioted the morning-glory-vines
On taut-set cotton-strings."
—*James Whitcomb Riley (A Child World).*

"The least of living things, I repeat, holds a more profound mystery than all our astronomy and our geology hold."
—*John Burroughs.*

"There is so much within our easy grasp
  For minds to know in radius of our eyes,
We only have to stretch our hands to clasp
  The 'Open Sesame' to a Paradise!"
—*Emily Selinger.*

"What is man, that Thou art mindful of him?
  And the son of man, that Thou visitest him?"

"Nature never did betray
  The heart that loved her; 'tis her privilege,
Through all the years of this our life, to lead
From joy to joy, for she can so inform
The mind that is within us, so impress
With quietness and beauty, and so feed
With lofty thoughts, that neither evil tongues,
Rash judgments, nor the sneers of selfish men,
Nor greetings where no kindness is, nor all
The dreary intercourse of daily life,
Shall e'er prevail against us, or disturb
Our cheerful faith, that all which we behold
Is full of blessings."
—*William Wordsworth.*

CHAPTER XLIV

The Teaching of Plant Diseases in the Grades

As the history of parallel movements may throw light upon our present problem it may be worth while, as well as interesting, to briefly review old acquaintances.

In searching the mouldy rolls of the past one finds that the fruit of that forbidden tree furnishes a frequent theme. As early as 1340 in the "Ayenbite of Inwyt," the title of a religious treatise by a monk, appears the well known saying that "a roted eppel amang the holen, maketh rotie the yzounde." Shakespeare in the Merchant of Venice (VI. iii, 102) mentions "a goodly apple rotten at the heart" and in the Taming of the Shrew (I. i. 139) says: "Faith (as you say) there's small choice in rotten apples." As these are suggestive of the manner of mentioning plant diseases in the past one must conclude that the purpose was not so much to teach the disease of the plant as to teach human morals.

The handing down of knowledge of moulds, blight, and rust has been in the folk-lore stage and is so today in most cases. The crediting of the failure of a crop to an east wind or some such force is the method of the Australian native. The medicine lore of the savage is but just disappearing along the fence row of the farmer and remains with the untutored, yet a scientific knowledge of these things is a practical necessity for a civilized community.

One might expect that the history of the hygiene movement in the grades would be suggestive as to the possibilities in garden hygiene but such is not the case. A little research shows that we are only in the initial stages of teaching the child how to care for himself. "In 1900 only eight cities in America had any organized health work in schools" (U. S. Bureau of Education Bulletin, 1913). The Greeks emphasized physical training and Locke and Rouseseau preached it, but clean hands as a prevention of disease was unheard of at that time. It was not until 1885 that physiology and hygiene were made a compulsory study by the laws of Massachusetts, and that was one of the first states to make the subject a part of the curriculum. When human hygiene is so recent that it scarcely has a history it would not seem over encouraging for the study of plant diseases.

We do not need to be reminded that the teaching of plant diseases in any organized way is very recent. The Yearbook of the Department of Agriculture for 1899 tells us that in 1885 there were only three institutions teaching this subject and that ten years later "50 colleges and stations engaged in the work and at least 100 special investigators were devoting their time to it."

Notwithstanding these preliminary remarks there are sufficient reasons for introducing the subject at this time. In Massachusetts alone for the summer of 1918 it is reported that 75,000 boys and girls not living on farms had gardens. If they are to be encouraged in this work they must know how to take care of these diseases which are so prevalent. Then, again, the potato blight is not a concern of Maine alone nor the black wart of the potato, in Pennsylvania, just a question of that state. There must be a Federal intelligence in regard to the things which concern its welfare. Likewise, a healthy garden is a community asset, and the ignorance of one gardener is a menace to the whole neighborhood. Other by-products of this knowledge will be a greater intelligence in regard to the causes of all diseases and their cure by patent medicines. Then, the consumer should have an appreciation of wholesome food and knowledge as to how it should be cared for in the home. He should have some notion as to what it costs to grow clean healthy produce and a greater respect for the farmer. A far more important result may be the plant physician for every community. He will work after the fashion of the Chinese doctors who are paid for the prevention rather than the cure of diseases. In this way the working power of the gardens of the community will be kept at top-notch efficiency.

If plant diseases should be taught in the grades what are some of the topics that may be presented and what method ought to be used? The following notes are merely suggestive as to a few projects.

Cut slices of raw potato with a sterilized knife and place each slice in a saucer under a glass. Place under different conditions, such as: warm, cold; dry, moist; sunny, dark; a slice that has been in contact with a dusty surface and one that has not; peeled and unpeeled; a healthy potato in contact with a decaying spot on another potato, and a healthy potato in contact with a healthy potato; a bruised and a scratched potato and a sound potato. A child can easily derive the following practical conclusions by observing the color changes and decay effects without the use of a microscope: Vegetables and fruits decay more readily in dark, moist, warm cellars. Unclean receptacles aid decay. The skin keeps out germs of decay. Decay is passed on by contact. Fruit should be handled carefully. Coldness, dryness, and sunshine are germ killers.

Another series of experiments is with dishes of agar. Make finger prints before and after washing the hands. Expose a dish for five minutes before and for five minutes after sweeping; before and after a thunder shower; early in the morning and late in the afternoon; a drop of distilled water, faucet water, and dish water. Try inoculation experiments with a sterile needle, as—removing bits of decaying potatoes, oranges, onions, apples, parsnips, etc., to a test tube with agar and stopping mouth of tube with absorbent cotton. Inoculation now becomes a visible thing to the child.

Visit a diseased potato field. Where is the disease thriving most,—in the rich or poor soil, at the low moist end of the garden or where it is well drained, where the foliage is crowded or where plants are far apart, on the sunny corner or the shady corner? Are there any varieties that appear to be more susceptible? The class will be impressed with the fact that the lack of proper food, over-crowding, and absence of sunshine lowers the vitality of the plants and makes them susceptible to disease. These conclusions must make the conditions of human hygiene seem real and not a matter of preaching.

A field trip to a woodlot is also of great value. Find trees being destroyed by fungi. What enabled the fungi to attack the tree? See how many kinds can be collected. Find leaves with blemishes and colored spots which are symptoms of disease. Symptoms in plant diseases are much more evident than with people.

A lesson on the higher parasites such as dodder, mistletoe, etc., may be made a basis of morals. The law of the use and disuse of parts is rather striking in these plants.

A lesson in history will also be instructive such as a report on the cause and effect of the potato blight in Ireland during the great famine of 1844.

Have an exhibition of moulds brought in from the homes. Include everything,—shoes, books, fruits, preserves, etc. In class discussion bring out the causes and emphasize the preventions. Try to get an estimate as to how much is destroyed each year by moulds in the home.

Some people may feel that a knowledge of these things leads to unhappiness. It reminds me of a story that I read a long time ago about an Arab who admonished a traveller for having stepped on a worm. The traveller asked the Arab if he did not know that he was destroying hundreds of living beings when he ate a fig. When the dusky inhabitant of the desert was shown the organisms through a microscope he took the microscope and dashed it against a rock. Let us not accept the philosophy that 'ignorance is bliss'.

It is an Ill Wind that Blows Nobody Good

A FABLE SEQUEL TO TEACHING PLANT DISEASES IN THE GRADES

"Abominable East Wind!" cried the farmer, as he gazed at his potatoes; "to what a woe-begotten end have you brought my winter food!"

"The same old story!" murmured the wind, in reply. "Always blame the weather for troubles you have brought upon yourself. What more could have been done for you? All this day have I brought water to your garden to save you from famine. If you did not spray your potatoes to prevent blight, when warned by the Farmers' Bulletin, who is to blame but yourself?"

"I am unfavored, indeed," rejoined the farmer. "I thought you were a friend, but have been deceived."

"Not by me," replied the wind, patiently. "I tend to my work every day. I bring the crops warmth. I bring them moisture. I mislead none but the superstitious and ignorant."

"Superstitious! ignorant!" cried the farmer. "How little do you know as to who I am. Trustee of the Academy—Superintendent of the Sunday School for thirty years—a leader of the community."

"A leader who cannot lead! Wise, perhaps, in the laws of the village—ignorant in the laws of nature. You have mistaken the friend that brings a good harvest for the pest that causes famine. Alas for your neighborhood, if no better leader can be found."

The farmer turned away, and the wind played across the field. The wind danced up and down the rows and mourned his luckless fate. "Yet," said he to himself, as he dried up a muddy pool about to decay a hill of potatoes, "I will keep on trying. What an ignorant farmer!"

$$*\quad*\quad*\quad*\quad*$$

Scene: The kitchen.

Characters: Mother Hubbard, a rich lady who had gone through the form of being patriotic by preserving peas without being intelligent as to the correct method.

Faith, the daughter of Mother Hubbard who had great faith in her fashionable mother.

"These here peas are moulding," observed Mother Hubbard to her daughter Faith, as she slowly took one jar after another from the shelf. "Such miserable weather!" Some would have said "just my luck" but it comforted the fashionable lady's heart to lay all the blame on the weather. Faith, however, took but little interest in the matter. Her mother was always grumbling about the east wind and her rheumatic pains which should have been called the gout.

The door banged with great violence. It was a pity that Mother Hubbard had placed some jars so near the edge of the table, for, when the door was blown too, they fell with a crash, and mouldy peas were strewn across the floor.

And, "Do we meet once again?" said the Jar Spore to the Floor Spore, in whose company he had traveled at preserving time. "Do we meet once again?" How pleasant indeed. "I have not seen you since Mother Hubbard locked me up with the peas. Well, well, well. Let me first ask how you are this morning?"

"Oh, pretty well," replied Floor Spore, "but very, very sad. You have little cause to be sad. You have had some fine peas to grow upon. But I! Alas, the cruel wind has dried me up and I never can grow again. Most of the merry little cousin spores that played with us have dried up and died. What are you smiling at?"

"I am smiling," said the Jar Spore, "at your calling the Wind a cruel being."

"And why shouldn't I? Do I not well know?" asked the Floor Spore? "I wonder, Floor Spore, what we *do know!* People are very sure as to what they know and then they find out that it is a mistake."

"What makes you think that?" inquired Floor Spore.

"I have learn't it," replied Jar Spore "from an acquaintance I have made here,—Mother Hubbard. She just said that the weather caused the peas to mould and now—"

Just at that moment the door opened. Faith came in and began to look around with wide staring eyes. "Why mother," cried the maiden, "What has happened?"

"That horrid wind!" wept the mother in despair, as she threw the dripping mass into the garbage pail.

"Whew-w-w," said the Wind angrily. "It is always some one else that is to blame. You called me horrid. Why did you open the window and invite anything 'horrid' to come in?"

"I thought that you would cool the room. I mistook your hateful temper. I know you now! Must I lose my preserves? Must my patriotism go for naught? Ay, whistle on in your joy."

"Fool! It is no joy to me to see your jars of peas spoilt nor your moldy views upset. It is my duty to help the peas to grow, to bring rain and warmth. I destroy germs. It is ignorant people like you that turn good into evil. You have turned me toward your ruin. What ignorant parents brought you up and did not teach you the laws of nature?"

"My poor mother!" wept Faith; "how unkindly you speak to her! But you are nothing but the wind. You know not what she does for me, her only child. She takes me into society, I have beautiful gowns, and fairy stories to read."

"Even so?" swayed the wind, "accomplished in the laws of fashion that changed but yesterday—unacquainted with the simple realities of life which have worked through the ages. Oh, that you knew the laws by which I live."

The Wind stole out of the window and across the garden. "I may be of service yet," said he. "What a foolish world."

* * * * *

Little Truth rambled about the fields gathering wild flowers and running after birds and insects. It was her mother who first taught her where to find the gentians and bluets and about the beauty of the hills.

Truth never wearied of watching the garden. She used to throw herself upon the ground and watch the bean plants. One day she spied rose colored spots on the bean pods. She had never seen them before. She thought she knew, and running to her mother, shouted, "Mother! there are roses on the beans!"

Truth's mother took the little girl on her knee, and tried to explain that the colored spots were accidental. Roses could not grow on beans. Truth was very silent, and then asked, "Why?" The mother sighed, as she did not understand these spots herself.

The next day was Sunday. Truth and her mother walked to church. Strange to relate the preacher talked about the colored spots on the beans. He called it blight. Truth heart beats' very fast for she was to hear about the roses. But, alas! The pastor told of the wicked beans and the Divine Wrath, and prayed that his congregation take warning. The little girl began to cry and the distressed mother had to get up and leave the church, leading Truth by the hand.

The next day found Truth in her favorite haunt. She was watching the beans with a look of pity on her face when she was interrupted by the voice of a stranger. The stranger smiled and said, "What are you doing little girl?"

"I am looking at the colored spots on the bean pods."

"And why are you looking so sadly at the bean plants?"

"I am so sorry for them!" cried Truth. "I am so sorry that God is angry with them."

"What makes you think that God is angry with the bean plants?"

"Why the preacher said so in his sermon."

The stranger nodded with a smile and placed his hand on Truth's head and said: "I will tell you a secret, little girl. I suspect that the preacher never studied the color spots on the bean and may not know very much about them." As he spoke he took a lens from his pocket and let Truth look through it at one of the colored spots. "That is as much a plant as the bean," continued the stranger.

Truth could hardly speak. A look of admiration came upon her face.

Then he touched the red spot with a needle and placed a tiny speck under the microscope. "These are like seeds but we call them spores," he went on to explain. "Do you wish to know more about them?" Truth eagerly nodded her head. The stranger now proceeded to explain. "Each of these little beads is a spore. They are so small that we can only see them through the microscope. The wind blows them around and when they land on the bean plant they send small threads into the pods or leaves. The red spots then appear and in a few hours there are thousands of more spores ready to be blown around. So now, little girl, you know why those red spots are on the bean pods."

* * * * *

Meanwhile the wind had heard the conversation between the kind-hearted stranger and the little girl. The wind was happy now, and said, "I have at last found some one who can face the truth and explain it in simple words. I am thankful that some people are searching out the wonders of nature instead of blaming her for what they do not know." With these words he whistled merrily and danced away to do his part in the world of natural laws.

CHAPTER XLV

Rhode Island Trees: A Type Study in Tree Geography

At the end of the Glacial Period there were no trees in Rhode Island. It is interesting to trace their source since that time. It is not mere accident but natural causes that have determined what trees selected this part of New England for their home. The character of the climate as tempered by Narragansett Bay, the sand plains along the coast, the granite hills of the west, the moisture in the soil—in short,

Fig. 1. A Black Spruce Swamp at North Scituate. A Labrador Scene in Rhod. Island. The Black Spruce Grows but a Few Feet in Height. Note the Leavee of the Leather Leaf or Cassandra Peering Above the Snow on the Lower Rights

all of the forces that make up the environment of a tree—have had to do with the migration of trees into our valleys and up our rugged slopes. A knowledge of the sources of our generous supply of trees awakens admiration and wonder.

I. Rhode Island the Meeting Place of Trees.

A. Trees of a Northern Range

The Black Spruce. In the new reservoir basin at North Scituate there is a black spruce moor. As this is the only accessible area in which the black spruce occurs in Rhode Island, tree lovers should visit it before the flooding of the reservoir destroys it forever. This cold, deep bog is a refrigerating spot where the roots are immersed in cold water until late in the spring. The roots cannot absorb nutrition at a freezing temperature and only certain plants are able to thrive there. Years ago the black spruce found this formation, gained a foothold on the margin, and later spread over the center. The black spruce is a northern tree and extends along the tributaries of the Yukon in Alaska. The tree is never more than 15 to 20 feet high, and bears fruit when it is two or three feet high. Sphagnum, the pitcher plant, and the creeping snowberry or moxie plum (Chiogenes hispidula), which is common in Labrador, grow beneath this spruce. Strange is it not that we should find a bit of Klondike scenery right here in Rhode Island.

The Canoe Birch. The canoe birch is also known as the white birch and the paper birch. We often call the gray birch the white birch, but this is not correct. The canoe birch is one of our most beautiful trees. It can be easily identified by its white bark, which is easily peeled. There are considerable numbers growing wild on Diamond Hill, and it is ornamental in our parks and private grounds. This tree comes to us from the wilderness region of Labrador. Every school child is sure to call to mind Hiawatha's request, "Give me of your bark, O Birch-Tree!" when he wishes to make a birch bark canoe.

> "Give me of your bark, O Birch Tree!
> Of your yellow bark, O Birch Tree!
> Growing by the rushing river,
> Tall and stately in the valley!
> I a light canoe will build me.
> —*From Longfellow's "Hiawatha."*

The Yellow Birch. Another essentially northern tree is the yellow birch. It has often made the members of the Maine Club feel at home in Little Rhody. Rhode Island boys and girls also have a chance to go to the "yellow birch swamp." They can recognize it by the silky lustre of the bark and by the way it curls into shreds. This tree may be seen in the Grotto of the Butler Asylum grounds and in the Metcalf Botanical Garden. This is an example of a tree that grows chiefly in a cold swamp, and like most swamp trees, is northern.

The American Beech. The beech is one of the characteristic trees of the northern woodlands. Poets and artists might find much of charm in the majestic beeches growing in the Butler Hospital ravines. The smooth steel-gray trunks make them easy to distinguish when walking through the Moses Brown Woods. And it gives a touch of the historic to know that some of our oldest beeches may have furnished shade for Roger Williams.

> "Oh, leave this barren spot to me!
> Spare, woodman, spare the beechen tree!
> —*Thomas Campbell.*"

B. Trees of a Southern Range

The Pin Oak. Rhode Island is the northern limit of the pin oak. The tree occurs wild in the border-land of the great Kingston Swamp and along the banks of the Pawcatuck River. The tree is more easily recognized in winter. It is a straight-trunked tree with slender branches, most of which are horizontal. It has been frequently used for ornament in Providence and may be seen in front of the Union Station, along the Blackstone Boulevard, and in Swan Point Cemetery.

The Red Birch. This is another southern tree that haunts the river borders. Its reddish, ragged bark makes it picturesque. It can be identified easily in the shrubbery of the Moses Brown School, where it forms a pleasing spectacle amongst the forsythias. There are several specimens in the Metcalf Garden.

The Black Walnut. There is some doubt as to whether the black walnut was ever native to Rhode Island. Several trees are thriving in Apponaug. There are two black walnuts on the southwest corner of Cypress and Ivy streets, in Providence. This tree grows as far south as Florida. It has been said that it takes a century for the tree to reach market size. The wood is so valuable and so much sought after that the tree has become almost extinct. The tree played an important part in the world war in furnishing wood for rifle stocks and airplane propellers. The boy who starts a black walnut farm will not only be making a good investment, but will be performing a patriotic duty, besides saving a most valuable tree for the future.

The Yellow Wood. The yellow wood is an ornamental tree, as in fact are all the other southern trees which we will mention. It grows wild in a very limited area between North Carolina and Alabama. Rhode Island people are very fortunate in being able to see this tree, especially when it is covered with its white blossoms. This rare tree may be seen in Roger Williams Park, on the Metcalf grounds, and west of the small pond in the Butler Hospital grounds.

The Red Bud. In early spring the red bud decorates the hillsides from New York to Florida. Its deep tint and profuseness suggest the peach orchard. The red bud may be seen in the shrubbery near the gymnasium of Pembroke Hall, Brown University.

The Kentucky Coffee Tree. This rather curious tree is related to our honey locust. As the name suggests, the fruit was used by the colonists of Kentucky as a substitute for coffee. This rare tree grows wild from New York south to Tennessee. Rhode Island people are fortunate in seeing this ornamental tree. One is located at Roger Williams Park, and another may be seen on the Normal School grounds, being the class tree in the spring of 1916. Another representative hangs over the wall on Power street, near Brown street.

The Catalpa Tree. The catalpa was once confined to the south, but is now naturalized in Rhode Island. It is quite common in our parks and along our streets. One street is called Catalpa Road. The tree is also known as the Indian Bean, Candle tree and Bean tree.

C. Trees from Abroad

When Roger Williams came, all our plants were native. The region was a wide expanse of forests, and beneath them were tender wood-loving plants such as the ladies' slipper and the Jack-in-the-pulpit. The coming of the colonists upset the flora, and as civilization advances trees disappear and along with them native plants. The killing off of native species gave chance for foreign plants. As the forests of Europe were largely destroyed ages ago, the plants are those of the open and their coarse hardy features have been inbred for years. The introduction and thrift of such European plants as dandelion, mallow, plantain, chickweed, burdock, mullein, sorrel, yarrow, and toadflax are indications of the passing of our forests.

Along with the weeds have come various trees. One may take a "tree trip abroad" in Providence. Some of the more commonly introduced species are mentioned.

Fruit Trees. The *apple, pear,* and *quince* were introduced to America from the Old World. They were brought on account of their fruit value. The apple frequently escaped from cultivation, and in old pastures has assumed a bushy character and is often protected from cattle by thorn-like branches. The fruit reverts to its ancestral, wild flavor when growing under these conditions.

> "High o'er the mead-flowers' hidden feet
> I bear aloft my burden sweet."
> —*The Pear Tree—William Morris.*

Rhode Island might well be called the cradle of the apple industry of America. The first horticulturist of this part of the country was William Blackstone. His orchard was planted in a region which is now a part of Rhode Island. He is said to have originated a new species of apple. The state is a natural orchard area, and produces finely flavored fruit, as proved by the Rhode Island Greening. The southern slopes of the hills and land suited for farming but not so used, should be covered with orchards. A generation ago no one would dream that apples would cost more than oranges in our markets. The demand is becoming greater. The population of Rhode Island has been vastly increased, but the production of apples is constantly decreasing. This is largely due to the neglect and destruction of old orchards. We can no longer rest on the reputation of past history. The time is ripe for progressive action in Rhode Island fruit growing.

> "Ho! The little-red-apple Tree!
>   Sweet as its juiciest fruit
> Spanged on the palate spicily,
>   And rolled o'er the tongue to boot,
> Is the memory still and the joy
>   Of the Little-red-apple Tree,
> When I was the little-est bit of a boy
>   And you were a boy with me!"
> —*James Whitcomb Riley.*

The Beeches, the Purple, Fern-Leaved, and Weeping Beech. These are all varieties of the European Beech. They may be seen at Roger Williams Park and elsewhere.

Maples from Europe, the Norway and Sycamore Maples. These introduced maples are common along our streets and in our drives. In winter one may distinguish them by the red buds of the Norway and the green buds of the syc-amore. The leaf-stem of the Norway maple has a milky juice.

European Poplars, the Lombardy and the Silver Poplars. The Lombardy Poplar is supposed to have originated in Lombardy. It also is native to the mountains of Afghanistan. It never produces seed in America, and has to be reproduced by cuttings just like a geranium. It may be readily distinguished by the vertical growth of its branches. The tree is very common in Providence. There is a conspicuous hedge or screen of Lombardy Poplars on the Blackstone Boulevard, and another on the State House grounds.

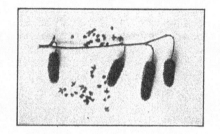

Grey Birch Catkins and Seed

The Silver or White poplar is so-named because of the white down on the under side of its leaves. The tree was brought to Rhode Island by the early colonists, and thickets of these trees often mark the yard of an old homestead which has long since disappeared. This tree is more apt to frequent the older villages and farms.

Other Tree Immigrants from Europe. Specimens of the beautiful *English Elm* may be seen on College Hill, near 54 College street, and on Benefit street, near Power street. The spring class of 1917 planted an *English White Oak* on the Normal School grounds. Another tree of this species is on Benefit street, near Star street. Several large *Yellow Willows*, a handsome winter tree, grow at Nayatt. *Austrian Pines* may be seen at Roger Williams Park, and in front of the former Morris Heights School building on Morris avenue, Providence.

D. *Tree Settlers from Asia.*

The Ginkgo. This oriental tree is said to have traveled from China to Japan, thence to England and thence to America. It may be recognized by its fan-like leaf, which has given it the name of Maiden-hair Tree. It is a sacred tree in Japan. There are several of these trees in front of the Providence Public Library and along the Black-stone Boulevard near Swan Point. It is a relative of the pines and spruces, but like the larch sheds its leaves.

The Chinese Chestnut. This tree has been introduced to take the place of our American Chestnut, which is threatened with extermination by the chestnut blight. There are several in the nursery at

Roger Williams Park, and two were planted in the spring of 1915 on the Normal School grounds.

> "The gray hoss-chestnut's leetle hands unfold,
> Softern a baby's be a three days old."
> —*Lowell.*

Horsechestnut. The horsechestnut tree is a native of southern Asia. Anna Botsford Comstock writes: "The wealth of children is, after all, the truest wealth in this world; and the horsechestnuts, brown and smooth, looking so appetizing and so belying their looks, have been used from time immemorial by boys as legal tender—a fit use, for these handsome nuts seemed coined purposely for boys' pockets." Every Providence boy knows the location of a horsechestnut tree. Bowen street is noted for its horsechestnut shade trees.

Ailanthus Tree. The ailanthus or Tree of Heaven came from China. It is quite extensively planted in parks and private grounds throughout the city. It probably has the largest leaf of any tree in Rhode Island, often reaching a length of three feet. The name Chinese Sumach tells the character of the leaf. The tree was first planted in 1820 on Long Island, and was probably literally blown into Rhode Island, as its winged seeds were wafted on a favoring wind.

The Peach Tree. This tree came from Asia. We have already seen how many of our fruit trees originated in Europe. In Japan cherries and peaches are cultivated as ornamental trees.

White Mulberry. The white mulberry is probably a native of China. It is interesting to know that the leaves are fed to silkworms and the tree has gone westward or in advance of the silkworm. The tree reached England in the early part of the 17th century, and America about 1830. It grows on Copley Lane in Providence and near the Arbor Vitae hedge in the Metcalf Botanical Garden.

II *Rhode Island Cross Roads of Trees.*

A. *The Sandy Trail.*

It is thought by some botanists that after the Glacial Period, Nova Scotia, Cape Cod, Southern Rhode Island, and New Jersey formed part of a sandy coastal plain. Georges Bank would have been dry land at that time. Since that time the New England coast has been slowly sinking and the plants of the ancient coastal plain have retreated to these sand area pockets. Consequently similar plants are isolated in these regions. The ancient coast area formed a north and south sand plain for the migration of plants that could adapt themselves to that kind of a soil. Many of our coastal trees, therefore, skip the rockbound coast of Cape Ann, New Hampshire and Maine to reappear in the land of Evangeline, or toward the south show up on the sandy stretch of Cape Cod, on Nantucket and Marthas Vineyard, east of New London, and in New Jersey.

The American Holly. In Rhode Island the holly is common in South Kingston and Little Compton It follows the coast into the sand areas south of Boston, to Quincy and Norwell. The tree's habitat extends south to Florida. There is danger of its being exterminated in Southern New England, as great quantities are sent to market in Boston and Providence during the Christmas season.

The White Cedar. The white cedar is very common in the Kingston Swamp. There are a few stands as far north in the state as Rumford. It is very rare north of Boston except at Cape Breton Island and in Nova Scotia, where it is found quite abundantly. The tree is found along the coast south to Florida. Associated with cedar bogs and ponds is the sedge spike rush (Eleocharis interstincta), which follows the coast into the tropics, actually growing in Cuba and the Isle of Pine.

The Post Oak. The post oak is found along the shore at Wickford. It grows on the same sort of sterile soil on Cape Cod as far as Brewster. It is not mere chance that the post oak and pitch pine do not grow on Block Island. The island has sand dunes in one part, but in the main it is a great block of clay. The tree is found southward to Florida, obtaining a greater size as it nears the more favorable climate of the south.

B. The Granite Way.

The granite hills of the western part of the state form a cross road to the south for the trees that prefer the granite soil. These same trees are found in the granite areas of Labrador, along the granite coast hills of northern New England and across Rhode Island, and then follow the Appalachian ridges to North Carolina. They find the same climate on the mountain tops of the south that prevails in the lowlands of the north.

Mountain Maple. This tree is found in Nova Scotia and along the mountains to Georgia. It is occasional in northern Rhode Island. We are not all fortunate enough to spend our vacations in mountainous New England or in Nova Scotia, but we still have the opportunity to admire this beautiful colored maple.

Mountain Ash. The mountain ash is not a relative of the ash, but belongs to the rose family. It is very beautiful in the autumn, when its fruit becomes a bright red. This tree grows in cool swamps and on the mountain slopes of the north, and is found occasionally in the northern section of Rhode Island, extending along the mountains to North Carolina.

The American Larch, and Fir Balsam. These two conifers prefer cool swamps and grow as far north as the Arctic Circle. From Labrador they extend along the mountains to New Jersey and Pennsylvania, being absent along the coast. The larch grows in the Moses Brown grounds and in the Metcalf Garden.

> "Give me of your roots, O Tamarack!
> Of your fibrous roots, O Larch Tree!
> My canoe to bind together,
> So to bind the ends together
> That the water may not enter,
> That the river may not wet me!"
>
> *—Longfellow, in "Hiawatha."*

The Hemlock, White Pine, Red Oak, American Hornbeam, and Black Birch are other trees which have crept in from the north. They prefer to occupy the cool spots and grow south along the Alleghanies to Georgia. All of these trees except the hornbeam may be seen at Roger Williams Park.

> "This is the forest primeval. The murmuring pines and the hemlocks,
> Bearded with moss, and in garments green, indistinct in the twilight,
> Stand like Druids of eld, with voices sad and prophetic,
> Stand like harpers hoar, with beards that rest on their bosoms."
> —*Longfellow.*

III. A Typical Tree Walk in Rhode Island.

We have just learned that Rhode Island has a cosmopolitan forest. Its conditions of soil and climate make it a natural assembly ground for trees. A tree walk in Providence will give us an idea of a tree census and the value of such a census to our citizens.

This tree walk will start at the corner of Barnes and Prospect streets and will continue along the south side of Barnes street. On the opposite corner is a broken down elm tree. Note the woodpecker's home. He learned that the tree was dead some years ago by tapping it and decided to set up housekeeping. Just back of this tree are two old cherry trees. The forester would say that they have been "dehorned". This is a harsh treatment and it is doubtful if they will recover, and it certainly does not improve their appearance.

1.   The large tree under which we stand is a *Norway Maple.* Look at the deep furrows running up and down the trunk. The fruit stalks in the top of the tree show that it had many "keys" last year. Can you find any seedlings which have come from the fallen seeds? Find a seedling two years old.

2.   This is a *sugar maple.* Compare the size of the twigs with those of the Norway maple. How does the bark differ?

3.   *Silver Maple.* The bark of the silver maple is scaly.

4.   *American Elm.* The flower buds make black spots against the sky.

5.   This is a *Maple.* Look back and see which one of the three maples it most nearly resembles. In back of this are *pear trees.*

6.   We have just seen this kind of tree. What is it?

7.   What kind of a maple is this?

8.   Examine the trunk and name this tree.

9.   *Yellow Wood.* What other trees have we seen that came from Europe? In the yard in back of this tree is an evergreen with large leaves. It is the rhododendron, which is the state flower of West Virginia. What is our state tree? To the left of the evergreen is a *magnolia.* Note the large buds. What will come from them? There are *pear trees* in back of the magnolia.

10.   *Horsechestnut.* From what continent were the horsechestnuts introduced? This is an example of tree dentistry. Why was an

Method of Presentation. Have the class collect several locusts. They should make the following observations during the collecting trip to report to the class. Describe the home of the locust. How does it move about? What is its color? How does it escape? Have the pupils bring the insects to school in a paste board box. Holes should be made in the box so that the insects can breathe. Have fresh clover, tumblers and blotters for covers on the front table. Let each student take a tumbler and a few pieces of clover. Place one or two locusts in a tumbler, cover the opening with blotting paper and invert on desks. Do not become alarmed at the escape of any of the locusts.

Agassiz tells in his notes, of giving a lecture to an audience, each one of which was holding a grasshopper. Occasionally one of the hoppers would escape and Agassiz would stop the lecture until it was caught. He considered this a part of the training. This is preparing the pupil for later work. There should be a chart drawing of the mouth parts, as shown in Comstock's Handbook of Nature Study, Pg. 368 or Linville and Kelly's Zoology, Pg. 3. The mouth parts should be greatly enlarged and labeled. Use common names, such as upper lip, lower lip, jaws, tongue and helpers.

After hearing observations made in the field prepare the students to observe the grasshopper eating. Bring out the fact that each animal has its own way of eating. When we eat corn we hold the cob horizontally, move the corn instead of our head and usually begin at the left end. We do it this way because it is the easiest. The class is now given some time to watch the grasshopper eat. Pupils should note the mouth parts and the manner of eating. (Holds food vertically because the jaws work from right to left like a pair of scissors. Begins at the top and eats down.) If the class cannot discover these points without some aid, ask a few questions, such as— Do the jaws move up and down or from right to left? Does it move its head or the food? Hold a grasshopper and place a pencil point in its mouth.

The Legs. Why does the locust have such large hind legs? (Jumping). For what does it use the forelegs? (Holding food.) Why does it need the middle legs when feeding? (To keep its balance.) Why does the locust need three pair of legs?

How does the grasshopper hold his hind legs just before jumping? (Have a boy take a position as though he were going to jump. Class note that he bends the knees and then straightens them.) How far can he jump? What is the advantage of this? What enables the grasshopper to climb the side of the tumbler? (Pads on bottom of the feet which secrete a sticky substance.) What else keeps the insect from slipping when climbing grasses? (Claws and spines.) Compare with men who climb telephone poles. Note the thread like affairs on top of the head. For what are they used? (Feel.) In how many directions can the grasshopper feel? (All.) What structure of the feelers enables the grasshopper to feel in all directions? (Jointed.)

operation performed on this tree? What material was used for filling? Why? Find some of last year's leaf stems on the ground. Carefully bend a limb down and find the places where the leaves were attached. What kind of a fruit tree is in back of the horse-chestnut?

11. *Sugar Maple.* Note something unusual about the large limbs in the top of this tree. This was done during the winter. What do you suppose did it? How many limbs were affected this way last winter?

12. *American Elm.*

13. What kind of a maple is this? What state is noted for this kind of tree?

14, 15. What are these trees? Find a bird box on the elm on the opposite side of the street.

Cross Brown Street, keeping on the same side of Barnes.

1. Note the young *elm* with a *forsythia* beneath it in the corner of the yard.

2. This is a large ——————? How many main trunks does it have? In what direction do the ends of the twigs point? This is one way to tell this tree. What makes the knot-like appearance on the twigs? Look at the small white birch or canoe birch in the back of the yard.

3, 4, 5, 6, 7. These trees are sometimes called rock maples. They grow in "orchards" in northern New England. What name have we already given these trees?

8. The buds will tell you the name of this maple. Observe the three clusters of Gypsy moth eggs on the northern side of the tree. What is the advantage of their being on the underside of the limb?

9. Name this maple.

10. Observe the buds and name this tree.

11, 12. What are these trees? Note the Norway maple across the street.

Moses Brown Walk. Just inside the gate is an elm. To the far left by the shrubs is a young blue spruce. To the right by the tennis court is a pignut hickory. Along the left of the path alternate sugar maples and Lombardy poplars. Which have branches curving sharply upward? Which are taller? These trees grow faster and when they become old and decrepit they will be cut down and the tops of the sugar maples will spread across the open spaces. Find an elm with a squirrel's winter home. On the right is a black oak with one main trunk. It has kept some of its leaves on all winter. This is characteristic of oaks. On what part of the tree have the leaves remained? Why are there none in the other part? Note the white oak with many wide-spreading limbs which are lighter than the black oak. In back of this tree in the shrubbery are several red or river birches. Suggest how they got their names? Next we come to an old apple tree. Note the pruned limbs which have commenced to heal. This is called tree surgery. Note the lines of small holes on the large limb which leans easterly toward the flag pole. These holes were made by the sap-sucker. This bird is a kind of woodpecker and is related to the flicker whose home we saw on the beginning of the trip. The large, dark tree over on the left is a larch. It sheds its leaves, but the cones show that it is a conifer.

IV. *The Rhode Island Tree University.*

You must now realize that every Rhode Island community is an arboretum with trees from the North, trees from the South, trees

JUNIPER   PIN OAK   UMBRELLA PINE   SPRUCE
(Native to R. I.)   (From the South)   (From Japan)   (From the North)
Fig. 3. Rhode Island is the Meeting Place of Trees. It might well be said that they come "From Greenland's icy mountains to India's coral strand."

from across the Atlantic and trees from the Far East. A great collection of trees is here. The next step in a well-organized arboretum is to label the trees. This spring many classes will make graduation gifts. Why not label a few trees? Let us make Arbor Day, 1920, a notable event by founding the Rhode Island Tree University. I know of nothing so inexpensive that will bring such satisfactory returns on the investment. To know our trees is the beginning of life-long acquaintances. To be interested, to be able to call them by name, to protect them—mean increasing civic pride. Such a gift will endow an educational system that not only works during school time but after hours and through vacations. It will afford an extension course for all the time for all the people.

CHAPTER XLVI

"GRASSHOPPER" A TYPE LESSON IN INSECT-STUDY

The Locust

Subject Matter. Both the name and the strong hind legs that this insect is a jumper. When it straightens its hind l sent for a remarkable distance through the air. The wings al in escaping from the many bird enemies. When the gras alights it is not easily seen as its color resembles the grasses

The grasshopper is well equipped for climbing. When it ing up the side of a tumbler we can see the small pads whic it to stick to the glass. Its claws are also useful in climbing The spines in the hind legs are used as a comb for cleaning oth

The two pair of wings make an interesting study. The fi are hard and thick forming a protective cover for the und The front wings are held erect when flying, the hind wings l only ones used in flight. The delicate flying wings are fold fan beneath the covers when the animal is walking through t thus preventing them from being torn by the sharp blad ability to conceal the bright colored back wings beneath the p color of the forewings, is an advantage when pursued by bi

The mouth consists of two lips, a pair of black bony jaws t sideways like scissors, and a pair of helpers or arrangers. T leaf or grass blade is held in a vertical position by means of feet. The insect then cuts the grass downward. It keeps up and cutting the blade downward until the food is devoui til its hunger is overcome. The "molasses" is partially dige used in defense. The great appetite of the grasshopper en to comprehend the locust plagues spoken of in the Bible Exodus) and of its recent damages in Kansas and Argentir

The grasshopper has two large eyes and three single one is a small eye in front of each big eye and one in the cen forehead. The feelers are also thought to be organs of sm joints enable the insect to feel in all directions. The breatl may be seen on each segment of the abdomen. The ears first segment of the abdomen and are well protected by t

The female grasshopper has four projections at the end o which are used for burrowing. The eggs are laid in the fa made by the female in the ground. The young hatch out in They resemble the adult except in size and in the absence Fall plowing exposes many eggs and is one way of keepir check.

Grasshoppers have different ways of singing. Some rub legs together, others rub the back leg against the first pai and still others rub the wings together.

The true grasshopper has feelers longer than the body. which we commonly call grasshoppers, are locusts.

The Wings. How many pair? Which pair does it use in flying? (Class watch closely as the grasshopper flies.) What does it do with the hind pair when walking? What are the advantages of this? How is it able to place the larger hind pair under the front wings? Show a mounted specimen with wings spread.

Other Observations for the Pupil. The movement of the abdomen is caused when the insect breathes. The breathing pores are along the sides of the abdomen. The ear is under the wings on the first ring. Note the two large eyes and three smaller ones.

In the grammar grades the questions may be written on the board and the pupil allowed to think out the answers at leisure. In the primary grades the teacher develops the lesson, one question at a time. In the grammar grades the results may be tabulated as the pupils give the answers.

As—*Description*   *Advantage*

Green or brown color. . . . . not easily seen by enemies.

Jointed antennae. . . . . to feel in all directions.

The general vivarium for the locusts should consist of a cage made of netting. Moistened earth could be placed on the bottom and fresh grass given to the insects daily. Wheat or grass could be raised in pots and placed in the cage.

Suggestions for Correlations

Drawing. An enlarged drawing, using a whole page of drawing paper is a good test as to whether the pupil sees all the parts. The side view and a top view with wings spread would be worth while.

Written Language. A description or a story of the life of a grass-hopper would give good opportunity for originality of expression. Written language in the school is for the purpose of expressing clearly what one already knows, and not for the purpose of originating new ideas. Most of us are incapable of the latter.

Oral Language. The pupil might describe himself as a grasshopper leaving out the name. This description might continue until every-one in the class recognized what was being described.

Story. Read or adapt some story for the class.

Literature Quotations

John Burroughs, Signs and Seasons—Page 24, speaks of seeing grasshoppers on a warm, thawy day in February. "The grass hatches out under the snow, and why should not the grasshopper. And yet, if a poet were to put grasshoppers in his winter poem, we should re-quire pretty full specifications of him or else fur to clothe them with. Nature will not be cornered, yet she does many things in a corner and surreptitiously."

Homer writes of the voices of old men, too old to be in the army, who

"In summer days like grasshoppers rejoice,
A bloodless race, that send a feeble voice."

"Green little vaulter in the sunny grass,
Catching your heart up at the feel of June,
Sole voice that's heard amid the lazy noon."
                                    —*Leigh Hunt.*

"The poetry of earth is never dead;
When all the birds are faint with the hot sun,
And hide in cooling trees, a voice will run
From hedge to hedge about the new-mown mead;
That is the grasshopper's. He takes the lead
    In summer luxury."
                                    —*John Keats.*

"Where the dusty highway leads,
High above the wayside weeds
They sowed the air with butterflies,
    like blooming flower seeds,
Till the dull grasshopper sprung
Half a man's height up, and hung
Tranced in the heat with whirring wings
And sung and sung and sung."
                                    —*James Whitcomb Riley.*

"The flying grasshopper clacked his wings
Like castanets gayly beating."
                                    —*Elizabeth Akers.*

References:

Boyden, A. C. 1900. Nature Study by Months. The New England Publishing Company. Excellent suggestions as to method.

Comstock, A. B. 1912. The Handbook of Nature-Study. The Comstock Publishing Company. The best all-around book for teachers.

Daulton, Agnes M. 1905. Autobiography of a Butterfly and other Stories. Rand McNally & Company. The Insects' Fiddle-Dee-Dee. An excellent story for children which is scientifically correct.

Fairchild, David. 1913. The Monsters of our Backyards. In the National Geographic Magazine, May 1913. Contains thirty-nine full-paged pictures of insects which are excellent for class work. This magazine can be obtained at 25 cents per copy by writing to the National Geographic Society, Washington, D. C.

Foot, Constance M. 1909. Insect Wonderland. John Lane Co., New York. Grasshoppers Lane, Pg. 139–157. A conversation between a grasshopper and a field-mouse. Good for oral reading in the Grammar Grades. Can be easily adapted to other grades.

Linville and Kelly. 1906. A Text-book in General Zoology. Ginn & Company. Written for secondary schools. An excellent, up-to-date text for the teacher with good suggestions as to method.

Morley, Margaret W. 1903. Insect Folk, Pg. 59–113. Ginn & Company. Written for children but has too much detail and expects the use of scientific names. Good for subject matter and in illustrating tact at certain points.

Morley, Margaret W. 1907. Grasshopper Land. A. C. McClurg & Company. A very complete treatise with an index. Chapter XVI, the Diary of a Locust, good for grammar grades.

Riley, C. V. Destructive Locusts. Bulletin 25. The U. S. Dept. of Agriculture, Washington, D. C.

Schwartz, Julia A. 1910. Grasshopper Green's Garden. Little, Brown & Company, Boston, Mass. An excellent story for grammar grades. Could be easily adapted for an oral story in the primary grades.

Wilson, Mrs. L. L. 1897. Nature-Study in Elementary Schools. The Macmillan Company. Suggestive as to method.

CHAPTER XLVII

Suggestive Lessons in Bird-Study: the Woodpecker

The following lessons are suggestive for an introduction to bird-study in the grades. The Flicker is taken as a type, since it is a permanent resident, at least as far north as Massachusetts, and may become an acquaintance before the arrival of other species. Moreover, the Flicker is a good bird to know. This woodland drummer is venturing into cities where it is adapting itself to civilization. One has taken up its abode in a telephone pole, within sight of my home, and its reveille on tin roofs may be heard nearly every morning. It seems as pleased with this new invention as a boy with a new drum. An old barn at home has been a Flicker hotel for years. These facts may be an indication of how other birds might fall into civilized habits if we should meet them half way. If we can develop an appreciative interest in these things in our boys and girls, we will have taken a long step toward gaining this end.

Lesson I. Field Observations.

The teacher should become acquainted with a Flicker rendezvous, or retreat, as the species is usually solitary, and take the class to visit the place. The pupils must approach on the alert, "all eyes and ears," for any secrets which the birds may divulge. Suddenly one flies up from the ground. What color did it show when it flew? (White rump.) What was the path of its flight? (A wavy, up-and-down motion. When the wings went down the bird went up, and *vice versa*.) Someone should make a drawing on the ground, to show the manner of flight. If the pupils do not observe these points, they must sharpen their eyes for another trial. What was the Flicker probably doing on the ground? (Feeding.) All birds do not eat the same food. If we would like to know what the Flicker was eating when we disturbed its feast, let us walk to the place where it was feeding and investigate. What do we find that might be eaten by the Flicker? (Weed seeds, bayberries, black alder, poison sumac, and poison ivy berries. An ant's hill might be present, as this is a favorite morsel of the Flicker.) The Flicker eats all of these things that we have found. We might think that it is a good thing for the Flicker to eat the seeds of these poisonous plants, but it has been found that after the waxy substance on the outside of the berry has been digested the seed is thrown out from the mouth. These seeds will germinate and, since the scattering of poisonous plants is not desirable, this cannot be placed on the credit side of our account with Mr. Flicker.

Who saw where our friend went? (To an old apple tree across the field.) Let us visit the home of the Flicker family. On our way we may hear the Flicker call to its mates. If we do, let us try to tell what it says. After interpretations by the class, tell them how other listeners have read the call.

"If-if-if-if-if-if-if," Burroughs; "Up, up, up, up, up, up, up," Thoreau; "Wick, wick, wick, wick," Mrs. Wright; "Wake-up, wake-up, wake-up, wake-up," Dr. Charles Conrad Abbott; "Kee-yer, kee-yer, kee-yer, kee-yer," Chapman; "Yarup! yarup! yarup-up-up-up!" Dallas Lore Sharp. Does anyone think that this Woodpecker sings? In which does it excel, instrumental or vocal music? What kind of a

Fig. 1. Specimens to show WHY and How Dr. Woodpecker performed an operation.

musician might we call it? (Drummer.) Investigate and describe its drum. (A hollow dead limb.) Sometimes it telegraphs a wireless message to its mate; at other times it is a sort of an anvil solo, and quite frequently a duller beat in the search for food. Try to learn these sounds in the Flicker's signal code. As we get nearer, let us make an effort to see some of the Flicker's colors. (Black crescent on breast, golden shaft of quill feathers, and spotted underparts.) In what position is the bird resting on the tree? (Perched on a limb or clinging to the trunk.) Remember this is a Woodpecker, and most of its kind cling to trees instead of perching. The class should observe the position of the tail (outer end braced against the trunk) and, if possible, note character of tail-feathers. (Sharp, pointed ends.) Of what use is such a tail? (Acts as a prop.) Since Mr. and Mrs. Flicker have not set up housekeeping, we may look in at the door. In what kind of limb are they building? (Dead limb. Knock on the limb with a stone.) Why? (Because it is easier to dig out the decaying particles of wood.) Fathom the hold, to find how far it extends. (One to three feet.) What is the advantage of so deep a hole? (To escape enemies and better protect inmates from the weather.) Let the class look for places on the tree where a Woodpecker has been drilling. What was it after? (Grubs.) We may call the Flicker a *tree surgeon*. Why? (The tree is the landlord and Dr. Flicker pays rent to his Treeship by removing undesirable insect visitors. These insect lodgers do not pay rent and are injurious to the health of the tree.) We have found that Dr. Flicker sometimes eats things which reflect upon his good character, and at other times he eats things which make him very useful.

Lesson II. Indoor Observation.

Use stuffed specimens and pictures. The class should collect illustrative material such as that shown in Figure 1. The teacher may exchange material with distant schools. The portion of a tree, for instance, illustrated in Figure 2 came from the Pacific Coast. It shows the work of the California Woodpecker, a red-headed Woodpecker on the western edge of our continent, which drills holes and stores acorns in them for future use.

Review the field-trip, asking about the Flicker's flight, colors, home, call and food. The class is now ready to make close observations, and to study some of the detailed structures which fit the Woodpecker for its life, which has been observed in the field.

Lead the class to discover the difference between the male and the female. Mr. Flicker has a moustache. Madame Flicker, of course, has not. If all of the colors of the plumage were not seen on the trip, they should be noted now.

Compare the arrangement of the toes with that of the Robin. The Flicker has two toes in front and two behind, the Robin has three in

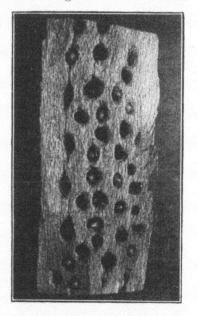

Fig. 2. The Carpenter Woodpecker stores acorns in trees. This is common in California. Material of this kind may be exchanged with Eastern Schools.

front and one behind. Who remembers something the Woodpecker was doing that it could not have done as well if its toes had been arranged like the Robin's? (Clinging to the side of the tree.) What was the position of its tail when it was clinging to the trunk? (It was bent under against the tree.) Look closely at the tail and tell how it differs from the Robin's tail. (It has sharp-pointed, stiff feathers.)

What use does the Flicker make of such a tail? (Helps hold itself on the trunk.) We call this kind of tail a prop. Tell the different ways in which the Woodpecker is fitted to cling to tree trunks. (The toes are arranged like ice-tongs for nipping, and the bird braces itself with its tail.) Why does the Flicker want to cling to the side of the

Fig. 3. Work of the Sapsucker and home of the Downy Woodpecker. As far as possible, material should be collected by the pupils.

tree? (To excavate for grubs, or to build a home.) What tool does the Flicker use for this work? (The bill.) In what way is its bill a good instrument for this work? (Sharp-pointed, stout and hard.)

The teacher may now tell the class the following story, using material such as is shown in Fig. 1 to illustrate the point. Yesterday,

we found places in the apple tree where Dr. Woodpecker had performed a surgical operation. (Open the sticks, which have been split.)
Inside of this tree were "worm tracks" such as are seen here. Worms
did not make these borings, but young beetles called grubs. They
correspond to the caterpillar stage of the butterfly. Dr. Woodpecker
came along and saw where Mr. Grub had broken entrance and decided
that here was a good meal. Now he did not start to get baby beetle
by boring in at the place where the grub entered, as perhaps you and
I would do. He held his head close to the trunk and listened. The
hard, dry wood is a good telephone, and he heard the grub clicking
away as he was digging his tunnel. Dr. Woodpecker, after his diag-
nosis, determined the nearest way to the worm and began to drill.
How could he get the worm out after drilling the hole? He has just
the right kind of an instrument for such work, his tongue. He thrust
his tongue through the white grub, drew him out and ate him. His
tongue is covered with a sticky substance which enables him to catch
ants. Three thousand ants have been found in the stomach of one
Flicker.

The Flicker is a carpenter, as well as a doctor. I am going to tell
you how he builds his home. First he outlines his doorway like this.
(Make a circle with dots.) He gets it just the right size. It is not so
large that cats can come in, and not so small that he cannot get in
himself. Could we draw a doorway just the right size for our house?
He then uses his bill as a pick and begins to chip away the wood, to
make a hole. He enjoys the work in the same way that we do when
we build a house. Fig. 3.

Lesson III. Comparisons.

Use stuffed specimens, pictures of other kinds of Woodpeckers,
and exchange material. Have the class discover points in which all
Woodpeckers are alike. How may we distinguish them? The Downy
and the Hairy Woodpeckers may often be attracted near school-
houses and homes by hanging pieces of beef fat in the trees. Fig. 4

SUGGESTIONS FOR CORRELATIONS

Lesson IV. Language.

Let the class suppose that they are Flickers, and tell about themselves. Ask each pupil to write a story on what one Woodpecker did
as he watched it for fifteen minutes. In schools where children
dramatize, it might be profitable and interesting to write a drama
with the Flicker, an apple tree, and a fat baby beetle as characters.

The Flicker affords an unusual opportunity for word study. Mr.
Colburn gave 36 common names of this species in the Audubon
Magazine for June, 1887. The Country Life in America, July, 1913,
says that there are 126 names. These names are nicknames, each
of which gives a hint of some characteristic of the bird. Have the
class determine which indicates the color, song, flight, and habits of
the bird: Yellow-hammer, Piquebois Jaune, Yellow-shafted Wood-

pecker, Yellow-winged Woodpecker, Crescent-bird, Clape, Cave-duc, Fiddler, Hittock, Hick-wall, Piute or Perrit, Wake-up, Yaffle, Yarrup, Yucker, Tapping-bird, High-hold, High-holder, and the High-hole.

The Woodpeckers have not attained the literary rank of the Bluebird, the Oriole, and some others. Walt Whitman speaks of "The High-hold flashing his golden wings."

Lesson V. Drawing.

Fill in outline drawings with colored pencils or water-colors. These outlines may be made on a hectograph. It is worth while to make different views, as a front view of the Flicker to show polka-dots and locket; side view, to show the moustache of Father Flicker or its absence in Madame Flicker, and the golden wing shafts; back view in flight, to show the white field mark, barred color scheme on the back, and the red patch on the back of the head. Simple drawings, to illustrate the story of the Flicker's activities, bring out skill and interest. Such a series of sketches might include the bird flying up from the ground; position on the trunk; head bent back for hammering; outline of a doorway; the completed mansion; the eggs in the nest; bringing food; the babies, with mouths wide-opened to receive the food, and the young on a limb receiving a lesson in flying. The food for the young, it should be explained, is invisible as it is partly digested in the alimentary canal. The process of feeding is peculiar since the food is literally pumped into the mouth of the young.

Lesson VI. Manual Training.

The construction of a home for the Flicker. Hollow out a small block of wood leaving the bark on the outside. The opening from the outside should have a diameter of two and a half inches. Modeling the home and eggs in clay is fascinating work for the younger grades. The Flicker does not build a nest. The eggs rest upon small chips, which probably fall to the bottom of the hole during the construction of the house.

Lesson VII. Music.

There are not many opportunities to correlate the study of the Flicker with music. The cry is rather difficult to imitate. The drumming is worthy of imitation in the elementary grades. Try to differentiate between the Flicker's drumming as a pastime and its picking for food. The noisiness of the Flicker may be contrasted with the music of some of our more accomplished feathered singers.

Fig. 1. Yellow-Crowned and Black-
Crowned Night Herons

CHAPTER XLVIII

THE BLACK-CROWNED NIGHT HERON
(*Nycticorax nycticorax nævius*)

A METHOD OF STUDY

I. Subject Matter.

The Black-crowned Night Heron is known more commonly when
called by one of its nicknames: Squawk, Quawk, or Qua Bird. In
some of the southern states it is known as Gros-bec, Indian Hen,
or Indian Pullet. The bird receives the first of these appellations
from its call as it flies to and from its hunting-ground late in the
afternoon or at night. Longfellow gives the proper setting, in
'Evangeline,' when he says,

"Deathlike the silence seemed, and unbroke, save by the herons
Home to their roosts in the cedar trees returning at sunset."

The Black-crowned is the most abundant and familiar of the Heron
family. A large colony of these birds has a breeding-ground on Cape
Cod, not far from Camp Chequesset, a girls' camp, where the writer
had frequent opportunity to visit the heronry and to experiment
upon the birds with the camera. This particular colony is in a
pitch-pine grove which is located near a marsh.

If one enters the rookery in daytime—which is usually bedtime
for this species—he finds things rather quiet until discovered. The
invader is then serenaded with a great din. The parents fly into the
air, squawking and cackling promiscuously. Blanchan likens it to
pandemonium, and Wilson compares the noise with that of two or
three hundred Indians "choking or throttling" each other. Such is
the heralding as one enters the *sanctum sanctorum of* herondom.

479

The housekeeping is no more inviting than the notes of greeting. The ground and trees are white with excrement, and a foul odor comes from decomposing pieces of fish which have fallen to the ground. Here and there are the remains of a young bird who did not meet the laws of arboreal life successfully. Should a visitor climb toward the nests, the young birds still further show their unsociability by disgorging their last meal. The Herons also have good allies in the mosquitos, whose method of attack might repel any human foe who attempted to harm the landlords of the settlement.

Fig. 2. A Banded Black-crowned Night Heron, two years old.

The nests of these Herons were ragged platforms of dead sticks built in the forks of trees about 10 to 20 feet from the ground. The birds, it should be noted, usually repair the old nests, which do not show the high degree of craftsmanship exhibited by other birds, while cleanliness is an unknown factor. In this latitude, the Black-crowned Night Heron returns about the second week of May, and as it is found breeding often into June, one may find the young in all stages of development during the midsummer months.

The baby Herons wear a coat of gray down and have a prominent head-crest. The one in the picture appeared to be dead, and I had to poke him several times before I learned that he was 'playing possum.' He would not hold up his chin, so appears as a fluffy ball in his photograph.

In less than a month these babies become nearly as large as their parents. This rapid growth is due to their enormous appetites. The old birds not only work nights but have to leave the homestead in the afternoon to keep the young well fed. They bring in fish, eels, and frogs and can ill afford to have 'fisherman's luck.' The food is softened and partially digested in the alimentary canal of the adult before it is served to the young. To have a family of three or four average 1.5 feet in height in their bare feet, in four weeks, is an undertaking that keeps both father and mother Heron working full time.

The young Heron has a dress more like the Bittern's than that of its parents. In this early period the color of the young blends well with the trees, making it more difficult to discover them. Sitting, day in and day out, like sentinels, except with a more expectant look, they await the return of the parents with food.

When I climbed a tree to take a picture of one of the young Herons, it began to climb away, and went rapidly to the end of a neighboring branch. If it lost its balance it regained it by using its bill. One unfortunate youngster fell to the ground and was allowed to pose on a limb. If disturbed on the perch the bird erected its crest, opened its cavernous beak, and spread its wings, presenting a terrifying appearance sufficient to drive away anyone having designs on its tender and plump makeup.

It was interesting to watch the adult birds feeding on the marshes and along the creek in front of the camp. They wade ankle-deep

Fig. 3. A Black-Crowned Night Heron in its Haunts.
(Photograph of Mounted Specimen)

(usually thought of as knee-deep), often standing still. When food is spied, the arched neck allows the bird to strike with great force. To a small fish the Heron's leg must resemble a stick, and the light ventral color must render its body imperceptible when looked at against the sky from below. The Heron's juvenile plumage enables the owner to escape becoming food for enemies, while the color of the adult is an aid in securing food.

The Black-crowned Night Heron is very widely distributed. It breeds from New Brunswick to Patagonia.* In this region it begins to migrate about the middle of October. Audubon says that the adults go farther south than the young.

---

*Distribution and Migration of North American Herons and Their Allies', Wells W. Cooke. 1913. Biological Survey, Bulletin No. 45.

II. *Method of Teaching.*

If there are Herons in the neighborhood, it is best to encourage pupils to observe these birds out-of-doors. Begin the work with an interesting description and ask a few questions to arouse the pupils' interest. If there is a rookery that can be visited it might be desirable to have a few pupils make a study of it. Observations of the adult feeding is an excellent training which is worth while for all the class. Pupils might have individual notebooks in which they could write answers to certain questions. These questions should be written on the blackboard when it is time to make the observations and the class be given at least a week for observation work.

A. *Observations.*

(a) *Questions for Observation at the Rookery.*

1. Try to enter the heronry without being discovered. If you are discovered: How did the bird discover you? How do you know that you were discovered?

2. What does the bird say? If you have heard the same call before, when did you hear it? Where did you hear it?

3. Where does the Heron carry its long legs when flying? How do they help it at this time?

4. What is the color of the underside of the Heron?

5. When the birds come back to the trees note the color of the legs, the eyes, the top of the head, the back.

6. Where are the nests placed? Of what material are they made? Where is the material obtained? How do you suppose it is obtained? Compare the nest with some familiar nest.

7. Do the young birds make any noise? What do the parents feed the young? How can you tell this?

8. What is the difference in color between the old bird and the young one?

9. Describe a baby bird.

(b) *Questions for Observation at the Marsh.*

1. Does the bird ever stand motionless? (Audubon says that this one never does.) Advantage of the habit?

2. Does this Heron prefer to walk or wade? If it wades, how deep does it go in the water? How is it adapted for its method of locomotion?

3. In what position does it hold the head? What is the advantage of this?

4. What does the bird do when it sees prey? Why does it do this?

5. In what position does it hold its head when flying? Why?

6. Describe its colors. Why would it be difficult for a fish to discover it?

7. Where does it place the legs when flying? How does this position help it? How would it be hindered if it did not do this?

8. What does the bird do when it hears a noise. (Note that birds differ in their response to a noise. The Bittern, for example, rather squats than flies.)

9. Try to discover what the bird eats.

(c) *Questions for Observation in the Laboratory.* (Preferably use a stuffed specimen, otherwise a picture.)

1. Describe the beak. What is the character of the edges of the bill? Advantage?

2. Compare the length of the tail with that of other birds. Disadvantage? How is it overcome?

3. What is characteristic of the legs? Why should they be so? Compare the growth of feathers on the legs with the growth of feathers on the legs of the Owl. Explain the difference.

4. Describe the wings. What does that tell you?

5. What is unusual about the toes? How does this help the bird?

6. What duck-like characteristic is found in the foot? How can this be of service?

7. Compare the length of the neck with that of the legs. Why should there be this relation?

8. What is the position of the neck? When would this poise be helpful?

9. What part of the eye is red? Look at the pupil of several birds. What color is the pupil in every case? How does the iris of the immature Night Heron differ from that of the adult?

Each pupil has now had opportunity to make careful observations, and each one has had the benefits of the training.

B. Organization.

Closely following the observation period should come the stage of organization. The teacher meets the class as a group. Questions about the observations made are asked and the results grouped somewhat as follows, the teacher writing down the facts on the blackboard as they are obtained from the pupils:

THE BLACK-CROWNED NIGHT HERON

Home: Marshes and Creeks.   *Food:* Frogs, eels, worms, fish, mice.

Color.....Black head................hence name.
　　　　　Light below...............not easily seen by animals it seeks.
　　　　　Dark above................not so easily seen from above.
　　　　　Red iris...................peculiar to some birds.
　　　　　Three white crown feathers...ornamentation at breeding-time.
Beak.....Large, lance-like...........killing prey.
　　　　　Sharp-edged................to hold slippery food.
Eyes.....Well forward...............for quick sight.
　　　　　Large pupils...............to see at night.
Neck.....S-shaped..................to dart at food.
　　　　　Length of legs.............to reach food.
　　　　　Drawn in when flying.......better balanced.
Legs.....Long.....................to keep body above the water.
　　　　　Most part unfeathered......not to get feathers wet.
　　　　　Straight back when flying....to steer.
Toes.....Long and wide spread......to distribute weight.
　　　　　Slightly webbed............to bear them up in soft mud.
　　　　　Back toe well developed......useful in perching as well as for bal-·
　　　　　　　　　　　　　　　　　　ancing
Tail......Short, weak feathers........steers with legs.
Wings....Long and broad............strong flight.

C. Conclusion

The pupil should be led to make the general conclusion, from facts gathered, that the Black-crowned Night Heron is a wading bird well adapted to its home and habits of getting food. Later the pupil will begin to see that to a great extent any bird is structurally adapted to its environment, that is to where it lives and what it eats. These two topics namely, home and habits, should come first, therefore, in considering an animal.

D. Comparison.

The Black-crowned Night Heron has been used as a type of Heron and of a larger group commonly known as wading birds. The pupil is now ready to understand related forms and to search out differences and possibly, to give reasons for variations. A visit to a museum, if convenient, may well be made or pictures of the various forms studied used. Questions for observations should be given for this lesson in the same manner as before. The following notes are subjoined for the benefit of the teacher who may add to them as occasion demands.

1. *Plovers.* Slender bill, to probe ground; wings long and acute; swallow-like flight; hind toe small, scarcely touching ground; builds nest on ground; toes not webbed, gathers food from upper part of beach in firm sand; each toe has lobes (water propellors), yet is suited to running along the beach.

2. *Avocet.* Long curved bill, to search out worms and snails in crevices and under stones.

3. *Woodcock.* Long bill, to thrust into mud for worms; end extremely sensitive, for the purpose of feeling for food; eyes far back; tongue secretes a sticky substance to help hold worms; brown-colored plumage; builds nest on ground in leaves. Wilson's Snipe is a close relative of the Woodcock.

4. *Jacana* has feet adapted for walking on floating lily-pads; spurs on wings used for fighting; found in South America.

E. Correlations.

Have the class investigate and make reports on the following:

1. *Literature.*
   Story of the Egret.
   Why our Shore Birds are Disappearing.
   The Stork.
   Child Stories: Hans Brinker, or The Silver Skates, by Mary
      Mapes Dodge, pp. 237–39.
      Wonder Stories Told for Children by Hans Christian Anderson,
      pp. 431–36.
      The Cranes' Express. An Old Tale. Adapted for Lower
      Grades.

2. *Drawing.* The Heron and its relatives have been used a great deal, especially in ornamentation, by the Japanese. The class might look over the chinaware and vases at home and bring in pieces that

illustrate the use of birds in art. Birds are picturesque whether wading or flying. A cover design might be made at this time for the note-books. Drawings to show the various forms of beaks and feet are interesting. Visit an art museum and study the use of the Heron in art.

THE DOWNY WOODPECKER
By Garrett Newkirk

The Downy is a drummer-boy, his drum a hollow limb;
   If people listen or do not, it's all the same to him.
He plays a Chinese melody, and plays it with a will,
   Without another drumstick but just his little bill;
And he isn't playing all for fun, nor just to have a lark,
   He's after every kind of bug or worm within the bark;
Or, if there is a coddling-moth, he'll get him without fail,
   While holding firmly to the tree with all his toes, and tail.
He is fond of every insect, and every insect egg;
   He works for everything he gets, and never has to beg.
From weather either cold or hot he never runs away;
   So, when you find him present, you may hope that he will stay.

"A thousand voices whisper it is spring;
   Shy flowers start up to greet me on the way,
And homing birds preen their swift wings and sing
   The praises of the friendly, lengthening day."
        —*Louise Chandler Moulton.*

"When Nature had made all her birds,
   With no more cares to think on,
She gave a rippling laugh, and out
   There flew a Bobolinkon."
        —*Christopher P. Cranch.*

"In the budding woods the April days,
   Faint with the fragrance from the life begun,
Where the early fluttering sunbeam plays
   Like a prisoned creature of the sun,
With sweet trill or plaintive note,
Quick pulsation of a throat,
With the life and light of Spring,
There the birds of April sing."
        —*Dora Read Goodale.*

"Oh, every year hath its winter,
  And every year hath its rain—
But a day is always coming
  When the birds go North again."

"STUPIDITY STREET"

I saw with open eyes
Singing birds sweet
Sold in the shops
For the people to eat,
Sold in the shops of
Stupidity Street.

I saw in vision
The worm in the wheat,
And in the shops nothing
For people to eat;
Nothing for sale in
Stupidity street.
            By Ralph Hodgson.

"The woods were filled so full of song
  There seemed no room for sense of wrong."
                    —Tennyson.

CHAPTER XLIX

SUGGESTIVE LESSONS IN BIRD-STUDY

THE BLUE JAY

I. *Field Observations*

There is only one practical use to which you can put these suggestions. Make them the purpose for wood excursions, not for the class, but for individuals and small groups. No one should try to teach what he does not know, but there is a great deal about a Blue Jay that one can know. You must catch the spirit before the lesson, and a single excursion into the woods of autumn or winter will give it, for the Blue Jay is a permanent resident. You ought to hear his notes ring through the silence of the October frost! Stand still and see if you can discover his business.

1. In what sort of a locality do you discover him?
2. Describe his method of flight.
3. Does he walk or hop?
4. What does he eat?
5. How do the other birds like him?
6. How does the Jay break off an acorn?
7. How does he open the acorns?
8. Where does he hide the acorns?

[Birds' nests are more easily found in winter than in summer, and this is really the time to study them, as one can collect and observe them carefully without disturbing the tenants.]

9. Where do you find the Blue Jay's nest?
10. In what kind of a tree?
11. How high is it from the ground?
12. Where is it in the tree, on a branch or in a fork?
13. Is the nest easy to find? Why?
14. Of what material is it built?
15. How is the material arranged?
16. What holds the nest together?
17. In the spring try to find a Blue Jay building his home. Do both parents work at the nest-building?
18. When do they commence to build their nest?
19. How does the Jay get twigs?
20. Where are the twigs obtained and how carried to the nest?

This is a kind of nature-test. It differs from most school studies in that the test comes right at the beginning of the subject. It is a test of the power to observe nature. Again, it gives the child an experience of his own. He has something interesting for conversation. His own experience is really the only kind of a subject for him to write about. It gives him an opportunity for self-expression, something different from the phonograph method by which some-

one else's ideas are repeated.   Do not let him put on smoked glasses
or stuff cotton into his ears after he has observed these twenty points.
It would be like planting twenty seeds in a garden and never looking
at them again.   Some naturalists have been observing the Blue Jay
for fourscore years or more, and there are still new Blue Jay sounds
and tricks to hear and see.   Here, again, is the difference between
book-study and nature-study.   A test in the former ends the study,
but in the latter it is simply opening the way for a lifelong examination, besides being a great deal more fun.   By the latter method,
one's failures are not proclaimed, and his successes are a point in
pedagogy for other subjects.

II.   *Blue Jay Experiences*.   (*A Character Study*)

As I do not know the Blue Jay experiences of other people, I
shall have to tell about mine.   They started on a farm in South
Scituate, Mass.   The Blue Jays were stealing the corn, and that was
an unpardonable sin on the farm.   There are four more chapters of
this story of which I will simply give the titles: An Old Shot Gun;
Concealed in the Bushes; Imitations of the Blue Jay's Call; A Dead
Blue Jay.   This paragraph would not have to be written had I
been given the opportunities for bird-study that boys and girls
have to-day.

Right here I want to say that I do not belittle the opportunities
of the farm.   One has to *know things* to succeed on the farm.   He
must plant, harvest, prepare, and use.   In the city it is a little
money, a store, and a can-opener.   If the city boy or girl wishes to
share in the experience of the great out-of-doors, he only needs to
step into the parks and use his senses.   Thus he may acquire some
real knowledge by observation, a fundamental principle in education.

As a farmer-boy I knew the Blue Jay, his haunts and his failings,
and could call him to any tree.   What I needed was a teacher,
someone to organize, direct, and guide (not stuff) my observations.

The next notable Blue Jay experience that I recall was when I had
a class on a field-trip.   We went to a field to watch some purple
Grackles.   One of the Grackles flew to a large elm tree, carrying a
white grub which he had excavated from the ground.   Just as the
Grackle landed, a Blue Jay flew down, snatched the grub, and flew
to another limb, where it proceeded to beat the worm against the tree.
When this juicy morsel had been devoured, the Jay flew again, this
time to where its nest was located.   This whole picture was run off
in about two minutes.   The incident showed the thieving instincts
and "cheek" of the bird, but at the same time his fondness for grubs.
We had his character in a nutshell.

The Blue Jay is also a big tease, at times a bully.   The house
across the street has a picket fence along the side of the lawn.   One
day in the fall we saw a cat sitting peacefully on the upper ledge of
the fence.   Suddenly, two Blue Jays appeared on the scene.   They
flew back of the cat and perched three or four feet away, from time to

time swooping down at it, being perhaps within a foot above it. The starting-point was a maple tree that shaded the fence. Now and then the birds would call *jay-jay-jay*. The whole performance seemed to be a game, and was seen at two different times and several months apart.

Fig. 1. The Moose-bird Caught on the Wing.

In September, 1916, I went on a trip to New Brunswick. It was a 'camera hunt,' which is much more fun than shooting with a gun. The cruise led twenty-four miles from the nearest house and settlement, right into the woods on the headwaters of the Miramichi. Our party found quarters at an old abandoned lumber camp. On a fishing-trip up the Little Dungavon one day, we cooked our noon meal at the junction of two streams. From our cornmeal allowance we had made some bannock. It was considered rather valuable, since we had 'toted' our provisions on our backs, carrying enough for a week which is quite a lug. I had forded one of the streams to get some dry wood for the fire, and, upon turning toward the place where our provisions were spread out, I saw a bird making away with our golden bannock. I decided that if it tasted as good to the bird as it did to me, he would return, so I hid in the tall grass and focussed my camera on a tin cup which held the disputed food. I did not have to wait long before he came back. Without following even woodsman etiquette, this feathered messmate tried to stand on the rim of the cup, which upset both of our plans, blurring the picture I tried to make. Such little unexpected or unplanned incidents, however, only add to the excitement . This was the first time that I had ever seen the bird, but I remembered its picture and knew that it was the Canada Jay. On returning to civilization (?) we learned

that the lumbermen call it the Moose-bird. In some parts it is called
Meat Hawk, Carrion-bird or Whiskey-Jack. Kennicott suggests
that its Indian name, Wiss-ka-chon, was probably contorted into
Whiskey-John and thence to Whiskey-Jack.

Many of the strange noises we heard in camp, near sundown, were
undoubtedly not bears or wildcats but the Moose-bird. We later

Fig. 2.   Canadian Jay Caller.

made friends at camp. I would place bait on one of the lumber-camp
stools and sit eight feet away on another, ready to shoot with the
camera. As the picture shows, the bird had no fear of the revolver.
The bird ate a little and then would carry off a large piece. He gave
a sort of whining tone as he returned from one tree and then another.

Picking up an acquaintance with city Blue Jays is easier than one
would suppose. Last spring one sunflower seed was planted near
our grape-arbor. The Blue Jays came regularly to get the sunflower
seeds. To take a picture I placed the camera near the grape-arbor
and had a thread leading into the house. When the Jays came I
pulled the thread. Next year we plan to have a row of sunflowers
by the arbor for the Blue Jays.

My last experience was in a Providence park, while taking the
picture of a Blue Jay's nest. An old gate was used for a ladder, and
after I had climbed up into the tree, a Jay came and perched overhead. Soon I saw another Jay coming down the path. Both Jays
had a sort of military bearing, with their blue uniforms, white
collars, and black belts. The patrol of the branches, however,
was more alert than his mate below, and I was not called upon to
explain my presence in the tree.

III. Blue Jay Economics. (*Debit and Credit Account*)

My early impression of Jay morals was that they were not as 'true blue' as the bird's dress. I am not so sure now but what the Jay had a right to some of the corn. Audubon pictures a Jay sucking an egg and writes: "I have seen it go its round from one nest to another every day, and suck the newly laid eggs." Barrows, however, in 'Michigan Bird-life,' says that these robberies are restricted to particular Jays and are not general. Forbush, in 'Useful Birds and Their Protection' says that "Jays eat the eggs of the tent caterpillar moth and the larvae of the gipsy moth and other hairy caterpillars." He concludes that it should not be allowed to increase at the expense of smaller birds. Prof. F. E. L. Beal, in the bulletin entitled, 'The Blue Jay and its Food' (published by U. S. Department of Agriculture), says: "Jays do not eat the seeds of the poison ivy (*Rhus radicans*) or poison sumac (*Rhus vernix*)." The Blue Jay helps in forestation by planting seeds of various trees, such as nuts and the like. Thus, on the whole, and aside from the enjoyment we get from his beautiful color, his neighborliness and cheery call, we may say that there is a great deal to be added to his credit account, and that he is a good friend to man.

IV. The Blue Jay in Literature

What facts do the different poets tell us about the Blue Jay?

Could you appreciate what they write if you had not heard and seen the Jay?

Pick out the words that describe him.

This is what a few writers think the Jay says:

Flagg- Dilly-lily.

Hoffman: *Djay djay, tee-ar tee-ar teerr, too-wheedle too-wheedle*, which suggests the creaking of a wheelbarrow.

Matthews: *J-aa-y j-aa-y, ge-rul-lup, ge-rul-lup, heigh-ho*

Samuels: *Wheeo-wheeo-wheeo.*

Seton: *Sir-roo-tle, sir-roo-tle, sir-roo-tle.*

"Blue Jay,
Clad in blue with snow-white trimmings."
—*Frank Bolles.*

The Blue Jay
"Blows the trumpet of winter."
—*Thoreau.*

"The brazen trump of the impatient Jay."
—*Thoreau.*

"The Robin and the Wren are flown, but from the shrub the Jay,
And from the wood-top calls the Crow through all the gloomy day."
—*Bryant.*

"Proud of cerulean stains
From heaven's unsullied arch purloined,
The Jay screams hoarse."

—Gisborne.

"He who makes his native wood
Resound his screaming, harsh and rude,
Continuously the season through;
Though scarce his painted wing you'll view
With sable barred, and white and grey,
And varied crest, the lonely Jay!"

—Bishop Mant.

ECONOMIC ASPECTS OF OUR NATIONAL PARKS POLICY

By Robert S. Yard, National Parks Association, Washington, D. C.

"The popular museums and school-rooms which constitute our national parks system, with their millions of waiting students, are not yet utilized. The system may be compared to a school equipped with every educational device, filled with eager pupils and with no teachers."

"But we are nearing a danger limit. So rapid is the increase of travel to the parks that it is none too early to anticipate the time when their popularity shall threaten their primary purpose."

"The reorganization of the administrative departments of the government which the president will urge upon the next Congress contemplates the creation of a Department of Education and Public Welfare. . . . Our national park service is in its best and fullest sense a service in the interest of education and public welfare, nothing less and nothing else."

Quoted from an article in "The Scientific Monthly," April, 1923.

Fig. 1. Camping in the heart of the Rockies is the real thing. Father and son made this poncho tent as a protection from rain and snow while they gunned along the peaks with their cameras.

CHAPTER L

The Diary of a Nature Guide's Son

IN THE YOSEMITE NATIONAL PARK, SUMMER OF 1924

Sunday, June 29.
Daddy and I went in tonight just before supper and washed up. The water was pretty cold but it was there anyway. I staid longest besides beating him in. He had to splash it up onto him for about five minutes after I got in. I didn't go right in, I'll admit that.

Monday, June 30.
Well I like it already. The birds and squirrels come right up into the outside tent, giving us a good chance for pictures. I'm going to try my luck tomorrow. I got up and put my gym suit on, because my trunk hasn't come, and went to the river and waited for my feet to get cold so I couldn't feel the water. Then I was all right. I went in for a good swim. Daddy can't get cold feet so he doesn't go in. He says he might go in Sunday afternoons if he is dirty enough. He just said now that he will write his own diary.

Tuesday, July 1st.
We saw 13 deer within five minutes on the trip with Mr. Nichols today. One of the deer comes around the camp and is very tame. I got a picture of Daddy feeding it from his hand. I got within 12

493

or 15 feet of one this afternoon but it always edged away. Dr. Bryant has a feeding table for the birds right in front of the tent. There are a lot that come, such as, black-headed grosbeak, hermit thrush, junco, chickadee, robin, and Western chipping sparrow. We put crumbs on a rock but intend to put a table out so that we can take pictures.

Wednesday, July 2.
Tomorrow there will be a special bird trip which both of us are intending to go on. These trips go at 8 and 4 o'clock each day, alternating with Camp Curry and The Lodge. On Saturday an all day trip is taken around different points on the Rim. Daddy is already snoring and I think that I will join him.

Thursday, July 3.
I fed Mabel Doe (The deer) green apples. She seemed to like them. Today she took bread with syrup from me, instead of bread with salt that Dr. Bryant's children offered him. I'm trying to take a picture of him (rather, her) with a natural background but she won't get in the right position or in the sun. I guess I'll have to have patience.

Friday, July 4.
Last night after the entertainment and Dr. Bryant talked on birds at Camp Curry we saw the Fire Fall. A man spends all day collecting wood after which he builds a large bon fire. After the entertainment a man hollers, "Hello Glacier!" (A point on the rim of the valley). After a few moments an answer comes back, "Hello Camp Curry!" Then our man hollers, "Let the Fire Fall!" To which the man proceeds to push coals over the cliff for a 150 foot drop to a cliff. This lasts about three minutes when the man on the point hollers, "The Fire has Fallen!" During this a male quartet sang. Without the color and light it looks just like a water fall. You can just barely hear the man on the point but they claim that he can hear better because of some currents. This is all very pretty.

Saturday, July 5.
Today, Dad and I went on a twenty mile hike, climb, and descendance all combined. We went to the Little Yosemite. On the way we passed Vernal and Nevada Falls. Nevada Falls is supposed to be the second best for beauty. It drops over 300 feet. Bridal Veil Falls is called the prettiest. The trip started off with over thirty. Eleven did the whole trip, and four of these were women. It was a pretty stiff trip. Even though I wouldn't want to do it again I am glad we did it.

Sunday, July 6.
Daddy expects to start in regular duties this week. He has been on all of the trips and at the museum between times this last week. Today we walked five miles to Mirror Lake. Daddy is writing to mother. He isn't so much of a writer as an artist so he draws most

of his words. He has just finished drawing the tent, stove etc. The thrushes have very pretty songs. They whistle like a coiled aerial wire looks. We sleep good after our day's occupations.

Monday, July 7.

Dad took his first trip alone. It was a children's trip which took the record with an attendance of 22. One little girl was taken around on her father's back. They all enjoyed it. Mabel hasn't been around for a few days but tonight she appeared while we were eating supper. I sat in my chair and fed her bread and apples while Dad tried to shoot her (with the camera). He did it from inside the tent. It was quite dark so the results are not promising. He took it with the bulb and would click it just before he thought Mabel would move but he thinks that she beat him every time.

It is hard to get many apples as there are two or three deer at the orchard most of the time eating the windfalls. That is where I saw thirteen in five minutes. We have seen a weasel twice on the path to the Lodge.

The same day that Daddy took the children's trip he gave a lecture at a boys' camp and a geology talk at the museum. This is pretty strong for his first day.

Fig. 2. "Mabel Doe" is a friendly visitor. This boy is seen attending to a social call. How to win the friendship of animals in the wilderness is intensely exciting and highly educational.

Tuesday, July 8.

Tonight we walked to the bear pits (Garbage disposal area). Just as we branched off the road a bear came running across the road in front of us. It was quite a large one and presented the funniest sight in the style it ran in.

We then reached the pits. There was one bear there. Some men tried to feed it but were scared to approach it. They would scale crackers up to it. Another man dumped some sugar on a board and pushed it up to within a dozen feet of the bear, when a braver man took up the board and walked up and reached out as far as he could and let the bear lick the sugar. A still braver man fed the bear with his outstretched arm. The bear was as scared as he was as he snatched the bone and ran away with it.

It came back and started to eat from the garbage but was pretty wary. He kept looking around on each side of him. He evidently didn't like to have anyone in back of him. I heard a man say that they would turn and woof at you if you approached from the rear but if you come in front of him he was all right except he might keep backing away.

We saw two others on the way back to the road. Probably a mother and quite old cub. These kept running or rather galloping away on our approach.

Wednesday, July 9.

Dad gave his first lecture before a crowd of 2000. He did well. Even the announcer complimented him.

Dr. Bryant is in the back country with a party of 22 on a six day trip around the Hiker's Camps. They receive meals and beds for 75c each. Pretty cheap as it has to be hauled up on pack trains.

I haven't found a job yet. I have asked in the store and answered an ad at the bakery but they wanted a truck driver. You have to be eighteen to do that but they advertised for a boy.

This afternoon at four o'clock Dad and I both have the privilege, although it is Dad's job, to go with some Rangers to distribute some trout fry in some streams.

Thursday, July 10.

Yesterday I sat here at the table and watched a Western Tanager come to the feeding table quite regular so I got my camera out and got ready for his next trip. It so happened that a black-headed grosbeak came at the same time. I think I got a good picture of them both. I could have duplicated the picture the next trip but didn't want the same picture. The female tanager lit on the tree beside it but would not step on the table. She had been on it in the morning but wasn't very hungry I guess.

Friday, July 11.

Dad started at two o'clock for a two day trip to Glacier Point. He took his party up the Ledge Trail which is a little over a mile but takes about four hours to ascend. He didn't have a very good brand of hikers so they dragged most of the way.

Saturday, July 12.

I staid here at the tent overnight alone and went up to Glacier Point today with the one day party. We went up the Ledge Trail also. We ate lunch at the Point after which a few of the ambitious ones, of which I was one, went on to Sentinel Dome. Mr. Nichols led our party.

Dad and I went in for a bath. Dad thought it was awful cold but I think it was the warmest it has been.

Sunday, July 13.

This morning Dad and I went to the Cascades with Dr. Lehenbauer and a friend to take pictures of some nutmeg trees. We walked about fifteen miles, stopping at the Bridal Veil Falls to eat lunch. We saw one little nutmeg tree.

Dad and I stopped an hour and a half at the Bear Pits on the way home. We saw six bears, three of whom were up in trees. Two of these were a brown bear chasing a black bear cub which would keep one branch away from the brown one whether it went up or down. I took two pictures but am in doubt as to the outcome as it was just before sunset and the second just after.

Monday, July 14.

This morning Mabel came while Dad was eating breakfast. He cut up apples for her as they are too hard otherwise and fed her as I watched from bed. When he wouldn't give her any more she coughed. She was a better shot than I as she was right in front of him and coughed over his plate and face. She coughs every once in a while after eating anyway.

We make oatmeal now with raisins in it. It is pretty good and goes good when we can't get home in time to get milk.

The river is getting awful low.

We need three blankets apiece. I wore two jerseys over my pajamas.

Sunday, July 20.

Haven't written for quite a while. I have built a table out of some old lumber which will take the place of the table we rented. We might as well bought it. While I am writing I have my camera set for any birds that come to the feeding table. I have just taken a russet backed thrush. A blue fronted jay just came to the table but was in the shade so I didn't get a chance at him. Female grosbeak on the table but I have a picture of her already. She has only two feathers on her tail.

Haven't written as I have been to Tuolomne Meadows. We packed all our blankets, ponchos, and sweaters in a duffle bag. We went out by the Big Oak Flat Road which takes at least 45 minutes to reach the top. The road from here on skun Bowker Street and its hill twenty times. We were going up and down 90% of the time. The road was awful rough again leaving Bowker Street in the dust. A big event on the way was when we went through the Tuolomne Grove of Big Trees. We stopped and ate dinner there.

We ate supper with the men putting up the telephone line. They were a very nice bunch of men, about eight carrying their own cook. We had dandy meals with a big variety. We expected to sleep out of doors but the ranger was on a trip so we occupied his tent. Jack Emert slept on the ranger's cot.

We ate breakfast with the men and started to climb Lambert's Dome. It looked all right but it wasn't. We started up easily but it began to get steeper. It was clean rock. There were very few places to cling to. Dad sat down and backed up. When we reached the top we could see Mt. Lyell and the glacier very plainly.

Monday, July 21.

Dad gave a ten minute lecture at Camp Curry last night on "Denatured Children." The following day he had as a result 30 children and five or six school teachers. Lucky the teachers were along as they helped a lot. I haven't seen Mabel for a week.

The bear came yesterday to Mrs. Nichols tent and was the cause of the vanishing of a five pound bag of sugar. I saw him as he came down to take a bath in the river near where we were swimming. Also a skunk visited Mrs. Bryant and relieved her of two or three melons. We have yet to see, hear, and feel the results of one of these robber visitors. Dad has quit for good, swimming, I guess.

Friday, August 1.

Indian Field Day starts today. Dad and Mr. Russell will dress as miners and march in the parade with some burros. I went down and watched some cowboys practice yesterday. One cowgirl was there practicing Roman riding. She works as a guide taking parties on horseback around the valley and rim. You wouldn't know her in civilian's clothing after seeing her at first in her cowgirl costume.

There was a very beautiful array of baskets covered with beads. The prize went to an old Indian who had only entered two large baskets. One was designed with a butterfly while the other had the old tribal design. They finally gave the decision to the basket with the old tribal design because the other design was taken from the white man.

Harry Tom won the saddle by scoring the most points. He had a dandy horse.

The cowboy's chaps and spurs race was a sight. None of them could run. They put their chaps, spurs, and hats in different piles and raced to put them on again. On the way back to the line they looked like cripples trying to gallop.

Sunday, August 3.

We started off for Tenaya Canyon at 8:30 a. m. They say only five people have been through. The canyon is awful narrow and some of the drops were so bad we had to find our way along ledges up above. This was rather ticklish in places. I caught my first two fish in a pool at the foot of a cascade. We hit the Tenaya Lake Trail at seven o'clock which gave us a half hour before dark. We had fourteen miles ahead of us. We had a walking stick and you will have to imagine us going through a dark forest and down the zigzag trail with over 100 turns. It wasn't very pleasant as you will imagine. Animals kept running off in the bushes and one, which was probably a coyote, followed us part way down the zigzag. We did think of plastering pitch on a stick which served us to good advantage in two very dark places in finding the trail. The last time we must have used it ten minutes through a dark wooded part of the canyon. The street lights were out when we came to the Valley. We got back after one a. m. Our neighbors were worried about us.

Monday, August 4.

I have been taking it easy today and do not feel any the worse for our experience. I would not like to do it again, no matter how many years between time. Once was hard enough but we are both glad we attempted and accomplished this feat.

Tuesday, August 5.

A large cub paid us a visit. I was coming home just before sunset. The bear walked out from behind a tree and scared me. I followed him over to the Bryant's tent, which he walked right into, mounted the bed and sniffed at the baby who was just awaking, and then went

Fig. 3. Negotiating with a Brown Bear in the
High Sierras.

out the side of the tent which was raised. I then thought of taking a picture. I had to put a film in the camera so I was delayed but the bear staid also. He came up and started to eat a graham cracker that I had in my hand but decided he didn't like it. Just after I had taken the picture he came up and sniffed of the camera in a hungry fashion, then opening his mouth around my leg, and got his claw caught in my shoe. I felt pretty comfortable. Yes, after he started away.

Wednesday, August 6.

Just before dinner the cub appeared again. I got one picture with a natural background with good sunlight. Dad came along then with some candy he had just got in the mail. He opened the box and the bear got a taste of it. He rose on his hind feet and walked after Dad into the sunlight where I got another picture of him in this pose.

We got a glimpse today, for the first time, of Mabel and her two fawns. One stopped in the middle of the river, while crossing, and began to nurse. The fawns are very small and awfully cute.

Saturday, August 9.

Dad took a party of 26 to Eagle Peak. The trail was particularly steep going up to Yosemite Falls. It was about fifteen miles for the round trip. We had a very good but not a fast party. Twenty three reached the top. One lady did not intend to do the whole trip so that leaves two who failed.

Sunday, August 10.

We started on a three day trip carrying food but sleeping at the Hiker's Camp at Merced Lake. We went right to Merced Lake the first day covering 17 miles. The next day we reached about 11,000 feet at Vogelsang Pass. This trail was very good though slightly used. We had wonderful views from here. All we could see was mountainous country. We took our time and lots of pictures and part of a bath at Vogelsang Lake. We covered at least 16 miles.

Tuesday, August 12.

We only had one pack going home so went much easier. Climbing Cloud's Rest was very steep. It was an altitude of 9,925 feet. We then descended and had our dinner at 4 o'clock. We had hidden some food on the first day out. We didn't want to eat before climbing Clouds Rest. We built a fire and had hot corn and beans and cocoa and washed the dishes all in 30 minutes. We could not see far as the smoke from fires dimmed the view. We had covered 71 miles in four days. This does not mean as much as the heights we went up and down. The last day we saw 40 deer. We also saw three broods of quail which were the first we had seen. We enjoyed ourselves very much the whole trip.

Friday, August 15.

We are having Dr. Branoun, whom we met at the Grand Canyon, to supper. I am chief cook for this occasion. We will have smashed potatoes, sausage, and peas with pie as desert. I suggested prunes but Dad wouldn't agree.

Saturday, August 16.

Dad had another trip into Little Yosemite. The party consisted of six men. I went with him as usual.

Sunday, August 17.

Dad was getting his breakfast this morning and I was still asleep when he called me and I looked out and there was Mabel with her two fawns. He got some apples and began to feed her. It was Sunday so nobody had gotten up in the other tents or she wouldn't have brought them. They were very cute. They got behind their mother and peeked around at Dad. One of them got scared and ran off into the bushes. The other looked at Mabel and at its brother (or sister) and finally trotted off with the other fawn. Mabel wouldn't stay long after that.

We went to see Mr. Sonn, who feeds the birds. He has a very neat tent. His profession is making funny animals out of pine cones and other pieces of wood.

We then went to Glacier Point via the Ledge Trail. We went into Illilouette Canyon where I caught two rainbow trout. We came down the eleven mile trail to Nevada and Vernal Falls getting home before dark for some beef stew.

We then went to a camp fire for the Nature Guides. Mr. Hall told about his experiences abroad.

Monday, August 18.

Mr. Russell is in the Tuolumne Meadows collecting material for the new $75,000 museum. He has asked me to come up and help him. I will send my bedding up in a truck and walk. I am anticipating a dandy time.

Dad will be rather alone the next week or so. He will have to eat some to get rid of all our supplies.

Monday, August 25.

I just got back from Tuolumne Meadows last night. Mr. West kindly took two or three miles off of my mind by giving me a ride in his machine to Mirror Lake where the trail starts.

We passed a mule train just before Mirror Lake. Mr. West wanted me to ask the two Indians taking it for a ride as they would probably pass me going up the 102 zigzags. I had my doubts as to whether I would or not. I didn't like to for one reason and I would rather walk anyway. As it happened I reached the top a few minutes before them and felt fine. The rest was mostly level ground except for a few ups and one down.

This was the trail we walked down in the dark so you can imagine how I enjoyed myself in the light of day. I hit the Tioga road about ten minutes ahead of the pack train. They passed me while I was eating my lunch on the shore of Lake Tenaya.

The next day we got up at 6:30 and went to the beginning of the Mono Pass Trail. We carried traps, two guns and our lunch. We set the traps and started for Mono Pass. We saw several deserted cabins along the trail. At the Pass we could look down on Mono Lake and forests of Pinon Pine (Which is a one needle pine).

On the way up we shot four conies. A cony is something like a Guinea Pig, and is found in most every rock slide. He would shoot and keep his eye on it while I would go and get it. They usually slide down behind a rock so you couldn't find them, but we didn't miss one. We got up to a patch of snow. Just below it Mr. Russell found a mountain sheep skull and I found a horn. These are both very rare.

Helen Lake was very pretty. There were thunder clouds all about us and we were afraid we would get wet. We could see where it had snowed on a shoulder way off. When we hit the trail we hit snow. It really started to snow but we were very low and getting lower so we had it for only a few minutes. It was small chunks instead of flakes. It had neither rained or snowed, or looked that way, back in the meadows.

The next day I trapped five picket pins which are something like a prairie dog, and a Tahoe Chipmunk. Mr. Russell skinned the conies while I trapped. In the afternoon I skinned about a half of a picket pin, and salted the skin down.

Mr. and Mrs. Harwood invited us to supper. He is the checking ranger for the autos coming into the park over that road. I had fresh corn that night for the first time this summer. It tasted mighty good.

The next morning we packed and started for the valley at 9 o'clock. We got home about six after a hard days trip over a thousand Bowker Hills and a worse road.

Tuesday, August 26.

The Nichols went the morning I left for Tuolumne Meadows. There are fewer people every day. Cannot write any more until I get to Estes Park to the Boy Scout Conference.

Fig. 4. "Thar she blows." The Timber Line is always a fairy land of enchantments.

Ourselves and Our Equipment.
Wᵐ G. Vinal and Raymond G. Vinal.
Nature-Guide Service Yosemite, California.

CHAPTER LI

Notes by a Yosemite Nature Guide

Well here we are at the bottom of a trench—nearly a mile down. Our tent is in a group of large evergreens that tower high overhead. We are in Camp 19 which is a nature guide colony. The Merced River is near by and we have had our first real bath for a week. The tent is 12 x 14 feet with a large fly extending out front as a veranda. It is here that we cook and eat. We spent today, which is the Lord's Day, in building a shallow tray with four legs, for our stove. The tray is filled with soil and the stove rests on top. This is a scheme to prevent fire. Through an opening in the trees we can see Yosemite Falls. It is getting rather small due to the lack of rain.

Nearly all the birds, trees and flowers are new so I am going on all the trips in order to get acquainted. The people ask so many questions about heights, trails, distances etc., that I have to study a great deal. We are learning a heap and being paid for it while the tourists are paying a heap and learning a little. It is their fault though as the Nature Guide service is free. There are several trips every day.

The thing that strikes me most forcefully about Yosemite is that it exhibits the whole history of the west in a nut shell. It was not discovered until 1851 and the first road entered the valley in 1874. The man who drove the old stage coach, that is now front of the museum, is chief here in the American Express Company. It has been so recently opened up to the automobilist (1913) that it possesses a primitive charm wanting in many places. Here nature is seen in her primitive moods.

The Indians have a reservation near the southern wall. On Indian Field Day they come over the mountain trails, often a buck, squaw, and papoose on one horse. It is very picturesque. The oldest Indian in the valley is Lucy. One of the favorite questions of tourists is, "How old is Lucy?" In front of the Museum there is a Chuckas or granary. It is a large twig woven basket to store acorns from mice and chipmunks. On our trips to the southern wall we always point out granite slabs which have mortar holes where the squaws pounded the meats. We often speculated as to what the conversation around these old mortar holes could have been. The tannin was leached out with hot water and then cooked in a basket. You will say, "How in the world could they cook acorn mush in a basket?" They dropped hot rocks, the size of my fist, into the basket with wooden tongs. I have often thought that this would be a fine thing for scouts to try. The necessities of the Indian, like the trail, canoe, wigwam, moccasin, medicinal plants, and wild foods are becoming the recreation of the present day camper.

On Indian Field Day we decided to represent the old mining days. It was due to the Indians killing some miners on the lower Merced that Yosemite was discovered. The soldiers chased the refugees into the valley. While talking about our costumes a man went by who looked like a miner. We hailed him and told our plans. He opened his shirt and showed a whole row of gold nuggets on his under shirt. There was not much more difficulty in finding a second miner. We fitted up two burros and they marched in the parade.

"How did Yosemite form?" is told in the museum twice a day in story form. The formation of any region is the key to all that comes after. It used to be said that the devil held his hands together and cleft the ground and then spread his hands apart to form the valley. It has always been characteristic of the race to blame the unexplainable onto the devil. A more recent explanation is that after the Merced had cut the valley down about a thousand feet the glacier began to move down from the high Sierras. Think of a snowflake alighting way up there in the mountain, hardly conceivable in weight and sound, and then another snowflake, and another until we get perhaps a foot of snow. In a few days there is another snow storm. When these have accumulated for several centuries they may get heavy enough to begin to move down the mountain. It is thought that this ice sheet came down the slopes of Mt. Lyell at the time of the glacier pushed over the face of northeastern United States. This was perhaps 10,000 years ago. This great ice plow polished the granite walls and plowed out the valley. The scenery of the Yosemite was made by the ice.

One of the great attractions of the park is the bear. They have been taught to come to a feeding ground every night. Automobiles park across the river and watch the bears eat. He is very timid and not ferocious as in story books where he is pictured as eating naughty children. Of course it would not do to offer him a piece of candy and then snatch it away. In such a case he might slap your face.

The deer are very tame and some of them come around the tents to be fed. Thousands of people get more sport feeding them and hunting with the camera than with the gun. The example of the national parks in the preservation of deer and other wild life might well be emulated by state parks. It is better to have many deer that are seen and enjoyed by the whole community than to have a few which are never seen.

We kept a rattle snake in the museum. He is very particular and will only eat live food. One day we put in a mouse. The mouse sat on the snake and began to wash. He did not realize that he was sitting on dynamite. The snake slowly moved his head to the right and then to the left to investigate. The mouse ran across the cage and one thrust of the snake's fangs caused the mouse to tremble and pass away. A great mass of tradition has grown up around the rattle snake. In Rhode Island there is a place called Snake Den. The farmers use to go out there each spring and take a jug of whiskey as a remedy for snake bites. They never saw any rattle snakes and never brought back any whiskey. Rattle snakes cannot jump from the ground and only roll like a hoop in folklore. The other day I heard a mother tell her little boy not to go into the reptile room as the snakes would get him. People still persist in spreading irrational ideas.

The Big Trees should be included in the seven wonders of Yosemite. They are the greatest of living things. They have not been visited by the glacier but all have been burnt and struck by lightning. At one time they were threatened with extinction but the Redwoods Association did much to create a public sentiment to save them. Sir Joseph Hooker well says that they are the "Noblest of a noble race."

Trout fishing is one of the great assets of Yosemite. The government thinks of this as bait for sportsmen. The park is a vast playground. It pays to lure people into the wilderness. It is not the number of fish that they catch, but the fascinations, reminiscences, and adventures that make the vacation worth while. In the six day trips in the high country the party is always given time to fish and in the evening the fish story is an important part of around the camp fire.

A man was fishing back of the museum lately and made as much fuss over his catch as a hen over a new laid egg. He held up his fish and said, "Get me a string quick, before he gets away. My gracious but he is a brute. Can you get me a string?" He was running around like a hen with her head cut off.

"My gracious, it's all I can do to hold him." I wondered if he thought the fish would bud legs and run for he was two hundred feet from the river. I helped him tie the prize. Then to be doubly sure he put the fish in a paper bag. About a third of the fish protruded. "My gracious I didn't come prepared to catch such a big fish." (It was 19 inches long by the yard stick but immensely bigger, and with promises of a good healthy growth, in the mind of the sports- man.) He was as tickled as a barefoot boy. About half an hour afterwards I heard him whistle to some men crossing by the museum, and he held up the fish for them to see. The next morning he was in early to find out about the kinds of trout. If anyone has any doubt about the policy of the government he should have seen how this fish saved both body and mind. We have a Loch Leven Trout at the museum that weighs 9 pounds and 15 ounces.

I. My First Nature Guide Party

This started Friday afternoon, July 11, from Camp Curry. Twelve people assembled for the mighty climb up the Ledge Trail to Glacier Point. Glacier Point is at an elevation of 7214 feet or about 3200 feet above the Valley Floor. This meant about a mile and a half walk but going nearly up, like a flight of stairs. One gentleman delayed the party a little, while he left his coat. He came around the corner, patted his bay window—which protruded somewhat for a mountain- eer. I smiled to myself—and a lady said, "Why are you smiling." My answer was: "For the same reason that you are." The gentle- man was most vivacious. His daughter accompanied him. She had hobble skirts that were hobbled at the knees. This was new to me but being a green guide I asked nothing but tried to look wise to it all. A doctor—who teaches in New York City was along. There was also an ex soldier with his French wife, a lady from California, a man and lady with daughter and two friends of the family.

Let me say that a Mountain Trail is a wonderful place to bring out the inner man. It soon developed that the stout gentleman was near sighted. He would say: "Which way do we turn now," and I would have to show him. I would say: "Look at that beautiful Yellow Mimulus." He never could see it. I pointed out two deer. He changed his glasses and slammed the case as though it was a wood shed door. Then I had to point dextrously and by sheer effort he saw the hulk of the deer

We came to a slope of granite which was rather smooth. There was no particular danger in crossing it. If one had slipped he would have slid about the length of the roof of an ordinary Cape Cod house. It would have sandpapered his trousers to a thin edge but the coaster would have passed down but not out. Well sir we had some time getting our party across. Mr. Russell held out his arm as one step- ping place. Miss ――― could not stride because of her hobble. It finally resulted in her pulling her dress up so that she could stride. Being a nature guide I was utterly impervious to anything unusual.

Said Father――had great difficulty in seating himself and reaching the proper foot over to the extended arm. He persisted in turning himself the other way up. Finally after gentle persuasion we got his frame over. He said afterwards that he was not a coward but that he *was* scared. He is going to complain to the government. He thinks it absurd not to have steps carved in the granite and an iron rail. I had a free lecture for half an hour on the subject the next day. Dr. ——— showed off in another way. He shook like a quaking aspen. I did not dare look at Mr. Russell for fear of smiling and a Guide should never smile at the wrong time. One lady balked. Said that she could not do it. She didn't have to do it the last time she visited the park which was 30 years ago. After each one had showed off his innerself we got the party under way again. Evidently none of these people had had the summer camp training. Do you think that they would be any better sailors? I don't. It took just four hours to bring up the rear to the summit.

At the top there is a hotel called Glacier Point Hotel. Mr. Russell told me that Nature Guides are given all they want for 50 cents. It is on the cafeteria plan so my mind was all made up to have a square meal. This after two weeks of my own cooking. My meals remind me of Mrs. Cram they are so different. I took roast pork and mashed potato, side dishes of succotash, corn, rolls, lemonade, pie, and pudding. It was some lay out. Then I saw in the middle of operations that I had a big contract before me. I just smiled at the man and said that I was awfully hungry. That was a darn lie at that stage but I was ashamed not to eat it. After tucking the meal away "en toto" the proprietor came around and invited me to lecture in the lobby for half an hour. Couldn't do anything then but say that I would be the happiest man on earth to talk to his guests.

They had a roaring fire and I backed up to the fire and talked about the geology of the region. I talked overtime as usual and the manager had to bring a clock out and show it to me. This is the point where they push the fire over for the fire fall. The hostess came along and wanted to know if I would like to have a wonderful place to see it. Of course I said YES. She disappeared. I wandered out to the rim and thought I saw her. I walked up and trying to be sociable said: On which rock were you going to drape me? The lady said; "W-H-A-T?" I repeated but believe me I did not repeat again. It was the wrong lady. Finally the real hostess came along and said: Come down to photographer's rock. You have to lie down and peer over the edge. (Note accompanying diagram of the way I peered and the others.)

It was some sight to look down into the valley and see the lights. Finally we heard: "Hello Glacier" rolling up from the Valley. "Hello Camp Curry" was sent back. Then: "Let the fire fall." They had built a huge bonfire and the man then pushed the coals over the rim. The hostess said to hang over the edge and watch the large coals until they hit the bottom. Did you see that one she said: Yes, said I. That was my second lie. Well I was stretched out on that rock lying galore for about ten minutes. My feet never felt so light.

This being over, the Nature Guide received more attention by being invited to see the bears. It was moonlight and four of us proceeded to the garbage pit. When almost there we heard a mad jangle of tin cans. A big bruiser heard us coming and dashed off into the wilderness. After waiting on a rock some twenty minutes we heard something that sounded like voices up in a large yellow pine. I was told that that was mammy bear and her cub. Then followed the descent of mammy and the cub. We could see the tree plainly but not the bears. It sounded like 14 bears. Never heard such a breaking of twigs and scratching. Soon we saw the old bear and the cub coming along the trail. The cub was certainly cute. Mammy kept looking all around and at the cub to see that all was well. The porter had brought along some gum drops. Mammy came up and took one from his hand. Then would not listen. Finally he hollered: MAMMY! She reared right up on her hind legs and he threw the whole of the candy at her and said: "Take 'em all." He was scared.

After this they wanted to know if I wanted to see the sun rise. I said: SURE. So I was called at 4:45 a. m. I got to bed at 11 p. m. after feeling all around for a button or string to put on the light—which I found not.

The sun coming up over the Sierras was wonderful. I tried several photographs and hope that they come out O. K. I then learned the names of as many of the 499 peaks as I could. Everyone comes along and asks the Guide questions and of course it does not pay not to know about them. One lady on the way up said: Have you been up here often. I said: "I am climbing all the time." The people eat on the porch and toss crumbs over to the squirrels and birds. There were California Ground Squirrels, the Golden Mantle Ground Squirrel, the Chickoree and three species of Chipmunks. Some Sierra Partridges came along. They are as large as a hen. There were all kinds of birds. You can imagine the education that I am getting by being on a hotel piazza and having the guests quizz me. It does not do to say that you know a thing if you don't because you will get caught. My students will swear to that. It keeps me on the hump to learn all this and to keep ahead of the multitude.

At 7:30 a. m. I had the party under way. Mr. Russell had left me the night before. We were to follow the Pohona Trail along the southern rim of the Valley. It was a 16 mile hike to the Bridal Veil checking Station where a buss was to meet us and bring us back to the Valley.

The trail led through wonderful isles of huge trees and upland meadows of beautiful flowers. On top of Sentinel Dome we saw a curious shaped Jeffrey Pine. We ate dinner on Bridal Veil Creek. I took many pictures which I shall make into lantern slides and hope to show them to you some day.

I had a contour map and compass. You remember that I have always told you how important they are. I had never been over the trail before and you can imagine the hundreds of questions, especially: "How much farther is it?" How far is it to Taft Point? Which way do we turn next? What peak is that? I was the guide and could not say: "I have never been here before. I come from the little state of R. I. and don't know." My knowledge of topographic maps came in handy. It was shaky business sometimes.

We arrived home footsore, dusty, and tired—at about 7 p. m. I took a swim or dip rather—in the Merced—and retired.

This will give you an idea of Nature Guiding. It is fascinating work.

I am enclosing a copy of my talk at Camp Curry. There were about 2,000 people there. One man came to the museum the next day and was talking with me. By and by he said: "Are you the man that lectured at Curry last night?" I assured him that I was. Well, he said, you look younger then you did last night. You may remember that in some lectures I imitate the way robins, herons, etc., walk. I did that in my Curry talk. The man said that he turned to his wife and said: "My God, but that man is limber!" I will let you guess whether that was complimentary or not. I do not know yet.

One meets some very interesting people guiding. At Glacier Point I met a lady with whom I had spent a half hour the day before trying to identify a flower that she described to me. It turned out to be the Sky Pilot or Polemonium which she found near Mt. Dana in the high altitudes. This was at the Museum and by chance we met the next day on the rim. I also met her husband. Their names were Knibbs and he is a famous western story writer. He went to Harvard where Barrett Wendell and Dean Briggs and others were interested in him. He never obtained a degree because at the end of three years they told him that he had better get out before they spoiled his style. He has a style his own and they did not want to destroy his individuality. I think that they were pretty broad minded to do that.

The French lady used to be in the French Red Cross. She knows Mme. Pellissier and knew Bob. She did not know that they were brother and sister. She knew about the Robert Pellissier ambulance and said some very fine things about Bob. It made tears come to my eyes to think that I should meet someone in the High Sierras that knew him. It was a remarkable coincidence.

Here endeth the first trip.

II. *A Nature Guide Party to Yosemite Falls and Eagle Peak*

A nature guide party usually consists of a large number of college and university graduates. Tourists are usually appreciative and very intelligent. On this trip there was a Professor of Economics from the University of Nebraska and his brother-in-law, who owns a big ranch. He said that he has a barn that holds 100 head of cattle and also had 40,000 tons of prunes ripening. There was a professor from Wellesley, an M.D. from the University of Chicago, a chiropractor, a nurse from Los Angeles, and three kitchen helpers from the east who had been at Yosemite Lodge. The lady with the red parasol, the porch hound who plays Mah Jong in the afternoon and Jazzes in the evening, and the scenic bus riders who tour the park like an art gallery with a rubber-neck megaphone artist were absent.   There were twenty-six of us all told.

A nature guide has to be a good psychologist to keep his party together.   There are those who want to make the peak in so much time and there is the saunterer who is unconscious of the miles ahead. In a summer camp the leader could command but here he must inveigle.   It was rather hot climbing but none of the party blamed the guide for the weather.   There is usually someone in every party who does.

In the upper country we passed through several wild flower gardens. In the midst of every glacial meadow there is usually a 2 foot stream

Fig. 2.   The Snow Plant with its ruddy complexion is one of the most fascinating inhabitants of the Sierras.

trimmed with moss and violets.   The water is always cool and the party enjoyed the drink.   The crisp weather was quite a contrast to the dusty mule tracks on the way up.   The garden spots were crowded with blue lupine, the Indian Paint Brush, and Rudbeckia.   When we came to the fire-red snow plant the party had to be reminded that the flowers are not to be picked.   Some people always persist in loving flowers by pulling them up.

At noon we stepped off the trail by a pool to eat.   We took stock of our experiences coming up the trail.   One member of the party was missing.   I sent my son on to find her but he did not have success.   It seems that she had been told some stories about bears and coyotes and was making such a high speed to catch us that she missed the stopping off place and had gone on up.   The naturalist must pacify people as to the imaginary ferocious animals. It was also up to the guide to entertain the party while they were resting. This he did by telling stories. In turn he had to hear about blisters, pay attention to some, and last to see that all fires were completely out. Some were advised to wait by the trailside for the return of the party. The silence of fatigue finely fell over the party and it was fully 30 minutes before the guide whistled them up.

At the summit of Eagle Peak the climbers sat down and reveled in the panorama outspread. When the last puffing member crawled wearily up, I was called upon to give a talk on the geology of the peaks. Literally hundreds of questions were answered. After feasting upon the scene and firing the blood with the idea that "A thing of beauty is a joy forever," the deepening shadows of the virgin forest below us beckoned that it was time to descend. A last click of the camera and a last look and we were soon beneath the darkening foliage of the evergreens.

III. As Viewed by an Eastern Naturalist

A Rhode Island Nature Guide in the Yosemite experiences a succession of sensations. He finds himself in a park larger than his own state. He is "shut in" by canyon walls, and every time he looks up he thinks that a thunder cloud is rising. He remembers that he left the folks at home complaining about the daily rains rotting the crops. But the complaint here is "no rain." The Belichened cliffs may resemble a thunder head but offer little in the way of moisture.

Then comes the humiliation of not knowing the life that he sees, for exceedingly few eastern plants and animals have crossed the Sierras. He may be greeted with a clear whistled song of three notes. How queer! It must be a chickadee. Yet, in the East his call is two notes, "Phoebe." Perhaps he knows that I am puzzled and so says, "Oh, dear me." Then from overhead comes a dry, wheezy drone. It must be a flycatcher, for the bird is perched on the outside of the tree to get a clear view. Yes, it darts off for an insect, snaps his mandibles together as though he meant business, and returns to his perch in true style. But it takes a little imagination to believe that he is saying, "Pewee." We certainly know Mr. Jay even if he has put on a blue front for he has his crest, noise, and distrust. Look behind us! While we are gazing about, someone is stealing our butter. His large beak gives him away. He is a Blackheaded Grosbeak. I wonder if the butter made his song clearer than his eastern cousin. The Western Tanager has caught a cold and sings like the Scarlet Tanager of the East, but how different with his yellow colors. Here comes a camp visitor that makes us feel at home,—the Western Robin. We now know how the Pilgrim exiles felt when they saw the robin. Quite different from their English robin, but being a home loving people they said, "Why! Here is robin red breast." And so he has been called ever since, not because of the color of his breast for it is not red, but because he was a friendly bird that reminded them of home. The Western Robin is larger than the New England Robin but just as "homely." Is that why we all love him? And thus it goes. As we take to the trail, we find birds that "remind us" like the Russet-backed Thrush, and the Band Tailed Pigeon, and then we find others that are real old friends like the Chipping Sparrow with his red cap and dry chips, or the Yellow Warbler seeking the willow banks.

Mark Twain says that the best place to collect weather for an exhibition is in New England. Weather is the most talked about subject. If we are introduced to a stranger, ten to one we start right in on weather gossip. The weather is never just right. No one apparently admits that a rainy day is a good day for all concerned. Naturers rightly maintain that we are all concerned. When this weather steeped naturer gets to Yosemite, he finds himself greeting everyone, not this morning but at all times, with "This is a fine day." Then he feels sort of like a prevaricator for he realizes that every day is sunny along the San Joaquin Valley. The announcement is unnecessary. Strangest of all, he soon discovers that Westerners are weather wise too and talk about it like real Cape Codders. Just at present it is too dry. The rainfall, or the lack of it, is of course unusual. I wonder if they and their New England ancestors brought the habit along in their covered wagons? So here is another week of Yosemite sunlight with which to reflect and write.

The fineness of the weather has a decided effect on the plant life. Between the moist banks of the Merced and the dry talus slopes there is an army of interesting flowers. A few are the same as in the East. Yarrow and sorrel have traveled here from their European homes. I saw one plant of our daisy or whiteweed near Camp Curry this week and wonder if it is a late arrival. At least the meadows are not dotted with it, as we know daisies back home. Then there are some plants which are the same but different. Queen Anne's Lace grows on a much smaller scale and has no central purple floweret. The natives are worrying about the disappearance of the Evening Primrose. It once grew in abundance, but the deer have developed a liking for it and nip off the flowers, thereby preventing seed formation. It is considered a pest east of the Appalachians. There are Columbines, White Violets, and Fire Weeds in blossom now; the Azaleas are just passing, all being a little later than our Atlantic flowers, probably because of the higher altitude. And then come the real gems of the Yosemite, the flowers that are different. The famous blood-red Snow Plant, the White Mariposa Lily, the Indian Paint Brush, the tinted Pussy's Paws, the handsome Sierra Primrose, and we might go on ad infinitum naming these remarkable flowers, upon which a stranger, in his mute ignorance, can only stand and gaze. It is here that the visitor can call upon the Nature Guide Service, which is furnished free by the Government. It is with deep appreciation that we welcome the Nature Guide who comes to our assistance to teach us the wonders of the Yosemite Trailside.

IV. Ah-Wi-Yah!

Ah-wi-yah or "Quiet Water" is much more musical than the white man's version "Mirror Lake," The word mirror is much in keeping with the chewing gum tokens and the paper trail of the kodak that fringes the lake. We need simplicity, quietness, and the natural in our language and life.

It is the quietness of the lake that enables us to see the reflection of Clouds' Rest. To see over a mile of granite ledge reflected in a small sheet of water is sufficient cause for musing. It is not until midday

Fig. 3. This Avalanche filled the whole Yosemite Valley with a cloud of rock dust. A whole forest was brushed oneside. This is only a passing show in the dramatic story of the Mountains.

that the image is brushed away by the gentle, canyon breeze, which ripples the surface. And this may give cause for further meditation.

Our reverie may take us back to the beginning of the lake. It was born long after the Glacier had scoured out the U-shaped walls of Tenaya Gorge. Geologists tell us that about three hundred years ago a severe earthquake shook the loose material from the walls of the cliffs of this region, forming the major part of our talus slopes. It is possible that during such a disturbance the gravel debris was thrown across Tenaya Creek to form Ah-wi-yah.

Ah-wi-yah was no sooner born than the creek set in to destroy it. Granite born soil, sand from the quartz, and clay from the feldspar, were dropped into the lake. As the deposits accumulated, the vegetation began to march into the shallow stretches of the upper end. A close study of this conquest of the lake will probably picture the filling of the post-glacial Yosemite Lake.

Chara and burr-reed are practically the only aquatics that have found this recent lake. This is probably due to the youthfulness of the lake and to the fact that there are very few aquatic plants in the lakes of the upper Tenaya Creek.

The sand plain, which has accumulated at the entrance of the creek into the lake, is being laid down faster than the vegetation can seed it. Sorrel is in the front ranks. This plant probably did not grow on the shores of Yosemite Lake, as it is a recent weed arrival from Europe. The next plant settler is a native Artemesia, which could have been the first colonist on the ancient lake. Willow is in the front ranks of the woody battalion. Behind it in close formation are marching the lodgepole pines. In the open spaces are lupine, mint, pussy paw, and manzanita—pioneer plants of dry areas and the chaparral, which have been well trained to meet the drought conditions of this sand desert. Incense Cedar is not quite so venturesome but appears in the back ranks of the leafy cover. As soon as the willows stabilize the stream banks, the black cotton wood and the creek dogwood take up the stand, each plant preparing the way for the one in back of it. Coming down to the shores of the old lake are the black oak, yellow pines, and the firs. They have not yet ventured onto the lake plain but are evidently preparing to do so. The tree colonization is so rapid that the meadow stage is very brief.

The animals which accompany lakes are very few in Ah-wi-yah. This, too, is probably due to its youthfulness. During this season it is unusually low. Along its clay margin are autographed the tracks of the spotted sandpiper, deer which come down to drink, and summer folks. In its deeper recessives are the native rainbow trout. A few water striders walk along its surface. Taken all in all, its life is a solitary one. Not a frog to break the silence nor a turtle to sun himself upon a log.

Herein is being enacted in a short space of time the history of the Yosemite Valley. V-shaped valley, Glacier, U-shaped valley, dam, lake, meadow, shrubs, forest march in succession like a great play— the sand plain a stage, and the trees the actors. Visitors to Yosemite should go to Ah-wi-yah, for therein is breathed the spirit of the Great Valley.

V.   The Flora of Yosemite and Cape Cod

It would seem that if one start with a fair knowledge of the plants in his backyard he would find either the same species or their counterparts the world over. Such at least is the experience of a nature guide venturing from the sand dunes of Cape Cod on Massachusetts Bay to the Sierras of the Yosemite National Park. The marked similarities have awakened so many pleasant surprises—so many trains of reminiscences along the trailside—the writer has been tempted to put his random notes down as a permanent reminder.

Our story must date back to the end of the glacial period, some 20,000 years ago, when both Yosemite and Cape Cod started with a clean slate. The soil in both places is of granite origin—in Yosemite from the granite mountains of the Sierras and on Cape Cod from the granite coast of Cape Ann. The glacial streams sorted out and deposited some of the materials in lowlands but left large masses of unsorted gravel in hills, those pushed up at the end of the glacier being known as terminal moraines and those parallel to its course as lateral moraines. If the terminal moraine dammed a river course, lakes were formed in back of it. There were often ice masses left in the moraine material; as they melted they left kettle holes which now hold lakes. The shallower ice basins became sphagnum bogs without inlets to cover the plant growth with sand. The vegetation in both Yosemite and Cape Cod is mostly on this moraine material, the cliffs and domes of Yosemite and the cliffs and dunes of Cape Cod being scarcely planted as yet.

The distribution of plants on the moraine deposits is determined by the moisture. Both regions have fine glacial meadow gardens. The dry sand areas of ancient lake bottoms and the recent talus slopes are populated by drought resistant plants. The polished domes and wind swept dunes scarcely hold moisture enough for pioneer plants. The comparison of plants in these similar areas merely scratches the surface of possibilities. These introductory notes may serve to remind others of how they have renewed plant acquaintances as they traveled from east to west or west to east. If the chain of evidences can some day be gathered into one story, it may make a fascinating chapter in plant annals.

Arctostaphylos: Arctostaphylos Uva-ursi or the Bearberry of Cape Cod covers whole hillsides and indeed serves as the lawn for many summer homes. It was almost a thrill to meet what at first appeared to be the same species at a 7,000 foot elevation on the rim of the Yosemite. We now know that the Sierra form is Arctostaphylos Nevadensis or the Dwarf Manzanita. Both plants form loose mats in dry sunny open ground. They are evergreens with zigzag branches and peeling red bark. Their arbutus like blossoms come early in the season and mature into red acid berries, which are scarcely edible because of their many large seeds. After sending for a Cape Cod specimen, we were still impressed by the similarities rather than the differences. The leaves of both are thick, entire, and glabrous. The main difference in leaves is the presence of a minute tooth on the apex of the Manzanita leaf. It is also broader in proportion to its length, being on the average about one-eighth inch longer than its Atlantic cousin. The Manzanita berry is about one-eighth inch smaller in diameter, the persistent sepals being acute rather than ovate as in the Bearberry. The differences in these two members of the Heath Family are quite minute when we think of one growing within ten feet of sea level and then suddenly appearing from 7,000 feet to timberline some 3,000 miles away. I have seen Bearberry in the White Mountains growing under similar conditions. As the Appalachians are older, I am wondering if Arctostaphylos appeared there first? If it did, how did it get to the high Sierras? Did it come

by a granite route or by a moraine road? Where is it found in between, and what are its variations? If we knew all about Bearberry, we might be able to answer other enigmas of the plant world.

Drosera Rotundifolia. Here is another plant that grows near sea
level on Cape Cod and is found in Tenaya Canyon at an elevation of
about 6,000 feet. Here it grows on the north wall of the gorge in the
seam of a moist granite cliff some thirty feet above the river bed.
A comparison of the two plants shows a persistent difference in
leaves. The alpine plant has a rotund blade averaging about one-
fourth inch in diameter, whereas the leaf of the coast plain plant is
broadly elliptic, being transversely 11/16 inches wide and 7/16
inches in length. An examination of the flowers of the two plants
may show them to be separate species. There is no consistent difference in the height of the flower stalk or in the number of flowers.
A study of this plant according to its geographic range across the
Great Lake region and Montana to California might give interesting
data in variation. When the glaciers swept it south, how did it march
back?

Heath-Like Plants. We have been speaking of plants that are most
closely related. It is also possible to have totally unrelated plants
meet the same conditions in a like manner. On Cape Cod there are
two heath-like plants covered with awl-shaped leaves—Hudsonia
ericoides and Hudsonia tomentosa. They are locally known as
Poverty Grass as they can grow on dry sandy hills. These low shrubs
are members of the Rockrose Family (*Cistaceae*). On the granite
rocks of the Sierra summits, above 6,000 feet, is found the beautiful
Alpine Phlox (*Phlox douglasii*), a member of the Gilia Family (*Pole-
moniaceae*). These homologous plants of the East and West form
dense mats in dry open situations and bear flowers on the upper part
early in the season before the soil is completely dried out. The reduced awl-shaped leaves prevent excessive loss of moisture, and the
matting of the leaves may tend to hold the dampness in the soil.

Sedum. Plants may also meet dry situations by storing water in
fleshy leaves. A Cape Cod representative of this genus is *Sedum
acre*, the Mossy Stonecrop. The Yosemite counterpart is *Sedum
obtusatum*. Both species spread moss-like on the ground with thick
leaves and yellow blossoms. The Sierra form is common on rocks.

Artemisia or Wormwood is a most bitter, aromatic plant. The
typical western representative is *Artemisia tridentata* or Sagebrush,
and the prevailing Beach Wormwood on Cape Cod is commonly
known as Dusty Miller (*Artemisia stelleriana*). The dense gray wool
which covers these plants assists in retaining moisture and enables
them to grow in exceedingly dry situations. The Dusty Miller is
often used as a border plant in Cape Cod gardens and, although introduced from Asia, has taken naturally to sand beaches. These
perennials are dominant wherever they grow and attract more attention by their whitened foliage than they do by their less showy yellow-

ish flowers. Bitter concoctions were made from these plants by the settlers.

Robinia Pseudo-Acacia. The Common Locust or False Acacia was named for John Robin and his son Vespasian, who first cultivated the Locust tree in Europe. The colonist loved the tree for its fragrant white flowers and its value as an ornamental tree. This may be the reason that it was carried to Yosemite and also to Cape Cod. We find it growing in both places wherever the early homesteaders located and since those times it has established itself in waste places. It proved to be durable as a fence post. One old farmer said that it would last a hundred years. He knew that it would as he had tried it several times. The coast people found it useful in ship building.

Ulmus Americana. If we think of the Common Locust as a homestead tree, we can think of the Elm as a street tree. The American Elm was introduced as a village shade tree both in Yosemite and on Cape Cod. Its wide-spreading branches make it particularly suitable for this purpose, and it seems to thrive in its new home. Why the tree did not find these localities of its own accord is rather puzzling. Possibly these glaciated areas were so remote that the tree had not had time to get there. Such trees as the hickories, beeches, birches, and the chestnut (*Castanea dentata*) are not native to the Yosemite or Cape Cod, and with the exception of the Gray Birch on Cape Cod have not found their way to these regions. Is it because they will not grow there or because they have not been introduced? It is certainly not because they are shy of the mountains or of the lowlands nor is it because they are not of general distribution.

Wind Blown Trees. One example in the Yosemite is the Jeffrey Pine (*Pinus ponderosa* var. *jeffreyi* Vasey) on top of Sentinel Dome. It grows at a higher and bleaker altitude than the true Western Yellow Pine. The typical Cape Cod evergreen, and only pine, is the Pitch Pine (*Pinus rigida*). It also is able to grow in dry, exposed areas and responds to the wind by growing to the leeward and at other times forms deep carpets. Both of these pines are three needled with the scales of the cones bearing a short prickle. It is the same force at work that distorts them. The prevailing wind blows up the San Joaquin, and a southwestern prevails from the Atlantic. Near timber line in Yosemite the Sierra Juniper (*Juniperus occidentalis*) becomes much gnarled and stubby. The only Juniper on Cape Cod is the Red Cedar (*Juniperus virginiana*), which occurs occasionally near old dwellings and was probably introduced to the lower part of the Cape. Unlike its western namesake, it does not venture into exposed situations but simulates it by never growing into a forest, instead remaining alone and independent. (It does form thickets in other parts of Massachusetts and New England, especially in old, abandoned pastures).

Populus Tremuloides. The Quaking Aspen is one of the first broadleaved trees to march out into the open meadows of the Yosemite rim.

Its trim greenish-white bark and leaves, which tremble in the slightest breeze are sure to attract attention. The tree is not found in the Valley yet grows in light, open areas on Cape Cod. Is the fact that so many Cape Cod plants are found in the High Sierras and not in the Valley where the elevation is about 4,000 feet due to the latitude? It is doubtful if such an explanation is sufficient, as this species ranges from Hudson Bay to Mexico. The two other poplars of Cape Cod grow around old houses and were brought in by the settlers, the Silver-leaved Poplar (*Populus alba*) being introduced for its shade and the Balm of Gilead (*Populus balsamifera*) for the medicinal qualities of its sticky buds. The Silver Poplar is the most wide-spread of the Cape Poplars and has spread widely by its roots about old house lots. The Black Cottonwood (*Populus trichocarpa*) is the most conspicuous poplar along the rivers in the Yosemite and neighboring valleys. It has a slight varnish on its buds, which reminds one of the fragrance of the Balm of Gilead. Black Cottonwood thickets are quite similar to the Silver Poplar groves of the Cape except that the latter grow in dry situations.

Alder. The common Alder of Cape Cod is *Alnus rugosa* or the Smooth Alder. The Yosemite relative is *Alnus rhombifolia* or the White Alder. The interesting surprise to a New Englander is to find an Alder which is a tree 30–80 feet high, for to him all alders are shrubs. However, in the yards of Dr. Fred Canady and Captain Howe at Wellfleet on Cape Cod there are Alder Trees. These trees are not native but came ashore in the wreck of the British ship Franklin about 1870.

Other Plants common to both localities are Brome Grass, which grows in exceedingly dry situations; the Eagle Fern or Brachen (*Ateris aquilina*) being commonest in both places; certain Bur-reeds (*Sparganium*) and Pond weeds (*Potamogeton*) of undetermined species; Fire Weed (*Epilobium augustifolium*); and Shadbush.

VI. *Impressions*

There are two kinds of impressions, as related to Yosemite visitors. One is the impression made on the visitor, and the other is the impression left on the Valley. This note refers to the impressions left by the visitors.

Most visitors come in an automobile. The machine stirs up a cloud of dust and is gone. The dust settles in a new place, and the record is soon rewritten by another machine. A hundred thousand tourists may leave no greater impression than this.

There is the Mariposa Batallion, which came up here in 1851— the first white men to visit the region. This was only about seventy-five years ago, yet any record of their visit is a curiosity and is on exhibition at the Yosemite Museum.

And the Yosemite Indians! What of them? The only evidences written in the Valley are the mortar holes where they used to grind acorns for meal. Here and there one may find a chip of obsidian that flaked off when they were fashioning arrow heads. The rest is handed down in myth and story.

May it not be said of Yosemite, "What is man that Thou art mindful of him?"

The visitor which made the greatest impression came some 20,000 years ago. It appeared silently without the sound of a trumpet or the glare of steel. There was a snow storm up on the summit of Mt. Lyell. Then there was another snow storm, and they kept adding to

Fig. 4. Reflections, with a Beaver Dam as a frame and a Beaver Pond as a Canvas, Nature presents a remarkable scene. Photographed by the author on a Rocky Mountain Trail.

the accumulation until a great snow field covered Lyell, Clark, Hoffman, and the others. It became nearly a thousand feet thick and so massive that it moved down the mountain sides to the valleys. This huge ice sheet picked up the loose granite debris from these summits and quarried the sides of Tenaya and Little Yosemite. It polished Liberty Cap and Mount Broderick and hewed out granite rock basins for Emerald Pool, Merced, and Boothe, and other summit lakes. All markings on the granite were erased, and, when the great glacier retreated, the walls were as new.

The first post glacial visitors were descendants of the old inhabitants. They marched into the valley in single file. At the head of the oak column came the huckleberry oak because it was more congenial for it to live nearest the glacier. Then up out of the dry country came the Golden Cup Oak and, when the talus formed, it took a stand near the cliffs. The Kellogg Black Oak came about the same time and stood sentinel-like on the open plain of the old lake bottom. The march of the oaks was up hill work and was accomplished with the assistance of squirrels who buried the acorns.

The pines came by the wind express usually on the afternoon up-valley schedule—each seed being a monoplane. The foremost in rank of the pine battalion was the white bark pine. Then came the mountain pine, with the Jeffrey close on its heels. The sugar, lodge pole, and yellow were later arrivals. The conifers are only now beginning to etch their record upon the granite cliffs and domes. With the acid secretion of their roots they make sketches upon the rocky walls and with the expansion of growth they flake off new pages. The story that they are writing is one of plant succession and today is but the first chapter of their impressions.

The glacier coaxed some plants, like the Alpine willow and the sorrel, up into the high spots and then went off and left them. The weather is still congenial to these cold-seeking plants, and they live in islands, so to speak, way up in the high Sierras. They do not visit and mingle back and forth because the valleys are too warm for cross roads. When the refrigeration was more general, these plants may have paved the valleys. Now they are marooned mountain high—stranded as it were—on a stern and rock bound coast.

And we must not go on with our description of plant impressions unmindful of the lichens and mosses. Their work goes back to the time of the first Yosemite plants, curious growths that crept upon the rocks for a long time before man appeared. The lichens may have remained like magic when only the top of Half Dome and a few others broke the ice waves. It must have been a cheerless period of cold, yet these tiny plants may have carried on during the age of the ice and rock giants, and their wee writings are still going on preparing the way for less hardy followers.

The latest arrivals have only lately ventured onto the valley floor. They take up the coarse soil, live their span of life, and then die. They add a little bit of humus and thereby are slowly darkening the top layer. The common locust and the elm are such and were brought by the colonist. The curly dock, sorrel, and knotweed of the yards stole in with the garden seeds. So came the common plantain and red clover. And there are others like the white daisy, chickory, and bouncing bet that have not reached here. Some like the burdock, beggars tick, and cocklebur are always hooking rides. They have not yet stolen into the Yosemite. Let it be hoped that they will not be allowed within the park limits. May Yosemite be one place where the native granites may be written upon by native plants in a native tongue. For herein may visitors come and see history being written in the American way—as it was before America was known to Europe.

CHAPTER LII

NATURE-STUDY ON THE PLAYGROUND*

How many playground leaders are country born? The country is a wonderful place to *come* from. Possibly some of you think that it is a great place to come *from*. Can you do all the things in a playground that you as a child could do in the fields and forests? I do not refer to stealing bird's eggs and shooting gray squirrels for that is out of date. Of course I realize that blueberrying is impossible in a city playground. A city park is saturated with *don'ts*. Don't spit on the walk, Don't walk on the grass, Don't pick the flowers. Many know the Park Dont's in Nature better than the do's. The caretaker can see no sense in watering the grass to make it grow and then having to cut it down again. I know of a teacher who recently had a class out in the park. She bent a limb down for the little tots to see the leaves. She received a call down from the Superintendent of Parks. I heard of some children over in our city who were climbing a pine tree and trying to swing in it. Those of you from the country know the fun of swinging in a birch tree. Two Boy Scouts saw them and immediately went over and told the children that they would break the tree. They must not do it. Those scouts had a forest sense. They were right. There is a long list of Don'ts for your playground and parks. If there were time and you were feeling in the mood I would like to have you make a list of "Do's" and "Don'ts" in nature experiences in your particular playground area. It would be enlightening to us all. I suspect that the liabilities would outnumber

---

*An address given at the Twelfth National Recreational Congress, Asheville, North Carolina.

the assets in the accounting of nature possibilities on playgrounds.

I want also to distinguish between the artificial playground and the natural playground. An artificial playground has ladders to climb instead of birches, awnings instead of shade trees, a dust laying preparation instead of green grass, an iron fence instead of shrubs, an Italian-tiled wading-pool instead of a frog pond. Up to this point the artificial playground might just as well be in a shed or basement. The artificial playground is a very uninteresting place for the enjoyment of nature. The tendency today is as much as possible toward the natural playground.

A natural playground is an outdoor neighborhood area on which are associated fresh air and sunshine; trees, flowers, and animal life; and play apparatus. There may be such a thing as too many swings and too much apparatus. When it means the removal of nature's furnishings it becomes an outdoor gymnasium. A playground is not an outdoor gymnasium. A playground is not a parking place. It is not simply an amusement area. It is an outdoor school room and should supplement but not duplicate school activities. Birds and flowers are as essentially a part of the leaders equipment as graveled areas and artificial equipment. He should utilize all the resources of the area. The playground which has been fortunate enough to have had its trees and weeds tolerated has therein a great natural resource in a crude state. The development and utilization of this resource is in a pioneer stage.

Mention has been made that this work is under an organized director. A playground leader is more than a nurse maid. He is not necessary to push someone in a swing. Children will play on apparatus without leadership. Give the boys a bat and ball and they will organize at once. Often times they know more about it than the leader. A playground leader is not a junior policeman. Children in healthy play do not need one. It is just as essential to train leaders for the outdoor school room as for the four-walled room, and the playground that counts most is going to be educational. The leader is going to give children a habit of seeing and doing. He is going to teach them to protect life. He is going to develop habits that will make for the enrichment of leisure time. He is going to bring out the best that is in them.

We have made a big sacrifice to get our parks and playgrounds and now we must make a big sacrifice to save them. It is just like getting democracy and then having to make a second struggle to make the world safe for democracy. One way to save a playground is to get iron apparatus that cannot be destroyed but this does not teach the finer contacts. The most attractive side of playground work includes the enjoyment of the natural life. Our parks are here but they represent undeveloped mines. They are in the hands of the people. An uneducated park republic can destroy the park. If people do not learn good park manners at home they will not know them in our state or national parks. The acts of people in our large recreation areas is a good indication of their early training. How will your community be represented?

What are some of the ways in which we may utilize the natural resources of our parks and playgrounds? This, I take it, is the reason for my being invited here—to suggest what may be done in this fertile field. If there is anything that I say which suggests discussion make a note of it and if there is a demand and time we will hold a discussion group, or possibly have a field demonstration. I am anxious to be of service while here.

First of all, do not publish a list of birds or trees in the area. That would be no more interesting than a list of dates in history. An ideal book on the natural history of a park is that published by Ansel F. Hall in 1921 on Yosemite National Park. If possible, get a similar book for your community.

Enlist the active support of your community bird or field club. Have them meet at your park. Have some of their members take children on field trips altho these leaders will have to be selected with great care as they will tend toward the catalogue style.

Get a naturalist champion for your park, as was John Muir for Yosemite and Enos Mills for the Rocky Mountain Park. Encourage this naturalist to get out nature publications written in a simple and interesting way.

Foresee the nature guide movement. Every community, along with its physician, teacher, and parson is going to have its nature guide. There is no better way to introduce that work than through the park or playground system. The nature guide is not only able to lead trips but has been trained to tell nature stories and to give nature talks. The field excursions will develop closer friendship and com-radeship than is possible with the swing or shute.

Nature guide trips should be given for adults. Playgrounds are usually thought of for children. Why not have a little play that can be carried over into old age. How many of you are now playing the games that you are making possible on the playground? What recreation do you take? A survey of the recreation of recreation leaders would be enlightening—possibly amusing. Here again, I would like to take time for paper and pencil. If it is true in your case—that there is no provision for the early enjoyment of the games that carry over into later life—how much more so may it be in the case of those not interested in recreation. Theirs is wreck-reation. What is sadder than an old person without capacity for recreation? The man who has forgotten how to play is a sad spectacle. He who gets into the great game of life with the plants and animals has met a lifetime of interests.

Have a Park Museum. One of the best city park museums is the Roger Williams Park Museum in Providence. In the past they have gotten out some very valuable publications. One of the best books on the subject is by Harlan I. Smith, called Park Museum Handbook of the Canadian Rocky Mountain Park. The museum should, of course, deal with local material—local historical material, minerals and rocks of the locality, have pictures or specimens of flowers and birds in season, monthly star maps, insects that are likely to appear, a model of the water system, etc. The children, themselves, will enjoy making the museum and many of their nature activities may center around this project. This is not the case in the artificial playground. The only geology that they then get is dust and possibly stone bruises. The only plant may be the green scum from the wading or swimming pool, and the only personal contact may be with the mosquito. They cannot get much appreciation from such acquaintances but if these be the nature opportunities I would at least have them understand that many. If possible, surround the children with the best friends of the wild that you can collect.

Some parks have lions and elephants. How much more wonderful is the grasshopper. He is much cheaper to feed. Pick him up and he spits tobacco. He is noted for his ability to jump. He has five eyes— one right in the middle of his forehead. He sings with his hind leg. He breathes through his abdomen and has ears on the sides of his abdomen. If any such individual were advertised for the next comedy you would surely get a ticket yet you pass by these shows every day because no one has pointed them out.

Another source of nature inoculation is through celebrating special days. Make a special program for Arbor Day and for May Day. The State of Rhode Island publishes an Arbor Day Bulletin for each one of its public school children, an enterprise worthy of emulating.

Nature-study as a form of play does not mean clearing the field for action but quite the contrary—keeping the wild places wild for nature activities. I fully realize that you have asked me to speak about Nature play. It seems to me, however, that we cannot play checkers until we get the checkerboard—in other words we cannot utilize this form of playground activity unless we prepare the way. I have suggested some ways of getting the entering wedge.

CHAPTER LIII

Winter Nature-Study: Was and Is

Some folks think that Nature-study retires for the winter with the ground hog or perhaps that it goes to Palm Beach along with the bird migration. This is just as imaginary as the belief that pussy willows are only here when they have *pussies*.

Many people eat a big turkey dinner and retire for the winter. They are said to be *housed up*. With the fear of pneumonia as an alibi, others take their last bath of the season. Some neighbors, southern European we are told, sew up their children in several layers of shirts topped off with a red sweater. These people believe that everyone else does the same thing. This also is a supposition.

Thanksgiving marks the retiring time of nature crops. The leaves have fallen, the insects have had their last medley and the beavers have gone to their winter cabins to live on aspen bark. Everything, in the style of Moby Dick, is stored down and cleared up. This again is not so. With the retiring of the chipping sparrow comes the junco. Although the American silkworm is hanging in a cocoon, the woolly bear still roams. It is spawning time for the codfish. Winter nature-study is as interesting as summer nature-study. There is every indication that there are those who are being aroused to the possibilities of winter interests.

December opens the season of unnatural winter. When business is poor with editors the *old timer* is made to observe squirrels storing an extra large crop of nuts and their fur is reported as unusually thick. On the strength of this the prophet predicts a hard winter. But large crops are due to past weather rather than future and a thick coat of fur is the result of good food rather than what is to come.

Winter Camping at Bear Mountain Park.

This annual display of current unnatural events is being censored by our young naturalists.

The classical Old Farmers' Almanac always predicts a snow storm along in the first two weeks in January. The writers assume that if the period was long enough there would be sure to be a snow storm. If a winter sport party is going to Jaffrey in the White Mountains, however, the members are apt to consult the weather-man as to whether there will be a snow storm over the week-end. Ground hog weather is giving away to the weather bureau.

The almanac has also been found to be a wonderful advertising medium for patent medicines, probably to prevent the ill effects of winter. Horse chestnuts and muskrat furs are still used to keep away rheumatics and the rabbit's foot is carried for good luck. The fear of winter has sentenced more people to close confinement than is commonly realized. But there is an uprising. Modern youth is showing an utter disregard for winter ailments. They are insisting in ever increasing numbers upon opportunities for winter sport.

Bear Mountain, the largest camping park in the world, is opening its annual season of winter camping. The commission has constructed an outdoor skating rink, two toboggan slides, and rents skis, sleds and snowshoes. The old fashioned straw ride is being revived. The winter hiker is getting a genuine thrill following snow clad streams and animal trails. They insist on seeing the tracks of the fox and the snowshoe rabbit which before have been limited to book nature.

The Girl Scouts of Rochester are interested in a plan suggested by the National Plant, Flower, and Fruit Guild, of distributing to

A BOY SCOUT IS PREPARED: These scouts have cut a long pole to use in case someone breaks thru the ice.

shut-in people small Christmas trees in pots. If these trees are kept alive in the winter they are to be transplanted in the spring. This project is being carried on in cooperation with the New York State College of Forestry. This shows not only a fine way of carrying out the scout laws, but the broad policy of the Forestry College in not discouraging the Christmas Tree. It is the Christmas Tree brought up to date and in harmony with all laws of conservation.

An interesting source of enjoyment with potted plants from the out-of-doors is with the winter rosettes of biennials. The mullein plant is sold as the American Velvet plant in London. Queen Ann's Lace suggests the beauty of the leaves of that plant and the cultivated carrot when grown in flower pots becomes a close rival of our ferns. Even the dandelion and primrose will blossom when brought to a warm room. The green colors of these weeds are as refreshing as that of the laurel, Prince's Pine, and Christmas fern. While getting an up-to-date winter view-point why not get better acquainted with our weeds and stop the extermination of these rarer plants of the woodlands?

The Massachusetts State Girl Scout Camp at Cedar Hill, Waltham, is getting ready for winter scouting parties. Early mornings will find merry girls hiking through snow flurries to hemlock hill or the cedar swamp to see the footprints of the partridge, or to watch the nuthatches and myrtle warblers. Many a rollicking group has decided that the lean-to is the favorite shelter in winter. They build their lean-tos of evergreen boughs and have a reflector fire built in front with a log or stone back to reflect the heat into the shelter. These scouts sleep inside as warm as toast, and with an absolutely clear conscience, for the snow eliminates the forest fire menace.

Another sign of a busy time outdoors this winter comes from the schools. The observer recently saw a group at the State School of Agriculture at Alfred, New York, on a nature trip in a heavy snow storm. Upon inquiry he learned that they were prospective teachers learning nature that they in turn might take their pupils into the open.

The forests and snowfields are our natural playgrounds in winter. The gap between play in summer and hibernating in winter is becoming remarkably narrow. People are going to the woods in winter in greater numbers. If a half million participated in winter play last year we may expect a million this season. Shall we uphold the American standards for recreation in the winter? Progressive cities are beginning to point with pride to their winter playgrounds.

———

"Clearly the place to seek for the nature-study that is true to human life is first of all in the historic development of man's relations toward nature.—*Hodges.*

"Nature-study not only educates, but it educates nature-ward; and nature is ever our companion, whether we will or no. Even though we are determined to shut ourselves in an office, nature sends her messengers. The light, the dark, the moon, the cloud, the rain, the wind, the falling leaf, the fly, the bouquet, the bird, the cockroach—they are all ours. Few of us can travel. We must know the things at home."—*L. H. Bailey in "The Nature-Study Idea."*

"As from a hidden organ-loft upsoaring,
　The rare song-rapture rises through the hush;
So from the topmost boughs outpouring
　Flows all the liquid silver of the thrush."
　　　　　　　　—*Mrs. Merrill E. Gates.*

CHAPTER LIV

Nature-Lore: A Selected Bibliography

There is an abundance of nature literature. The following list has been arranged that there may be economy of effort and time. Books have been selected to represent the different phases of nature-study. It is obvious that no two would select the same list. Special attention has been given to Nature Stories. The author believes that all books included in the list are accurate. Wherever it has been known that the story personifies or has the animals doing unheard of things it has been omitted. The list is not complete. In the collection of dog stories, for example, the list has been limited to the generally recognized standard stories. It is also true that there may be many other dog stories that should be added. There are many nature books that are wholly informational and these also are arranged alphabetically according to subject. It is often difficult to say that a book is either informational or in story form but they have been limited to one list to save space. It is also well known that prices frequently change.

I. Methods

Bailey, L. H., Nature-study Idea, Doubleday..........................$1.00
Berry, James B., Teaching Agriculture, World Book Co................. 2.00
Bigelow, E. F., Spirit of Nature-study, A. S. Barnes................... 1.00
Burbank, Luther, The Training of the Human Plant....................
Comstock, Anna B., Handbook of Nature-study, Comstock Pub. Co...... 5.00
Comstock and Vinal, Field and Camp Notebook, Comstock Pub. Co.....
Cooper, Lane, Louis Agassiz as a Teacher...........................
Hodge, C. F., and J. Dawson, Civic Biology, Ginn.................... 1.75
Keller, Helen, The Story of My Life.................................
MacCaughey, The Natural History of Chautauqua, H. W. Huebsch 1.00
Mills, Enos, The Adventures of a Nature Guide, Doubleday............ 3.50
Palmer, E. L., Nature Landscapes and Life History Charts, Comstock....
Roosevelt, Theo., Letters to His Children..........................
Rousseau, Emile, Appleton.. 1.50
Skilling, W. T., Nature-study Agriculture, World Book Co............. 1.68
Skinner, M. P., The Yellowstone Nature Book, A. C. McClurg.......... 2.50
Spillman, W. J., Farm Science (Teaching), World Book Co............. 1.68
Wiggam, Albert E., The New Decalogue of Science, Bobbs-Merrill.......

II. Stories

Animals in General (See Mammals).

Austin, Mary, The Trail Book, Houghton Mifflin.....................
Beard, Dan C., American Boys' Book of Wild Animals, Lippincott.......$3.00
Beebe, Wm., Jungle Peace (Guiana), Henry Holt.....................
Bostock, F. C., Training of Wild Animals, Century.................... 1.00
Brearley, Harry C., Animal Secrets Told............................
Brunner, Josef, Tracks and Tracking, Macmillan...................... 1.00
Burroughs, John, Camping and Tramping with Roosevelt.............. 1.50
    Winter Sunshine... 2.00
    Ways of Nature... 2.00
    Songs of Nature, McClure Phillips...............................
    Wakerobin, Houghton...

Chapman, W., Green Timbered Trails, Century.........................
Cooper, Courtney R., Under the Big Tops (Circus Animals), Little....... 2.00
Creighton, Kath., Nature Songs and Stories, Comstock.................
Dixon, R., F. A. Stokes, Human Side of Animals, Stokes...............
Donald, C. H., Companions, Feathered, Furred, and Scaled, John Lane...
DuChaillu, Paul, Wild Life Under the Equator, Harper................. 1.75
Eaton, Walter P., On the Edge of the Wilderness, Wilde................ 1.75
Fabre, J. H., The Story Book of Science, Century..................... 1.75
Frentz, Edward, Uncle Zeb and His Friends, Atl. Mo. Press...........
Gould, A. W., Mother Nature's Children, Ginn....................... 1.00
Groos, Karl, The Play of Animals...................................
Hawkes, Clarence, The Way of the Wild, Jacobs...................... 1.60
Hornaday, Minds and Manners of Wild Animals...................... 2.50
Houssay, Frederick, The Industries of Animals......................
Hudson, W. H., The Book of a Naturalist, Dutton................... 3.00
Ingersoll, Ernest, Wit of the Wild, Dodd, Mead...................... 2.00
Jenkins, Oliver, Interesting Neighbors, Blakiston.................... 1.50
Long, W. J., Ways of Wood Folk, Ginn.............................. .50
       Wood Folk Comedies, Harper.................................. 3.00
Maeterlink, Maurice, Mountain Paths.............................. 2.00
       Unknown Guests.. 2.00
       Our Eternity... 2.00
McNally, Geo. M., Babyhood of Wild Beasts, Doran.................. 2.00
Mills, Enos, Wild Life of the Rockies, Houghton.....................
       Wild Animal Homesteads, Doubleday......................... 2.50
Mukerji, Dhan G., Jungle Beasts and Men, Dutton.................. 2.00
Muir, John, Steep Trails... 3.25
       The Cruise of the Corwin.................................. 3.25
       A Thousand Mile Walk to the Gulf.......................... 3.25
       Travels in Alaska.. 3.25
       The Story of My Boyhood and Youth......................... 3.25
       My First Summer in the Sierra............................. 3.25
Murrill, Wm. A., Billy the Boy Naturalist, Bronxwood Park, N. Y. City..
Pellett, Frank C., Our Backdoor Neighbors, The Abingdon Press........ 1.50
Roberts, C. G. D., The Backwoodsman, Macmillan.................... .50
       Neighbors Unknown.. 1.00
       Secret Trails.. 1.00
Rogers, J. E., Wild Animals Every Child Should Know, Doubleday...... 1.00
Seton, E. T., Wild Animals at Home, Grosset........................ .75
Sharp, Dallas Lore, Whole Year Round, Houghton...................
Sqnier, Emma L., On Autumn Trails, Cosmopolitan.................... 2.00
Thompson, E. S., Wild Animals I Have Known, Scribners.............. 2.00
       Animal Heroes, Grosset.................................... 1.00
Thoreau, H. D., Walden, Houghton, 2 vol. each...................... 1.00
Ward, F., Animal Life Under Water, Funk and Wagnalls...............
Washburne, S. W., and H. C., Story of the Earth, Century...........
Weed, C. M., Seeing Nature First, Lippincott....................... 1.25
White, Gilbert, Natural History of Selbourne....................... 4.00
Wiggam, Decalog of Science......................................
Wilcox, Alice Wilson, Treasured Nature Lyrics, R. G. Badger..........
Wright, Mabel Osgood, Four Footed Americans, Macmillan............. 2.50
Zavarziger, Animal Kingdom, Macmillan...........................

Ants.

McCook, Henry M., Ant Communities...............................
Vamba, The Prince and His Ants..................................

Bears

Carter, M. H., Bear Stories (Retold from St. Nicholas), Century......... .65
Hittell, Theo., Adventures of James C. Adams......................
Major, Chas., The Bears of Blue River, Macmillan................... 1.50

Mills, Enos, The Grizzly, Houghton Mifflin..........................
Seton, E. T., The Biography of a Grizzly........................... 1.50
Wright, W. H., The Black Bear, Scribners........................... 1.00

Beaver

Dugmore, A. R., The Romance of the Beaver, J. B. Lippincott..........
Hawkes, Clarence, Shaggycoat, Jacobs................................ 1.25
Mills, Enos, In Beaver World, Trail Bkstore, Long's Pk.............. 2.25

Bee

Fabre, Jean Henri, The Mason Bees, Dodd Mead...................... 2.50
     Bramble Bees and Others, Dodd Mead........................... 2.50
Lovell, John H., The Flower and the Bee............................
Maeterlink, Maurice, Life of the Bee, Dodd Mead................... 1.50
Morley, M. W., The Bee People, McClurg............................ 1.35

Birds

Bralliar, Floyd, Knowing Birds through Stories, Funk and Wagnalls...... 2.00
Burgess, Thornton, The Birdbook for Children, Little.................. 2.50
Burroughs, John, Bird and Bough (Poems)............................
     Birds and Poets.. 2.00
Durand, Herbert, Taming the Wildings, Putnam....................... 3.50
Eckstrom, Fannie H., The Bird Book, Audubon Society................ 1.40
Grinnell, E. and J., Birds of Song and Story......................
Hudson, W. H., Adventures among Birds, Dutton..................... 4.00
Lanier, Sidney, Bob, The Story of Our Mockingbird, Scribner.......... .75
Mathews, F. Schuyler, The Book of Birds for Young People, Putnam.....
Miller, Olive Thorne, True Bird Stories from my Notebooks, Houghton... .60
Myers, Harriet W., The Birds Convention, Out West Magaz., Los Angeles..
Pearson, T. Gilbert, Tales from Birdland, Mass., Aud. Soc. 66 Newbury,
     Boston, Mass.. 1.10
     Stories from Bird Life, Aud. Soc. 66 Newbury, Boston, Mass.......... .80
Porter, Gene Stratton, Homing with the Birds......................
Rolt-Wheeler, Francis, The Boy with the U. S. Naturalist, Lothrop........ 1.75
Stratton-Porter, Gene, The Fire Bird.............................. 1.75
     Friends in Feathers... 3.00
     Homing with the Birds....................................... 2.00
     Song of the Cardinal.. 1.75
Torrey, Bradford, Birds in the Bush..............................
Walker, M. C., Our Birds and Their Nestlings, Am. Bk. Co............ .85

Buffalo

Dimock: Wall Street and the Wilds................................

Butterflies and Moths

Porter, Gene Stratton, Moths of the Limberlost....................
Scudder, S. H., Life of a Butterfly..............................

Cats

Carter, M. H., Cat Stories (Retold from St. Nicholas), Century.......... .65
Jackson, Cat Stories, Little..................................... 2.00
Kipling, Just So Stories, The Cat That Walked by Himself............. 1.20

Conservation

Fairbanks, Harold W., Conservation Reader, World Book Co............

Dogs

Atkinson, Eleanor, Greyfriars Bobby, Harper....................... 2.00
Baynes, Ernest H., The Story of an Eskimo Dog, Macmillan............ 1.50
Caldwell, Frank, Wolf, The Storm Leader, Dodd..................... 2.50
Carter, M. H., ed., Stories of Brave Dogs, Century................. .65
Darling, Esther B., Baldy of Nome, Penn.......................... 2.50

Davis, Richard Harding, The Bar Sinister.............................
Derieux, Samuel, Frank of Freedom Hill, Doubleday................... 1.75
Fitspatrick, James P., Jock of the Bushweld, Longsman................ 1.75
Gask, L., True Stories about Dogs, Crowell.......................... 1.50
London, Jack, Call of the Wild, Grossett............................ 1.00
Maeterlinck, M., Our Friend the Dog................................
Mills, Enos, The Story of Scotch, Longs Pk., Bkstore.................. 1.25
Muir, John, Stickeen...
Olivant, Alfred, Bob, Son of Battle, Doubleday...................... 2.00
Ramee (Ouida), The Dog of Flanders.................................
Roberts, C. G. D., Jim—Story of the Backwoods Police Dog, Macmillan... 1.00
Saunders, Marshall, Beautiful Joe, Various Publ.....................
    The Wandering Dog, Doran....................................... 1.50
Terhune, Albert P., Buff, A Collie, Doran........................... 2.00
    Lad, A Dog, Dutton.. 2.00

Dunes

Reed, Earl H., The Dune Country....................................
Townsend, Chas. W., Sand Dunes and Salt Marshes...................

Elephants

Eardley-Wilmot, S., Life of an Elephant, London, Ed. Arnold..........
Mukerji, Dhan G., Kari the Elephant, Dutton........................ 2.00

Fish

Baskett, James N., The Story of the Fishes.........................
Jordan, David Starr, Science Sketches (Stories about Fish)..............

Flowers

Mathews, F. Schuyler, The Book of Wild Flowers for Young People, Putnam 3.00
Stratton-Porter, Gene, Freckles....................................

Forests (See Trees)

Noyes, Wm., Wood and Forests......................................
Pack, A. N., Our Vanishing Forests................................
Rolt-Wheeler, Francis, The Boys with the U. S. Foresters, Lothrop....... 1.75
Schwartz, G. H., Forest Trees and Forest Scenery...................

Fox

Roberts, C. G. D., Red Fox, L. C. Page.............................
Seton, E. T., Biography of a Silver Fox............................. 1.90

Horses

Baldwin, Story of Horses..
Saunders, Marshall, Bonnie Prince Fetlar, Doran..................... 2.00
Sewell, Black Beauty..

Insects

Badenock, L. N., Romance of the Insect World.......................
Brailliar, Floyd, Knowing Insects through Stories, Funk and Wagnalls...
Clark, G. Glenwood, Tiny Toilers and Their Works, Century............ 1.75
Comstock, Anna B., The Way of the Six-footed, Comstock.............. .75
Cragin, Our Insect Friends and Foes, Putnam........................ 1.75
Fabre, Jean Henri, Life of the Fly, Dodd Mead....................... 2.50
    Hunting Wasps, Dodd Mead.....................................
    Life of the Caterpillar, Dodd Mead...........................
    Life of the Grasshopper, Dodd Mead...........................
    The Sacred Beetle and others, Dodd Mead......................
    The Mason Wasps, Dodd Mead...................................
    Glow-worm and other Beetles, Dodd Mead.......................
    More Hunting Wasps, Dodd Mead................................
    Life of the Weevil, Dodd Mead................................

More Beetles, Dodd Mead...
Insect Adventures, World Book.................................... 1.00
Fairchild, David and Marian, The Book of Monsters, Nat'l Geog. Soc....
Fell, E. P., Insects on the Farm, Cuting..............................
Kellogg, Vernon, Insect Stories, D. Appleton........................ 1.50
McCook, H. C., Nature's Craftsman.................................
Morley, M. W., Will o' the Wasps, McClurg......................... 1.35
Patch, Edith M., Hexapod Stories, Atl. Mo. Press....................
Selous, E., The Romance of Insect Life.............................
Stratton-Porter, Gene, Moths of Limberlost......................... 3.50
Swartz, Julia A., Grasshopper Green's Garden, Little Brown........... .40

Fossils

Lucas, F. A., Animals of the Past..................................
Rolt-Wheeler, Francis, The Monster Hunters, Lothrop................. 1.75

Lion and Tiger

Carter, M. H., ed., About Animals (Retold from St. Nicholas), Century... .65

Man

Baitsell, G. A., ed., The Evolution of Man.........................
Conklin, E. G., The Direction of Human Evolution...................
Fiske, John, The Destiny of Man..................................
Patten, Wm., The Grand Strategy of Evolution......................
Shaler, N. S., Man and the Earth.................................
Thomson, J. Arthur, What is Man..................................

Minerals

Kelley, Jay G., The Boy Mineral Collectors, Lippincott............... 1.75
Rolt-Wheeler, Francis, The Boy with the U. S. Miners, Lothrop......... 1.75

Mammals

Burgess, T. W., Burgess Animal Book for Children, Little.............. 3.00
Ingersoll, Ernest, The Life of Animals: the Mammals, MacMillan........ 2.00

Mountains

Abraham, Geo. F., The Complete Mountaineer.......................
Jeffers, LeRoy, The Call of the Mountains, Dodd, Mead...............
King, Clarence, Mountaineering in the Sierras.......................
Maeterlinck, Maurice, Mountain Paths............................. 2.00
Meany, Edmund S., Mount Rainer.................................
Mills, Enos, Wild Life of the Rockies, Houghton..................... 1.75
Muir, John, My First Summer in the Sierra.......................... 3.25

Pebbles

Hawksworth, Hallam, The Strange Adventures of a Pebble, Scribner....... 1.60

Pigeons

Seaman, Augusta H., Jacqueline of the Carrier Pigeons, Macmillan....... 1.50

Sea

Crowder, Wm., Dwellers of the Sea and Shore, Macmillan.............. 2.25
Duncan, F. Martin, Wonders of the Shore..........................
Hardy, Mrs. A. S., Sea Stories for Wonder Eyes, Ginn................ .40
Verrill, A. Hyatt, The Ocean and its Mysteries......................

Seal

Jordan, David Starr, The Story of the Seal.........................

Squirrels

Burroughs, John, Squirrels and Other Fur Bearers, Houghton........... 1.65
Morley, M. W., Little Mitchell, McClurg........................... 1.25

Stars

Johnson, Gaylord, The Star People, Macmillan........................ 1.50
 The Sky Movies, Macmillan.. 1.50
Warren, G. C., Star Stories for Little Folks, Pilgrim Press............... .60

Trees

Dixon and Fitch, The Human Side of Trees...........................
Huntington, Studies of Trees in Winter..............................
McFarland, J. H., Getting Acquainted with the Trees..................
Mills, Enos, The Story of a Thousand Year Pine, Houghton.............
Thoreau, Henry D., The Maine Woods................................ 2.00
White, Stewart Edward, The Magic Forest, The Juvenile Library........ .50

Whales

Beddard, F. E., Book of Whales, Putnam............................. 2.50
Bullen, Cruise of the Cachalot round the World after Sperm Whale, Dodd,
 Mead.. 1.50
Melville, Herman, Moby Dick or the Great White Whale, Dodd, Mead... .70
Verrill, A., Hyatt, The Real Story of the Whaler.Appletons............. 1.50

Wolf

Wallace, Dillon, The Gaunt Grey Wolf..............................

Sources of Information

Animals (See Mammals)

Adams, C. C., Guide to the Study of Animal Ecology, Macmillan.... ... 1.90
Champlin, J. D., Young Folks' Cyclopedia of Natural History, Holt........ 3.00
Gibson, W. H., Sharp Eyes, Harper.................................. 4.00
Hornaday, W. T., The American Natural History, Scribner......... .... 5.00
Shelford, Victor E., Animal Communities in Temperate America.........
Wood, Theodore, Natural History for Young People, Dutton............ 3.00
———, Gold Fish Breeds and Other Aquarium Fishes, Innes and Sons...

Aquaria

Brind, W. L., The Practical Fish Fancier, W. L. Brind, N. Y............
Osborne, Raymond C., Care of Home Aquaria, N. Y. Zool. Soc.........
Smith, Eugene, The Home Aquarium and How to Care for it, Dutton....

Bacteria

Conn, H. W., Bacteria, Yeasts and Molds, Ginn...................... 1.48

Birds

National Assoc. of Audubon Societies, 1974 Eroadway, N. Y., Secure list
 of Public...
Birds of New York, State Museum, Albany, bound..................... 2.50
Book of Birds, 250 colored Illust. Nat. Geog. Mag................... 3.40
Bailey, Florence M., Handbook of Birds of Western U. S., Aud. Soc...... 4.15
 Birds of Village and Field, Aud. Soc............................... 1.19
Baynes, Ernest Harold, Wild Bird Guests, Dutton.................... 2.00
Blanchan, Neltje, Birds Worth Knowing, Doubleday Page.............. 1.60
 Bird Neighbors, Doubleday....................................... 4.00
Bowdish, B. S., Putting up Bird Boxes, Aud. Society.................15c Doz.
Chapman, F. M., What Bird is That? Appleton........................ 3.00
Dugmore, A. R., Bird Homes, Doubleday, Page....................... 5.00
Eaton, E. H., Birds of New York, 2 Vol., State Museum................ 6.00
Finley, W. L., American Birds, Scribners............................ 1.50
Forbush, Edward Howe.......................................
Grinnell, Bryant, and Storer. Game Birds of California, Univ. of Cal. Press. 6.00
Henshaw, H. W., Book of Birds, Nat. Geog. Soc...................... 3.00
Hoffman, Ralph, Guide to Birds of N. Eng. and E. New York, Houghton. 3.00
Job, Herbert K., Propogation of Wild Birds, Doubleday, Page..........
 The Sport of Bird Study, Macmillan............................... 2.50

Keeler, C., Bird Notes Afield, Paul Elder Co., San Fr..................
Mathews, F. Schuyler, Wild Birds and Their Music, Putnam............ 3.50
Miller, Olive Thorne, First Book of Birds, Houghton................... 2.00
Miner, Manley F., Jack Miner and the Birds, M. F. Miner, Kingsville,
Myers, H. W., Western Birds, Macmillan.............................. 4.00
    Ontario, Canada...................................... 3.00
Pearson, T. Gilbert, The Bird Study Book, Doubleday, Page........... 1.25
Reed, Chester A., Pocket Handbooks, Each, Cloth..................... 1.25
    Land Birds East of Rockies, Doubleday...........................
    Water Birds East of Rockies, Doubleday..........................
    Land Birds West of Rockies, Doubleday...........................
Siepert, Albert F., Bird Houses Boys can Build, Bradley Institute, Peori..
Trafton, Gilbert H., Bird Friends, Houghton, Mifflin.................. 2.00
Wright, Mabel O., Birdcraft, Macmillan.............................. 2.00

Butterflies (See Moths)

Coloration
Beddard, F. E., Animal Coloration, Macmillan........................
Poulton, E. B., The Colours of Animals, Appleton....................
Thayer, G. H., Concealing-coloration in the Animal Kingdom, Macmillan. 7.00

Domestic Animals
Burkett, C. W., ed., Our Domestic Animals, Ginn..................... 3.50

Earth's Beginning
Ball, Sir R. S., The Earth's Beginning...............................
Chamberlain, T. C., The Origin of the Earth........................
Gregory, J. W., The Making of the Earth............................

Evolution
Bickerton, A. W., The Romance of the Earth.........................
Clodd, Edward, The Story of Creation...............................
Metcalf, M., Evolution...
Schmucker, S. C., The Meaning of Evolution.........................
Thomson, J. Arthur, The Bible of Nature............................

Ferns
Beecroft, W. I., Who's Who Among the Wild Flowers, and Ferns, Moffat. 1.50
Clute, W. N., Our Ferns and Their Haunts, Stokes.................... 3.00
Eastman, Helen, New England Ferns.................................
Parsons, F. T., How to Know the Ferns, Scribners................... 1.50
Tilton, Fern Lovers' Companion, Little, Brown....................... 3.00
Underwood, L. M., Our Native Ferns................................
Woolson, G. A., Ferns and How to Grow Them, Doubleday, Page........ 1.10

Fishes (See Aquaria)
Holder, C. F., The Fishes of the Pacific Coast, Dodge Pub. Co..........
Innes, W. T., Goldfish Varieties, Innes.............................. 3.00
Jordan and Evermann, American Food and Game Fishes, Doubleday,
    Page.. 4.00
Nichols, J. T., Fishes of the Vicinity of N. Y. City., Am. Mus. Nat. Hist. .75
Rhead, L. J., Book of Fish and Fishing, Scribner.................... 2.00

Flowers
Armstrong, Margaret, Field Book of Western Wild Flowers, Putnam...... 3.50
Blanchan, Neltje, Nature's Garden, Doubleday....................... 5.00
Burgess, Thornton, Flower Book for Children, Lothrop............... 3.50
Creevey, Caroline, Guide to Wild Flowers, Harper................... 1.75
Dana, Mrs. W. S., How to Know the Wild Flowers, Scribners.......... 2.00
House, H. D., Wild Flowers of New York, State Museum, Albany, 2 Vol.. 7.50
Keeler, H. L., Wild Flowers, 3 Vols., Scribners, each............... 1.75
Lounsberry, Alice, A Guide to the Wild Flowers, Stokes............. 1.90
    Southern Wild Flowers and Trees, Stokes.......................... 3.75

Maeterlinck, M., Old Fashioned Flowers..................................
Mathews, F. Schuyler, Field Book of American Wild Flowers, Putnam.... 3.50
Reed, C. A., Flower Guide (Pocket Edition), Doubleday............... 1.25
Saunders, Western Flower Guide...................................... 1.75
Walton, G. L., The Flower Finder, Lippincott....................... 2.00

Forestry
Berry, James B., Farm Woodlands, World Book Co.................... 2.00
Gifford, J. C., Practical Forestry, Appleton......................... 2.50
Moon, F. F., The Book of Forestry, D. Appleton..................... 2.00
Pack, C. L., The School Book of Forestry, Amer. Nature Assoc., Wash....
Pinchot, G., The Primer of Forestry, U. S. Bureau Forestry.......2 pts. ea. 5c
Roth, Filebert, First Book of Forestry, Ginn........................ 1.00

Frogs
Bureau of Fisheries, Document 888, Bulletin on Frogs.................
Dickerson, Mary E., The Frog Book, Doubleday Page.................. 5.00

Fruits
Peterson, M. G., How to Know the Wild Fruits, Macmillan............ 1.50
Walton, Geo. L., Wild Flowers and Fruits............................

Fungi
Duggar, B. M., Fungous Diseases of Plants, Ginn.................... 2.00
McCubbin, W. A., Fungi and Human Affairs, World Book Co........... 1.00

Gardens
Burkett, Stevens and Hill, Agriculture for Beginners, Ginn.............. 1.08
Croy, M. S., Putnam's Garden Handbook, Putnam..................... 1.90
Duncan, Frances, When Mother Lets us Garden, Dodd, Mead........... 1.25
Durand, Herbert, Taming the Wildings, Putnam...................... 3.50
Keeler, H. L., Our Garden Flowers, Scribner........................ 3.00
Meier, School and Home Gardens, Ginn.............................. 1.28
Shaw, E. E., Garden Flowers of Spring, Summer, Autumn and Winter,
    4 Vol., Doubleday.. 1.50

Geology
Burroughs, John, Time and Change...................................
Fabre, Jean Henri, This Earth of Ours, Century..................... 2.50
Geikie, James, Geology (Small Book)................................
Hopkins, T. C., Elements of Physical Geography, Sanborn............. 1.35
Salisbury, Barrows, Tower, Elements of Geography, Holt..............
Seers, A. W., Earth and Its Life, World Book Co....................
Shaler, N. S., First Book in Geology, Heath........................ .45

Grasses
Francis, Mary E., Book of Grasses................................. 5.00
Knobel, Edward, Grasses, Sedges, and Rushes, Bradlee Whidden........

Insects (*See Moths*)
Beard, Dan C., American Boys' Book of Bugs, Lippincott.............. 2.00
Comstock, J. H., Manual of the Study of Insects, Comstock Pub. Co..... 3.75
    Insect Life, Comstock.. 1.75
Howard, L. O., The Insect Book, Doubleday Page..................... 3.00
Kellogg, V. L., American Insects, H. Holt.......................... 5.00
Lutz, Frank E., The Field Book of Insects, Putnam................... 3.50
Miall, L. C., The Natural History of Aquatic Insects, Macmillan......... 1.75
Washburn, F. L., Injurious Insects and Useful Birds, Lippincott.........

Invertebrates
Holder, C. F., Half Hours with the Lower Animals, American Book Co.... .60

Landscape
Hamblin, Stephen, Man's Spiritual Contact with the Landscape, Badger.. 2.50
Waugh, Frank, The Natural Style in Landscape Gardening, Badger...... 3.00

Mammals

Knight, C. R., Animals of the World for Young People, Stokes...........
Miller, Garret S., Key to Land Mammals of N. E., N. Amer., N. Y. State
    Museum, Albany.. .35
Nelson, E. W., Wild Animals of N. America, Nat. Geog. Soc............. 3.00
Seton, E. T., Life History of Mammals...............................
Stone, W. & Cram, W. E., American Animals, Doubleday............... 5.00
Westell, W. P., The Book of the Animal Kingdom: Mammals, Dutton.... 2.50

Meteorology

Archibald, E. D., The Story of the Earth's Atmosphere................
Buckley, A. B., The Fairy Land of Science...........................
Gilberne, Agnes, The Ocean of Air...................................
Huntington, Ellsworth, Civilization and Climate.....................
Moore, W. L., The New Air World.....................................
Thompson, J. M., Water Wonders.....................................

Minerals

Clapp, Observation Lessons in Minerals, Heath....................... .30
Dana, E. S., Minerals and How to Study Them, Wiley.................. 1.50
Fairbanks, H. W., Rocks and Minerals, Educ. Publ. Co............... .75
Loomis, F. B., Field Book of Common Rocks and Minerals, Putnam...... 3.50
Pirsson, Louis V., Rocks and Rock Minerals..........................
Spencer, L. J., The World's Minerals, Stokes........................ 5.00

Mosses

Dunham, Mrs. E. M., How to Know Mosses, Houghton................. 2.50
Grout, A. J., Mosses with Hand Lens and Microscope, Grout........... 7.00
Marshall, N. L., Mosses and Lichens, Doubleday, Page. 4.00

Moths and Butterflies

Ballard, Julia P., Among the Moths and Butterflies, Putnam........... 1.75
Dickerson, M. C., Moths and Butterflies, Ginn....................... 1.80
Holland, W. J., The Butterfly Guide, Doubleday...................... 1.25
    The Moth Book, Doubleday....................................... 4.00
Miller, E. R., Butterfly and Moth Book, Scribners...................
Weed, C. M., Butterflies Worth Knowing, Doubleday.................. 1.75

Mountains

Geikie, James, Mountains ....

Mushrooms

Atkinson, Geo., Mushrooms, Henry Holt..............................
Marshall, N. L., The Mushroom Book, Doubleday, Page............... 3.00
Patterson and Charles, Mushrooms and Other Common Fungi, Bulletin 175,
    U. S. Dept. of Agric., Washington..............................

National Parks

Mills, Enos, Your National Parks....................................
Muir, John, Our National Parks..................................... 1.75

Natural Philosophy

Thomson, J. Arthur, The Outline of Science ....

Pets

Comstock, Anna B., The Petbook, Comstock Pub. Co.................. 2.50
Crandall, Lee S., Pets: Their History and Care, Henry Holt...........
Johnston, Constance, When Mother Lets us Keep Pets, Dodd, Mead..... 1.25
MacSelf, A. J., Pets for Boys and Girls, Dutton...................... 2.00

Photography

Dimock, Julian, Outdoor Photography, Macmillan....................
Dugmore, A. R., Nature and the Camera, Doubleday.................. 1.35
Jenks, Tudor, Photography for Young People, F. A. Stokes............. 1.50
Snell, F. C., The Camera in the Fields, T. Fisher Unwin, London........

onds
Needham and Lloyd, Life of Inland Waters, Comstock Pub. Co..........

Reptiles
Baskett, J. N. and Ditmars, R. L., The Story of Amphibians and Reptiles,
    Appleton... 1.00
Ditmars, Raymond L., The Reptile Book, Doubleday Page.............. 3.50
Stejnegr, Leonhard, Poisonous Snakes of N. Amer., Gov't Printing Office...

Sea-Shore (See Shells)
Arnold, A. F., Sea-Beach at Ebb Tide, Century...................... 4.50
Duncan, F. Martin, The Sea-shore, Stokes...........................
Dycraft, W. P., The Sea-shore, Macmillan........................... 1.75
Flattely, F. W., Biology of the Seashore, Macmillan................ 5.00
Holder, C. F., Half Hours with Lower Animals, Amer. Book Co........ .75
Mayer, A. G., Sea-shore Life, Laidlaw.............................. 1.20
Wood, Theodore, Sea-shore Shown to the Children, Dutton........... 1.25

Sea-weeds
Murray, G., Sea Weeds..

Sex
Cady, B. C. and V. M., The Way Life Begins, American Social Hygiene,
    105 W. 48 St., N. Y. City...................................... 1.00

Shells
Baker, F. C., Shells of Land and Water, Mumford................... 2.50
Keep, J., West Coast Shells, Whitaker and Ray-wiggin, San Francisco.... 2.00
Rogers, Julia, The Shell Book, Doubleday, Page................... 4.00

Shrubs
Apgar, Ornamental Shrubs of the U. S., Amer. Book Co..............
Keeler, Harriet L., Our Northern Shrubs, Scribners............... 3.00
Newell, C. S., Shrubs of Northeastern America....................

Spiders
Comstock, J. H., Spiders, Comstock Pub. Co....................... 5.00
Emerton, Spiders, Ginn...
Fabre, Life of the Spider.. 1.75
Patterson, A. J., The Spinner Family, McClurg.................... 1.50

Stars
Brown, Louise, Nature-Study Pamphlets, Woman's Press, Y. W. C. A., 600
    Lexington Ave., N. Y. City....................................
Clarke, E. C., Astronomy from a Dipper, Houghton................. 1.25
Collins, R. F., The Book of Stars, D. Appleton................... 1.50
Irving, Edward, How to Know the Starry Heavens, Stokes........... 2.00
Kippax, J. R., Call of the Stars (Poems and Myths)..............
Martin, M. E., Friendly Stars, Harper........................... 2.00
McReady, Kelvin, A Beginners Star Book, Putnam..................
Newcomb, Simon, Astronomy for Everybody.........................
Olcott, Wm. T., A Field Book of Stars, Putnam................... 3.00
Serviss, Around the Year with the Stars, Harpers................ 1.00
    Astronomy with an Opera Glass, Appleton...................... 2.50
Sundell, E. W., Radium Star Map, E. W. Sundell, 643 Ontarioa St., Oak
    Park, Illinois.. 8.50

Trees
Blakeslee and Jarvis, Trees in Winter, Macmillan................ 2.00
Britton, N. L., North American Trees, Holt...................... 2.50
Brown, H. P., Trees of N. Y. State, N. Y. State College of Forestry...... 2.00
Collins and Preston, Key to the Trees, Holt.....................
Dame and Brooks, Handbook of Trees in N. Eng., Ginn.............
Emerson and Weed, Our Trees, How to Know Them, Lippincott........ 3.50
Hough, R. B., Handbook of Trees, Hough.......................... 8.00

Illic, Joseph, Trees of Pennsylvania, Dep't of Forestry, Harris..........
Keeler, Our Native Trees, Scribners.................................... 3.00
Mathews, F. S., Fieldbook American Trees and Shrubs, Putnam......... 3.50
Pettis, C. R., Practical Tree Planting, Outing..........................
Rogers, Julia, The Tree Guide, Doubleday 1.25
Sargent, C. S., Manual of Trees of N. Amer., Houghton, Mifflin.........
Webster, A. P., Tree Wounds and Diseases, Lippincott.................. 2.50

Weather

Bulletins, U. S. Weather Bureau......................................
The Earth and Its Weather, Cornell Rural School Leaflet, Vol. xv, Jan.
   1922, No. 3..
Barnard, Charles, Tales About the Weather, Funk, Wagnalls...........
Longstreth, T. M., Reading the Weather, Outing......................
Martin, Edward C., Our Own Weather, Harper.......................
McAdie, Alexander, Wind and Weather, Macmillan...................
Moore, W. L., The New Air World, Little, Brown....................

Weeds
Georgia, Ada E., A Manual of Weeds, Macmillan..................... 2.00

Woodcraft
Beard, Dan, Amer. Boys Handy Book of Camp Lore and Woodcraft...... 3.00
Seton, E. T., The Woodcraft Manual for Boys, Lippincott.............
   The Woodcraft Manual for Girls, Lippincott......................

IV.   Myth and Humor

Branner, How and Why Stories..
Farrar, F. A., Old Greek Nature Stories.............................
Hawthorne, N., The Great Stone Face...............................
Holmes, O. W., The Dorchester Giant...............................
Judson, Katharine B., Old Crow and His Friends (Indian Stories), Little,
   Brown... 1.35
Kipling, Jungle Stories... 1.50
   Qui Quern (Eskimo Dog)
   Ricki Ticki Tarvi
   Story of the Sea Cow
Linderman, Indian Why Stories, Scribners........................... 2.00
Muller, J. W., First Aid to Naturers, The Platt and Peck Co...........
Toogood, Hector B., The Outline of Everything, Little, Brown..........
Wells, Carolyn, Nonsense Anthology, Scribners......................

V.   Philosophy

Burroughs, John, The Summit of the years.
Carman, Bliss, The Breath of Life..................................
   The Kinship of Nature.......................................
Dixon, Royal, The Human Side of Plants............................
Long, Wm. J., Briar Patch Philosophy..............................
Mabie, H. W., Essays on Nature and Culture.......................
Shaler, N. S., The Interpretation of Nature........................
Smith, Mary O., The Autobiography of a Tree.....................
Thoreau, H. D., Excursions.......................................
Whitman, W., In Leaves of Grass.................................

VI.   Poems

Badger, Clark, Sun and Saddle Leather (Cowboy Poems and Songs), Badger 2.50
Bartlett, Brooks, Pine Tree Verse, Badger........................... 2.00
Browning, Pippa Passes...
Burroughs, John, Songs of Nature.................................
Carman, Bliss, The Joys of the Road..............................

Chapman, Arthur, Out Where the West Begins.......................
Clark, Badger, Sun and Saddle Leather...........................
Drummond, Habitant Poems in Canada............................
Emerson, R. W., Rhodora and Other Poems......................
Herford, Oliver, A Child's Primer of Natural History and More Animals,
    Scribners..
Holmes, O. W., The Chambered Nautilus, etc.......................
Longfellow, H. W., Flowers.....................................
Lowell, J. R., Vision of Sir Launfal............................
Neihardt, John, The Song of Hugh Glass.........................
Richards, Mrs. Waldo, High Tide...............................
Service, Robert W., The Spell of the Yukon......................
Smith, Horace, Hymn to the Flowers............................
Tennyson, Flower in the Cranied Wall, etc., Houghton, Mifflin..........
Thompson, James, The Seasons................................
Van Dyke, Henry, God of the Open Air.........................
Wordsworth, The World is too Much with us, etc....................

VII. *Religions*

Burr, Around the Fire Stories, Assoc. Press.......................
Burroughs, John, Accepting the Universe........................
    The Light of Day..
Clodd, Edw., The Childhood of Religions.........................
Cobb, Cora S., God's Wonder World, Beacon Press, Boston.............
Fiske, John, The Idea of God..................................
    Through Nature to God......................................
Grosser, Scripture Natural History.............................
Huntington, Ellsworth, Palestine and its Transformation, Houghton,
    Mifflin..
Jefferies, Richard, The Story of My Heart.......................
Jefferson, Chas. E., Nature Sermons, Revell...................... 1.50
Mattoon and Brayton, Services for the Open, Century ....
Quayle, Wm. A., Out-of-doors with Jesus, Abingdon Press.............
Smith, Geo. Adams, The Historical Geography of the Holy Land.........
Tristram, The Natural History of the Bible.......................

VIII. *Nature-study Magazines*

$1.00 per year

Bird-Lore (Bi-monthly); National Association of Audubon Societies, N. Y.
Cornell Rural School Leaflets (Quarterly); N. Y. State College of Agriculture at
    Cornell University. Free to rural teachers N. Y. State only.
The Guide to Nature (Monthly); The Agassiz Association, Sound Beach, Con-
    necticut. $1.50 per year.
Natural History (Bi-Monthly); American Museum of Natural History; New York
    City. $3.00 per year.
Nature Magazine (Monthly); American Nature Association, Washington, D. C.
    $2.00 per year.
Our Dumb Animals (Monthly); Massachusetts Society for Prevention of Cruelty
    to Animals, Boston, Mass. $1.00 per year.

IX. *Suggested $10.00 Nature Library*

Comstock—Handbook of Nature-study
Reed—Bird Guide and Flower Guide
Rogers—Trees
Blakeslee and Jarvis—Trees in Winter
Pinchot—Primer of Forestry
Hodge and Dawson—Civic Biology

X. Suggested Additions for a $25.00 Nature Library

Nature Magazine
Bigelow's Spirit of Nature-study
Parsons—How to Know the Ferns
Burgess—Bird Book
Mills—Story of Scotch
    Story of a 1000 Year Pine
Fabres—Insect Adventures
Dickerson—Moths and Butterflies
Muir—Our National Parks
Serviss—Around the Year with Stars
Seton—Woodcraft Manual
Branner—How and Why Stories
Comstock—Petbook

INDEX

Lightning Source UK Ltd.
Milton Keynes UK
UKHW020724150719
346096UK00014B/263/P